Moses Foster Sweetser

The Maritime provinces

A handbook for travellers

Moses Foster Sweetser

The Maritime provinces
A handbook for travellers

ISBN/EAN: 9783337210830

Printed in Europe, USA, Canada, Australia, Japan

Cover: Foto ©Andreas Hilbeck / pixelio.de

More available books at **www.hansebooks.com**

THE

MARITIME PROVINCES:

A

HANDBOOK FOR TRAVELLERS.

A GUIDE TO

THE CHIEF CITIES, COASTS, AND ISLANDS OF THE MARITIME PROVINCES OF CANADA, AND TO THEIR SCENERY AND HISTORIC ATTRACTIONS; WITH THE GULF AND RIVER OF ST. LAWRENCE TO QUEBEC AND MONTREAL; ALSO, NEWFOUNDLAND AND THE LABRADOR COAST.

With Four Maps and Four Plans.

BOSTON:
JAMES R. OSGOOD AND COMPANY,
LATE TICKNOR & FIELDS, AND FIELDS, OSGOOD, & CO.
1875.

PREFACE.

THE chief object of the Handbook to the Maritime Provinces is to supply the place of a guide in a land where professional ⸱ides cannot be found, and to assist the traveller in gaining Lie greatest possible amount of pleasure and information while passing through the most interesting portions of Eastern British America. The St. Lawrence Provinces have been hitherto casually treated in books which cover wider sections of country (the best of which have long been out of print), and the Atlantic Provinces have as yet received but little attention of this kind. The present guide-book is the first which has been devoted to their treatment in a combined form and according to the most approved principles of the European works of similar purpose and character. It also includes descriptions of the remote and interesting coasts of Newfoundland and Labrador, which have never before been mentioned in works of this character. The Handbook is designed to enable travellers to visit any or all of the notable places in the Maritime Provinces, with economy of money, time, and temper, by giving lists of the hotels with their prices, descriptions of the various routes by land and water, and maps and plans of the principal cities. The letter-press contains epitomes of the histories of the cities and the ancient settlements along the coast, statements of the principal scenic attractions, descriptions of the art and architecture of the cities, and statistics of the chief industries of the included Provinces. The brilliant and picturesque records and traditions of the early French and Scottish colonies, and the heroic exploits of the Jesuit missionaries, have received special attention in connection with the localities made famous in those remote days; and the remarkable legends and mythology of the Micmac Indians are incorporated with the accounts of the places made classic by them. The naval and military operations of the wars which centred on Port Royal, Louisbourg, and Quebec have been condensed from the best authorities, and the mournful events which are commemorated in "Evangeline" are herein analyzed and recorded. The noble coast-scenery and the favorite summer-voyages with which the northern seas abound have been described at length in these pages.

The plan and structure of the book, its system of treatment and forms of abbreviation, have been derived from the European Handbooks of Karl Baedeker. The typography, binding, and system of city plans also resemble those of Baedeker, and hence the grand desiderata of compactness and portability, which have made his works the most popular in Europe, have also been attained in the present volume. Nearly all the facts concerning the routes, hotels, and scenic attractions have been framed or verified from the Editor's personal experience, after many months of almost incessant travelling for this express purpose. But infallibility is impossible in a work of this nature, especially amid the rapid changes which are ever going on in America, and hence the Editor would be grateful for any *bona fide* corrections or suggestions with which either travellers or residents may favor him.

The maps and plans of cities have been prepared with the greatest care, and will doubtless prove of material service to all who may trust to their directions. They are based on the system of lettered and numbered squares, with figures corresponding to similar figures, attached to lists of the chief public buildings, hotels, churches, and notable objects. The hotels indicated by asterisks are those which are believed by the Editor to be the most comfortable and elegant.

M. F. SWEETSER,
Editor of Osgood's American Handbooks,
131 Franklin St., Boston.

CONTENTS.

	PAGE
I. PLAN OF TOUR	1
II. NEWFOUNDLAND AND LABRADOR	2
III. MONEY AND TRAVELLING EXPENSES	4
IV. RAILWAYS AND STEAMBOATS	4
V. ROUND-TRIP EXCURSIONS	5
VI. HOTELS	7
VII. LANGUAGE	7
VIII. CLIMATE AND DRESS	8
IX. FISHING	8
X. MISCELLANEOUS NOTES	9

NEW BRUNSWICK.

ROUTE
General Notes	13
1. ST. JOHN	15
2. THE ENVIRONS OF ST. JOHN	22
1. Lily Lake. Marsh Road	22
2. Mispeck Road. Suspension Bridge	23
3. Carleton	24
3. ST. JOHN TO EASTPORT AND ST. STEPHEN. PASSAMAQUODDY BAY	25
1. Eastport	26
4. GRAND MANAN	28
5. ST. JOHN TO ST. ANDREWS AND ST. STEPHEN. PASSAMAQUODDY BAY	30
1. St. George. Lake Utopia	32
2. St. Andrews. Chamcook Mountain	33
3. St. Stephen. Schoodic Lakes	35
6. ST. ANDREWS AND ST. STEPHEN TO WOODSTOCK AND HOULTON .	36
7. ST. JOHN TO BANGOR	37
8. ST. JOHN TO FREDERICTON. THE ST. JOHN RIVER . . .	39
1. Kennebecasis Bay	40
2. Belleisle Bay	42
3. Fredericton	44
4. Fredericton to Miramichi	46
9. WASHADEMOAK LAKE	47
10. GRAND LAKE	48
11. FREDERICTON TO WOODSTOCK	49
12. FREDERICTON TO WOODSTOCK, BY THE ST. JOHN RIVER . .	51
13. WOODSTOCK TO GRAND FALLS AND RIVIÈRE DU LOUP . . .	53

ROUTE	PAGE
1. Tobique to Bathurst	54
2. The St. John to the Restigouche	56
3. The Madawaska District	57
4. The Maine Woods. Temiscouata Lake	58
14. ST. JOHN TO SHEDIAC	59
15. THE BAY OF CHALEUR AND THE NORTH SHORE OF NEW BRUNSWICK	60
1. Chatham to Shippigan	61
2. Shippigan. Bay of Chaleur	64
3. Bathurst to Caraquette	66
4. Campbellton to St. Flavie	69
16. ST. JOHN TO AMHERST AND HALIFAX	70
1. Quaco. Sussex Vale	71
2. Albert County. Moncton to Quebec	72
3. Dorchester. Sackville	73

NOVA SCOTIA.

General Notes	75
17. ST. JOHN TO AMHERST AND HALIFAX	78
1. Tantramar Marsh. Chignecto Peninsula	79
2. North Shore of Nova Scotia	81
18. ST. JOHN TO HALIFAX, BY THE ANNAPOLIS VALLEY	83
1. Annapolis Royal	85
2. The Annapolis Valley	88
3. Kentville to Chester	90
19. HALIFAX	93
20. THE ENVIRONS OF HALIFAX	100
1. Bedford Basin. Point Pleasant	100
21. THE BASIN OF MINAS. HALIFAX TO ST. JOHN	101
1. Advocate Harbor and Cape d'Or	103
2. The Basin of Minas	104
22. THE LAND OF EVANGELINE	107
23. ANNAPOLIS ROYAL TO CLARE AND YARMOUTH	112
1. The Clare Settlements	113
2. The Tusket Lakes and Archipelago	115
24. DIGBY NECK	116
25. HALIFAX TO YARMOUTH. THE ATLANTIC COAST OF NOVA SCOTIA	117
1. Cape Sambro. Lunenburg	118
2. Liverpool	120
3. Shelburne	121
4. Cape Sable	123
26. HALIFAX TO YARMOUTH, BY THE SHORE ROUTE	126
1. Chester. Mahone Bay	127
2. Chester to Liverpool	128
27. THE LIVERPOOL LAKES	129
28. HALIFAX TO TANGIER	131
29. THE NORTHEAST COAST OF NOVA SCOTIA	133
30. SABLE ISLAND	134

CONTENTS. vii

ROUTE	PAGE
31. ST. JOHN AND HALIFAX TO PICTOU	136
32. ST. JOHN AND HALIFAX TO THE STRAIT OF CANSO AND CAPE BRETON	138

CAPE BRETON.

General Notes	141
33. THE STRAIT OF CANSO	142
34. ARICHAT AND ISLE MADAME	145
35. THE STRAIT OF CANSO TO SYDNEY, CAPE BRETON	146
36. HALIFAX TO SYDNEY, CAPE BRETON	148
37. THE EAST COAST OF CAPE BRETON. THE SYDNEY COAL-FIELDS	152
38. THE FORTRESS OF LOUISBOURG	154
39. THE NORTH SHORE OF CAPE BRETON	158
1. St. Anne's Bay	158
2. St. Paul's Island	160
40. THE BRAS D'OR LAKES	161
1. Baddeck	162
2. Great Bras d'Or Lake	164
3. The Bras d'Or to Halifax	166
41. BADDECK TO MABOU AND PORT HOOD	167
1. St. Patrick's Channel. Whycocomagh	167
42. THE WEST COAST OF CAPE BRETON	168
1. Port Hood. Mabou	169
2. Margaree. The Lord's Day Gale	170

PRINCE EDWARD ISLAND.

General Notes	172
43. SHEDIAC TO SUMMERSIDE AND CHARLOTTETOWN	174
1. The Northumberland Strait	174
44. PICTOU TO PRINCE EDWARD ISLAND	175
45. CHARLOTTETOWN	175
1. Environs of Charlottetown	177
46. CHARLOTTETOWN TO SUMMERSIDE AND TIGNISH. THE WESTERN SHORES OF PRINCE EDWARD ISLAND	177
1. Rustico. Summerside	178
47. CHARLOTTETOWN TO GEORGETOWN	180
48. CHARLOTTETOWN TO SOURIS	182
49. THE MAGDALEN ISLANDS	183
50. ST. PIERRE AND MIQUELON	185

NEWFOUNDLAND.

General Notes	187
51. HALIFAX TO ST. JOHN'S, NEWFOUNDLAND	188
52. ST. JOHN'S, NEWFOUNDLAND	189
53. THE ENVIRONS OF ST. JOHN'S	195
1. Portugal Cove. Logie Bay. Torbay	195
54. THE STRAIT SHORE OF AVALON. ST. JOHN'S TO CAPE RACE	196

ROUTE	PAGE
1. The Grand Banks of Newfoundland	199
55. ST. JOHN'S TO LABRADOR. THE NORTHERN COAST OF NEWFOUNDLAND	200
1. Bonavista Bay	203
2. Twillingate. Exploits Island	205
56. ST. JOHN'S TO CONCEPTION BAY	206
57. TRINITY BAY	208
58. THE BAY OF NOTRE DAME	210
59. PLACENTIA BAY	212
60. THE WESTERN OUTPORTS. ST. JOHN'S TO CAPE RAY	213
1. Fortune Bay	214
2. Hermitage Bay	215
61. THE FRENCH SHORE. CAPE RAY TO CAPE ST. JOHN	216
1. The Interior of Newfoundland	218
2. The Strait of Belle Isle	220

LABRADOR.

General Notes	223
62. THE ATLANTIC COAST, TO THE MORAVIAN MISSIONS AND GREENLAND	224
1. The Moravian Missions	226
63. THE LABRADOR COAST OF THE STRAIT OF BELLE ISLE	227
64. THE LABRADOR COAST OF THE GULF OF ST. LAWRENCE	229
1. The Mingan Islands	231
2. The Seven Islands	232
65. ANTICOSTI	234

PROVINCE OF QUEBEC.

General Notes	235
66. PICTOU TO QUEBEC. THE COASTS OF GASPÉ	238
1. Paspebiac	240
2. Percé	242
3. Gaspé	244
67. THE LOWER ST. LAWRENCE	246
1. Father Point. Rimouski	250
2. Bic. Trois Pistoles	251
3. St. Anne de la Pocatière. L'Islet	253
68. QUEBEC	255
1. Durham Terrace	259
2. Jesuits' College. Basilica	261
3. Seminary	262
4. Laval University. Parliament Building	263
5. Hôtel Dieu. Around the Ramparts	266
6. The Lower Town	271
69. THE ENVIRONS OF QUEBEC	276
1. Beauport. Montmorenci Falls	276
2. Indian Lorette	278
3. Château Bigot. Sillery	280
4. Point Levi. Chaudière Falls	282

CONTENTS.

ROUTE	PAGE
70. Quebec to La Bonne Ste. Anne	283
1. The Falls of St. Anne	283
71. The Isle of Orleans	288
72. Quebec to Cacouna and the Saguenay River	291
1. St. Paul's Bay	292
2. Murray Bay	294
3. Cacouna	296
73. The Saguenay River	297
1. Tadousac	299
2. Chicoutimi	300
3. Ha Ha Bay. Lake St. John	301
4. Eternity Bay. Cape Trinity	303
74. Quebec to Montreal. The St. Lawrence River	305
75. Montreal	309
1. Victoria Square. Notre Dame	311
2. The Gesù. St. Patrick's Church	313
3. Cathedral. McGill University. Great Seminary	314
4. Hôtel Dieu. Mount Royal. Victoria Bridge	316
76. The Environs of Montreal	318
1. Around the Mountain. Sault au Recollet	318
2. Lachine Rapids. Caughnawaga	319
3. Belœil Mt. St. Anne	320
Index to Localities	321
Index to Historical and Biographical Allusions	332
Index to Quotations	333
Index to Railways and Steamboats	334
List of Authorities Consulted	334

MAPS.

1. Map of the Maritime Provinces: before title-page.
2. Map of Newfoundland and Labrador: after the index.
3. Map of the Acadian Land: between pages 106 and 107.
4. Map of the Saguenay River: between pages 296 and 297.
5. Map of the Lower St. Lawrence River: between pages 296 and 297.

PLANS OF CITIES.

1. St. John: between pages 14 and 15.
2. Halifax: between pages 92 and 93.
3. Quebec: between pages 254 and 255.
4. Montreal: between pages 308 and 309.

ABBREVIATIONS.

N. — North, Northern, etc.
S. — South, etc.
E. — East, etc.
W. — West, etc.
N. B. — New Brunswick.
N. S. — Nova Scotia.
N. F. — Newfoundland.
Lab. — Labrador.

P. E. I. — Prince Edward Island.
P. Q. — Province of Quebec.
M. — mile or miles.
r. — right.
l. — left.
ft. — foot or feet.
hr. — hour.
min. — minute or minutes.

Asterisks denote objects deserving of special attention.

INTRODUCTION.

I. Plan of Tour.

THE most profitable course for a tourist in the Lower Provinces is to keep moving, and his route should be made to include as many as possible of the points of interest which are easily accessible. There are but few places in this region where the local attractions are of sufficient interest to justify a prolonged visit, or where the accommodations for strangers are adapted to make such a sojourn pleasant. The historic and scenic beauties are not concentrated on a few points, but extend throughout the country, affording rare opportunities for journeys whose general course may be replete with interest. The peculiar charms of the Maritime Provinces are their history during the Acadian era and their noble coast scenery, — the former containing some of the most romantic episodes in the annals of America, and the latter exhibiting a marvellous blending of mountainous capes and picturesque islands with the blue northern sea. And these two traits are intertwined throughout, for there is scarce a promontory that has not ruins or legends of French fortresses, scarce a bay that has not heard the roaring broadsides of British frigates.

The remarkable ethnological phenomena here presented are also calculated to awaken interest even in the lightest minds. The American tourist, accustomed to the homogeneousness of the cities and rural communities of the Republic, may here see extensive districts inhabited by Frenchmen or by Scottish Highlanders, preserving their national languages, customs, and amusements unaffected by the presence and pressure of British influence and power. Of such are the districts of Clare and Madawaska and the entire island of Cape Breton. The people of the cities and the English settlements are quaintly ultra-Anglican (in the secular sense of the word), and follow London as closely as possible in all matters of costume, idiom, and social manners.

All these phases of provincial life and history afford subjects for study or amusement to the traveller, and may serve to make a summer voyage both interesting and profitable.

Travelling has been greatly facilitated, within a few years, by the establishment of railways and steamship routes throughout the Provinces. From the analyses of these lines, given in the following pages, the tourist will be able to compute the cost of his trip, both in money and in time. The following tour would include a glimpse at the chief attractions of the country, and will serve to convey an idea of the time requisite: —

Boston to St. John	1½ days.
St. John	1 "
St. John to Annapolis and Halifax	2 "
Halifax	1 "
Halifax to Sydney	1½ "
The Bras d'Or Lakes	1 "
Port Hawkesbury to Pictou, Charlottetown, and Shediac	2 "
Shediac to Quebec (by steamer)	4 "
Quebec	3 "
Quebec to Boston	1 "
Failures to connect	3 "
	21 days.

To this circular tour several side-trips may be added, at the discretion of the traveller. The most desirable among these are the routes to Passamaquoddy Bay, the St. John River, the Basin of Minas (to Parrsboro'), from Halifax to Chester and Mahone Bay, Whycocomagh, or Louisbourg (in Cape Breton), and the Saguenay River. Either of these side-trips will take from two to four days.

If the tourist wishes to sojourn for several days or weeks in one place, the most eligible points for such a visit, outside of St. John and Halifax, are St. Andrews, Grand Manan, or Dalhousie, in New Brunswick; Annapolis, Wolfville, Parrsboro', or Chester, in Nova Scotia; Baddeck, in Cape Breton; and, perhaps, Summerside, in Prince Edward Island. At each of these villages are small but comfortable inns, and the surrounding scenery is attractive.

II. Newfoundland and Labrador.

Extended descriptions of these remote northern coasts have been given in the following pages for the use of the increasing number of travellers who yearly pass thitherward. The marine scenery of Newfoundland is the grandest on the North Atlantic coast, and here are all the varied phenomena of the northern seas, — icebergs, the aurora borealis, the herds of seals, the desolate and lofty shores, and the vast fishing-fleets from which France and the United States draw their best seamen. English and American yachtsmen grow more familiar every year with these coasts, and it is becoming more common for gentlemen of our Eastern cities to embark on fishing-schooners and make the voyage to Labrador or the Banks.

The tourist can also reach the remotest settlements on the Labrador coast by the steamship lines from Halifax to St. John's, N. F., and thence to Battle Harbor. This route takes a long period of time, though the expense is comparatively light; and the accommodations on the steamships beyond St. John's are quite inferior. A shorter circular tour may be made by taking the steamer from Halifax to St. John's, and at St. John's embarking on the Western Outports steamship, which coasts along the entire S. shore of the island, and runs down to Sydney, C. B., once a month. From Sydney the tourist can return to Halifax (or St. John, N. B.) by way of the Bras d'Or Lakes. The Western Outports steamship also visits the quaint French colony at St. Pierre and Miquelon fortnightly, and the traveller can stop off there and return directly to Halifax by the Anglo-French steamship, which leaves St. Pierre fortnightly.

Sea-Sickness. The chief benefit to be derived on these routes is the invigoration of the bracing air of the northern sea. Persons who are liable to sea-sickness should avoid the Newfoundland trip, since rough weather is frequently experienced there, and the stewards are neither as numerous nor as dexterous as those on the transatlantic steamships. The Editor is tempted to insert here a bit of personal experience, showing how the results of early experiences, combined with the advice of veteran travellers, have furnished him with a code of rules which are useful against the *mal du mer* in all its forms. During 28 days on the Mediterranean Sea and 45 days on the Canadian waters, the observance of these simple rules prevented sickness, although every condition of weather was experienced, from the fierce simoom of the Lybian Desert to the icy gales of Labrador. The chief rule, to which the others are but corollaries, is, Don't think of your physical self. Any one in perfect health, who will busy himself for an hour in thinking about the manner in which his breath is inhaled, or in which his eyes perform their functions, will soon feel ill at ease in his lungs or eyes, and can only regain tranquillity by banishing the disturbing thoughts. Avoid, therefore, this gloomy and apprehensive self-contemplation, and fill the mind with bright and engrossing themes, — the conversation of merry companions, the exciting vicissitudes of card-playing, or the marvellous deeds of some hero of romance. Never think of your throat and stomach, nor think of thinking or not thinking of them, but forget that such conveniences exist. Keep on deck as much as possible, warmly wrapped up, and inhaling the salty air of the sea. Don't stay in the lee of the funnel, where the smell of oil is nauseating. And if you are still ill at ease, lie down in your state-room, with the port-hole slightly opened, and go to sleep. The tourist should purchase, before leaving Halifax, two or three lively novels, a flask of fine brandy, a bottle of pickled limes, and a dozen lemons.

III. Money and Travelling Expenses.

The tourist will experience great inconvenience from the lack of a uniform currency in the Provinces. If he carries New-Brunswick money into Nova Scotia or Quebec, it can only be passed at a discount; and the same is true with Nova-Scotia or Quebec bills in either of the other Provinces. There appears to be no standard currency in circulation. To save frequent discounts, it is best for the tourist to carry U. S. money, changing it, in each Province, for the amount of local currency that he will be likely to need there. Respectable shop-keepers in the cities take U. S. money in payment for their goods, valuing it at the rate at which it is quoted on the local exchange. It is, however, more economical and convenient to take the U. S. money to an exchange office and buy as much of the local currency as will be needed during the sojourn. The shop-keepers are apt to charge at least full prices to people who have American money.

The silver coins of this country could only be defined in a lengthy numismatical treatise. There are half-crowns, two-shilling pieces, florins, shillings, and several smaller grades of English coins, independent and varying silver and copper tokens of each of the Lower Provinces, the money of Newfoundland, and large quantities of American silver. The latter is very unstable in its valuation, since a 25-cent piece goes for from 20 to 24 cents in the same city and on the same day, the rate of exchange apparently depending on the time of day and the mood of the shop-keeper. Nova-Scotian or Canadian money is held at a heavy discount in Newfoundland, and it is better to carry greenbacks there.

IV. Railways and Steamboats.

The new-born railway system of the Maritime Provinces is being extended rapidly on all sides, by the energy of private corporations and the liberality of the Canadian Government. The lines are generally well and securely constructed, on English principles of solidity, and are not yet burdened by such a pressure of traffic as to render travelling in any way dangerous. The cars are built on the American plan, and are sufficiently comfortable. On most trains there are no accommodations for smokers, and, generally, when any such convenience exists, it is only to be had in the second-class cars. Pullman cars were introduced on the Intercolonial Railway in 1874, and will probably be retained there during the summer seasons. They have been used on the European and North American road for three years. There are restaurants at convenient distances on the lines, where the trains stop long enough for passengers to take their meals. The narrow-gauge cars on Prince Edward Island and on the New Brunswick Railway will attract the attention of travellers, on account of their singular construction. The tourist has choice of three grades of accommodation on the chief railways, — Pullman car, first class, and second class. The latter mode of travelling is very uncomfortable.

The steamships which ply along these coasts afford material for a naval museum. At least two vessels of the Quebec and Gulf Ports fleet were captured blockade-runners; the *Edgar Stuart* was one of the most daring of the Cuban supply-ships, and was nearly the cause of a battle between the Spanish steamer *Tornado* and the U. S. frigate *Wyoming*, in the harbor of Aspinwall; the *M. A. Starr* was built for a British gunboat; it is claimed that the *Virgo* was intended for a U. S. man-of-war; and there are several other historic vessels now engaged in these peaceful pursuits. Good accommodations are given on the vessels which ply between Boston and St. John and to Halifax and Prince Edward Island. The cabins of the Quebec and Gulf Ports steamships are elegantly fitted up, and are airy and spacious. The Annapolis, Minas, Prince Edward Island, and Newfoundland lines have comfortable accommodations, and the Yarmouth and North Shore vessels are also fairly equipped. The lines to the Magdalen Islands, St. Pierre, and along the Newfoundland and Labrador coasts are primarily intended for the transportation of freight, and for successfully encountering rough weather and heavy seas, and have small cabins and plain fare. The Saguenay steamers resemble the better class of American river-boats, and have fine accommodations. Since the Canadas are under the English social system and have retained the Old-World customs, it will be found expedient, in many cases, to conciliate the waiters and stewards by small gifts of money. As the results thereof, the state-rooms will be better cared for, and the meals will be more promptly and generously served.

The Mail-Stages. — The remoter districts of the Provinces are visited by lines of stages. The tourist will naturally be deceived by the grandiloquent titles of "Royal Mail Stage," or "Her Majesty's Mail Route," and suppose that some reflected stateliness will invest the vehicles that bear such august names. In point of fact, and with but two or three exceptions, the Provincial stages are far from corresponding to such expectations; being, in most cases, the rudest and plainest carriages, sometimes drawn by but one horse, and usually unprovided with covers. The fares, however, are very low, for this class of transportation, and a good rate of speed is usually kept up.

V. Round-Trip Excursions.

During the summer and early autumn the railway and steamship companies prepare lists of excursions at greatly reduced prices. Information and lists of these routes may be obtained of George F. Field, General Passenger Agent of the Eastern R. R., 134 Washington St., Boston; T.

Edward Bond, Ticket Agent of the Central Vermont R. R., 148 Washington St., Boston; and from Stevenson and Leve, Passenger Agents of the Quebec and Gulf Ports S. S. Co., Quebec. Small books are issued every spring by these companies, each giving several hundred combinations of routes, with their prices. They may be obtained on application, in person or by letter, at the above-mentioned offices. The excursion tickets are good during the season, and have all the privileges of first-class tickets. The following tours, selected from the books of the three companies (for 1874), will serve to convey an idea of the pecuniary expense incurred in a trip through the best sections of the Maritime Provinces.

The Central Vermont R. R. — (Excursion 139.) International steamship, Boston to St. John; St. John to Halifax, by the Annapolis route; Halifax to Pictou, by the Intercolonial Railway; Pictou to Quebec, by the Q. & G. P. steamships (meals and state-room extra); Quebec to Montreal, by the Richelieu steamer, or the Grand Trunk Railway; Quebec to Boston, by the Central Vermont R. R. Fare, $34.50; or if the Eastern Railroad is preferred between Boston and St. John, $36.50.

Boston to Portland, by Eastern R. R.; N. E. & N. S. S. S. Co. to Halifax; Halifax to Point du Chene, by the Intercolonial Railway; Point du Chene to Quebec, by Q. & G. P. S. S. Co.; Quebec to Montreal, by railway or steamer; Montreal to Boston, by the Central Vermont R. R. Fare, $33.35.

Boston to Montreal, by Central Vt. R. R. and connections; Montreal to Quebec, by railway or steamer; Quebec to Point du Chene, by Q. & G. P. steamship; Point du Chene to St. John, by Intercolonial Railway; St. John to Boston, by International steamship. Fare, $29.15.

Eastern R. R. — Boston to St. John, by rail; St. John to Point du Chene, by Intercolonial Railway; Point du Chene to Quebec, by Quebec and Gulf Ports S. S. Co.; Quebec to Boston, by Grand Trunk and Eastern Railways. Fare, $35.65.

Boston to St. John and Shediac, by rail; Shediac to Summerside, Charlottetown, and Pictou, by steamship; Pictou to Halifax, by rail; Halifax to St. John, by the Annapolis route; St. John to Boston, by rail. Fare, $34.10.

Boston to Portland, by rail; Portland to St. John, by steamer; St. John to Halifax, by Annapolis route; Halifax to St. John, by Intercolonial Railway; St. John to Boston, by rail. Fare, $26.50.

Quebec and Gulf Ports S. S. Co. — Boston to Pictou, by the Boston and Colonial S. S. Co.; Pictou to Quebec, by the Q. & P. S. S. Co. Fare, $21; fare from Quebec to Boston, $10.

Boston to Halifax, by Boston and Colonial S. S. Co.; Halifax to St. John, by the Annapolis route; St. John to Point du Chene, by Intercolonial Railway; Point du Chene to Quebec, by Q. & G. P. S. S. Co. Fare, $26.50.

Boston to Portland, by Eastern R. R.; Portland to St. John, by International S. S. Co.; St. John to Point du Chene, by Intercolonial Railway; Point du Chene to Quebec, by Q. & G. P. S. S. Co.   Fare, $19.

VI. Hotels.

The Hotels of the Maritime Provinces are far behind the age. The Victoria Hotel, at St. John, is the only first-class house in the four Provinces, though the two chief hotels at Halifax are comfortable. The Island Park Hotel, at Summerside, P. E. I., is the only summer resort of any consequence. The general rates at the better hotels of the second class is $2 a day; and the village inns and country taverns charge from $1 to $1.50, with reductions for boarders by the week.

VII. Language.

The English language will be found sufficient, unless the tourist desires to visit the more remote districts of Cape Breton, or the Acadian settlements. The Gaelic is probably the predominant language on Cape Breton, but English is also spoken in the chief villages and fishing-communities. In the more secluded farming-districts among the highlands the Gaelic tongue is more generally used, and the tourist may sometimes find whole families, not one of whom can speak English.

In the villages along the Lower St. Lawrence, and especially on the North Shore, the French language is in common use, and English is nearly unknown. The relation of this language to the polite French speech of the present day is not clearly understood, and it is frequently stigmatized by Americans as "an unintelligible *patois*." This statement is erroneous. The Canadian French has borrowed from the English tongue a few nautical and political terms, and has formed for itself words describing the peculiar phenomena and conditions of nature in the new homes of the people. The Indians have also contributed numerous terms, descriptive of the animals and their habits, and the operations of forest-life. But the interpolated words are of rare occurrence, and the language is as intelligible as when brought from the North of France, two centuries ago. It is far closer in its resemblance to the Parisian speech than are the dialects of one fourth of the departments of France. Travellers and immigrants from Old France find no difficulty in conversing with the Lower-Canadians, and the aristocracy of Quebec speak as pure an idiom as is used in the Faubourg St. Germain. Among others whose testimony has been given in support of this fact, the Editor would adduce a gentleman whom he recently met in Canada, and who was an officer in the Imperial Guard until its capture in the Franco-Prussian war. He stated that neither he nor any of his compatriots, who came over after the triumph of Germany, had ever had any difficulty with the Canadian language, and that he had not yet learned a word of English.

This language has an extensive and interesting literature, which includes science, theology, history, romance, and poetry. It has also numerous newspapers and magazines, and is kept from adulteration by the vigilance of several colleges and a powerful university. It is used, co-ordinately with the English language, in the records and journals of the Dominion and Provincial Parliaments, and speeches and pleadings in French are allowable before the Parliaments and courts of Canada.

Thus much to prove the substantial identity of the Lower-Canadian and French languages. The tourist who wishes to ramble through the ancient French-Canadian districts will, therefore, get on very well if he has travelled much in Old France. But if the language is unknown to him, he will be subjected to many inconveniences and hardships.

VIII. Climate and Dress.

The more northerly situation of the Maritime Provinces and their vicinity, on so many sides, to the sea, render the climate even more severe and uncertain than that of New England. The extremes of heat and cold are much farther apart than in the corresponding latitudes of Europe, and, as Marmier expresses it, this region "combines the torrid climate of southern regions with the severity of an hyperborean winter." During the brief but lovely summer the atmosphere is clear and balmy, and vegetation flourishes amain. The winters are long and severe, but exercise no evil effect on the people, nor restrain the merry games of the youths. Ever since Knowles sent to England his celebrated dictum that the climate of Nova Scotia consisted of "nine months of winter and three months of fog," the people of Britain and America have had highly exaggerated ideas of the severity of the seasons in the Provinces. These statements are not borne out by the facts; and, though Nova Scotia and New Brunswick have not the mild skies of Virginia, their coldest weather is surpassed by the winters of the Northwestern States. The meteorological tables and the physical condition of the people prove that the climate, though severe, is healthy and invigorating. The time has gone by for describing these Provinces as a gloomy land of frozen Hyperboreans, and for decrying them with pessimistic pen.

The worst annoyance experienced by tourists is the prevalence of dense fogs, which sometimes sweep in suddenly from the sea and brood over the cities. In order to encounter such unwelcome visitations, and also to be prepared against fresh breezes on the open sea, travellers should be provided with heavy shawls or overcoats, and woollen underclothing should be kept at hand.

IX. Fishing.

"Anglers in the United States who desire to fish a salmon-river in the Dominion of Canada should club together and apply for the fluvial parts of rivers. . . . . The government leases the rivers for a term of nine years, and rivers unlet on the first day of each year are advertised by the government to be let to the highest bidders. The places of residence of those tendering for fishings are not considered in letting a river; and if a gentleman from the States overbids a Canadian, the river will be declared as his. Rivers are therefore hired by Europeans as well as by Canadians and citizens of the States. . . . . Rivers are either let in whole or parts, each part permitting the use of a given number of rods, generally four. Parties who desire to lease a Canadian river should address a letter to the Minister of Marine and Fisheries, at Ottawa, stating how many rods they have, and the district which they prefer to fish. He will forward them a list of the leasable rivers, and a note of information, upon which they should get some Canadian to make the tender for them. The leases of fluvial parts of rivers vary from two to six hundred dollars a year for from three to eight rods, and the price for guides or gaffers is a dollar a day." (This subject is fully discussed in Scott's "Fishing in American Waters.")

"The Game Fish of the Northern States and British Provinces," by Robert B. Roosevelt (published by Carleton, of New York, in 1865), contains an account of the salmon and sea-trout fishing of Canada and New Brunswick. The pursuit of sea-trout on the Lower St. Lawrence and Laval is described in pages 50-88 and 315-321; the Labrador rivers, pages 107-111; the Miramichi and Nepisiguit Rivers, pages 111-145; the Schoodic Lakes, pages 145-147.

"Fishing in American Waters," by Genio C. Scott (published by Harper and Brothers, 1869), contains practical directions to sportsmen, and graphic descriptions of fishing in the rivers of New Brunswick and Lower Quebec.

"Frank Forester's Fish and Fishing of the United States and British Provinces of North America," by H. W. Herbert (New York, 1850), is to a large extent technical and scientific, and contains but a few incidental allusions to the provincial fisheries.

"The Fishing Tourist," by Charles Hallock (published by Harper and Brothers, 1873), contains about 100 pages of pleasant descriptions relating to the Schoodic Lakes, the best trout and salmon streams of Nova Scotia, New Brunswick, and Cape Breton, the Bay of Chaleur, the Saguenay and Lower St. Lawrence, Anticosti, and Labrador.

IX. Miscellaneous Notes.

The times of departure of the provincial steamships are liable to change every season. The tables given in the ensuing routes are based on those of 1874, and the changes for 1875 are indicated so far as the Editor has been able to learn them. The tourist can find full particulars of the days of sailing, etc., on arriving at St. John, from the local and the Halifax newspapers. The names of the agents of these lines have also been given hereinafter, and further information may be obtained by writing to their addresses.

The custom-house formalities at the national frontiers depend less upon the actual laws than upon the men who execute them. The examination of baggage is usually conducted in a lenient manner, but trunks and packages are sometimes detained on account of the presence of too many Canadian goods. It is politic, as well as gentlemanly, for the tourist to afford the officers every facility for the inspection of his baggage.

Travellers are advised to carefully inspect the prices of goods offered them by shop-keepers, since the lavish and unquestioning extravagance of American tourists has somewhat influenced the tone of commercial morality.

The people of the Provinces are generally courteous, and are willing to answer any civilly put questions. The inhabitants of the more remote districts are distinguished for their hospitality, and are kindly disposed and honest.

ROUTES FROM BOSTON TO THE MARITIME PROVINCES.

1. *By Railway.*

The *Eastern and Maine Central R. R. Lines* afford the best mode of approach by land. Their trains leave the terminal station on Causeway St., Boston, and run through to Bangor, without change of cars. Pullman cars are attached to the through trains, and tickets are sold to nearly all points in the Eastern Provinces. At Bangor passengers change to the cars of the European & North American R. R., which runs E. through the great forests of Maine and New Brunswick to the city of St. John. Between Boston and Portland this route traverses a peculiarly interesting country, with frequent glimpses of the sea; but the country between Bangor and St. John is almost devoid of attractions.

The *Boston & Maine R. R.* may also be used as an avenue to the Eastern Provinces, though the Editor does not know what connections (if any) it makes at Portland with the lines to the Eastward.

2. *By Steamship.*

The *International Steamship Company* despatches vessels three times weekly from June 15 to October 1, leaving Commercial Wharf, Boston, at 8 A. M., on Monday, Wednesday, and Friday. They touch at Portland, which is left at 6 P. M.; and afterwards they run along the Maine coast, calling at Eastport and traversing Passamoquoddy Bay. Fares, — from Boston to Eastport, $ 5; to St. John, $ 5.50.

The steamers of the *Portland Steam Packet Company* leave India Wharf, Boston, every morning, running along the New England coast to Portland. At that city they connect with the fine steamship *Falmouth*, which leaves Portland every Saturday at 5.30 P. M., stretching out over the open sea, and, beyond Cape Sable, following the Nova-Scotia coast to Halifax.

Clements' Line affords the most convenient route to visit the famous hunting and fishing grounds of the western counties of Nova Scotia. The steamship *Dominion* leaves Lewis Wharf, Boston, every Tuesday noon, for Yarmouth and St. John, giving an exhilarating voyage across the open sea.

The *Boston, Halifax, and Prince Edward Island Steamship Line* despatch vessels from T Wharf, Boston, every Saturday at noon. After reaching Halifax these steamships run N. E. along the Nova-Scotia coast, round Cape Canso, and traverse the picturesque Gut of Canso. They call at Pictou and then run across to Charlottetown. By leaving the vessel at Port Hawkesbury, the tourist can easily reach the Bras d'Or and other parts of the island of Cape Breton.

3. *Routes by way of Montreal and Quebec.*

Montreal may be reached by either the Central Vermont R. R., the Montreal & Boston Short Line (Passumpsic R. R.), or the Eastern and Grand Trunk lines. These routes are all described in Osgood's *New England: a Handbook for Travellers* (revised up to 1875). The most picturesque route from Quebec to the Maritime Provinces is by the vessels of the Quebec & Gulf Ports Steamship Company, which leave every week for the eastern ports of Quebec, New Brunswick, and Nova Scotia, connecting with the local lines of travel. It now seems improbable that the Intercolonial Railway can be opened to travel from (Quebec and) Rivière de Loup to Moncton and Halifax this year.

Further particulars about these lines and their accommodations, the days on which they depart for Boston, etc., may be found in their advertisements, which are grouped at the end of the book. There, also, may be found the names and addresses of the agents of the lines, from whom other information may be obtained, by letter or by personal application. The main question for the summer tourist will naturally be whether he shall go eastward by rail or by a short sea-voyage. The Editor has travelled on each of the above-mentioned lines (with one exception) and on some of them several times, and has found them well equipped and comfortable.

MARITIME-PROVINCES HANDBOOK.

NEW BRUNSWICK.

THE Province of New Brunswick is situated nearly in the centre of the North Temperate Zone, and is bounded by Maine and Quebec on the W., Quebec and the Bay of Chaleur on the N., the Gulf of St. Lawrence and the Northumberland Strait on the E., and Nova Scotia and the Bay of Fundy on the S. It is 140 M. long from E. to W., and 190 M. from N. to S., and contains 27,105 square miles. The direct coast-line (exclusive of indentations) is 410 M., which is nearly equally divided between the S. and E. shores, and is broken by many fine harbors. The Bay of Fundy on the S., and the Bay of Chaleur on the N., are of great size and commercial importance, — the former being 140 M. long by 30 – 50 M. wide; the latter being 90 M. long by 10 – 25 M. wide. The fisheries in the great bays and in the Gulf are of immense value, employing many thousand men, and attracting large American fleets. They have furnished sustenance to the people of the maritime counties, and have been the occasion of developing a race of skilful mariners. During the past 50 years 6,000 vessels have been built in this Province, valued at nearly $80,000,000. The lumber business is conducted on a vast scale on all the rivers, and the product amounts to $4,000,000 a year.

The country is generally level, and is crossed by low ridges in the N. and W. There are numerous lakes, whose scenery is generally of a sombre and monotonous character. The interior is traversed by the rivers St. John, Restigouche, Miramichi, Petitcodiac, Nepisiguit, and Richibucto, which, with their numerous tributaries, afford extensive facilities for boat-navigation. The river-fisheries of New Brunswick are renowned for their variety and richness, and attract many American sportsmen.

There are 14,000,000 acres of arable land in the Province, a great portion of which has not yet been brought into cultivation. The intervales of the rivers contain 60,000 acres, and are very rich and prolific, being fertilized by annual inundations. The chief agricultural products are wheat, buckwheat, barley, oats, potatoes, butter, and cheese ; but farming operations are still carried on in an antiquated and unscientific manner.

The climate is less inclement on the Bay of Fundy than farther inland. The mean temperature for the last ten years at St. John was, for the winter, $17\frac{1}{2}°$; spring, $37\frac{1}{2}°$; summer, $58°$; autumn, $44\frac{2}{3}°$. The thermometer ranges between —22° and 87° as the extremes marked during the past ten years.

The present domain of New Brunswick was formerly occupied by two distinct nations of Indians. The Micmacs were an offshoot of the Algonquin race, and inhabited all the sea-shore regions. They were powerful and hardy, and made daring boatmen and fishermen. The Milicetes were from the Huron nation, and inhabited the St. John valley and the inland forests, being skilful in hunting and all manner of woodcraft. They were less numerous and warlike than the Micmacs. Both tribes had a simple and beautiful theology, to which was attached a multitude of quaint mythological legends.

This region was included in the ancient domain of Acadie (or Acadia) which was granted to the Sieur De Monts by King Henri IV. of France in 1603. De Monts explored the St. John River, and planted an ephemere colony on the St. Croix, in 1604. From 1635 until 1645 the St. John Rive was the scene of the feudal wars between La Tour and Charnisay. Olive Cromwell sent an expedition in 1654, which occupied the country; bu it was restored to France by Charles II. in 1670. After the war of 1689 97, this region was again confirmed to France, and its W. boundary wa located at the St. George River, W. of Penobscot Bay. Meantime th shores of the Bay of Chaleur and the Gulf of St. Lawrence had been se tled by the French, between 1639 and 1672. The New-Englanders invad the Province in 1703, and in 1713 Acadia was ceded to England.

The French limited the cession to Nova Scotia, and fortified the line c the Missignash River, to protect the domains to the N. In 1755 a nav expedition from Boston took these forts, and also the post at St. Joh and in 1758 the whole Province was occupied by Anglo-American troop In 1763 it was surrendered to England by the Treaty of Versailles.

The Americans made several attacks on northern Acadia during tl Revolutionary War, but were prevented from holding the country by t British fleets at Halifax. At the close of the war many thousands American Loyalists retired from the United States to this and the adjoi ing countries. In 1784 New Brunswick was organized as a Provinc having been previously dependent on Nova Scotia; and in 1788 the cap tal was established at Fredericton. Immigration from Great Britain no commenced, and the forests began to give way before the lumbermen. 1 1839 the Province called out its militia on the occasion of the boundar disputes with Maine; and in 1861 it was occupied with British troops o account of the possibility of a war with the United States about the *Tren* affair. In 1865 New Brunswick refused, by a popular vote, to enter th Dominion of Canada, but it accepted the plan the next year, and becam a part of the Dominion in 1867.

The population of New Brunswick was 74,176 in 1824, 154,000 in 1840, and 285,777 in 1871.

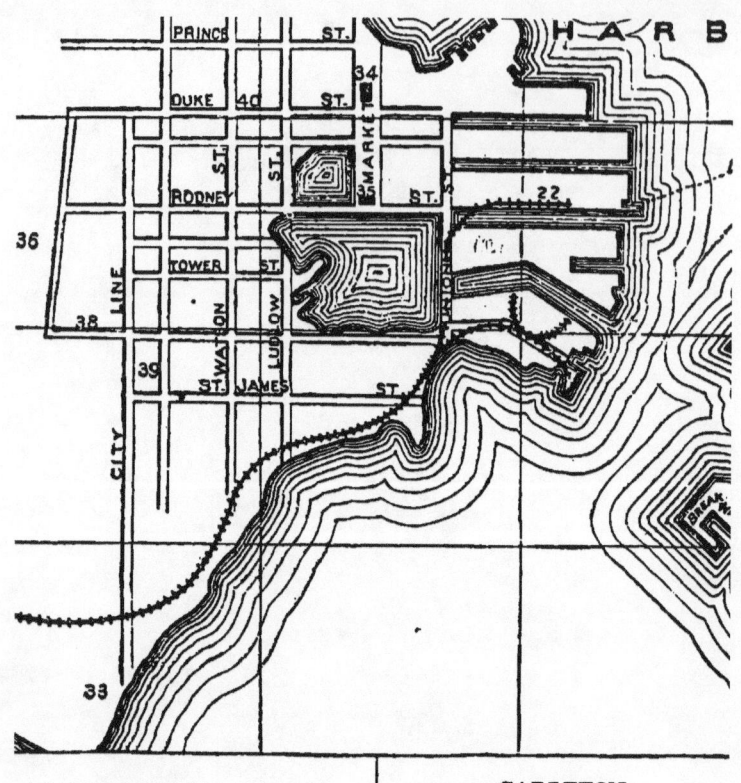

29. Portland, ...D. 1.	34. City Hall, ...B. 2.
30. Marsh Bridge, ...F. 2.	35. Market, ...B. 3.
31. Suspension Bridge, ...A. 1.	36. Martello Tower, ...A. 3.
32. Reed's Point, ...C. 4.	37. Lunatic Asylum, ...A. 1.
33. Negrotown Point, ...A. 5.	38. Church of the Assumption, ...A. 3.
	39. St. Jude's, ...A. 4.
	40. St. George's, ...A. 2

CARLETON.

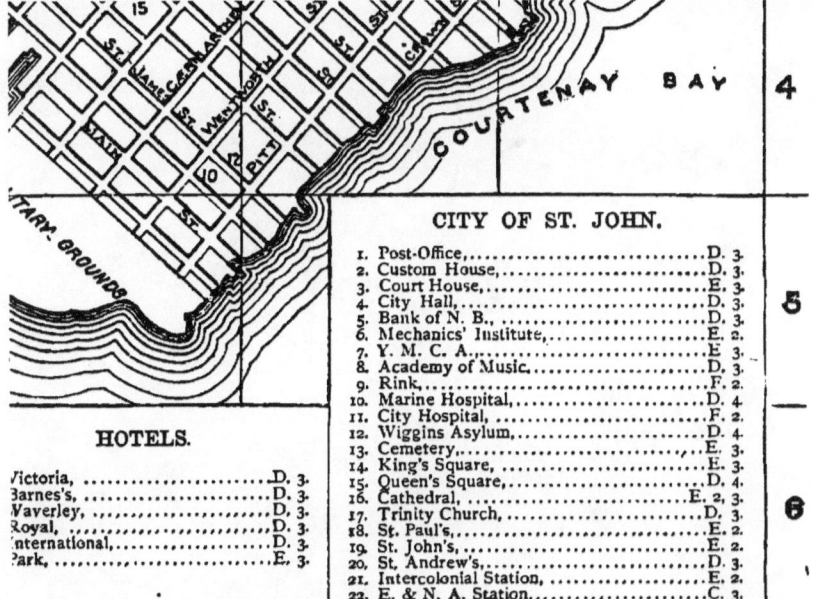

CITY OF ST. JOHN.

1. Post-Office,...........................D. 3.
2. Custom House,........................D. 3.
3. Court House,..........................E. 3.
4. City Hall,.............................D. 3.
5. Bank of N. B.,.......................D. 3.
6. Mechanics' Institute,................E. 2.
7. Y. M. C. A..........................E. 3.
8. Academy of Music...................D. 3.
9. Rink.................................F. 2.
10. Marine Hospital,...................D. 4.
11. City Hospital, ...F. 2.
12. Wiggins Asylum,....................D. 4.
13. Cemetery,..........................E. 3.
14. King's Square, ...E. 3.
15. Queen's Square,....................D. 4.
16. Cathedral, ...E. 2, 3.
17. Trinity Church, ...D. 3.
18. St. Paul's,.........................E. 2.
19. St. John's,.........................E. 2.
20. St. Andrew's,......................D. 3.
21. Intercolonial Station, ...E. 2.
22. E. & N. A. Station,................C. 3.

HOTELS.

Victoria, ...D. 3.
Barnes's, ...D. 3.
Waverley, ...D. 3.
Royal, ...D. 3.
International,....................D. 3
Park, ...E. 3.

1. St. John.

Arrival from the Sea. — Soon after passing Negro Head, the steamer runs in by *Partridge Island*, the round and rocky guard of the harbor of St. John. Its precipitous sides are seamed with deep clefts and narrow chasms, and on the upland are seen the Quarantine Hospital, the buildings of the steam fog-horn and the lighthouse, and the ruins of a cliff battery. On the l. is the bold headland of *Negrotown Point*, crowned by dilapidated earthworks. The course now leads in by the Beacon-light (l. side), with the Martello Tower on Carleton Heights, and the high-placed St. Jude's Church on the l. In front are the green slopes and barracks of the Military Grounds, beyond which are the populous hills of St. John.

Hotels. — The * Victoria, corner of Duke and Germain Sts., is the best hotel in the Maritime Provinces. It is centrally located, and accommodates 300 guests; terms, $3 a day. Barnes's Hotel, Prince William St., near Princess St; the Waverley, King St., near King Square ($2 a day); the Royal, 146 Prince William St.; the International, corner of Prince William and Duke Sts. The Park and the Continental are comfortable hotels fronting on King Square, near which are several smaller houses. The American is a second-class hotel on lower King St.; the Bay View is on Prince William St., near Reed's Point.

Amusements. — Theatrical performances and other entertainments are frequently given at the Academy of Music, on Germain St., near Duke St. The Academy can accommodate 2,000 people. Lectures and concerts are given in the hall of the Mechanics' Institute, near the head of Germain St. Varieties and minstrels at Lee's Opera-House, on Dock St.

Reading-Rooms. — The Young Men's Christian Association, on Charlotte St., near King Square; open from 9 A. M. until 10 P. M. The Mechanics' Institute, near the head of Germain St., has an extensive variety of British papers on file.

Carriages. — For a course within the city, 30c. for one passenger, 10c. for each additional one. For each half-hour, 50c. If the river is crossed the passenger pays the toll, which is, for a double carriage, 15c. each way by ferry, 20c. by the bridge.

Horse-cars run from Market Square through Dock and Mill Sts., to the terminus of the river steamboat-lines, at Indiantown (fare, 5c.).

Railways. — The European and North American Railway runs W. to Bangor in 206 M., connecting there with the Maine Central and Eastern lines for Boston, 449 M. from St. John. The same road also has a branch to Fredericton. The Intercolonial Railway runs E. to Shediac, Truro, and Halifax (276 M.).

Steamships. — The Temperley and other lines run steamships occasionally between St. John and Liverpool, or London. The steamship *Linda* leaves St. John every Friday evening for Boston, touching at Yarmouth, N. S. The International Steamship Company despatches one of their vessels every Monday, Wednesday, and Friday morning, at 8 o'clock, for Boston, touching at Eastport and Portland, and connecting with a steamer for St. Andrews and St. Stephen. A steamer leaves the Reed's Point Wharf, at 8 A. M., every Thursday and Saturday, for St. George, St. Andrews, and St. Stephen (calling at Beaver Harbor on Saturdays). The *Empress* crosses the Bay of Fundy to Digby and Annapolis, on Monday, Wednesday, Friday, and Saturday, at 8 A. M., connecting at Annapolis with the railway for Halifax. A steamer leaves the Reed's Point Wharf every Tuesday evening for Parrsboro', Windsor, and the ports on the Basin of Minas.

St. John River Lines. — The *David Weston*, of the Union Line, leaves Indiantown on Tuesday, Thursday, and Saturday, at 9 A. M., for Fredericton and the intermediate landings. The *Rothesay*, of the Express Line, leaves Indiantown Monday, Wednesday, and Friday, at 9 A. M., for Fredericton and the intermediate landings. The *May Queen* leaves Indiantown on Wednesday and Saturday, at 8 A. M., for Gagetown and Grand Lake. The *Star* leaves Indiantown on Tuesday, Thursday, and Saturday, at 10 A. M., for Cole's Island and the Washademoak Lake.

The Carleton ferry-steamers leave the foot of Princess St. every 15 minutes until 9.30 P. M. Fare, 3c.; for one-horse carriages, 9c.; for two-horse carriages, 15c.

ST. JOHN, the chief city of the Province of New Brunswick and the commercial metropolis of the Bay of Fundy, occupies a commanding position at the mouth of the St. John River. From its favorable situation for the purposes of commerce it has been termed "the Liverpool of America" (a claim, however, which Halifax stoutly combats, and which should be limited at least to "the Liverpool of Canada"). The city has 28,805 inhabitants (census of 1871), and the contiguous suburb of Portland has 12,520 more. The ridge upon which it is built is composed of solid rock, through which streets have been cut at great expense; and the plan of the streets is regular, including a succession of rectangular squares. The general appearance of the city is, however, somewhat uneven and dingy, owing to the difference in the size of the buildings and to the absence of paint. The harbor is good, and is kept free from ice by the high tides of the Bay of Fundy and the sweeping current of the St. John River. It is usually well filled with shipping, and the shores are lined with wharves and mills. The hill-country in the vicinity is barren but picturesque, and affords a variety of pleasing marine views. Since 1853 the water supply of the city has been drawn from Little River, and the works have a daily capacity of 5,500,000 gallons. The fire department has 3 steam-engines, but is seldom called into service. There are 26 churches in St. John and Portland, of which the Baptists claim precedence in point of numbers. There are 4 banks, and 4 daily and several weekly newspapers.

King Street is the main business street of the city, and runs from the harbor across the peninsula to Courtenay Bay. All the principal shops are on this street, between the harbor and King Square, and along *Prince William St.*, which intersects it near the water. At the foot of the street is the *Market Slip*, into which the light packet-boats and produce-vessels from the adjacent rural counties bring wood and provisions for the use of the city. At low tide, these vessels are, for the most part, left to hold themselves up on the muddy flats. At this point landed the weary and self-exiled American Loyalists, in 1783, and founded the city of St. John. The rather dreary breadth of King St. is occupied in its lower part by wagoners and unemployed workmen. From this point the street ascends a steep hill, passing the telegraph-office, police-court, and several banks and hotels. **King Square** is entered through a pretentious "triumphal arch" of wood, which was erected in honor of Prince Arthur's visit, and has since been utilized for sustaining the fire-alarm bell. The Square is an open space of about 3 acres in area, studded with young trees, and adorned in the centre with a small fountain. To the E. is the County Market, a narrow street filled with rude stalls. A few steps N. W. of the Square (on Charlotte St.) is the new and handsome building of the Young Men's Christian Association, containing a large hall, gymnasium, parlors, and class-rooms. The library and reading-room are open daily (except Sunday) from 9 A. M. to 10 P. M., and strangers are welcomed.

The building cost $38,000, and was dedicated in 1872, but subsequently gave signs of instability, and has since been strengthened at considerable expense. The County Court House and Jail are at the S. E. corner of King Square, and are antiquated and homely stone buildings. To the E. is the *Old Burying-Ground*, containing the graves of the pioneers of the Province, with epitaphs in many cases quaint and interesting. *Trinity Church* is on Germain St., near Princess St., and is a large and plain wooden building with a spire and clock-tower. It was built in 1788, and has had subsequent additions and enlargements. The roof is sustained by two lines of wooden columns, of the Doric order ; and the walls are adorned with mural tablets and with the Royal Arms which formerly belonged to Trinity Church in New York, and was brought here by the Loyalists in 1783, having been rescued from the New York church during the great fire of 1776. Beyond Trinity is *St. Andrew's Church* (Presbyterian), with its quaint interior, by the side of which rise the lofty walls of the Victoria Hotel. By ascending the next street (Queen) to the l., *Queen Square* is reached,—a carelessly kept park surrounded with dwelling-houses. A short distance to the E., on St. James Street, is the *Wiggins Male Orphan Institution*, a new building in Gothic architecture, of red and gray sandstone. It is the most elegant and symmetrical structure of its size in the Province, and cost over $100,000, but is only adequate to the accommodation of 30 orphans. The Marine Hospital is in this vicinity.

A short walk out Sydney St. or Caermarthen St. leads to the **Military Grounds**, on the extreme S. point of the peninsula. Here is a spacious parade-ground, which is now used only by the cricket and base-ball clubs, and barracks for the accommodation of 2,000 soldiers. These grounds were formerly occupied by large detachments from the British army, whose officers were a desired acquisition to the society of the city, while the military bands amused the people by concerts on Queen Square. From the Military Grounds is obtained a series of views of the harbor and bay, with Partridge Island near at hand in the foreground.

Prince William Street runs S. from Market Square to Reed's Point, and is one of the chief thoroughfares of the city, containing several hotels and some of the largest shops. Where it crosses Princess St., the Carleton ferry is seen to the r., and on the l. is *Ritchie's Building*, the headquarters of lawyers and Freemasons. At the S. W. corner of Prince William and Princess Sts. is the new * **Post-Office**, an elegant building of gray sandstone, ornamented with columns of the polished red granite of St. George. It is surmounted by a clock tower 100 ft. high. The next building, with a classic front and one wing, is occupied by the Bank of New Brunswick, beyond which is the **Custom House**, a plain and massive stone building, which dates from 1842. It is 250 ft. long, and contains several of the provincio-national offices, and a storm-signal station which receives warnings from "Old Probabilities" at Washington. The street ends at *Reed's Point*, the headquarters of several lines of coasting-steamers, whence may be seen the Breakwater, W. of the Military Grounds.

At the N. end of Germain St. is the old *Stone Church*, a sanctuary of the Episcopalians under the invocation of St. John. Its square stone tower is visible for a long distance, on account of the elevation of the site on which it stands. Nearly opposite is the brick *Calvin Church* (Presbyterian); and in the same vicinity is the classic wooden front of the *Mechanics' Institute*, which has a large hall, and is the domicile of one of the city schools. The reading-room is supplied with Canadian and British newspapers, and the library contains about 7,000 volumes (open from 2½ to 5 o'clock). From this point roads descend to the water-side and to the railway station in the Valley.

The Roman Catholic *****Cathedral** is situated on Waterloo St., and is the largest church in the Province. It is constructed of marble and sandstone, in pointed architecture, and has a tall and graceful stone spire. The interior is in a style of the severest simplicity, the Gothic arches of the clere-story being supported on plain and massive piers. The windows are of stained glass, and are very brilliant and rich. The chancel and transept windows are large and of fine design; a rose window is placed over the organ-loft; and the side windows represent Saints Bernard, Dominic, Ambrose, Jerome, Mark, Matthew, Andrew, Benedict, Francis, John, Luke, Augustine, and Gregory. The building is 200 ft. long, and 110 ft. wide at the transepts. The *Bishop's Palace* is the fine sandstone building towards Cliff St., beyond which is the extensive building of the *Orphan Asylum*, fronting on Cliff St. On the other side of the Cathedral is the plain brick building of the Nunnery. The visitor should notice, over the Cathedral portal adjacent to the Nunnery, the great marble bas-relief of the Last Supper (after Leonardo Da Vinci's painting at Milan).

From this point Waterloo St. descends to the Marsh Bridge, at the head of Courtenay Bay. By ascending Cliff St. for a short distance, a point may be reached from which are seen the Valley, with its churches and streets, and the embowered villas on Portland Heights, over which Reed's Castle is prominent.

The *General Public Hospital* is situated on a bold rocky knoll which overlooks the Marsh Valley, and is entered from Waterloo St. It consists of a large brick building with one wing, and accommodates 80 patients. The structure pertains to the city, and was erected in 1865 at a cost of $54,000. Directly below the precipitous sides of the knoll on which it is built is the broad Marsh, covered with houses, and extending on the r. to Courtenay Bay. The geologists entertain a plausible theory that in remote ages the St. John River flowed down this valley from the Kennebecasis to the sea, until finally the present channel through the Narrows was opened by some convulsion of nature.

That suburb which is known as the Valley lies between the rocky hills of the city proper and the line of the Portland Heights. It is reached from King Square by Charlotte and Cobourg Sts., and contains the tracks and station of the Intercolonial Railway. The most prominent object in the Valley is *St. Paul's Church* (Episcopal), a graceful wooden edifice with transepts, a clere-story, and a tall spire. The windows are of stained glass. The brick church of St. Stephen and the white Zion Church (Reformed Episcopal) are also situated in the Valley, and the road to Lily Lake diverges to the r. from the latter. Farther to the E., on the City Road, is the *Skating Rink*, a round wooden building, 160 ft. in diameter, covered with a domed roof. This is the favorite winter resort of the aristocracy of St. John, and strangers can gain admission only by introduction from one of the directors.

The site of St. John was the *Menagwes* of ancient Micmac tradition, where the divine Glooscap once had his home. Hence, during his absence, his attendants were carried away by a powerful evil magician, who fled with them to Grand Manan, Cape Breton, and Newfoundland, where he was pursued by Glooscap, who rode much of the way on the backs of whales which he called in from the deep sea. Passing through Cape Breton, he at length reached the dark Newfoundland shores, where he assumed such a stature that the clouds rolled about his head. The evil-doing wizard was soon found and put to death and the servants of Glooscap were set free.

The site of St. John was discovered by Champlain and De Monts, on St. John's Day (June 24), 1604, but was not occupied for 30 years after.

Claude de la Tour, a Huguenot noble, was one of the earliest of the French adventurers in this region, and received a grant of all Acadia from Charles I. of England. After his repulse and humiliation (see Route 25), the French government divided Acadia into three provinces, placing there as governors, M. Denys, Razilly, and the young and chivalrous Charles de St. Estienne, Lord of La Tour (son of Claude). Denys contented himself with the ocean-fisheries from Canso and Cape Breton. Razilly soon died, leaving his domain to his kinsman Charles de Menou, Sieur d'Aulnay Charnisay, who was also related to Cardinal Richelieu. D'Aulnay and La Tour began to quarrel about the boundaries of their jurisdictions, and the former employed a powerful influence at the Court of France to aid his cause. Louis XIII. finally ordered him to carry La Tour to France, in chains, and open war ensued between these patrician adventurers. La Tour had erected a fort at St. John in 1634, whence he carried on a lucrative fur-trade with the Indians. In 1643 this stronghold was attacked by D'Aulnay with six vessels, but La Tour escaped on the ship *Clement*, leaving his garrison to hold the works. He entered Boston Harbor with 140 Huguenots of La Rochelle, and sought aid from Massachusetts against the Catholic forces which were besieging him. The austere Puritans referred to the Bible to see if they could find any precedent for such action, but found no certain response from that oracle. "On the one hand, it was said that the speech of the Prophet to Jehoshaphat, in 2d Chronicles xix. 2, and the portion of Solomon's Proverbs contained in chap. xxvi, 17th verse, not only discharged them from any obligation, but actually forbade them to assist La Tour; while, on the other hand, it was agreed that it was as lawful for them to give him succor as it was for Joshua to aid the Gibeonites against the rest of the Canaanites, or for Jehoshaphat to aid Jehoram against Moab, in which expedition Elisha was present, and did not reprove the King of Judah." But when they had assured themselves that it would be allowable for them to aid the distressed nobleman, they sent such a fleet that D'Aulnay's forces were quickly scattered, and the siege was raised. Two years later, while La Tour was absent, D'Aulnay again attacked the fort, but was handsomely repulsed (with a loss of 33 men) by the little garrison, headed by Madame La Tour. Some months later he returned, and opened a regular siege on the landward side (the fort was in Carleton, near Navy Island). After three days of fighting a treacherous Swiss sentry admitted the enemy into the works; and even then Madame La Tour led her troops so gallantly that the victor gave her her own terms. These terms, however, were shamefully violated, and the garrison was massacred before her face. Three weeks afterward, she died of a broken heart. La Tour came back to St. John some years later, and found that D'Aulnay was dead, whereupon he effectually recaptured his old domain by marrying the widow of the conqueror (1653). D'Aulnay died in 1650, having spent 800,000 livres in Acadia, and built 5 fortresses, 2 seminaries, and several churches. He had several sons, all of whom entered the French army, and were slain in the service.

In 1690 a sharp engagement took place in St. John harbor, between the French frigate *Union* and two English vessels. The former had entered the harbor bearing the Chevalier de Villebon, and was taken at a disadvantage. After a severe cannonade, the *Union* hauled down her colors. Villebon soon descended the river with a party of Indians and attacked the ships, but without success. In 1696, while the Chevalier de Villebon governed Acadia from the upper St. John and hurled destructive Indian bands upon New England, Massachusetts sent three men-of-war to blockade the mouth of the river and cut off his supplies. They were soon attacked by D'Iberville's French frigates, and made a desperate resistance. But the *Newport*, 24, was unable to withstand the heavy fire of the *Profond*, and soon lay dismasted and helpless. After her surrender the other American vessels escaped under cover of a thick fog. A new fleet from Boston soon afterwards overhauled the French frigates, cruising between Mount Desert and St. John, and captured the *Profond*, with M. de Villebon, the Governor of Acadia, on board. In 1701 the fort of St. John was dismantled by Brouillan; but in 1708 it was rebuilt, and had 4 bastions and 24 pieces of artillery.

In July, 1749, H. B. M. sloop-of-war *Albany* entered the harbor and drove away the French troops, lowering also the standard of France. The frigates *Hound* and *York* had a skirmish with the French here in 1750, and were ordered out of the port by Boishébert, the commandant of the fort. In 1755, four British war-vessels entered the harbor, and the French garrison demolished the fort, blew up the magazine, and retreated into the country. In 1758 Fort La Tour was still garrisoned by French soldiers, but, after a short siege by an Anglo-American force, the post was surrendered at discretion. Two years later, the place was visited by James Simonds, an adventurous New-Englander, who was, however, soon driven away by the Indians, "Catholics and allies of France." In 1764 he returned with a party of Massachusetts fishermen, and settled on the present site of the city, erecting defensive works on Portland Heights, under the name of Fort Howe. In 1775 a naval expedition of Americans from Machias entered the harbor and destroyed the old French fortifications (then called Fort Frederick), completing their work by plundering and bombarding the village. May 18, 1783, a British fleet arrived in the port bringing 5,000 of the self-styled "United Empire Loyalists," Americans who were loyal to King George and could not or would not remain in the new Republic of the United States. From this day may be dated the growth of the city of St. John.

New Brunswick was set off from Nova Scotia as a separate Province the next year, and in 1786 its first Legislative Assembly was convened here. In 1787 Trinity Church was founded; in 1788 harbor-lights were established on Partridge Island, and in 1799 the *Royal Gazette* was started. In 1837 one third of the commercial portion of the city was burned, involving a loss of £250,000. During the boundary dispute with the State of Maine (1839-42) the citizens were all enrolled and drilled in military exercises, in preparation for a war on the borders. Large fortunes were made by the merchants during the Crimean war, when the British timber-market, which had depended largely on the Baltic ports for its supply, was by their closing forced to draw heavily on the American Provinces. The last historic event at St. John was its occupation, in the winter of 1861, by several of the choicest regiments of the British army, among which were the Grenadier Guards, the Scotch Fusiliers, and other *élite corps*. After the peaceful solution of the *Trent* affair this formidable garrison was removed, and the city has since been left to prosper in the arts of peace and industry.

"Here is picturesque St. John, with its couple of centuries of history and tradition, its commerces, its enterprise felt all along the coast and through the settlements of the territory to the northeast, with its no doubt charming society and solid English culture; and the summer tourist, in an idle mood regarding it for a day, says it is naught." (WARNER's *Baddeck*.)

St. John. 1647.

"To the winds give our banner!
 Bear homeward again!"
Cried the Lord of Acadia,
 Cried Charles of Estienne;
From the prow of his shallop
 He gazed, as the sun,
From its bed in the ocean,
 Streamed up the St. John.

O'er the blue western waters
 That shallop had passed,
Where the mists of Penobscot
 Clung damp on her mast.
St. Savior had looked
 On the heretic sail,
As the songs of the Huguenot
 Rose on the gale.

The pale, ghostly fathers
 Remembered her well,
And had cursed her while passing,
 With taper and bell,
But the men of Monhegan,
 Of Papists abhorred,
Had welcomed and feasted
 The heretic Lord.

They had loaded his shallop
 With dun-fish and ball,
With stores for his larder,
 And steel for his wall.
Pemequid, from her bastions
 And turrets of stone,
Had welcomed his coming
 With banner and gun.

And the prayers of the elders
 Had followed his way,
As homeward he glided
 Down Pentecost Bay.
O, well sped La Tour!
 For, in peril and pain,
His lady kept watch
 For his coming again.

O'er the Isle of the Pheasant
 The morning sun shone,
On the plane-trees which shaded
 The shores of St. John.
"Now why from yon battlements
 Speaks not my love?
Why waves there no banner
 My fortress above?"

Dark and wild, from his deck
 St. Estienne gazed about,
On fire-wasted dwellings,
 And silent redoubt;
From the low shattered walls
 Which the flame had o'errun,
There floated no banner,
 There thundered no gun.

But beneath the low arch
 Of its doorway there stood
A pale priest of Rome,
 In his cloak and his hood.
With the bound of a lion
 La Tour sprang to land,
On the throat of the Papist
 He fastened his hand.

"Speak, son of the Woman
 Of scarlet and sin!
What wolf has been prowling
 My castle within?"
From the grasp of the soldier
 The Jesuit broke,
Half in scorn, half in sorrow,
 He smiled as he spoke:

"No wolf, Lord of Estienne,
 Has ravaged thy hall,
But thy red-handed rival,
 With fire, steel, and ball!
On an errand of mercy
 I hitherward came,
While the walls of thy castle
 Yet spouted with flame.

"Pentagoet's dark vessels
 Were moored in the bay,
Grim sea-lions, roaring
 Aloud for their prey!"
"But what of my lady?"
 Cried Charles of Estienne.
"On the shot-crumbled turret
 Thy lady was seen:

"Half veiled in the smoke-cloud,
 Her hand grasped thy pennon,
While her dark tresses swayed
 In the hot breath of cannon!
But woe to the heretic,
 Evermore woe!
When the son of the church
 And the cross is his foe!

"In the track of the shell,
 In the path of the ball,
Pentagoet swept over
 The breach of the wall!
Steel to steel, gun to gun,
 One moment, — and then
Alone stood the victor,
 Alone with his men!

"Of its sturdy defenders,
 Thy lady alone
Saw the cross-blazoned banner
 Float over St. John."
"Let the dastard look to it!"
 Cried fiery Estienne,
"Were D'Aulnay King Louis,
 I'd free her again!"

"Alas for thy lady!
 No service from thee
Is needed by her
 Whom the Lord hath set free:
Nine days, in stern silence,
 Her thraldom she bore,
But the tenth morning came,
 And Death opened her door!"

As if suddenly smitten,
 La Tour staggered back;
His hand grasped his sword-hilt,
 His forehead grew black.
He sprang on the deck
 Of his shallop again.
"We cruise now for vengeance!
 Give way!" cried Estienne.

"Massachusetts shall hear
 Of the Huguenot's wrong,
And from island and creekside
 Her fishers shall throng!
Pentagoet shall rue
 What his Papists have done,
When his palisades echo
 The Puritan's gun!"

O, the loveliest of heavens
 Hung tenderly o'er him,
There were waves in the sunshine,
 And green isles before him:
But a pale hand was beckoning
 The Huguenot on;
And in blackness and ashes
 Behind was St. John!
   JOHN G. WHITTIER.

2. The Environs of St. John.

* **Lily Lake** is about 1 M. from King Square, and is reached by crossing the Valley and ascending Portland Heights. The road which turns to the r. from the white (Zion) church conducts past several villas and rural estates. From its end a broad path diverges to the r., leading in a few minutes to the lake, a beautiful sheet of water surrounded by high rocky banks. The environs are thickly studded with clumps of arbor-vitæ and evergreens, among which run devious rambles and pathways. No houses or other signs of civilization are seen on the shores, and the citizens wish to preserve this district in its primitive beauty by converting it into a public park. The water is of rare purity, and was used for several years to supply the city, being pumped up by expensive machinery. This is a favorite place for skating early in the season, and at that time presents a scene of great activity and interest. A pleasant pathway leads on one side to the *Lily Lake Falls*, which are attractive in time of high water.

The **Marsh Road** is the favorite drive for the citizens of St. John, and presents a busy scene on pleasant Sundays and during the season of sleighing. It is broad, firm, and level, and follows the (supposed) ancient bed of the St. John River. At 1½ M. from the city the *Rural Cemetery* is reached (only lot-owners are admitted on Sunday). This is a pleasant ground occupying about 12 acres along a cluster of high, rocky knolls, and its roads curve gracefully through an almost unbroken forest of old (but small) evergreen trees. The chief point of interest is along Ocean Avenue, where beneath uniform monuments are buried a large number of sailors. 1½ M. beyond the Cemetery the Marsh Road passes the Three-Mile House and *Moosepath Park*, a half-mile course which is much used for horse-racing, especially during the month of August. 3 - 4 M. farther on (with the Intercolonial Railway always near at hand) the road reaches the *Torryburn House*, near the usual course for boat-racing on the broad **Kennebecasis Bay**. The course of this estuary is now followed for 2 M., with the high cliff called the *Minister's Face* on the farther shore. Passing several country-seats, the tourist arrives at **Rothesay**, prettily situated on the Kennebecasis. This village is a favorite place of summer residence for families from the city, and has numerous villas and picnic grounds. The facilities for boating and bathing are good. Near the railway station is *Rothesay Hall*, a summer hotel, accommodating 30 - 40 guests ($ 8 - 10 a week). There are pleasant views from this point, including the broad and lake-like Kennebecasis for many miles, the palisades of the Minister's Face, and the hamlet of Moss Glen.

Loch Lomond is about 11 M. N. E. of St. John, and is a favorite resort for its citizens. Many people go out to the lake on Saturday and remain there until Monday morning. The road crosses the Marsh Bridge and passes near the *Silver Falls*, a pretty cascade on Little River (whence the city draws its water supply). There are two small hotels near Loch Lomond, of which Bunker's is at the lower end and Dalzell's is 3-4 M. beyond, or near the head of the First Lake. These waters are much resorted to by trout-fishers, and the white trout that are found near Dalzell's Lake House are considered a delicacy. Boats and tackle are furnished at the hotels; and there is good shooting in the vicinity. The shores consist, for the most part, of low rolling hills, covered with forests. The First Lake is 4 × ½ M. in area, and is connected by a short stream with the Second Lake, which is nearly 2 M. long, and very narrow. The Third Lake is smaller than either of the others.

" An elevated ridge of hard-wood land, over which the road passes near the narrowest part, afforded me from its summit a view of the lower lake, which would not suffer in comparison with many either of our English or our Scottish lakes. Its surface was calm and still; beyond it rose a wooded ridge of rounded hills, purpled by the broad-leaved trees which covered them, and terminated at the foot of the lake by a lofty, so-called Lion's Back, lower considerably than Arthur's Seat, yet still a miniature Ben Lomond." — PROF. JOHNSTON.

Ben Lomond, Jones, Taylor's, and other so-called lakes (being large forest-ponds) are situated in this neighborhood, and afford better fishing facilities than the much-visited waters of Loch Lomond. Both white and speckled trout are caught in great numbers from rafts or floats on these ponds; and Bunker's or Dalzell's affords a favorable headquarters for the sportsman, where also more particular information may be obtained.

The *Penitentiary* is a granite building 120 ft. long, situated in an in-walled tract of 18 acres, on the farther side of Courtenay Bay. The *Poor House* is a spacious brick building in the same neighborhood. The road that passes these institutions is prolonged as far as *Mispeck*, traversing a diversified country, and at times affording pretty views of the Bay of Fundy. Mispeck is a small marine hamlet, 10 M. from St. John.

4 M. N. of the city is the estate of the *Highland Park Company*, an association of citizens who have united for the purpose of securing rural homes in a beautiful and picturesque region. There are three lakes on the tract (which includes 500 acres), the chief of which is *Howe's Lake*, a small but pretty forest-pond.

The * Suspension Bridge is about 1½ M. from King Square, and most of the distance may be traversed by horse-cars, passing through the town of Portland and under Fort Howe Hill (whence a good view of the city is afforded). The bridge crosses the rocky gorge into which the wide waters of the St. John River are compressed, at a height of nearly 100 ft. above low water. The rush of the upward tide, and the falls which become visible at low tide, fill the stream with seething eddies and whirls and render navigation impossible. At a certain stage of the flood-tide, and for a few minutes only, this gorge may be passed by vessels and rafts.

The St. John River is over 450 M. long, and, with its many tributaries, drains a vast extent of country. Yet, at this point, where its waters are emptied into the harbor, the outlet of the river is narrowed to a channel which is in places but 450 ft. wide, with cliffs of limestone 100 ft. high hemming it in on either side. The stream rushes through this narrow pass with great impetuosity, and its course is further disturbed by several rocky islets. The tides in the harbor rise to a height of 22-26 ft., and rush up the river with such force as to overflow the falls and produce level water at flood-tide. The bridge was built in 1852 by an American engineer, and cost $80,000. It is 640 ft. long and contains 570 M. of wire, supported on 4 slender but solid towers. One-horse carriages pay 13c. toll; 2-horse carriages, 20c.

Over the head of the bridge, on the Carleton shore, is the *Provincial Lunatic Asylum*, an extensive brick building with long wings, situated in pleasant grounds. Its elevated situation renders it a prominent object in approaching the city from almost any direction. The building was erected in 1848, and accommodates 200 patients. From this vicinity, or from the bridge, are seen the busy manufacturing villages about Indiantown and Point Pleasant, most of which are engaged in the lumber business.

On the summit of the highest hill in Carleton is a venerable and picturesque stone tower, which gives an antique and feudal air to the landscape. It is known as the Martello Tower, and was built for a harbor-defence at the time when this peculiar kind of fortification was favored by the British War Office. Many of these works may be seen along the shores of the British Isles, but they are now used (if used at all) only as coast-guard stations. The tower in Carleton is under the charge of a sub-officer, and near by are seen the remains of a hill-battery, with a few old guns still in position. The * view from this point is broad and beautiful, including St. John, with the Victoria Hotel and the Cathedral most prominent, Portland and the Fort Howe Hill, the wharves of Carleton and its pretty churches, the harbor and shipping, the broad Bay of Fundy, extending to the horizon, and in the S. the blue shores of Nova Scotia (the North Mt.), with the deep gap at the entrance to the Annapolis Basin, called the Digby Gut.

The streets of Carleton are as yet in a transition state, and do not invite a long sojourn. On the hill near the Martello Tower is the tall and graceful *Church of the Assumption*, with pleasant grounds, in which is the fine building of the presbytery. Below this point is the Convent of St. Vincent, S. of which is seen the spire of St. Jude's Episcopal Church.

The **Fern Ledges** are about 1 M. from Carleton, on the shore, and are much visited by geologists. They consist of an erratic fragment of the Old Red Sandstone epoch, and are covered with sea-weed and limpets. On clearing away the weeds and breaking the rock, the most beautiful impressions of ferns and other cryptogamous plants are found.

The **Mahogany**[1] **Road** affords a fine drive along the Bay shore, with a succession of broad marine views. It is gained by crossing the Suspension Bridge and passing the Insane Asylum. About 4 M. from the city is the *Four-Mile House*, a favorite objective point for drives. The road is often followed as far as *Spruce Lake*, a fine sheet of water 5 M. long, and situated about 7 M. from St. John. Perch are found here in great numbers, but the facilities for fishing are not good. The water supply of the suburb of Carleton is drawn from this lake.

---

[1] Mahogany, a popular adaptation of the Indian word *Manawagonish*, applied to the neighboring bay.

3. St. John to Eastport and St. Stephen. — Passamaquoddy Bay.

The commodious vessels of the International Steamship Company leave the Reed's Point Wharf, at St. John, every Monday, Wednesday, and Friday, at 8 A. M., and reach Eastport (60 M. distant) a little after noon. A connection is made there with the light steamboat *Belle Brown*, which ascends Passamaquoddy Bay and the St. Croix River to St. Andrews and St. Stephen.

Travellers who wish to gain a thorough idea of the quaintly picturesque scenery of Passamaquoddy Bay would do well to go to St. Stephen by Route 3 and return to St. John by Route 5, or *vice versa*. Except during very stormy weather the waters of Passamaquoddy Bay are quiet and without much swell.

After leaving St. John, the steamer runs S. W. into the Bay of Fundy, and soon passes Split Rock, and stretches across to Point Lepreau. The peculiarities of the coast, which is always visible (in clear weather) on the N., are spoken of in Route 5, and are thus epitomized by Mr. Warner: "A pretty bay now and then, a rocky cove with scant foliage, a lighthouse, a rude cabin, a level land, monotonous and without noble forests,— this was New Brunswick as we coasted along it under the most favorable circumstances."

After passing the iron-bound islets called the Wolves (where the *New England* was wrecked in 1872), the steamer runs in towards the *West Isles*, whose knob-like hills rise boldly from the blue waters. Sometimes she meets, in these outer passages, great fleets of fishing-boats, either drifting over schools of fish, or, with their white and red sails stretched, pursuing their prey. If such a meeting occurs during one of the heavy fogs which so often visit this coast, a wonderfully weird effect is caused by the sudden emergence and disappearance of the boats in the dense white clouds.

Soon after passing the White Horse islet, the steamer enters the Eastern Passage, and runs to the S. W. into Friar's Road. On the r. is *Deer Isle*, a rugged island, 7 M. long by 3 M. wide, with a poor soil and no good harbors. There are about 1,000 inhabitants on this island, and it is surrounded by an archipelago of isolated rocky peaks. The shores attain an elevation of 300 ft., and from some of the higher hills are gained beautiful panoramic views of the Passamaquoddy Bay, on one side, and the Bay of Fundy, on the other.

Campobello Island lies on the left side of the course, with bold and rocky shores. It is 8 M. long by 3 M. wide, and contains numerous profitable farms. On its N. point is a lighthouse, below which is the entrance to the fine harbor of *Welchpool*, where there is a pretty marine village. Wilson's Beach is a populous fishing-settlement on the S. shore; and the island contains over 1,000 inhabitants. The surrounding waters are rich in fisheries, especially of herring and haddock, which are followed by the island flotillas; and the hills are said to yield copper, lead, and plaster. The proximity of the lower shores to the American towns of Lubec and Eastport affords favorable opportunities for smuggling, which was formerly practised to a considerable extent. The island is frequently visited by summer tourists, on account of the fine marine scenery on its ocean front and for the sport afforded by the deep-sea fishing. Some years ago there was much talk of erecting a first-class hotel on the east shore, but the project now lies in abeyance. The view from the abrupt heights of *Brucker's Hill* embraces a wide expanse of blue waters, studded with an archipelago of islets. On the W. shore is the singular group of rocks known as the *Friar's Face*, which has been a favorite target for marine artillery.

The earliest settlement on the Bay was established about 1770, by the Campobello Company, and was located at Harbor de Lute, on Campobello Island. It was named Warrington, but the Welchpool settlement has long since surpassed it. The island was for some time the property of Capt. Owen, of the Royal Navy, to whom the residents paid tenants' dues. At certain stages of the tide, Eastport can only be approached by passing around Campobello, concerning which Mr. Warner indulges in the following pleasantry: "The possession by the British of the island of Campobello is an insufferable menace and impertinence. I write with a full knowledge of what war is. We ought to instantly dislodge the British from Campobello. It entirely shuts up and commands our harbor, — one of our chief Eastern harbors and war stations, where we keep a flag and cannon and some soldiers, and where the customs officers look out for smuggling. There is no way to get into our own harbor, except in favorable circumstances of the tide, without begging the courtesy of a passage through British waters. Why is England permitted to stretch along down our coast in this straggling and inquisitive manner? She might almost as well own Long Island. It was impossible to prevent our cheeks mantling with shame as we thought of this, and saw ourselves, free American citizens, landlocked by alien soil in our own harbor. We ought to have war, if war is necessary to possess Campobello and Deer Islands, or else we ought to give the British Eastport. I am not sure but the latter would be the better course."

Eastport (**Passamaquoddy House*, $2.50 a day; *Tuttle's Hotel*, $2) is an American border-town, on the coast of Maine, and has 3,738 inhabitants and 8 churches. It is built on the slope of a hill at the E. end of Moose Island, in Passamaquoddy Bay, and is engaged in the fisheries and the coasting-trade. Over the village are the ramparts of **Fort Sullivan**, a garrisoned post of the United States, commanding the harbor with its artillery. Eastport is much visited in summer for the sake of the saltwater fishing and the unique marine scenery in the vicinity, and has several reputable boarding-houses. It is connected with the mainland by a bridge, over which lies the road to the Indian village. Eastport is the most convenient point from which to reach Campobello, Grand Manan (see Route 4), and the adjacent islands. A steam-ferry runs hence in 3 M. to **Lubec** (*Lubec House, Cobscook Hotel*), a picturesque marine village towards Quoddy Head, with advantages for summer residents. This pleasant little place is decaying slowly, having lost over 400 inhabitants between 1860 and 1870. The present population is a little over 2,000. Lubec is 1 M. farther E. than Eastport, and is therefore the easternmost town of the United States. The purple cliffs of Grand Manan are seen from Quoddy Head.

In 1684 the Passamaquoddy islands were granted by the King of France to Jean Sarreau de St. Aubin. In the summer of 1704 the few French settlers about Passamaquoddy Bay were plundered by an expedition under Col. Church, consisting of 600 Massachusetts soldiers, escorted by the men-of-war *Jersey*, 48, and *Gosport*, 32. They ascended the St. Croix as far as the head of navigation, then returned and crossed the bay to ravage the Minas settlements. They visited Moose Island and the adjacent main, and carried off all the settlers as prisoners. Eighteen years later a Boston ship was captured by the Indians among these islands, but was retaken by its crew when a fair wind arose. In 1744 Massachusetts declared war against the Indians on this bay and on the St. John River; and in 1760 the tribes sued for peace, sending hostages to Boston. In 1734 Gov. Belcher (of Mass.) visited the bay, and in 1750 and 1762 its shores and islands were regularly surveyed.

During the War of the Revolution the Passamaquoddy Indians were loyal to the United States, and declined all offers from the British agents. The boundary question began to assume great importance after the close of the war. The treaty stipulated that the St. Croix River should form the frontier; but Massachusetts, supported by the Indians, claimed that the Magaguadavic was the true St. Croix; while Great Britain asserted and proved that the outlet of the Schoodic Lakes was the veritable river. The islands were surrendered to Britain; but Moose, Dudley, and Frederick Islands were restored to the United States in 1818.

Eastport was founded about 1784, by fishermen from the coast of Essex County, Mass., who settled here on account of the facilities for catching and curing fish. In 1808 the walls of Fort Sullivan were raised, and a detachment of troops was stationed there. In 1813 the valuable British vessel, the *Eliza Ann*, was captured by the privateer *Timothy Pickering* and sent into Eastport. She was followed by H. M. S. *Martin*, whose commander demanded her surrender, on pain of destroying the town. The citizens refused to release the prize, and the *Martin* opened fire on Eastport, but was soon driven away by the guns of the fort. July 11, 1814, a British fleet appeared off the town, and informed the commander that if he did not haul down his flag within five minutes they would bombard the town. The flag came down, the garrison laid down their arms, and the hostile fleet, headed by the *Ramilies*, 74, anchored off the town. British martial law was enforced here for the next four years, after which the place was restored to the United States.

The steamer *Belle Brown*, in ascending the bay, runs for some distance between Deer Isle and Moose Island. At about 5 M. from Eastport, Pleasant Point (known to the Indians as *Sybaik*) is seen on the l. Here is the chief settlement of the Passamaquoddy Indians, who were driven from the peninsula of St. Andrews nearly a century ago, and received their present domain from the American government. They are about 400 in number, and draw an annuity and a school-fund from the Republic.

They are the remnant of the ancient Openango tribe of the Etchemin nation, and they cling tenaciously to the faith delivered unto them of old by the Jesuits. Their church is dedicated to St. Anne, and is served by Indian deacons; and the picturesque cemetery is in the same vicinity. They support themselves by hunting, fishing, and basket-making, and their favorite amusement is dancing, for which they have built a hall. There are scarcely any pure-blooded Indians here, but the adulteration has been made with a choicer material than among the other tribes, since these are mostly French half-breeds, in distinction from the negro half-breeds of the lower coasts. Many years ago there was a controversy about the chieftaincy, in consequence of which a portion of the tribe seceded, and are now settled on the Schoodic Lakes.

The name *Passamaquoddy* is said to be derived from *Pesmo-acadie*, "pollock-place." Others say that *Quoddy* means "pollock"; but Father Vetromile, the scholarly Jesuit missionary, claims that the whole word is a corruption of the Indian *Peskamaquontik*, derived from *Peskadaminkkanti*, a term which signifies "it goes up into the open field."

As the bay is entered, above Pleasant Point, the West Isles are seen opening on the r., displaying a great variety of forms and combinations. On the l. are the pleasant shores of Perry, and far across, to the r., are the highlands about the Magaguadavic River. After passing Navy Island, the boat rounds in at St. Andrews.

St. Andrews, the St. Croix River, and St. Stephen, see pages 33 – 36.

4. Grand Manan.

This "paradise of cliffs" is situated off Quoddy Head, about 7 M. from the Maine coast, and pertains to the Province of New Brunswick. It is easily reached from Eastport (during fair winds), with which it has a mail communication. The summer climate would be delicious were it not for the fogs; and it is claimed that invalids suffering from gout and dyspepsia receive much benefit here (very likely from the enforced abstinence from rich food). The brooks and the many freshwater ponds afford fair trouting and bird-shooting, and a few deer and rabbits are found in the woods. There are no bears nor reptiles on the island. There is a small inn at Grand Harbor, but the sojourner will prefer to get board in some of the private houses. Neat rooms and simple fare may there be obtained for $4 – 7 a week.

"As we advanced, Manan gradually rose above the waves and changed its aspect, the flat-topped purple wall being transmuted into brown, rugged, perpendicular cliffs, crowned with dark green foliage. Passing, as we did, close in by the extreme northern point, we were impressed by its beauty and grandeur, which far exceeds even that of the cliffs at Mount Desert.

"As a place of summer resort, Grand Manan is in some respects unequalled. At certain seasons the fog is abundant, yet that can be endured. Here the opportunities for recreation are unequalled, and all persons fond of grand sea-shore views may indulge their taste without limit. The people are invariably kind and trustworthy, and American manners and customs prevail to such an extent that travellers at once feel at home." (DE COSTA.)

The island of Grand Manan is 22 M. long and 3 – 6 M. wide, and lies in the mouth of the Bay of Fundy, whose powerful tides sweep impetuously by its shores. It has about 1,500 inhabitants, who dwell along the road which connects the harbors on the E. shore, and are famous for their daring and expertness in the fisheries. They have 3 schools, 5 churches (mostly Baptist), and a military organization; while the advantages of free-trade, insignificant taxation, government-built roads, and complete self-legislation, give reason for the apostrophe, "Happy Mananites, who, free from grinding taxation, now rove out from rock-bound coves, and quarry at will in the silvery mines of the sea!" The harbors on the E. shore afford safe shelter for small vessels, and are connected with the great cliffs on the W. by narrow roads through the woods. The fisheries of cod, herring, and haddock are very extensive in this vicinity, and form the chief resource of the people, who are distinguished for the quaint simplicity which usually pertains to small and insulated maritime communities. Grand Manan has been for many years a favorite resort for American marine painters, who find excellent studies in its picturesque cliffs and billowy seas. It was visited by Champlain in 1605, but was occupied only by the Indians for 180 years after. Col. Allan, the American commander in E. Maine during the Revolution, held the island with his Indian auxiliaries, but it was finally ceded to Great Britain. After the war it was settled by several Loyalists from Massachusetts, chief among whom was Moses Gerrish. A recent writer demands that the island be fortified and developed, claiming that its situation, either for commerce or war, is strategically as valuable as those of the Isle of Man, Guernsey, and Jersey, and that it would make a fine point of attack against Portland and the coast of Maine.

Grand Harbor is the chief of the island hamlets, and is situated on the safe and shallow bay of the same name. It has an Episcopal church of stone and two or three stores, besides a small inn. Off shore to the S. E. lie Ross, Cheyne, and White Head Islands, on the latter of which Audubon studied the habits of the herring-gulls, in 1833. To the E. are the rock-bound shores of Nantucket Island, and on the S. are the Grand Ponds.

The South Shore is reached by a good road leading down from Grand Harbor. At 5 M. distance is the narrow harbor of *Seal Cove*, beyond which the road lies nearer to the sea, affording fine marine views on the l., including the Wood Islands and the Gannet Rock Lighthouse, 9 - 10 M. at sea. 4 M. beyond Seal Cove the road reaches *Broad Cove*, whence a path leads across the downs for about 2 M. to the high and ocean-viewing cliffs of S. W. Head. Among the rugged and surf-beaten rocks of this bold promontory is one which is called the Old Maid, from its rude resemblance to a colossal woman. About the S. W. Head is a favorite resort and breeding-place of the gulls, whose nests are made in the grass. A forest-path leads N. to Bradford's Cove, on the W. shore, a wide bight of the sea in which the ship *Mavourneen* was wrecked.

The North Shore. The road from Grand Harbor to Whale Cove is 7 - 8 M. long, and is firm and well-made. 3 M. N. of Grand Harbor, Woodward's Cove is passed, with its neat hamlet, 4 M. beyond which is Flagg's Cove. *Sprague's Cove* is a pretty fishing-hamlet on the S. side of Swallow-Tail Head, where "everything appears to have been arranged for artistic effect. The old boats, the tumble-down storehouses, the picturesque costumes, the breaking surf, and all the miscellaneous paraphernalia of such a place, set off as they are by the noble background of richly-colored cliffs, produce an effect that is as rare as beautiful." *Swallow-Tail Head* is a fan-shaped peninsula, surrounded by wave-worn cliffs, and swept by gales from every quarter. On its outer point is a lighthouse which holds a fixed light (visible for 17 M.) 148 ft. above the sea.

Whale Cove is on the N. E. shore, and is bordered by a shingle-beach on which are found bits of porphyry, agate, jasper, and other minerals. " Here the view is surprisingly fine, the entire shore being encircled by immense cliffs that rise up around the border of the blue waves, with a richness of color and stateliness of aspect that cannot fail to impress the beholder..... On the E. side is Fish Head, and on the W. Eel Brook and Northern Head, the latter extending out beyond its neighbor, and between are the blue sky and water." On the melancholy cliffs at Eel Brook Cove the ship *Lord Ashburton* was wrecked, and nearly all on board were lost (21 of them are buried at Flagg's Cove). Beyond this point, and near the extreme northern cape, is the *Bishop's Head*, so called because of a vague profile in the face of the cliff.

The W. coast of Grand Manan is lined with a succession of massive cliffs, which appear from West Quoddy like a long and unbroken purple wall. These great precipices are 3–400 ft. high (attaining their greatest elevation at the N. end), and form noble combinations of marine scenery. A cart-track leads across the island from near Woodward's Cove to the romantic scenery about *Dark Cove;* near which is Money Cove, so named because search has been made there for some of Capt. Kidd's buried treasures. To the N. is *Indian Beach*, where several lodges of the Passamaquoddy tribe pass the summer, attending to the shore fishery of porpoises. Still farther N. are the rocky palisades and whirling currents of Long's Eddy.

"When the cliff is brought out on such a stupendous scale as at Grand Manan, with all the accessories of a wild ocean shore, the interest becomes absorbing. The other parts of the island are of course invested with much interest. The low eastern shore, fringed with small islands and rocks, affords many picturesque sights. In a pleasant day a walk southward has many charms. The bright sky, the shingle beach, the picturesque boats, and blue land-locked bays continually enforce the admiration of an artistic eye, and allure the pedestrian on past cape, cove, and reach, until he suddenly finds that miles of ground intervene between him and his dinner." (DE COSTA.)

"Grand Manan, a favorite summer haunt of the painter, is the very throne of the bold and romantic. The high precipitous shores, but for the woods which beautify them, are quite in the style of Labrador." (L. L. NOBLE.)

Charlevoix speaks of an old-time wonder which seems to have passed away from these shores: "It is even asserted that at ¾ of a league off Isle Menane, which serves as a guide to vessels to enter St. John's River, there is a rock, almost always covered by the sea, which is of lapis-lazuli. It is added that Commander de Razilli broke off a piece, which he sent to France, and Sieur Denys, who had seen it, says that it was valued at ten crowns an ounce."

5. St. John to St. Andrews and St. Stephen.—Passamaquoddy Bay.

The steamer leaves the Reed's Point Wharf every Thursday and Saturday, at 8 A. M., and reaches St. Stephen before dark. She returns from St. Stephen every Monday and Friday morning. Fares, St. John to St. George, $1.75; to St. Andrews, $1.50; to St. Stephen, $1.75. This route was served in 1874 by the famous Cuban blockade-runner *Edgar Stuart*, but another vessel will run here in 1875.

St. John to St. Andrews by stage.

The Royal Mail traverses this route daily over roads which are rugged and tiresome. Distances: St. John to Fairville, 2½ M.; Spruce Lake, 7; Prince of Wales, 11; Musquash, 14; Lepreau, 25; New River, 33; Pennfield, 39; St. George, 45; Bocabec, 55; St. Andrews, 65. Fare, $4. The Bay Shore Railway is a new line which was recently projected, and is intended to follow the direction of this mail-route.

After leaving the harbor of St. John the steamer runs S. W. by W. 9½ M., passing the openings of Manawagonish Bay and Pisarinco Cove. The course is laid well out in the Bay of Fundy, which "wears a beautiful aspect in fine summer weather, — a soft chalky hue quite different from the stern blue of the sea on the Atlantic shores, and somewhat approaching the summer tints of the channel on the coast of England." Beyond the point of Split Rock, *Musquash Harbor* is seen opening to the N. It is a safe and beautiful haven, 2 M. long and very deep, at whose head is the pretty Episcopal village of *Musquash* (Musquash Hotel), with several lumber-mills. About two centuries ago a French war-vessel was driven into this harbor and destroyed by a British cruiser. From Split Rock the course is W. ½ S. for 11½ M. to Point Lepreau, passing the openings of Chance Harbor and Dipper Harbor, in which are obscure marine hamlets. In the latter, many years ago, the frigate *Plumper* was wrecked, with a large amount of specie on board. The harbor is now visited mostly by lobster-fishers. *Point Lepreau* is a bold and tide-swept promontory, on which are two fixed lights, visible for 18 and 20 M. at sea.

The traveller will doubtless be amazed at the rudeness and sterility of these frowning shores. "Two very different impressions in regard to the Province of New Brunswick will be produced on the mind of the stranger, according as he contents himself with visiting the towns and inspecting the lands which lie along the sea-board, or ascends its rivers, or penetrates by its numerous roads into the interior of its more central and northern counties. In the former case he will feel like the traveller who enters Sweden by the harbors of Stockholm and Gottenburg, or who sails among the rocks on the western coast of Norway. The naked cliffs or shelving shores of granite or other hardened rocks, and the unvarying pine forests, awaken in his mind ideas of hopeless desolation, and poverty and barrenness appear necessarily to dwell within the iron-bound shores..... But on the other hand, if the stranger penetrate beyond the Atlantic shores of the Province and travel through the interior, he will be struck by the number and beauty of its rivers, by the fertility of its river islands and intervales, and by the great extent and excellent condition of its roads." (PROF. J. F. W. JOHNSTON, F. R. S.)

From Point Lepreau the course is laid nearly W. for 16½ M. to Bliss Island, crossing the bight of Mace's Bay, a wide and shallow estuary in which are two fishing-hamlets. The Saturday steamer stops on this reach at *Beaver Harbor*, a place of 150 inhabitants. S. of this harbor, and seen on the l. of the course, are the five black and dangerous islets called the *Wolves*, much dreaded by navigators. A vessel of the International Steamship Company was wrecked here two or three years ago. One of the Wolves bears a revolving light, 111 ft. high, and visible for 16 M.

The steamer now rounds Bliss Island (which has a fixed red light), and to the N. is seen the entrance to *L'Etang Harbor*, a deep and picturesque inlet which is well sheltered by islands, the largest of which is called Caitiff. A few miles S. W. are seen the rolling hills of Campobello; Deer Island is nearer, on the W.; and the bay is studded with weird-looking hummocky islands, — the Nubble, White, and Spruce Islands, the grim trap-rock mamelon of White Horse, and many other nameless rocks. They are known as the *West Isles*, and most of them are inhabited by hard-working fishermen. The course is laid to the N. W. through the *Letite Passage*, between MacMaster Island and the Peninsula of Mascarene, and *Passamaquoddy Bay* is entered. Sweeping up to the N., along and close to a bold shore 150 – 225 ft. high, the steamer rounds the Mijic Bluff on the r. and enters the harbor at the mouth of the *Magaguadavic River*. To the N. are the wooded slopes of Mount Blair, and some distance up the estuary is the hamlet of Mascarene. The vessel drifts about in the harbor while passengers and freight are transferred to the dingy little steamer that ascends to St. George.

St. George (three inns) is a town of 600 inhabitants, devoted to the lumber-trade, and situated about 10 M. from the mouth of the river. It has 4 churches, a masonic hall, and a custom-house. It is at the head of tide, and ships can load, in the deep water below, all the year round. This district has recently become celebrated for its production of a fine granite, of a rose-red color, which receives a high polish, and is being introduced for ornamental columns and monuments. It resembles the beautiful Scotch granite of Peterhead (popularly called " Aberdeen granite"). At St. George are the * *Lower Falls* of the Magaguadavic, where the river is compressed into a chasm 30 ft. wide, and falls 100 ft. in five successive steps. Along the sides of the gorge are several powerful sawmills, clinging to the rocks like eagles' nests, and sluicing their lumber into the deep pools below. Geologists have found, in this vicinity, marked evidences of the action of icebergs and glaciers.

"The village, the cataract, the lake, and the elevated wilderness to the N., render this part of the country peculiarly picturesque; indeed, the neighborhood of St. George, the Digdeguash, Chamcook, and the lower St. Croix, present the traveller with some of the finest scenery in America." (DR. GESNER.)

Lake Utopia is picturesquely situated in a deep and sheltered depression among forest-covered hills, along whose slopes ledges of red granite crop out here and there. It is about 4 M. from St. George, and is 6 M. long by 1 – 2 M. wide. The road from Beaver Harbor to Gagetown follows its E. shore through an almost unbroken solitude. On a bluff over this lake the earliest pioneers found the remains of an ancient and mysterious temple, all traces of which have now passed away. Here also was found a slab of red granite, bearing a large bas-relief of a human head, in style resembling an Egyptian sculpture, and having a likeness to Washington. This remarkable medallion has been placed in the Natural History Museum at St. John. For nearly 40 years the Indians and lumbermen near the lake have told marvellous stories of a marine prodigy called " the Monster of Utopia," which dwells in this fair forest-loch. His last appearance was in 1867, when several persons about the shores claimed to have seen furious disturbances of the waters, and to have caught momentary glimpses of an animal 10 ft. thick and 30 ft. long. The lake abounds in silvery-gray trout, and its tributary streams contain many brook-trout and smelt.

Among the hills along the valley of the Magaguadavic River are the favorite haunts of large numbers of Virginian deer. Moose were formerly abundant in this region, and it is but a few years since over 400 were killed in one season, for the sake of their hides. This noble game animal has been nearly exterminated by the merciless settlers, and will soon become extinct in this district.

The **Magaguadavic River** (an Indian name meaning " The River of the Hills ") rises in a chain of lakes over 80 M. N. W., within a short portage of the Sheogomoc River, a tributary of the upper St. John. Traversing the great Lake of Magaguadavic it descends through an uninhabited and barren highland region, tersely described by an early pioneer as " a scraggly hole." Much of its lower valley is a wide intervale, which is supposed to have been an ancient lake-bottom. The river is followed closely by a rugged road, which leads to the remote Harvey and Magaguadavic settlements.

After leaving the port of St. George, the steamer runs S. W. across Passamaquoddy Bay, with the West Isles and the heights of Deer Island on the S., and other bold hummocks on either side. On the N. are the estuaries of the Digdeguash and Bocabec Rivers, and the massive ridge of the Chamcook Mt. Large fleets of fishing-boats are sometimes met in these waters, following the schools of herring or pollock. In about an hour, the steamer approaches St. Andrews, passes its great summer hotel, and lands between Navy Island and the peninsula.

St. Andrews (*Central Exchange*, $1.50 a day), the capital of Charlotte County, is finely situated on a peninsula at the mouth of the St. Croix River, which is here 2 M. wide. It has about 1,800 inhabitants, and a few quiet old streets, surrounded by a broad belt of farms. The town was founded about a century ago, and soon acquired considerable commercial importance, and had large fleets in its harbor, loading with timber for Great Britain and the West Indies. This era of prosperity was ended by the rise of the town of St. Stephen and by the operation of the Reciprocity Treaty, and for many years St. Andrews has been retrograding, until now the wharves are deserted and dilapidated, and the houses seem antiquated and neglected. It has recently attracted summer visitors, on account of the pleasant scenery and the facilities for boating, fishing, and excursions among the adjacent islands. A large and handsome summer hotel has been erected near the shore, but the enterprise of the town has not been able to furnish it, so that it is not eligible to tourists, who must therefore dwell at the village inns.

Steamboats run daily between St. Andrews and Eastport, Calais, and St. Stephen. There is a ferry to the American village of Robbinston, 2 M distant. The New Brunswick and Canada Railway runs thence to Houlton and Woodstock, 90 and 93 M. N. (See Route 6.)

The **Chamcook Mt.** is about 6 M. N. of St. Andrew, and its base is reached by a good road (visitors can also go by railway to the foot of the mountain). It is often ascended by parties for the sake of the beautiful view, which includes "the lovely Passamaquoddy Bay, with its little islands and outline recalling recollections of the Gulf of Naples as seen from the summit of Vesuvius, whilst the scenery toward the N. is hilly, with deep intervening troughs containing natural tarns, where the *togue* or gray-spotted trout is plentiful." The bright course of the St. Croix River is visible for a long distance, and numerous pretty frontier-villages are seen on either shore. "The glacial rounded top" of Chamcook is scored with the long scratches which indicate that at some remote age a glacier from the northern highlands has grated and ground its way across the mountain. The views of the Chamcook Lake and Harbor, and of the numerous conical hills to the N., are of much interest.

As the steamer swings out into the river, the little ship-building village of Robbinston is seen, on the American shore. On the r. the bold bluffs of

Chamcook Mt. are passed, and occasional farm-houses are seen along the shores. 5-6 M. above St. Andrews, the steamer passes on the E. side of **Doucet's Island**, on which a lighthouse has been erected by the American government. W. of the island is the village of *Red Beach*, with its plaster-mills, and on the opposite shore is the farming settlement of *Bay Shore*.

In the year 1604 Henri IV. of France granted a large part of America to Pierre du Guast, Sieur de Monts, and Governor of Pons. This tract extended from Philadelphia to Quebec, and was named *Acadie*, which is said to be derived from a local Indian word. De Monts sailed from Havre in April, with a motley company of impressed vagabonds, gentlemen-adventurers, and Huguenot and Catholic clergymen, the latter of whom quarrelled all the way over. After exploring parts of Nova Scotia and the Bay of Fundy, the voyagers ascended the Passamaquoddy Bay and the river to St. Croix Isle, where it was determined to found a settlement. Batteries were erected at each end, joined by palisades, within which were the houses of De Monts and Champlain, workshops, magazines, the chapel, and the barracks of the Swiss soldiery. But the winter soon set in with its intense cold, and the ravages of disease were added to the miseries of the colonists. 35 out of 79 men died of the scurvy during the winter; and when a supply-ship arrived from France, in June, the island was abandoned.

"It is meet to tell you how hard the isle of Sainte Croix is to be found out to them that never were there; for there are so many isles and great bays to go by (from St. John) before one be at it, that I wonder how one might ever pierce so far as to find it. There are three or four mountains imminent above the others, on the sides; but on the N. side, from whence the river runneth down, there is but a sharp pointed one, above two leagues distant. The woods of the main land are fair and admirable high, and well grown, as in like manner is the grass..... Now let us prepare and hoist sails. M. de Poutrincourt made the voyage into these parts, with some men of good sort, not to winter there, but as it were to seek out his seat, and find out a land that might like him. Which he having done, had no need to sojourn there any longer." Late in the year, "the most urgent things being done, and hoary snowy father being come, that is to say, Winter, then they were forced to keep within doors, and to live every one at his own home. During which time our men had three special discommodities in this island: want of wood (for that which was in the said isle was spent in buildings), lack of fresh water, and the continual watch made by night, fearing some surprise from the savages that had lodged themselves at the foot of the said island, or some other enemy. For the malediction and rage of many Christians is such, that one must take heed of them much more than of infidels." (LESCARBOT's *Nouvelle France*.)

In 1783 the river St. Croix was designated as the E. boundary of Maine, but the Americans claimed that the true St. Croix was the stream called the Magaguadavic. It then became important to find traces of De Monts's settlement of 180 years previous, as that would locate the true St. Croix River. So, after long searching among the bushes and jungle, the boundary-commissioners succeeded in finding remnants of the ancient French occupation on Neutral (Doucet's) Island, and thus fixed the line.

About 10 M. above St. Andrews the river deflects to the W., and to the N. is seen the deep and spacious * **Oak Bay**, surrounded by bold hills, and forming a beautiful and picturesque prospect. It is supposed that the French explorers named the St. Croix River from the resemblance of its waters at this point to a cross, — the upright arm being formed by the river to the S. and Oak Bay to the N., while the horizontal arm is outlined by the river to the W. and a cove and creek on the E. At the head of the bay is the populous farming-village of *Oak Bay*, with three churches.

Rounding on the l. the bold bluff called *Devil's Head* (from one Duval, who formerly lived there), the course is laid to the N. W., in a narrow channel, between sterile shores. 2-3 M. above is the antiquated marine hamlet called *The Ledge* (1. bank), most of whose inhabitants are dependent on the sea for their living. 4 M. above this point the steamer reaches her dock at St. Stephen.

St. Stephen (*Watson House*) is an active and enterprising provincial town, situated at the head of navigation on the St. Croix River, opposite the American city of Calais. The population is about 5,000, with 6 churches, 2 newspapers, and 2 banks. The business of St. Stephen is mostly connected with the manufacture and shipment of lumber. The falls of the river at this point give a valuable water-power, which will probably be devoted to general manufacturing purposes after the lumber supply begins to fail. A covered bridge connects St. Stephen with **Calais** (*International Hotel; St. Croix Exchange*), a small city of the State of Maine, with 6,000 inhabitants, 7 churches, 2 weekly papers, and 2 banks. Although under different flags, and separated by lines of customs-officers, St. Stephen and Calais form practically but one community, with identical pursuits and interests. Their citizens have always lived in perfect fraternity, and formed and kept an agreement by which they abstained from hostilities during the War of 1812. At that time the authorities also restrained the restless spirits from the back country from acts of violence across the borders. 2-3 M. above is another Canado-American town, with large lumber-mills at the falls, which is divided by the river into Milltown-St. Stephen and Milltown-Calais. Travellers who cross the river either at Calais or Milltown will have their baggage looked into by the customs-officers, squads of whom are stationed at the ends of the bridges.

The New Brunswick & Canada Railway runs N. from St. Stephen to Houlton and Woodstock (see Route 6). Calais is connected with the Schoodic Lakes by railway, and with Eastport by stages. The U. S. Mail-stage runs daily to Bangor, 95 M. W. (fare, $ 7.50), passing through a wide tract of unoccupied wilderness. The steamboat *Belle Brown* leaves Calais or St. Stephen tri-weekly for St. Andrews and Eastport, where it connects with the International steamships for Portland and Boston (see also Route 3, and Osgood's *New England*). Fares, Calais to Portland, $ 4.50; to Boston, by water, $ 5.50; to Boston, by rail from Portland, $ 7.

The Schoodic Lakes.

A railway runs 21 M. N. W. from Calais to *Lewey's Island* (2 inns), in Princeton, whence the tourist may enter the lovely and picturesque Schoodic Lakes. The steamer *Gipsey* carries visitors 12 M. up the lake to **Grand Lake Stream**, one of the most famous fishing-grounds in America. The trout in Lewey's Lake have been nearly exterminated by the voracious pike, but the upper waters are more carefully guarded, and contain perch, pickerel, land-locked salmon, lake-trout, and fine speckled-trout. The Grand Lake Stream is 3-4 M. long, and connects the Grand and Big Lakes with its rapid waters, in which are found many of the famous silvery salmon-trout. The urban parties who visit these forest-lakes usually engage Indian guides to do the heavy work of portages and camp-building, and to guide their course from lake to lake. There is a large village of the Passamaquoddy tribe near the foot of Big Lake. A two hours' portage leads to **Grand Lake**, a broad and beautiful forest-sea, with gravelly shores, picturesque islets, and transparent waters. The cry of the loon is often heard here, and a few bear and deer still lurk along the shores. From Grand Lake a labyrinth of smaller and yet more remote lakes may be entered; and portages conduct thence to the navigable tributaries of the Machias and Penobscot Rivers.

" One of the most picturesque portions of the western Schoodic region is Grand Lake. This noble sheet of water is broken here and there by islets, and surrounded, even to the water's edge, with forests of pine and hard wood, whilst its bottom is covered with granitic bowlders, which, in combination with drift, are spread far and wide among the arboreal vegetation around."

"While the fog is lifting from Schoodic Lake,
And the white trout are leaping for flies,
It's exciting sport those beauties to take,
Jogging the nerves and feasting the eyes."
GENIO C. SCOTT.

6. St. Andrews and St. Stephen to Woodstock and Houlton.

By the New Brunswick & Canada Railway. Fare from St. Stephen to Woodstock, $2.90.

Distances. — St. Andrews to Chamcook, 5 M.; Bartlett's, 11; Waweig, 13; Roix Road, 15; Hewitt's, 19; Rolling Dam, 20; Dumbarton, 24; Watt Junction, 27 (St. Stephen to Watt Junction, 19); Lawrence, 29; Barber Dam, 34; McAdam Junction, 43; Deer Lake, 59; Canterbury, 65; Eel River, 75; Wickham, 80; Debec Junction, 90 (Houlton, 98); Hodgdon, 98; Woodstock, 101.

The country traversed by this line is one of the most irredeemably desolate regions in North America. The view from the car-windows presents a continual succession of dead and dying forests, clearings bristling with stumps, and funereal clusters of blasted and fire-scorched tree-trunks. The traces of human habitation, which at wide intervals are seen in this gloomy land, are cabins of logs, where poverty and toil seem the fittest occupants; and Nature has withheld the hills and lakes with which she rudely adorns other wildernesses. The sanguine Dr. Gesner wrote a volume inviting immigration to New Brunswick, and describing its domains in language which reaches the outer verge of complaisant optimism; but in presence of the lands between the upper St. John and St. Stephen his pen lost its hyperbolical fervor. He says: "Excepting the intervales of the stream, it is necessary to speak with circumspection in regard to the general quality of the lands. Many tracts are fit for little else but pasturage." This district is occupied, for the most part, by the remains of soft-wood forests, whose soils are always inferior to those of the hard-wood districts.

For a short distance beyond St. Andrews the railway lies near the shores of Passamaquoddy Bay, affording pleasant views to the r. Then the great mass of Chamcook Mt. is passed, with its abrupt sides and rounded summit. *Waweig* is between Bonaparte Lake and Oak Bay (see page 34). About 7 M. beyond, the line approaches the Digdeguash

River, which it follows to its source. At *Watt Junction* the St. Stephen Branch Railway comes in on the l., and the train passes on to McAdam Junction, where it intersects the European & North American Railway (page 38). There is a restaurant at this station, and the passenger will have time to dine while the train is waiting for the arrival of the trains from Bangor and from St. John.

The forest is again entered, and the train passes on for 16 M. until it reaches the lumber-station at Deer Lake. The next station is *Canterbury* (small inn), the centre of extensive operations in lumber. Running N. W. for 10 M., the Eel River is crossed near Rankin's Mills, and at *Debec Junction* the passenger changes for Woodstock.

A train runs thence 8 M. N. W. to **Houlton** (*Snell House, Buzzell House*), the shire-town of Aroostook County, in the State of Maine (see Osgood's *New England*, Route 50). The other train runs N. E. down the valley of the South Brook, and in about 6 M. emerges on the highlands above the valley of the St. John River. For the ensuing 5 M. there are beautiful views of the river and its cultivated intervales, presenting a wonderful contrast to the dreary region behind. The line soon reaches its terminus at the pretty village of **Woodstock** (see Route 11).

7. St. John to Bangor.

By the European & North American Railway, in 10-12 hours.

Distances. — St. John; Carleton, ½ M.; Fairville, 4; South Bay, 7; Grand Bay, 12; Westfield, 16; Nerepis, 20; Welsford, 26; Clarendon, 30; Gaspereaux, 33; Enniskillen, 36; Hoyt, 39; .Blissville, 42; Fredericton Junction, 46; Tracy, 49; Cork, 61; Harvey, 66; Magaguadavic, 76; McAdam Junction, 85; St. Croix, 91; Vanceboro', 92; Jackson Brook, 112; Danforth, 117; Bancroft, 126; Kingman, 139; Mattawamkeag, 147; Winn, 150; Lincoln Centre, 159; Lincoln, 161; Enfield, 170; Passadumkeag, 175; Olamon, 179; Greenbush, 182; Costigan, 187; Milford, 192; Oldtown, 193; Great Works, 194; Webster, 196; Orono, 197; Basin Mills, 198; Veazie, 201; Bangor, 205. (Newport, 233; Waterville, 260; Augusta, 281; Brunswick, 315; Portland, 343; Portsmouth, 395; Newburyport, 415; Boston, 451.)

The traveller crosses the Princess St. ferry from St. John to Carleton, and takes the train at the terminal station, near the landing. The line ascends through the disordered suburb of Carleton, giving from its higher grades broad and pleasing views over the city, the harbor, and the Bay of Fundy. It soon reaches *Fairville*, a growing town near the Provincial Lunatic Asylum and the Suspension Bridge. There are numerous lumber-mills here, in the coves of the river. The train sweeps around the South Bay on a high grade, and soon reaches the Grand Bay of the St. John River, beyond which is seen the deep estuary of the Kennebecasis Bay, with its environment of dark hills. The shores of the *Long Reach* are followed for several miles, with beautiful views on the r. over the placid river and its vessels and villages (see also page 41). To the W. is a sparsely settled and rugged region in which are many lakes, — Loch Alva, the Robin Hood, Sherwood, and the Queen's Lakes.

The line leaves the Long Reach, and turns to the N. W. up the valley of the Nerepis River, which is followed as far as the hamlet of *Welsford* (small inn). The country now grows very tame and uninteresting, as the Douglas Valley is ascended. Clarendon is 7 M. from the Clarendon Settlement, with its new homes wrested from the savage forest. From Gaspereaux a wagon conveys passengers to the *South Oromocto Lake*, 10-12 M. S. W., among the highlands, a secluded sheet of water about 5 M. long, abounding in trout. Beyond the lumber station of Enniskillen, the train passes the prosperous village of Blissville; and at *Fredericton Junction* a connection is made for Fredericton, about 20 M. N.

Tracy's Mills is the next stopping-place, and is a cluster of lumber-mills on the Oromocto River, which traverses the village. On either side are wide tracts of unpopulated wilderness; and after crossing the parish of New Maryland, the line enters Manners Sutton, passes the Cork Settlement, and stops at the *Harvey Settlement*, a rugged district occupied by families from the borders of England and Scotland. To the N. and N. W. are the Bear and Cranberry Lakes, affording good fishing. A road leads S. 7-8 M. from Harvey to the **Oromocto Lake**, a fine sheet of water nearly 10 M. long and 3-4 M. wide, where many large trout are found. The neighboring forests contain various kinds of game. Near the N. W. shore of the lake is the small hamlet of Tweedside. The *Bald Mountain*, "near the Harvey Settlement, is a great mass of porphyry, with a lake (probably in the crater) near the summit. It is on the edge of the coal measures, where they touch the slate."

Magaguadavic station is at the foot of Magaguadavic Lake, which is about 8 M. long, and is visited by sportsmen. On its E. shore is the low and bristling Magaguadavic Ridge; and a chain of smaller lakes lies to the N.

The train now runs S. W. to **McAdam Junction** (restaurant in the station), where it intersects the New Brunswick and Canada Railway (see Route 6). 6 M. beyond McAdam, through a monotonous wilderness, is *St. Croix*, on the river of the same name. After crossing the river the train enters the United States, and is visited by the customs-officers at **Vanceboro'** (*Chiputneticook House*). This is the station whence the beautiful lakes of the upper Schoodic may be visited.

The **Chiputneticook Lakes** are about 45 M. in length, in a N. W. course, and are from ¼ to 10 M. in width. Their navigation is very intricate, by reason of the multitude of islets and islands, narrow passages, coves, and deep inlets, which diversity of land and water affords beautiful combinations of scenery. The islands are covered with cedar, hemlock, and birch trees; and the bold highlands which shadow the lakes are also well wooded. One of the most remarkable features of the scenery is the abundance of bowlders and ledges of fine white granite, either seen through the transparent waters or lining the shore like massive masonry. "Universal gloom and stillness reign over these lakes and the forests around them."

Beyond Vanceboro' the train passes through an almost unbroken wilderness for 55 M., during the last 16 M. following the course of the Mattawamkeag River. The station of *Mattawamkeag* is at the confluence of the Mattawamkeag and Penobscot Rivers; and the railway from thence follows the course of the latter stream, traversing a succession of thinly populated lumbering towns. 45 M. below Mattawamkeag, the Penobscot is crossed, and the train reaches **Oldtown** (two inns), a place of about 4,000 inhabitants, largely engaged in the lumber business. The traveller should notice here the immense and costly booms and mills, one of which is the largest in the world and has 100 saws at work cutting out planks.

On an island just above Oldtown is the home of the Tarratine Indians, formerly the most powerful and warlike of the Northern tribes. They were at first well-disposed towards the colonists, but after a series of wrongs and insults they took up arms in 1678, and inflicted such terrible damage on the settlements that Maine became tributary to them by the Peace of Casco. After destroying the fortress of Pemaquid to avenge an insult to their chief, St. Castin, they remained quiet for many years. The treaty of 1720 contains the substance of their present relations with the State. The declension of the tribe was marked for two centuries; but it is now slowly increasing. The people own the islands in the Penobscot, and have a revenue of $6-7,000 from the State, which the men eke out by working on the lumber-rafts, and by hunting and fishing, while the women make baskets and other trifles for sale. The island-village is without streets, and consists of many small houses built around a Catholic church. There are over 400 persons here, most of whom are half-breeds.

Below Oldtown the river is seen to be filled with booms and rafts of timber, and lined with saw-mills. At *Orono* is the State Agricultural College; and soon after passing Veazie the train enters the city of **Bangor.**

For descriptions of Bangor, the Penobscot River, and the route to Boston, see Osgood's *New England.*

8. St. John to Fredericton.—The St. John River.

The steamer *Rothesay*, of the Express Line, leaves St. John (Indiantown) at 9 A. M. on Monday, Wednesday, and Friday. The steamer *David Weston*, of the Union Line, leaves Indiantown at 9 A. M. on Tuesday, Thursday, and Saturday. See also Routes 9 and 10. These vessels are comfortably fitted up for passengers, in the manner of the smaller boats on the Hudson River. Dinner is served on board ; and Fredericton is usually reached late in the afternoon.

The scenery of the St. John River is pretty, and has a pleasing pastoral quietness. The elements of the landscapes are simple ; the settlements are few and small, and at no time will the traveller find his attention violently drawn to any passing object. There are beautiful views on the Long Reach, at Belleisle Bay, and during the approach to Fredericton, but the prevalent character of the scenery is that of quiet and restful rural lands, by which it is pleasant to drift on a balmy summer-day. Certain provincial writers have done a mischief to the St. John by bestowing upon it too extravagant praise, thereby preparing a disappointment for such as believed their report. One calls it "the Rhine of America," and another prefers it to the Hudson. This is wide exaggeration ; but if the traveller would enjoy a tranquillizing and luxurious journey through a pretty farming country, abounding in mild diversity of scenery, he should devote a day to this river.

Distances. — (The steamboat-landings bear the names of their owners, and the following itinerary bears reference rather to the villages on the shores than to the stopping-places of the boats.) St. John; Brundage's Point, 10 M. ; Westfield, 17 ; Greenwich Hill, 19 ; Oak Point, 25 ; Long Reach, 26 ; Tennant's Cove (Belleisle Bay), 29 ; Wickham, 32 ; Hampstead, 36 ; Otnabog, 41 ; Gagetown, 50 ; Upper Gagetown, 58 ; Maugerville, 72 ; Oromocto, 75 ; Glasier's, 81 ; Fredericton, 86.

Fares. — St. John to Gagetown, $1 ; to Fredericton, $1.50.

Route 8.   KENNEBECASIS BAY.

This river was called *Looshtook* (Long River) by the Etchemin Indians, and *Ouangondie* by the Micmacs. It is supposed to have been visited by De Monts, or other explorers at an early day, and in the commission of the year 1598 to the Lieut.-General of Acadia it is called *La Rivière de la Grande Baie*. But no examination was made of the upper waters until St. John's Day, 1604, when the French fleet under De Monts and Poutrincourt entered the great river. In honor of the saint on whose festival the exploration was begun, it was then entitled the St. John. After spending several weeks in ascending the stream and its connected waters, the discoverers sailed away to the south, bearing a good report of the chief river of Acadia. De Monts expected to find by this course a near route to Tadousac, on the Saguenay, and therefore sailed up as far as the depth of water would permit. " The extent of this river, the fish with which it was filled, the grapes growing on its banks, and the beauty of its scenery, were all objects of wonder and admiration." At a subsequent day the fierce struggles of the French seigneurs were waged on its shores, and the invading fleets of New England furrowed its tranquil waters.

The St. John is the chief river of the Maritime Provinces, and is over 450 M. in length, being navigable for steamers of 1,000 tons for 90 M., for light-draught steamers 270 M. (with a break at the Grand Falls), and for canoes for nearly its entire extent. It takes its rise in the great Maine forest, near the sources of the Penobscot and the Chaudière; and from the lake which heads its S. W. Branch the Indian *voyageurs* carry their canoes across the Mejarmette Portage and launch them in the Chaudière, on which they descend to Quebec. Flowing to the N. E. for over 150 M. through the Maine forest, it receives the Allagash, St. Francis, and other large streams; and from the mouth of the St. Francis nearly to the Grand Falls, a distance of 75 M., it forms the frontier between the United States and Canada. It is the chief member in that great system of rivers and lakes which has won for New Brunswick the distinction of being " the most finely watered country in the world." At Madawaska the course changes from N. E to S. E., and the sparsely settled N. W. counties of the Province are traversed, with large tributaries coming in on either side. During the last 50 M. of its course it receives the waters of the great basins of the Grand and Washademoak Lakes and the Belleisle and Kennebecasis Bays, which have a parallel direction to the N. E., and afford good facilities for inland navigation. The tributary streams are connected with those of the Gulf and of the Bay of Chaleur by short portages (which will be mentioned in connection with their points of departure).

Immediately after leaving the dock at St. John a fine retrospect is given of the dark chasm below, over which is the light and graceful suspension-bridge. Running up by Point Pleasant, the boat ascends a narrow gorge with high and abrupt banks, at whose bases are large lumber-mills. On the r. is *Boar's Head*,' a picturesque rocky promontory, in whose sides are quarries of limestone; 3 – 4 M. above Indiantown the broad expanse of *Grand Bay* is entered, and South Bay is seen opening on the l. rear.

The **Kennebecasis Bay** is now seen, opening to the N. E. This noble sheet of water is from 1 to 4 M. wide, and is navigable for large vessels for over 20 M. It receives the Kennebecasis and Hammond Rivers, and contains several islands, the chief of which, *Long Island*, is 5 M. long, and is opposite the village of Rothesay (see page 22). The E. shore is followed for many miles by the track of the Intercolonial Railway.

The testimony of the rocks causes scientists to believe that the St. John formerly emptied by two months,— through the Kennebecasis and the Marsh Valley, and through South Bay into Manawagonish Bay,— and that the breaking down of the present channel through the lofty hills W. of St. John is an event quite recent in geological history. The Indians still preserve a tradition that this barrier of hills was once unbroken and served to divert the stream.

On the banks of the placid Kennebecasis the ancient Micmac legends locate the home of the Great Beaver, "feared by beasts and men," whom Glooscap finally conquered and put to death. In this vicinity dwelt the two Great Brothers, GLOOSCAP and MALSUNSIS, of unknown origin and invincible power. Glooscap knew that his brother was vulnerable only by the touch of a fern-root; and he had told Malsunsis (falsely) that the stroke of an owl's feather would kill him. It came to pass that Malsunsis determined to kill his brother (whether tempted thus by Mik-o, the Squirrel, or by Quah-beet-e-sis, the son of the Great Beaver, or by his own evil ambition); wherefore with his arrow he shot Koo-koo-skoos, the Owl, and with one of his feathers struck the sleeping Glooscap. Then he awoke, and reproached Malsunsis, but afterwards told him that a blow from the root of a pine would kill him. Then the traitorous man led his brother on a hunting excursion far into the forest, and while he slept he smote him with a pine-root. But the cautious Glooscap arose unharmed, and drove Malsunsis forth into the forest; then sat down by the brookside and said to himself, "Naught but a flowering rush can kill me." Musquash, the Beaver, hidden among the sedge, heard these words and reported them to Malsunsis, who promised to do unto him even as he should ask. Therefore did Musquash say, "Give unto me wings like a pigeon." But the warrior answered, "Get thee hence, thou with a tail like a file; what need hast thou of pigeon's wings?" and went on his way. Then the Beaver was angry, and went forth unto the camp of Glooscap, to whom he told what he had done. And by reason of these tidings, Glooscap arose and took a root of fern and sought Malsunsis in the wide and gloomy forest; and when he had found him he smote him so that he fell down dead. "And Glooscap sang a song over him and lamented."

Now, therefore, Glooscap ruled all beasts and men. And there came unto him three brothers seeking that he would give them great strength and long life and much stature. Then asked he of them whether they wished these things that they might benefit and counsel men and be glorious in battle. But they said, "No; we seek not the good of men, nor care we for others." Then he offered unto them success in battle, knowledge and skill in diseases, or wisdom and subtlety in counsel. But they would not hearken unto him. Therefore did Glooscap wax angry, and said: "Go your ways; you shall have strength and stature and length of days." And while they were yet in the way, rejoicing, "lo! their feet became rooted to the ground, and their legs stuck together, and their necks shot up, and they were turned into three cedar-trees, strong and tall, and enduring beyond the days of men, but destitute alike of all glory and of all use."

Occasional glimpses of the railway are obtained on the l., and on the r. is the large island of Kennebecasis, which is separated from the Kingston peninsula by the Milkish Channel. Then the shores of Land's End are passed on the r.; and on the l. is the estuary of the Nerepis River. At this point the low (but rocky and alpine) ridge of the *Nerepis Hills* crosses the river, running N. E. to Bull Moose Hill, near the head of Belleisle Bay.

The steamer now changes her course from N. W. to N. E., and enters the **Long Reach**, a broad and straight expanse of the river, 16 M. long and 1-3 M. wide. The shores are high and bold, and the scenery has a lakelike character. Beyond the hamlets of Westfield and Greenwich Hill, on the l. bank, is the rugged and forest-covered ridge known as the *Devil's Back*, an off-spur of the minor Alleghany chain over the Nerepis Valley. Abreast of the wooded Foster's Island, on the E. shore, is a small hamlet clustered about a tall-spired church. Caton's Island is just above Foster's, and in on the W. shore is seen the pretty little village of *Oak Point* (Lacey's inn), with a lighthouse and the spire of the Episcopal church of St. Paul. Farther up is the insulated intervale of Grassy Island, famous for its rich hay, which may be seen in autumn stacked all along the shore. The steamer now passes through the contracted channel off Mistaken Point, where the river is nearly closed by two narrow peninsulas which project towards each other from the opposite shores.

Belleisle Bay turns to the N. E. just above Mistaken Point. The estuary is nearly hidden by a low island and by a rounded promontory on the r., beyond which the bay extends to the N. E. for 12-14 M., with a uniform width of 1 M. It is navigable for the largest vessels, and is bordered by wooded hills. On the S. shore near the mouth is Kingston Creek, which leads S. in about 5 M. to **Kingston** (two inns), a sequestered village of 200 inhabitants, romantically situated among the hills in the centre of the peninsular parish of Kingston. This peninsula preserves an almost uniform width of 5-6 M. for 30 M., between the Kennebecasis Bay and river on the S. E. and the Long Reach and Belleisle Bay on the N. W. The scenery, though never on a grand scale, is pleasant and bold, and has many fine water views. A few miles E. of Kingston is the remarkable lakelet called the *Pickwaakert*, occupying an extinct crater and surrounded by volcanic rocks. This district was originally settled by American Loyalists, and for many years Kingston was the capital of Kings County. The village is most easily reached from Rothesay (see page 22).

Tennant's Cove is a small Baptist village at the N. of the entrance to the bay; whence a road leads in 5 M. to the hamlet of *Belleisle Bay* on the N. shore (nearly opposite Long Point village); from which the bay road runs in 3-4 M. to the larger Baptist settlement at Spragg's Point, whence much cord-wood is sent to St. John. 4 M. beyond is *Springfield* (small inn), the largest of the Belleisle villages, situated near the head of the bay, and 7 M. from Norton, on the Intercolonial Railway (Route 16).

At the head of the Long Reach a granite ridge turns the river to the N. and N. W. and narrows it for several miles. 4-5 M. above Belleisle Bay Spoon Island is passed, above which, on the r. bank, is the shipbuilding hamlet of *Wickham*. A short distance beyond, on the W. bank, is *Hampstead*, with several mills and a granite-quarry. The shores of the river now become more low and level, and the fertile meadows of *Long Island* are coasted for nearly 5 M. This pretty island is dotted with elm-trees, and contains two large ponds. On the mainland (W. shore), near its head, is the hamlet of *Otnabog*, at the mouth of a river which empties into a lake 3 M. long and 1-2 M. wide, connected with the St. John by a narrow passage. The boat next passes the Lower Musquash Island, containing a large pond, and hiding the outlet of the *Washademoak Lake* (see Route 9).

"This part of the Province, including the lands around the Grand Lake and along the Washademoak, must become a very populous and rich country. A great proportion of the land is intervale or alluvial, and coal is found in great plenty, near the Grand Lake..... No part of America can exhibit greater beauty or more luxuriant fertility than the lands on each side, and the islands that we pass in this distance." (MCGREGOR's *British America*.)

After passing the Upper Musquash Island, the steamboat rounds in at *Gagetown* (2 inns), a village of 300 inhabitants, prettily situated on the W. bank of the river. It is the shire-town of Queen's County, and is the shipping-point for a broad tract of farming-country. After leaving this point, the steamer passes between Grimross Neck (l.) and the level shores of Cambridge (r.), and runs by the mouth of the **Jemseg River.**

About the year 1640 the French seigneur erected at the mouth of the Jemseg a fort, on whose ramparts were 12 iron guns and 6 " murtherers." It was provided with a court of guard, stone barracks and magazines, a garden, and a chapel "6 paces square, with a bell weighing 18 pounds." In 1654 it was captured by an expedition sent out by Oliver Cromwell; but was yielded up by Sir Thomas Temple to the Seigneur de Soulanges et Marson in 1670. In 1674 it was taken and plundered by " a Flemish corsair." The Seigniory of Jemseg was granted by the French Crown to the ancient Breton family of Damour des Chaffour. In 1686 it was occupied by the seigniorial family, and in 1698 there were 50 persons settled here under its auspices. In 1739 the lordship of this district was held by the Marquis de Vaudreuil, who had 116 colonists in the domain of Jemseg. In 1692 it was made the capital of Acadia, under the command of M. de Villebon; and after the removal of the seat of government to Fort Nashwaak (Fredericton), the Jemseg fort suffered the vicissitudes of British attack, and was finally abandoned. About the year 1776, 600 Indian warriors gathered here, designing to devastate the St. John valley, but were deterred by the resolute front made by the colonists from the Oromocto fort, and were finally appeased and quieted by large presents.

The Jemseg River is the outlet of Grand Lake (see Route 10). Beyond this point the steamer runs N. W. by Grimross Island, and soon passes the hamlets of Canning (r.) and Upper Gagetown (l.). Above Mauger's Island is seen the tall spire of Burton church, and the boat calls at *Sheffield*, the seat of the Sheffield Academy.

" The whole river-front of the parishes of Maugerville, Sheffield, and Waterborough, an extent of nearly 30 M., is a remarkably fine alluvial soil, exactly resembling that of Battersea fields and the Twickenham meadows, stretching from the river generally about 2 M. This tract of intervale, including the three noble islands opposite, is deservedly called the Garden of New Brunswick, and it is by far the most considerable tract of alluvial soil, formed by fresh water, in the Province."

Above Sheffield the steamer passes Middle Island, which is 3 M. long, and produces much hay, and calls at *Maugerville*, a quiet lowland village of 300 inhabitants. On the opposite shore is *Oromocto* (two inns), the capital of Sunbury County, a village of 400 inhabitants, engaged in shipbuilding. It is at the mouth of the Oromocto River, which is navigable for 22 M.

The settlement of Maugerville was the first which was formed by the English on the St. John River. It was established in 1763 by families from Massachusetts and Connecticut, and had over 100 families in 1775. In May, 1776, the inhabitants of Sunbury County assembled at Maugerville, and resolved that the colonial policy of the British Parliament was wrong, that the United Provinces were justified in resisting it, that the county should be attached to Massachusetts, and that men and money should be raised for the American service: saying also, " we are Ready with our Lives and fortunes to Share with them the Event of the present Struggle for Liberty, however God in his Providence may order it." These resolutions were signed by all but 12 of the people; and Massachusetts soon sent them a quantity of ammunition. At a later day Col. Eddy, with a detachment of Mass. troops, ascended the St. John River to Maugerville, where he met with a warm welcome and was joined by nearly 50 men.

Oromocto was in early days a favorite resort of the Indians, one of whose great cemeteries has recently been found here. When the hostile tribes concentrated on the Jemseg during the Revolutionary War, and were preparing to devastate the river-towns, the colonists erected a large fortification near the mouth of the Oromocto, and took refuge there. They made such a bold front that the Indians retired and disbanded, after having reconnoitred the works.

" The rich meadows are decorated with stately elms and forest trees, or sheltered by low coppices of cranberry, alder, and other native bushes. Through the numerous openings in the shrubbery, the visitor, in traversing the river, sees the white fronts of the cottages, and other buildings; and, from the constant change of position, in sailing, an almost endless variety of scenery is presented to the traveller's eye. During the summer season the surface of the water affords an interesting spectacle. Vast rafts of timber and logs are slowly moved downwards by the current. On them is sometimes seen the shanty of the lumberman, with his family, a cow, and occasionally a haystack, all destined for the city below. Numerous canoes and boats are in motion, while the paddles of the steamboat break the polished surface of the stream and send it rippling to the shore. In the midst of this landscape stands Fredericton, situated on an obtuse level point formed by the bending of the river, and in the midst of natural and cultivated scenery." (GESNER.)

Fredericton.

Hotels. Barker House, Queen St., $2 a day; Queen's Hotel, Queen St., $1.50 a day.

Stages leave tri-weekly for Woodstock (62 M.; fare, $2.50); and tri-weekly for Boiestown and the Miramichi (105 M.; fare, $6).

Railways. The European & North American (branch line) to St. John, in about 64 M.; fare, $2. The New Brunswick Railway (narrow gauge), to Woodstock and Florenceville; fare to Woodstock, $1.75 (page 50).

Steamboats. Daily to St. John, stopping at the river-ports. Fare, $1.50. In the summer there are occasional night-boats, leaving Fredericton at 4 P. M. When the river has enough water, steamboats run from Fredericton, 65 - 70 M. N. W. to Woodstock and Grand Falls. Ferry-steamers cross to St. Mary's at frequent intervals.

FREDERICTON, the capital of the Province of New Brunswick, is a small city pleasantly situated on a level plain near the St. John River. In 1871 it had 6,006 inhabitants, with 4 weekly newspapers and a bank. It is probably the quietest place, of its size, north of the Potomac River. The streets are broad and airy, intersecting each other at right angles, and are lined with fine old shade trees. The city has no manufacturing interests, but serves as a shipping-point and depot of supplies for the young settlements to the N. and W. Its chief reason for being is the presence of the offices of the Provincial Government, for which it was founded.

Queen St. is the chief thoroughfare of the city, and runs nearly parallel with the river. At its W. end is the *Government House*, a plain and spacious stone building situated in a pleasant park, and used for the official residence of the Lieutenant-Governor of New Brunswick. Nearly in the middle of the city, and between Queen St. and the river, are the Military Grounds and Parade-ground, with the large barracks (accommodating 1,000 men), which were formerly the headquarters of the British army in this Province. They are now deserted, and are falling into dilapidation. Near the E. end of Queen St. are the *Parliament Buildings*, a group of inferior wooden structures, where the legislative bodies of the Province hold their sessions. The Library is in the brick building on the E., and contains about 13,000 volumes. It is, however, open only on Wednesdays. The Council Chamber and Chamber of Commons are comfortable, but small and plain, halls; and the Law Library is also contained in this building.

* **Christ Church Cathedral** is a short distance beyond the Parliament Buildings, and is embowered in a grove of fine old trees near the river (corner of Church and Queen Sts.). It is under the direct care of the Anglican Bishop of Fredericton, and its style of construction is modelled

after that of Christ Church Cathedral at Montreal. The beauty of the
English Gothic architecture, as here wrought out in fine gray stone, is
heightened by the picturesque effect of the surrounding trees. A stone
spire, 178 ft. high, rises from the junction of the nave and transepts. The
interior is beautiful, though small, and the chancel is adorned with a
superb window of Newcastle stained-glass, presented by the Episcopal
Church in the United States. It represents, in the centre, Christ cruci-
fied, with SS. John, James, and Peter on the l., and SS. Thomas, Philip,
and Andrew on the r. In the cathedral tower is a chime of 8 bells, each
of which bears the inscription :

"Ave Pater, Rex, Creator,	Ave Simplex, Ave Trine,
Ave Fili, Lux, Salvator,	Ave Regnans in Sublime,
Ave Spiritus Consolator,	Ave Resonet sine fine,
Ave Beata Unitas.	Ave Sancta Trinitas."

The *Provincial Exposition Building* is a spacious edifice on Westmoreland St., constructed in a singular variety of Saracenic architecture. It
is used for great industrial and agricultural fairs every 3 or 4 years. In
this vicinity is the skating-rink, and the railway-station is but a little way
beyond, on York St.

The University of New Brunswick is a substantial freestone building, 170
ft. long and 60 ft. wide, occupying a fine position on the hills which sweep
around the city on the S. It was established by royal charter in 1828,
while Sir Howard Douglas ruled the Province ; and was for many years
a source of great strife between the Episcopalians and the other sects, the
latter making objection to the absorption by the Anglicans of an institution which had been paid for by the whole people. It was fairly endowed,
but has not yet reached an era of prosperity, probably because there are
too many colleges in the Maritime Provinces. The view from the University is pleasant, and is thus described by Prof. Johnston :

" From the high ground above Fredericton I again felt how very delightful it is to
feast the eyes, weary of stony barrens and perpetual pines, upon the beautiful river
St John..... Calm, broad, clear, just visibly flowing on ; full to its banks, and re-
flecting from its surface the graceful American elms which at intervals fringe its
shores, it has all the beauty of a long lake without its lifelessness. But its accessories are as yet chiefly those of nature, — wooded ranges of hills varied in outline,
now retiring from and now approaching the water's edge, with an occasional clearing, and a rare white-washed house, with its still more rarely visible inhabitants,
and stray cattle..... In some respects this view of the St. John recalled to my
mind some of the points on the Russian river (Neva): though among European
scenery, in its broad waters and forests of pines, it most resembled the tamer portions of the sea-arms and fiords of Sweden and Norway."

St. Mary's and *Nashwaaksis* are opposite Fredericton, on the l. bank of
the St. John, and are reached by a steam-ferry. Here is the terminus of
the New Brunswick Railway (to Woodstock) ; and here also are the great
lumber-mills of Mr. Gibson, with the stately church and comfortable
homes which he has erected for his workmen. Nearly opposite the city
is seen the mouth of the Nashwaak River, whose valley was settled by
disbanded soldiers of the old Black Watch (42d Highlanders).

In the year 1690 the French government sent out the Chevalier de Villebon as Governor of Acadia. When he arrived at Port Royal (Annapolis), his capital, he found that Sir William Phipps's New-England fleet had recently captured and destroyed its fortifications, so he ascended the St. John River and soon fixed his capital at Nashwaak, where he remained for several years, organizing Indian forays on the settlements of Maine.

In October, 1696, an Anglo-American army ascended the St. John in the ships *Arundel, Province,* and others, and laid siege to Fort Nashwaak. The Chevalier de Villebon drew up his garrison, and addressed them with enthusiasm, and the detachments were put in charge of the Sieurs de la Côte, Tibierge, and Clignancourt. The British royal standard was displayed over the besiegers' works, and for three days a heavy fire of artillery and musketry was kept up. The precision of the fire from La Côte's battery dismounted the hostile guns, and after seeing the Sieur de Falaise reinforce the fort from Quebec, the British gave up the siege and retreated down the river.

The village of St. Anne was erected here, under the protection of Fort Nashwaak. Its site had been visited by De Monts in 1604, during his exploration of the river. In 1757 (and later) the place was crowded with Acadian refugees fleeing from the stern visitations of angry New England on the Minas and Port Royal districts. In 1784 came the exiled American Loyalists, who drove away the Acadians into the wilderness of Madawaska, and settled along these shores. During the following year Gov. Guy Carleton established the capital of the Province here, in view of the central location and pleasant natural features of the place. Since the formation of the Canadian Dominion, and the consequent withdrawal of the British garrison, Fredericton has become dormant.

7 M. above Fredericton is *Aukpaque,* the favorite home-district of the ancient Indians of the river. The name signifies " a beautiful expanse of the river caused by numerous islands." On the island of Sandous were the fortifications and quarters of the American forces in 1777, when the St. John River was held by the expedition of Col. Allan. They reached Aukpaque on the 6th of June, and saluted the new American flag with salvos of artillery, while the resident Indians, under Ambrose St. Aubin, their " august and noble chief," welcomed them and their cause. They patrolled the river with guard-boats, aided the patriot residents on the banks, and watched the mouth of St. John harbor. After the camp on Aukpaque had been established about a month it was broken up by a British naval force from below, and Col. Allan led away about 500 people, patriot Provincials, Indians, and their families. This great exodus is one of the most romantic and yet least known incidents of the American borders. It was conducted by canoes up the St. John to the ancient French trading-post called Fort Meductic, whence they carried their boats, families, and household goods across a long portage ; then they ascended the rapid Eel River to its reservoir-lake, from whose head another portage of 4 M. led them to North Pond. The long procession of exiles next defiled into the Grand Lake, and encamped for several days at its outlet, after which they descended the Chiputneticook Lake and the St. Croix River, passed into the Lower Schoodic Lake, and thence carried their families and goods to the head-waters of the Machias River. Floating down that stream, they reached Machias[1] in time to aid in beating off the British squadron from that town.

From Fredericton to the Miramichi. Through the Forest.

The Royal Mail-stage leaves on Monday, Wednesday, and Friday, at a very early hour, and the passenger gets breakfast at Eastman's, and sleeps at Frazer's. The trip requires 2 days, and costs $6 (exclusive of hotels), and the distance from Fredericton to Newcastle is 105 M. By far the greater part of the route leads through an unbroken forest, and the road leaves much to be desired. After crossing the ferry at Fredericton

[1] *Machias* is said to be derived from the French word *Mages* (meaning the Magi), and it is held that it was discovered by the ancient French explorers on the Festival of the Magi.

the route lies due N. and is as straight as an arrow for 9 M., when it reaches Nashwaak Village (small inn); thence it follows the Nashwaak River for 5 M., to the hamlet of *Nashwaak*, above which it enters a wild country about the head-waters of the river. To the W. are the immense domains of the New Brunswick Land Company, on which a few struggling settlements are located. In the earlier days there was a much-travelled route between the St. John valley and the Miramichi waters, by way of the Nashwaak River, from whose upper waters a portage was made to the adjacent streams of the Miramichi (see "Vacation Tourists," for 1862-3, pp. 464-474). At about 40 M. from Fredericton the stage reaches *Boiestown* (small inn), a lumbering-village of 250 inhabitants, on the S. W. Miramichi River. This place was founded in 1822, by Thomas Boies and 120 Americans, but has become decadent since the partial exhaustion of the forests. The road now follows the course of the S. W. Miramichi, passing the hamlets of Ludlow, 52 M. from Fredericton ; Doaktown, 55 M.; Blissfield, 62; Dunphy, 73; Blackville, 79; Indiantown (Renous River), 87; Derby, 96; and Newcastle, 105 (see Route 15).

9. Washademoak Lake.

The steamer *Star* leaves St. John (Indiantown) on Tuesday, Thursday, and Saturday, at 10 A. M. for Cole's Island and the intermediate landings. The distance is about 60 M.; the fare is $1. The boats leave Cole's Island on the return trip at 7.30 A. M., on Wednesday, Friday, and Monday.

The steamboat ascends the St. John River (see page 39) to the upper end of Long Island, where it turns to the N. E. in a narrow passage between the Lower Musquash Island and the shores of Wickham. On either side are wide rich intervales, over which the spring inundations spread fertilizing soil; and the otherwise monotonous landscape is enlivened by clusters of elms and maples. After following this passage for 1½ M., the steamer enters the **Washademoak Lake**, at this point nearly 2 M. wide. The Washademoak is not properly a lake, but is the broadening of the river of the same name, which maintains a width of from ½ M. to 2 M. from Cole's Island to its mouth, a distance of 25-30 M. It is deep and still, and has but little current. In the spring-time and autumn rafts descend the lake from the upper rivers and from the head-waters of the Cocagne, and pass down to St. John. The scenery is rather tame, being that of alluvial lowlands, diversified only by scattered trees. There are 10 small hamlets on the shores, with from 150 to 250 inhabitants each, most of them being on the E. shore. The people are engaged in farming and in freighting cord-wood to St. John. About 6 M. above McDonald's Point, Lewis Cove opens to the S. E., running down for about 3 M. into the parish of Wickham; and 4-5 M. farther on are *the Narrows*, where the lake is nearly cut in two by a bold bluff projecting from the E. shore. *Cole's Island* has about 200 inhabitants, and a small hotel. It is 20 M.

from Apohaqui, on the Intercolonial Railway. Roads run across the peninsula on the N. W. to Grand Lake in 5-7 M. It is 38 M. from Cole's Island to Petitcodiac, on the Intercolonial Railway, by way of Brookvale, The Forks, and New Canaan. The Washademoak region has no attractions for the summer tourist.

10. Grand Lake.

The steamer *May Queen* leaves St. John (Indiantown) on Wednesday and Saturday at 8 A. M., for Grand Lake and the Salmon River. The distance is 85 M.; the fare is $1.50. She leaves Salmon River on Monday and Thursday mornings; and touches at Gagetown in ascending and descending.

Grand Lake is 30 M. long and from 3 to 9 M. wide. It has a tide of 6 inches, caused by the backwater of the St. John River, thrown up by the high tides of the Bay of Fundy. The shores are low and uninteresting, and are broken by several deep coves and estuaries. There are numerous hamlets on each side, but they are all small and have an air of poverty. It is reasonably hoped, however, that these broad alluvial plains will become, in a few decades, the home of a large and prosperous population.

The lands in this vicinity were granted at an early date to the Sieur de Freneuse, a young Parisian, the son of that Sieur de Clignancourt who was so active in settling the St. John valley and in defending it against the New-Englanders. On Charlevoix's map (dated 1744) Grand Lake is called *Lac Freneuse*, and a village of the same name is indicated as being a few miles to the N. These shores were a favorite camping-ground of the ancient Milicete Indians, whose descendants occasionally visit Grand Lake in pursuit of muskrats. The lumber business, always baneful to the agricultural interests of a new country, has slackened on account of the exhaustion of the forests on the Salmon River; and it is now thought that a farming population will erelong occupy the Grand Lake country.

The steamer ascends the St. John River (see page 39) as far as *Gagetown*, where it makes a brief stop (other landings on the lower river are sometimes visited). She then crosses to the mouth of the Jemseg (see page 43), where the Jemseg River is entered, and is followed through its narrow, tortuous, and picturesque course of 4 M. This is the most interesting part of the journey. When nearly through the passage the boat stops before the compact hamlet of *Jemseg*, occupying the slope of a hill on the r. On entering the lake, a broad expanse of still water is seen in front, with low and level shores denuded of trees. On the l. is *Scotchtown* (150 inhabitants), near which is a channel cut through the alluvium, leading (in 2 M.) to *Maquapit Lake*, which is 5 M. long and 2-3 M. wide. This channel is called *the Thoroughfare;* is passable by large boats; and leads through groves of elm, birch, and maple trees. 1 M. from the W. end of Maquapit Lake is *French Lake*, accessible by another "Thoroughfare," and 3-4 M. long, nearly divided by a long, low point. This lake is 5-6 M. from Sheffield, on the St. John River.

The channel is marked out by poles rising from the flats on either side. (The course of the steamer is liable to variation, and is here described as followed by the Editor.) Robinson's Point is first visited, with its white lighthouse rising from the E. shore; and the steamer passes around into *White's Cove*, where there is a farming settlement of 200 inhabitants. Thence the lake is crossed to the N. to *Keyhole*, a curious little harbor near the villages of Maquapit and Douglas Harbor. After visiting Mill Cove and Wiggin's Cove, on the E. shore, and Young's Cove (2 inns), the boat rounds Cumberland Point and ascends the deep *Cumberland Bay*, at whose head is a populous farming settlement. On the way out of the bay Cox's Point is visited, and then the narrowing waters at the head of the lake are entered. At *Newcastle* and other points in this vicinity, attempts have been made at coal-mining. The coal district about the head of Grand Lake covers an area of 40 square miles, and the coal is said to be of good quality and in thick seams. But little has yet been done in the way of mining, owing to the difficulty of transporting the coal to market.

Soon after passing Newcastle Creek the steamer ascends the N. E. arm, rounds a long, low point, and enters the **Salmon River.** This stream is ascended for several miles, through the depressing influences of ruined forests not yet replaced by farms. Beyond Ironbound Cove and the Coal Mines, the boat ties up for the night at a backwoods settlement, where the traveller must go ashore and sleep in a room reserved for wayfarers in an adjacent cottage.

Brigg's Corner is at the head of navigation, and a road runs thence N. E. across the wilderness to Richibucto, in 50-60 M. It is stated by good authority that the fishing in the Salmon River has been ruined by the lumber-mills; but that very good sport may be found on the Lake Stream, 15-20 M. beyond Brigg's Corner. Visitors to this district must be provided with full camp-equipage. A road also leads N. W. from Brigg's Corner (diverging from the Richibucto road at Gaspereau) to *Blissville*, on the S. W. Miramichi, in about 40 M.

11. Fredericton to Woodstock.

By the *New Brunswick Railway*, a new line which has been but recently opened to trade. It is a narrow-gauge road, and travellers who are not familiar with that principle of railway-building will be interested in observing the comparatively low and narrow, but comfortable cars; the small locomotives; and the construction of the bridges, the sharpness of the curves, and the steepness of the grades.

The New Brunswick Railway is now completed to Florenceville, and is being graded to Tobique, whence it is proposed to construct a branch to Cariboo, 13 M. up the rich valley of the Aroostook. The company hopes that the line will be carried through to Rivière du Loup, on the St. Lawrence, at no distant date.

Stations. Gibson; St. Mary's, 1 M.; Douglas, 3; Springhill, 5¾; Rockland, 10; Keswick, 12; Cardigan, 16¼; Lawrence, 17¼; Zealand, 20; Stoneridge, 22½; Burnside, 25; Upper Keswick, 28¼; Burt Lake, 32; Haynesville, 36½; Millville, 38¼; Nackawic, 43; Falls Brook, 48; Woodstock Junction, 52; Newburgh, 57; Riverside, 60; Northampton, 61¾. Fare from Fredericton to Woodstock, $1.75.

Beyond Woodstock Junction the New Brunswick Railway runs N. to Hartland (61 M. from Fredericton) and to Florenceville (71 M.). The trains make connections with stages for Tobique and the upper St. John valley.

The traveller crosses the St. John River by the steam ferry-boat (5c.), from Fredericton to Gibson; and the terminal station of the railway is near the ferry-landing. As the train moves out, pleasant views are afforded

Route 11. FREDERICTON TO WOODSTOCK.

of the prosperous and happy settlements which have been founded here by Mr. Gibson, the lumber-merchant. Glimpses of Fredericton are obtained on the l., and beyond St. Mary's the Nashwaaksis River is crossed. Then follows a succession of beautiful views (to the l.) over the wide and placid St. John, dotted with numerous large and level islands, upon which are clusters of graceful trees. On the farther shore is seen the village of *Springhill* (see page 51); and the broad expanse of Sugar Island crosses the river a little way above. At about 10 M. from Fredericton the line changes its course from W. to N. W., and leaves the St. John valley, ascending the valley of the Keswick, — a district which is beginning to show the rewards of the arduous labors of its early pioneers. The Keswick Valley was settled in 1783, by the disbanded American-loyalist corps of New York and the Royal Guides, and their descendants are now attacking the remoter back-country. The Keswick flows through a pleasant region, and has bold features, the chief of which is the escarped wall of sandstone on the l. bank, reaching for 8 – 10 M. from its mouth. From Cardigan station a road leads into the old Welsh settlement of *Cardigan*.

The line next passes several stations on the old domain of the New Brunswick Land Company, an association which was incorporated by royal charter before 1840, and purchased from the Crown 550,000 acres in York County. They established their capital and chief agency at the village of *Stanley*, opened roads through the forest, settled a large company of people from the Isle of Skye upon their lands, and expended $500,000 in vain attempts to colonize this district.

The country now traversed by the line seems desolate and unpromising, and but few signs of civilization are visible. This forest-land is left behind, and the open valley of the St. John is approached, beyond *Newburgh*. For the last few miles of the journey beautiful views are given from the high grades of the line, including the river and its intervales and surrounding hills. The terminal station is, at present, in a field about 1½ M. from Woodstock, on the opposite shore of the St. John, which is here crossed by a primitive steam ferry-boat.

Woodstock (*American House*, comfortable), the capital of Carleton County, is situated at the confluence of the St. John and Meduxnekeag Rivers, in the centre of a thriving agricultural district. The population is over 2,000, and the town is favorably situated on a high bluff over the St. John River. The Episcopal Church of St. Luke and the Catholic Church of St. Gertrude are on Main St., where are also the chief buildings of the town. The academy called Woodstock College is located here. The country in this vicinity is very attractive in summer, and is possessed of a rich rural beauty which is uncommon in these Provinces. The soil is a calcareous loam, producing more fruit and cereal grains than any other part of New Brunswick. The bold bluffs over the St. John are generally well-wooded, and the intervales bear much hay and grain. There are large sawmills at the mouth of the Meduxnekeag, where the timber which is cut on its upper waters, in Maine, is made into lumber. 12 M. from Woodstock is the American village of *Houlton*, the capital of Aroostook County, Maine; and the citizens of the two towns are in such close social relations that Woodstock bears great resemblance to a Yankee town, both in its architecture and its society.

"Of the quality of the Woodstock iron it is impossible to speak too highly, especially for making steel, and it is eagerly sought by the armor-plate manufacturers in England. On six different trials, plates of Woodstock iron were only slightly indented by an Armstrong shot, which shattered to pieces scrap-iron plates of the best quality and of similar thickness. When cast it has a fine silver-gray color, is singularly close-grained, and rings like steel on being struck. A cubic inch of Woodstock iron weighs 22 per cent more than the like quantity of Swedish, Russian, or East Indian iron." (HON. ARTHUR GORDON.) The mines are some distance from the village, and are being worked efficiently, their products being much used for the British iron-clad frigates.

The N. B. & C. Railway runs S. from Woodstock to St. Stephen and St. Andrews (see page 36); fare, $2.90. The N. B. Railway goes S. E. to Fredericton; fare, $1.75. Steamers run to Fredericton and to Grand Falls, when the river is high enough. Stages pass by the river-road to Fredericton semi-weekly, and daily stages run N. to Grand Falls, and also W. to Houlton.

12. Fredericton to Woodstock, by the St. John River.

During the spring and autumn, when there is enough water in the river, this route is served by steamboats. At other times the journey may be made by the mail-stage. The distance is 62 M.; the fare is $2.50. The stage is uncovered, and hence is undesirable as a means of conveyance except in pleasant weather. Most travellers will prefer to pass between Fredericton and Woodstock by the new railway (see Route 11). The stage passes up the S. and W. side of the river. The ensuing itinerary speaks of the river-villages in their order of location, without reference to the stations of the stages and steamboats.

Distances. — Fredericton to Springhill, 5 M.; Lower French Village, 9; Bristol (Kingsclear), 16; Lower Prince William, 21; Prince William, 25; Dumfries, 32; Pokiok Falls, 39; Lower Canterbury, 44; Canterbury, 51; Lower Woodstock; Woodstock, 52.

On leaving Fredericton, pleasant prospects of the city and its Nashwaak suburbs are afforded, and successions of pretty views are obtained over the rich alluvial islands which fill the river for over 7 M., up to the mouth of the Keswick River. *Springhill* (S. shore) is the first village, and has about 250 inhabitants, with an Episcopal church and a small inn. The prolific intervales of Sugar Island are seen on the r., nearly closing the estuary of the Keswick, and the road passes on to the Indian village, where reside 25 families of the Milicete tribe. A short distance beyond is the *Lower French Village* (McKinley's inn), inhabited by a farming population descended from the old Acadian fugitives. The road and river now run to the S. W., through the rural parish of Kingsclear, which was settled in 1784 by the 2d Battalion of New Jersey Loyalists. Beyond the hamlet of *Bristol* (Kingsclear) Burgoyne's Ferry is reached, and the scattered cottages of Lower Queensbury are seen on the N. shore. After crossing Long's Creek the road and river turn to the N. W., and soon reach the village of Lower Prince William (Wason's inn). 9 M. S. W. of this point is a settlement amid the beautiful scenery of *Lake George*, where an antimony-mine is being worked; 3 M. beyond which is Magundy (small inn), to the W. of Lake George.

FORT MEDUCTIC.

The road passes on to Prince William, through a parish which was originally settled by the King's American Dragoons, and is now occupied by their descendants. On the N. shore are the hilly uplands of the parish of Queensbury, which were settled by the disbanded men of the Queen's Rangers, after the Revolutionary War. Rich intervale islands are seen in the river between these parishes. Beyond *Dumfries* (small hotel) the hamlet of Upper Queensbury is seen on the N. shore, and the river sweeps around a broad bend at whose head is *Pokiok*, with large lumber-mills, 3 M. from Allandale. There is a fine piece of scenery here, where the River Pokiok (an Indian word meaning "the Dreadful Place "), the outlet of Lake George, enters the St. John. The river first plunges over a perpendicular fall of 40 ft. and then enters a fine gorge, 1,200 ft. long, 75 ft. deep, and 25 ft. wide, cut through opposing ledges of dark rock. The Pokiok bounds down this chasm, from step to step, until it reaches the St. John, and affords a beautiful sight in time of high water, although its current is often encumbered with masses of riff-raff and rubbish from the saw-mills above. The gorge should be inspected from below, although it cannot be ascended along the bottom on account of the velocity of the contracted stream. About 4 M. from Pokiok (and nearer to Dumfries) is the pretty highland water of Prince William Lake, which is nearly 2 M. in diameter.

Lower Canterbury (inn) is about 5 M. beyond Pokiok, and is near the mouth of the Sheogomoc River, flowing out from a lake of the same name. At *Canterbury* (Hoyt's inn) the Eel River is crossed; and about 5 M. beyond, the road passes the site of the old French works of *Fort Meductic.*

This fort commanded the portage between the St. John and the route by the upper Eel River and the Eel and North Lakes to the Chiputneticook Lakes and Passamaquoddy Bay. Portions of these portages are marked by deep pathways worn in the rocks by the moccasons of many generations of Indian hunters and warriors. By this route marched the devastating savage troops of the Chevalier de Villebon to many a merciless foray on the New England borders. The land in this vicinity, and the lordship of the Milicete town at Meductic, were granted in 1684 to the Sieur Clignancourt, the brave Parisian who aided in repelling the troops of Massachusetts from the fort on the Jemseg. Here, also, during high water, the Indians were obliged to make a portage around the Meductic Rapids, and the command of this point was deemed of great importance and value. (See also the account of Allan's retreat, on page 46.)

Off this point are the *Meductic Rapids,* where the steamboats sometimes find it difficult to make headway against the descending waters, accelerated by a slight incline. The road now runs N. through the pleasant valley of the St. John, with hill-ranges on either side. *Lower Woodstock* is a prosperous settlement of about 500 inhabitants, and the road soon approaches the N. B. & C. Railway (see page 37), and runs between that line and the river.

"The approach to Woodstock, from the old church upwards, is one of the pleasantest drives in the Province, the road being shaded on either side with fine trees, and the comfortable farm-houses and gardens, the scattered clumps of wood, the windings of the great river, the picturesque knolls, and the gay appearance of the pretty straggling little town, all giving an air of a long-settled, peaceful, *English-looking* country." (GORDON.)

13. Woodstock to Grand Falls and Rivière du Loup.

The pleasanter route to Grand Falls is by the steamboats,—small, light-draught craft, which scuttle up the rapids and over the shallows as long as there is enough water in the river (usually only during the springtime and autumn).

The Royal mail-stages leave Woodstock at 6 P. M. daily; supper at Middle Simonds (Mills's), 15 M. out; breakfast at Tobique, at 4 A. M.; reach Grand Falls at 8 A. M., and remain one hour; dinner at Belyea's, 18 M. beyond; supper at Edmundston, and remain one hour; breakfast at La Belle's, at 1 A. M., and reach Rivière du Loup in time for the morning train for Quebec or Montreal. The time between Woodstock and Rivière du Loup is 36–40 hours. The New Brunswick Railway has been extended beyond Woodstock Junction to Florenceville and Muniac, and stages connect with the trains at the latter station and run through to Tobique. The railway will probably reach the latter point this year. Passengers leave Woodstock (Northampton) at 8 A. M., change cars at Woodstock Junction, and reach Muniac about 3.20 P. M.

Distances. — Woodstock to Victoria, 11 M.; Florenceville, 24; Tobique, 50; Grand Falls, 75; Edmundston, 113; Rivière du Loup, 193.

Fares. — By stage, Woodstock to Florenceville, $1.50; Tobique, $3; Grand Falls, $4.25; Grand Falls to Edmundston, $2.50; Edmundston to Rivière du Loup, $5.

The road from Woodstock to Florenceville is pleasant and in an attractive country. "It is rich, English, and pretty. When I say English, I ought, perhaps, rather to say Scotch, for the general features are those of the lowland parts of Perthshire, though the luxuriant vegetation—tall crops of maize, ripening fields of golden wheat, and fine well-grown hard-wood—speaks of a more southern latitude. Single trees and clumps are here left about the fields and on the hillsides, under the shade of which well-looking cattle may be seen resting, whilst on the other hand are pretty views of river and distance, visible under fine willows, or through birches that carried me back to Deeside." (HON. ARTHUR GORDON.)

Soon after leaving Woodstock the stage-road takes a direction to the N. E., keeping along the W. bank of the St. John River. Victoria and Middle Simonds (Mills's Hotel) are quiet hamlets on the river, centres of agricultural districts of 5–800 inhabitants each. *Florenceville* (large hotel) is a pretty village, "perched, like an Italian town, on the very top of a high bluff far over the river." The road now swings around to the N. W. and traverses the settlements of Wicklow. The district between Woodstock and Wicklow was settled after the American Revolution by the disbanded soldiers of the West India Rangers and the New Brunswick Fencibles.

"Between Florenceville and Tobique the road becomes even prettier, winding along the bank of the St. John, or through woody glens that combine to my eye Somersetshire, Perthshire, and the green wooded part of southwestern Germany." There are five distinct terraces along the valley, showing the geological changes in the level of the river, and the banks of the stream are composed of sand and gravel. The intervale is usually narrow, and is broken frequently by intrusive highlands.

5 M. S. W. of the river is Mars Hill, a steep mountain about 1,200 ft. high, which overlooks a vast expanse of forest. This was one of the chief points of controversy during the old border-troubles, and its summit was cleared by the Commissioners of 1794. The road now crosses the River des Chutes, at whose mouth are large saw-mills, near the site of an ancient waterfall which has disappeared on account of the erosion of the rocks. Above this point the country is less thickly settled, and the road passes up near the river. Perth village is seen on the E. shore, and the narrowing valleys of Victoria County are traversed.

Tobique (Newcomb's inn), otherwise known as Andover, is pleasantly situated on the W. bank of the St. John, nearly opposite the mouth of the Tobique River. It has 400 inhabitants and 2 churches, and is the chief depot of supplies for the lumbering-camps on the Tobique River. Nearly opposite is a large and picturesque Indian village, containing about 150 persons of the Milicete tribe, and situated on the bluff at the confluence of the rivers. They have a valuable reservation here, and the men of the tribe engage in lumbering and boating.

Fort Fairfield (*Fort Fairfield House*) is 7 M. N. W. of Tobique, and is an American border-town, with 900 inhabitants, 5 churches, and several small factories. This town was settled by men of New Brunswick in 1816, at which time it was supposed to be inside the Provincial line. A road runs from Fort Fairfield S. W. to **Presque Isle** (*Presque Isle Hotel*), a village of about 1,000 inhabitants, with 4 churches, an academy, several factories, and a newspaper (the "Presque Isle Sunrise"). This town is 42 M. N. of Houlton, on the U. S. military road which runs to the Madawaska district, and is one of the centres of the rich farming lands of the Aroostook Valley, parts of which are now occupied by Swedish colonists.

From Tobique to Bathurst. Through the Wilderness.

Guides and canoes can be obtained at the Indian village near Tobique. About 1 M. above Tobique the voyagers ascend through the *Narrows*, where the rapid current of the Tobique River is confined in a winding cañon (1 M. long, 150 ft wide, and 50 – 100 ft. deep) between high limestone cliffs. Then the river broadens out into a pretty lake-like reach, with rounded and forest-covered hills on either side. The first night-camp is usually made high up on this reach. Two more rapids are next passed, and then commences a stretch of clear, deep water 70 M. long. Near the foot of the reach is the settlement of *Arthurette*, with about 400 inhabitants. The *Red Rapids* are 11 M from the mouth of the river, and descend between high shores. Occasional beautifully wooded islands are passed in the stream; and by the evening of the second day the voyagers should reach the high red cliffs at the mouth of the broad Wapskehegan River. This Indian name signifies "a river with a wall at its mouth," and the stream may be ascended for 20 M., through a region of limestone hills and alluvial intervales. The Wapskehegan is 81 M. above the mouth of the Tobique.

Infrequent clearings, red cliffs along the shore, and blue hills more remote, engage the attention as the canoe ascends still farther, passing the hamlet of *Foster's Cove* on the N. bank, and running along the shores of Diamond and Long Island, 44 M. up river is the Agulquac River, coming in from the E., and navigable by canoes for 25 M. As the intervales beyond this confluence are passed, occasional glimpses are gained (on the r.) of the Blue Mts. and other tall ridges. At 80 M. from the mouth of the river, the canoe reaches *The Forks* (4 – 5 days from Tobique).

The Campbell River here comes in from the E. and S. E., from the great Tobique Lake and other remote wilderness-waters: the Mamozeket descends from the N., and from the N. W. comes the Nictor, or Little Tobique River. It is a good day's journey from the Forks to Cedar Brook, on the Nictor; and another day conducts to the *Nictor Lake, "possessing more beauty of scenery than any other locality I have seen in the Province, except, perhaps, the Bay of Chaleur. Close to its southern edge a granite mountain rises to a height of nearly 3,000 ft., clothed with wood to its summit, except where it breaks into precipices of dark rock or long gray shingly slopes. Other mountains of less height, but in some cases of more picturesque forms, are on other sides; and in the lake itself, in the shadow of the mountain, is a little rocky islet of most inviting appearance." It takes 2-3 hours to ascend the mountain (Bald, or Sagamook), whence "the view is very fine. The lake lies right at our feet, — millions of acres of forest are spread out before us like a map, sinking and swelling in one dark mantle over hills and valleys, whilst Katahdin and Mars Hill in Maine, Tracadiegash in Canada, the Squaw's Cap on the Restigouche, and Green Mountain in Victoria, are all distinctly visible." (GORDON.) From the head of Nictor Lake a portage 3 M. long leads to the Nepisiguit Lake, on whose E. shore is the remarkable peak called Mount Teneriffe. Near the outlet is a famous camping-ground, where the fishing is good and in whose vicinity deer and ducks are found

It takes about six days to descend the *Nepisiguit River* to the Great Falls, the larger part of the way being through forests of fir and between distant ranges of bare granite hills.

There is a Provincial highway which follows the W. shore of the Tobique River, and touches the lower end of Nictor Lake, whence it runs N. and N. E. across the uninhabited valley of the Upsalquitch to Campbellton, on the Restigouche. (See Route 14.)

6 M. above Tobique is the mouth of the Aroostook River, which traverses a great area of northern Maine, and for the last 5 M. of its course is in New Brunswick. It is not easily navigable on account of several rapids and the falls near Fort Fairfield; yet great quantities of lumber are floated down its current. There is a thriving village near the mouth of the river. 7 M. farther N. the hamlet of Grand Falls Portage is passed, and the road leaves the St. John, which here begins a broad bend to the W. About 10 M. above the Portage the steamboat or stage reaches *Grand Falls* (2 inns), otherwise known as Colebrooke. This town has about 700 inhabitants, and is picturesquely situated on a narrow peninsula near the cataract. It was formerly a fortified post of the British army, and is now the capital of Victoria County. It is hoped that large manufacturing interests will be developed here when the railway is completed from Woodstock to Rivière du Loup. Daily stages leave for Woodstock and for Rivière du Loup; and steamboats descend the river during the brief seasons of navigation. The environs of the village are remarkable for their picturesque beauty, and the view from the Suspension Bridge over the gorge of the St. John is worthy of notice.

The **Grand Falls are near the village, and form the most imposing cataract in the Maritime Provinces. The river expands into a broad basin above, affording a landing-place for descending canoes; then hurries its massive current into a narrow rock-bound gorge, in which it slants down an incline of 6 ft., and then plunges over a precipice of calcareous slate

58 ft. high. The shape of the fall is singular, since the water leaps from the front and from both sides, with minor and detached cascades over the outer ledges. Below the cataract the river whirls and whitens for ¾ M. through a rugged gorge 250 ft. wide, whose walls of dark rock are from 100 to 240 ft. high. "It is a narrow and frightful chasm, lashed by the troubled water, and excavated by boiling eddies and whirlpools always in motion; at last the water plunges in an immense frothy sheet into a basin below, where it becomes tranquil, and the stream resumes its original features." Within the gorge the river falls 58 ft. more, and the rugged shores are strewn with the wrecks of lumber-rafts which have become entangled here. The traveller should try to visit the Falls when a raft is about passing over. 3 - 4 M. below the Falls is the dangerous *Rapide de Femme*. Small steamers have been placed on the river above the Falls, and have run as far as the mouth of the St. Francis, 65 M. distant.

It is a tradition of the Micmacs that in a remote age two families of their tribe were on the upper St. John hunting, and were surprised by a war-party of the strange and dreaded Northern Indians. The latter were descending the river to attack the lower Micmac villages, and forced the captured women to pilot them down. A few miles above the falls they asked their unwilling guides if the stream was all smooth below, and on receiving an affirmative answer, lashed the canoes together into a raft, and went to sleep, exhausted with their march. When near the Grand Falls the women quietly dropped overboard and swam ashore, while the hostile warriors, wrapped in slumber, were swept down into the rapids, only to awaken when escape was impossible. Their bodies were stripped by the Micmacs on the river below, and the brave women were ever afterward held in high honor by the tribe.

Crossing the St. John at Grand Falls, the stage ascends the E. bank of the stream, and soon enters the Acadian-French settlements and farming-districts. 8 - 10 M. up the road is the village of *St. Leonard*, nearly all of whose people are French; and on the American shore (for the St. John River is for many leagues the frontier between the nations) is the similarly constituted village of *Van Buren* (two inns). This district is largely peopled by the Cyr, Violette, and Michaud families.

The Hon. Arthur Gordon thus describes one of these Acadian homes near Grand River (in 1863): "The whole aspect of the farm was that of a *metairie* in Normandy; the outer doors of the house gaudily painted, the panels of a different color from the frame, — the large, open, uncarpeted room, with its bare shining floor, — the lasses at the spinning-wheel, — the French costume and appearance of Madame Violet and her sons and daughters, — all carried me back to the other side of the Atlantic."

Grand River (Tardiff's inn) is a hamlet about 4 M. beyond St. Leonard, at the mouth of the river of the same name.

The St. John River to the Restigouche.

A rugged wilderness-journey may be made on this line, by engaging Acadian guides and canoes at the Madawaska settlements. 3 - 4 weeks will be sufficient time to reach the Bay of Chaleur, with plenty of fishing on the way. On leaving the St. John the voyagers ascend the Grand River to its tributary, the Waagansis. A portage of 5 - 6 M. from this stream leads to the Waagan, down whose narrow current the canoes float through the forest until the broad Restigouche is entered (see Route 15; see also Hon. Arthur Gordon in "Vacation Tourists" for 1862 - 63, p. 477).

6 M. above Grand River is *St. Basil* (two inns), which, with its back settlements, has over 1,400 inhabitants. A few miles beyond are some islands in the St. John River, over which is seen the American village of *Grant Isle* (Levecque's inn), a place of 700 inhabitants, all of whom are Acadians. This village was incorporated in 1869, and is on the U. S. mail-route from Van Buren to Fort Kent. Beyond the populous village of Green River the road continues around the great bend of the St. John to the Acadian settlement which is variously known as Madawaska, Edmundston, and Little Falls. There are about 400 inhabitants here, most of whom are engaged in lumbering and in agriculture. The town occupies a favorable position at the confluence of the Madawaska and St. John Rivers, and it is to be the objective point of the New Brunswick Railway (see page 50) during the year 1875. This is the centre of the Acadian-French settlements which extend from the Grand Falls to the mouth of the St. Francis, and up the Madawaska to Temiscouata Lake. This district is studded with Roman Catholic chapels, and is divided into narrow farms, on which are quaint little houses. There are rich tracts of intervale along the rivers, and the people are generally in a prosperous and happy condition. The visitor should ascend to the top of the loftily situated old block-house tower, over Edmundston, for the sake of the wide prospect over the district.

This people is descended from the French colonists who lived on the shores of the Bay of Fundy and the Basin of Minas at the middle of the 18th century. When the cruel edict of exile was carried into effect in 1755 (see Route 21), many of the Acadians fled from the Anglo-American troops and took refuge in the forest. A portion of them ascended the St. John to the present site of Fredericton, and founded a new home; but they were ejected 30 years later, in order that the land might be given to the refugee American Loyalists. Then they advanced into the trackless forest, and settled in the Madawaska region, where they have been permitted to remain undisturbed. When the American frontier was pushed forward to the St. John River, by the sharp diplomacy of Mr. Webster, the Acadians found themselves divided by a national boundary; and so they still remain, nearly half of the villages being on the side of the United States. It is estimated that there are now about 8,000 persons in these settlements.

" It was pleasant to drive along the wide flat intervale which formed the Madawaska Valley; to see the rich crops of oats, buckwheat, and potatoes; the large, often handsome, and externally clean and comfortable-looking houses of the inhabitants, with the wooded high grounds at a distance on our right, and the river on our left, — on which an occasional boat, laden with stores for the lumberers, with the help of stout horses, toiled against the current towards the rarely visited headwaters of the tributary streams, where the virgin forests still stood unconscious of the axe. This beautiful valley, with the rich lands which border the river above the mouth of the Madawaska, as far almost as that of the river St. Francis, is the peculiar seat of the old Acadian-French." (PROF. JOHNSTON.)

The American village of **Madawaska** (two inns) is opposite Edmundston, and has over 1,000 inhabitants. The U. S. mail-stages run from this point up the valley of the St. John for 10 M. to another Acadian village, which was first named Dionne (in honor of Father Dionne, who founded here the Church of St. Luce); in 1869 was incorporated as Dickeyville, in honor of some local statesman; and in 1871 received the name of Frenchville, " as describing the nationality of its settlers." From near Frenchville a portage 5 M. long leads to the shores of *Lake Cleveland*, a fine sheet of water 9 M. long, connected by Second Lake and Lake Preble with *Lake Sedgwick*, which is nearly 10 M. long.

16 M. S. W. of Madawaska is *Fort Kent*, an old border-post of the U. S. Army. It has two inns and about 1,000 inhabitants (including the adjacent farming settlements), and is the terminus of the mail-route from Van Buren. From this point stages run W. 20 M. to the Acadian village of *St. Francis*, near the mouth of the St. Francis River. The latter stream, flowing from the N. W., is the boundary of the United States for the next 40 M., descending through the long lakes called Welastookwaagamis, Pechtaweekaagomic, and Pohenegamook. Above the mouth of the St. Francis, the St. John River is included in the State of Maine, and flows through that immense and trackless forest which covers "an extent seven times that of the famous Black Forest of Germany at its largest expanse in modern times. The States of Rhode Island, Connecticut, and Delaware could be lost together in our northern forests, and still leave about each a margin of wilderness sufficiently wide to make the exploration without a compass a work of desperate adventure." Its chief tributary in the woods is the Allagash, which descends from the great Lakes Pemgockwahen and Chamberlain, near the Chesuncook and Moosehead Lakes and the head-waters of the Penobscot.

The U. S. mail-stages also run S. from Fort Kent to *Patten*, about 100 M. S., near Mount Katahdin; whence another stage-line runs out to *Mattawamkeag*, on the E. & N. A. Railway (see page 39), in 38 M. 8 – 10 M. S. of Fort Kent, by this road, is Lake Winthrop (15 M. long by 1 – 3 M. wide), the westernmost of the great **Eagle Lakes**, famous for their white-fish and burbot.

At Edmundston the Royal mail-route leaves the St. John River, and ascends the W. shore of the Madawaska. But few settlements are passed, and at 12 M. from Edmundston the Province of Quebec is entered. About 25 M. from Edmundston the road reaches the foot of the picturesque **Temiscouata Lake**, where there is a small village. The road is parallel with the water, but at a considerable distance from it, until near the upper part, and pretty views are afforded from various points where it overlooks the lake.

Temiscouata is an Indian word meaning " Winding Water," and the lake is 30 M. long by 2 – 3 M. wide. The scenery is very pretty, and the clear deep waters contain many fish, the best of which are the tuladi, or great gray trout, which sometimes weighs over 12 pounds. There are also whitefish and burbot. Visitors to the lake usually stop at Fournier's old inn, where canoes may be obtained. From the W., Temiscouata receives the Cabineau River, the outlet of Long Lake (15 by 2 M.); and on the E. is the Tuladi River, which rises in the highlands of Rimouski and flows down through a chain of secluded and rarely visited lakelets. The chief settlement on Temiscouata Lake is the French Catholic hamlet of *Notre Dame du Lac*, which was founded since 1861 and has 180 inhabitants. The military works of Fort Ingalls formerly commanded the lake, and had a garrison of 200 men as late as 1850.

"Temiscouata Lake is a fine large sheet of water, 20 M. long; it is deep, contains plenty of fish, and there are hills about it, down the valleys and ravines of which rush winds which occasion sudden and dangerous agitation in the dark waters."

The road from Temiscouata Lake to Rivière du Loup is 40 – 50 M. long, and descends through a wild region into which a few settlers have advanced within fifteen years.

14. St. John to Shediac.

Distances. — St. John to Moncton, 89 M.; Painsec Junction, 97; Dorchester Road, 102; Shediac, 106; Point du Chêne, 108.

St. John to Painsec Junction, see Route 16.

Passengers for Shediac and Point du Chêne change cars at Painsec Junction, and pass to the N. E. over a level and unproductive country.

Shediac (*Kirk Hotel*) is a marine village of 500 inhabitants, with 3 churches, — Baptist, the Catholic St. Joseph de Shediac, and St. Andrew's, the head of a rural deanery of the Anglican church. The town is well situated on a broad harbor, which is sheltered by Shediac Island, but its commerce is inconsiderable, being limited to a few cargoes of lumber and deals sent annually to Great Britain. The small oysters (*Ostrea canadensis*) of the adjacent waters are also exported to the provincial cities. Shediac was occupied by a French garrison in 1750, to protect the borders of Acadia, and in 1757 there were 2,000 French and Acadian troops and settlers here. The French element is still predominant in this vicinity, and its interests are represented by a weekly paper called "*Le Moniteur Acadien.*"

Point du Chêne (Schurman's Point du Chêne House) is 2 M. N. E. of Shediac, and is the E. terminus of the railway and the St. Lawrence port nearest to St. John. It has a village of about 200 inhabitants, with long piers reaching out to the deep-water channels. From this point passengers embark on the steamers for Prince Edward Island, the N. shore of New Brunswick (see Route 15), and Quebec and the Gulf Ports. Daily steamers run from Shediac to Summerside, P. E. I., where they make connections with the trains of the P. E. I. Railway (see Route 43). The Gulf Ports steamers ply between Point du Chêne and Pictou, the time of transit being about 12 hours, and the route being down the Northumberland Strait, with the red shores of Prince Edward Island on the l. In the time-tables and circulars of the steamships and railways, the term *Shediac* is generally used for Point du Chêne.

The Westmorland Coast. Infrequent mail-stages run E. from Shediac by Point du Chêne to Barachois, 8 M.; Tedish, 17; Great Shemogue (Avard's Hotel), 22; and Little Shemogue, 24. These settlements contain about 1,500 inhabitants, most of whom are Acadians. Capes Jourimain (fixed white light, visible 14 M.) and Tormentine are respectively 15 M. and 20 M. E. of Little Shemogue.

10–12 M. N. of Shediac (mail-stage daily) are the large and prosperous Acadian settlements of the *Cocagnes* (three inns), having about 1,500 inhabitants, seven eighths of whom are of French descent. These people are nearly all farmers, engaged in tilling the level plains of Dundas, although a good harbor opens between the villages. 21 M. from Shediac is *Buctouche* (two inns), a prosperous Acadian village of 400 inhabitants, engaged in shipbuilding and in the exportation of lumber and oysters.

15. The Bay of Chaleur and the North Shore of New Brunswick.

The vessels of the Quebec and Gulf Ports Steamship Line, the *Secret* and the *Miramichi*, leave Pictou every Tuesday morning at 7 o'clock, and Shediac (Point du Chêne) every Tuesday evening at 7 (after the arrival of the St. John train). They then ascend the coast, leaving Chatham at 7 A.M. on Wednesday, Newcastle at 8 A.M. on Wednesday, and Dalhousie at 4 A.M. on Thursday (for Quebec). Returning, they leave Dalhousie at 9 P.M. on Thursday, Chatham at 4 P.M. on Friday, Newcastle at 6 P.M. on Friday, Shediac at 3 A.M. on Saturday (connecting with the morning train to St. John), and arrive at Pictou at 1 P.M. on Saturday (connecting with the afternoon train to Halifax). These hours are liable to variation on account of the weather, or if heavy freights are landed or taken at any port. The Gulf Ports vessels are larger and more commodious than that of the North Shore Line, but they do not visit Richibucto, Bathurst, or Campbellton. (See also Route .)

The North Shore steamer *City of St. John* leaves Shediac (Point du Chêne) every Thursday, on the arrival of the morning train from St. John, and calls at Richibucto, Chatham, Newcastle, Bathurst, Dalhousie, and Campbellton. Chatham is reached on Thursday evening, the Bay-of-Chaleur ports on Friday. The steamer leaves the Bay-of-Chaleur ports on Monday, and the Miramichi ports on Tuesday, arriving at Shediac Tuesday evening, and connecting with a late train for St. John.

Fares (North Shore Line). — St. John (by railway and steamship) to Richibucto, $5; to Chatham and Newcastle, $6.50; to Bathurst, $9.50; to Dalhousie, $10; to Campbellton, $10.50.

Distances from Shediac along the N. shore: To Richibucto, by sea, 38 M., by land, 34 M.; to Chatham, by sea, 80 M., by land, 74 M.; to Bathurst, by land, 122 M.; to Dalhousie, by sea, 220 M., by land, 175 M. Daily mail-stages run N. by Cocagne and Buctouche to Richibucto, Chatham, and Newcastle.

The steamship leaves the long railway wharf at Point du Chêne, and passes the low shores of Shediac Island on the l. The course is laid well out into the Northumberland Strait. Between Shediac Point and Cape Egmont (on Prince Edward Island) the strait is nearly 20 M. wide. On the l. the harbors of Cocagne and Buctouche (see page 59) are soon passed. 14½ M. N. of Buctouche are the low cliffs and lighthouse of *Richibucto Head*, beyond which (if the weather permits) the steamer takes a more westerly course, and enters the great Richibucto River, which empties its stream through a broad lagoon enclosed by sand-bars.

Richibucto (*Kent Hotel*) is the capital of Kent County, and occupies a favorable position for commerce and shipbuilding, near the mouth of the Richibucto River. It has about 800 inhabitants and 3 churches, and is engaged in the exportation of fish and lumber. The river is navigable for 20 M., and has been a great highway for lumber-vessels, although now the supply of the forests is wellnigh exhausted. The rubbish of the saw-mills has destroyed the once valuable fisheries in this river. In the region about Richibucto are many Acadian farmers, and the hamlet of *Aldouin River*, 4 M. from the town, pertains to this people. Daily stages run from Richibucto to Shediac and to Chatham (see page 61). A road leads S. W. through the wilderness to the Grand Lake district (Route 10).

The name *Richibucto* signifies "the River of Fire," and the shores of the river and bay were formerly inhabited by a ferocious and bloodthirsty tribe of Indians. So late as 1787, when the American Loyalist Powell settled here, there were but four Christian families (and they were Acadians) in all this region (the present county of Kent). The power of the Richibuctos was broken in 1724, when all their warriors, under command of Argimoosh ("the Great Wizard"), attacked Canso and captured 17 Massachusetts vessels. Two well-manned vessels of Boston and Cape Ann were sent after them, and overtook the Indian fleet on the coast. A desperate naval battle ensued between the Massachusetts sloops and the Indian prize-ships. The Richibuctos fought with great valor, but were finally disconcerted by showers of hand-grenades from the Americans, and nearly every warrior was either killed or drowned.

After emerging from Richibucto harbor, the steamer runs N. across the opening of the shallow Kouchibouguac Bay, whose shores are low sand-bars and beaches which enclose shoal lagoons. 5 M. above Point Sapin is *Escuminac Point*, on which is a powerful white light, visible for 25 M. The course is now laid more to the W., across the Miramichi Bay, and on the l. are seen the pilots' village and the lighthouses on Preston's Beach. The entrance to the Inner Bay of Miramichi is between Fox Island and Portage Island, the latter of which bears a lighthouse. The Inner Bay is 13 M. long and 7-8 M. wide, and on the S. is seen Vin Island, back of which is the *Bay du Vin*. Two centuries ago all this shore was occupied by French settlements, whose only remnant now is the hamlet of Portage Road, in a remote corner of the bay.

When about 9 M. from the entrance, the steamer passes between Point Quart and Grand Dune Island (on the r.), which are 3½ M. apart. 3-4 M. farther on, the course is between Oak Point, with its two lighthouses (on the r.), and Cheval Point, beyond which is the populous valley of the Napan River, on the S. The hamlet of Black Brook is visible on the l., and off Point Napan is *Sheldrake Island*, a low and swampy land lying across the mouth of the river. The vessel now enters the **Miramichi River**, and on the r. is the estuary of the Great Bartibog, with the beacon-lights on Malcolm Point. The Miramichi is here a noble stream, fully 1 M. wide, but flowing between low and uninteresting shores.

Chatham (*Canada Hotel; Bowser's Hotel*) is the chief town on the North Shore, and has a population of nearly 3,000, with 5 churches, a weekly newspaper, and a Masonic hall. It is 24 M. from the sea, and is built along the S. shore of the river for a distance of 1½ M. On the summit of the hill along which the town is built is seen a great pile of Catholic institutions, among which are the Cathedral of St. Michael, the convent and hospital of the Hôtel Dieu de Chatham, and St. Michael's College. These buildings, like all the rest of the town, are of wood. The chief industries of Chatham are shipbuilding and the exportation of fish and lumber, and the river here usually contains several large ships, which can anchor off the wharves in 6-8 fathoms.

Daily stages run N. from Chatham to Batburst, in 45 M., over a road which traverses one of the dreariest regions imaginable. About 22 M. beyond Chatham it crosses the head-waters of the **Tabusintac River**, "the sportsman's paradise," a narrow and shallow stream in which an abundance of trout is found.

Semi-weekly stages run from Chatham N. E. to Oak Point, 11 M.; Burnt Church, 20; Neguac, 25; Tabusintac, 37; Tracadie, 52; Pockmouche, 64; Shippigan, 70; and Caraquette (Lower), 73. The first 30 M. of this road are along (or near) the N. shore of the Miramichi River and the Inner Bay, by the hamlets of Oak Point and Burnt Church.

62   *Route 15.*   THE MIRAMICHI.

Burnt Church is still the capital of the Micmac Indians of the Province, and here they gather in great numbers on St. Anne's Day and engage in religious rites and athletic sports and dances. Hon. Arthur Gordon says: "I was surprised by the curious resemblance between these dances and those of the Greek peasantry. Even the costumes were in some degree similar, and I noticed more than one short colored-silk jacket and handkerchief-bound head that carried me back to Ithaca and Paxo." (VACATION TOURISTS, 1863.)

Tabusintac (small inn) is near the mouth of the Tabusintac River, and is a Presbyterian village of about 400 inhabitants, most of whom are engaged in the fisheries. Many large sea-trout are caught near the mouth of the river, and in October immense numbers of wild geese and ducks are shot in the adjacent lagoons.

Tracadie is a settlement which contains 1,200 French Acadians, and is situated near a broad lagoon which lies inside a line of sand-bars. Salmon, cod, and herring are found in the adjacent waters, and most of the people are engaged in the fisheries. The *Tracadie Lazaretto* is devoted to the reception of persons afflicted with the leprosy, which prevails to some extent in this district, but has diminished since the government secluded the lepers in this remote hospital. There is an old tradition that the leprosy was introduced into this region during the last century, when a French vessel was wrecked on the coast, some of whose sailors were from Marseilles and had contracted the true *elephantiasis graecorum* (Eastern leprosy) in the Levant. Its perpetuation and hereditary transmission is attributed to the closeness of the relation in which intermarriage is sanctioned among the Acadians (sometimes by dispensations from the Church)

Pockmouche is a settlement of 800 Acadian farmers, and here the mail-route forks, — one road running 6 M. N. E. to Shippigan (see page 64), the other running 9 M. N. to Lower Caraquette (see page 66).

Daily stages run from Chatham to Shediac (see page 59), also twice weekly to Fredericton and to Bathurst. There are two steamers weekly to Shediac, and one to Quebec. The river-steamer *New Era* runs up the river four times daily to Newcastle (6 M.), touching at Douglastown, a dingy village on the N. bank, where much lumber is loaded on the ships which take it hence to Europe. This village contains about 400 inhabitants, and has a marine hospital, built of stone.

Newcastle (*Waverley Hotel*) is the capital of Northumberland County, and is situated at the head of deep-water navigation on the Miramichi River. It has about 1,500 inhabitants, and is engaged in shipbuilding and the exportation of fish and lumber, oysters, and preserved lobsters. One of the chief stations of the Intercolonial Railway will be located here, and a branch line is to be built to Chatham.

A short distance above Newcastle, and beyond the Irish village of Nelson, is the confluence of the great rivers known as the N. W. Miramichi and the S. W. Miramichi. These streams are crossed by the largest and most costly bridges on the line of the Intercolonial Railway. The name *Miramichi* signifies "Happy Retreat," and signifies the love that the Indians entertained for these fine hunting and fishing grounds. The upper waters of the rivers traverse wide districts of unsettled country, and are visited by hardy and adventurous sportsmen, who capture large numbers of trout and salmon. This system of waters is connected by portages with the Nepisiguit, the Restigouche, the Upsalquitch, the Tobique, and the Nashwaak Rivers. The best salmon-pools are on the S. W. Miramichi, beyond Boiestown, at the mouths of the Salmon, Rocky, Clearwater, and Burnt Hill Brooks. A tri-weekly stage runs from Newcastle to Boiestown and Fredericton (see page 46), traversing 105 M. of a rude and sparsely settled country.

SHIPPIGAN ISLAND.   *Route 15.*   63

Beaubair's Island is off upper Nelson, and was formerly occupied by a prosperous French town, but few relics of which are now to be seen. It was destroyed by a British naval attack in 1759. A colony was planted here in 1722, under Cardinal Fleury's administration, and was provided with 200 houses, a church, and a 16-gun battery.

In 1642-44 the Miramichi district was occupied by Jean Jaques Enaud, a Basque gentleman, who founded trading-posts on the islands and entered also upon the walrus fisheries. But a contention soon arose between Enaud's men and the Indians, by reason of which the Basque establishments were destroyed, and their people were forced to flee to Nepisiguit. In 1672, after the Treaty of Breda, several families from St. Malo landed on this coast and founded a village at Bay du Vin. From 1740 to 1757 a flourishing trade was carried on between the Miramichi country and France, great quantities of furs being exported. But the crops failed in 1757, and the relief-ships from France were captured by the British. In the winter of 1758 the transport *L'Indienne*, of Morlaix, was wrecked in the bay, and the disheartened colonists, famished and pestilence-stricken, were rapidly depleted by death. Many of the French settlers died during the winter, and were buried on Beaubair's Point. Those who survived fled from the scene of such bitter suffering, and by the arrival of spring there were not threescore inhabitants about the bay.

In 1759 a British war-vessel entered the bay for wood and water, and the first boat's-crew which landed was cut off and exterminated by the Indians. The frigate bombarded the French Fort batteries, and annihilated the town at Canadian Cove. Then sailing to the N. E., the commander landed a force at Neguac, and burnt the Catholic chapel, the inhabitants having fled to the woods. Neguac is known to this day only by the name of Burnt Church. After this fierce foray all the N. coast of New Brunswick was deserted and relapsed into a wilderness state.

In 1775 there was an insignificant Scotch trading-post on the S. W. Miramichi, where 1,500 - 1,800 tierces of salmon were caught annually. This was once surprised and plundered by the Indians in sympathy with the Americans, but in 1777 the river was visited by the sloop-of-war *Viper* and the captured American privateer *Lafayette*. The American flag was displayed on the latter vessel, and it was given out that her crew were Bostonians, by which means 35 Indians from the great council at Bartibog were decoyed on board and carried captive to Quebec.

In 1786 the Scottish settlers opened large saw-mills on the N. W. Miramichi, and several families of American Loyalists settled along the shore. Vast numbers of masts and spars were sent hence to the British dock-yards, and the growth of the Miramichi was rapid and satisfactory. In 1793 the Indians of the hills gathered secretly and concerted plans to exterminate the settlers (who had mostly taken refuge in Chatham), but the danger was averted by the interposition of the French Catholic priests, who caused the Indians to disperse.

In October, 1825, this district was desolated by the great Miramichi Fire, which swept over 3,000,000 acres of forest, and destroyed $ 1,000,000 worth of property and 160 human lives. The town of Newcastle was laid in ashes, and all the lower Miramichi Valley became a blackened wilderness. The only escape for life was by rushing into the rivers while the storm of fire passed overhead; and here, nearly covered by the hissing waters, were men and women, the wild animals of the woods, and the domestic beasts of the farm.

On leaving the Miramichi River and Bay the vessel steams out into the Gulf, leaving on the N. W. the low shores of Tabusintac and Tracadie, indented by wide and shallow lagoons (see page 62). After running about 35 M. the low red cliffs of **Shippigan Island** are seen on the W. This island is 12 M. long by 8 M. wide, and is inhabited by Acadian fishermen. On the S. W. shore is the hamlet of Alexander Point, on Alemek Bay, opposite the populous village and magnificent harbor of *Shippigan*. There are valuable fisheries of herring, cod, and mackerel off these shores, and the deep triple harbor is well sheltered by the islands of Shippigan and Pocksuedie, forming a secure haven of refuge for the American and Canadian fleets.

64   *Route 15.*   BAY OF CHALEUR.

Shippigan Harbor, though still surrounded by forests, has occupied a prominent place in the calculations of commerce and travel. It has been proposed that the Intercolonial Railway shall connect here with a transatlantic steamship line, thus withdrawing a large portion of the summer travel from Halifax and New York. The distance from Shippigan to Liverpool by the Straits of Belleisle is 148 M. less than the distance from Halifax to Liverpool, and Shippigan is 271 M. nearer Montreal than is Halifax.

The Ocean Ferry.—The following plan is ingeniously elaborated and powerfully supported, and is perhaps destined to reduce the transatlantic passage to 100 hours. It is to be carried out with strong, swift express-steamers on the Ocean and the Gulf, and through trains on the railways. The itinerary is as follows: London to Valentia, 640 M., 16 hours; Valentia to St. John's, N. F., 1,640 M., 100 hours; St. John's to St. George's Bay (across Newfoundland by railway), 250 M., 8¼ hours; St. George's Bay to Shippigan (across the Gulf), 250 M., 15½ hours; Shippigan to New York, 906 M., 31 hours; London to New York, 171 hours, or 7⅛ days. It is claimed that this route would escape the dangers between Cape Race and New York; would give usually quiet passages across the Gulf; would diversify the monotony of the long voyage by three transfers, and would save 4-6 days on the recorded averages of the steamships between New York and Liverpool (see maps and details in Sandford Fleming's "Intercolonial Railway Survey").

The steamer now crosses the Miscou Banks, and approaches **Miscou Island,** which is 20 M. in circumference and contains about 300 inhabitants. On its S. shore is a fine and spacious harbor, which is much used as a place of refuge in stormy weather by the American fishing-fleets.

Settlements were formed here early in the 17th century by the French, for the purpose of hunting the walrus, or sea-cow. Such an exterminating war was waged upon this valuable aquatic animal that it soon became extinct in the Gulf, and was followed into the Arctic Zone. Within five years a few walruses have been seen in the Gulf, and it is hoped that they may once more enter these waters in droves. At an early date the Jesuits established the mission of St. Charles de Miscou, but the priests were soon killed by the climate, and no impression had been made on the Indians. It is claimed that there may still be seen the ruins of the post of the Royal Company of Miscou, which was founded in 1635 for the pursuit of fish and walruses, and for a time derived a great revenue from this district. Fortifications were also erected here by M. Denys, Sieur de Fronsac.

The steamer alters her course gradually to the W. and passes the fixed red light on Birch Point, and Point Miscou, with its high green knoll. Between Point Miscou and Cape Despair, 25 M. N., is the entrance to the Bay of Chaleur.

The **Bay of Chaleur** was known to the Indians by the name of *Ecketuam Nemaache,* signifying "a Sea of Fish," and that name is still applicable, since the bay contains every variety of fish known on these coasts. It is 90 M. long and from 10 to 25 M. wide, and is nearly free from shoals or dangerous reefs. The waters are comparatively tranquil, and the air is clear and bracing and usually free from fog, affording a marked contrast to the climate of the adjacent Gulf coasts. The tides are regular and have but little velocity. The length of the bay, from Point Miscou to Campbellton, is about 110 M. These waters are visited every year by great American fleets, manned by the hardy seamen of Cape Cod and Gloucester, and valuable cargoes of fish are usually carried back to the Massachusetts ports.

This bay was discovered by Jaques Cartier in the summer of 1535, and, from the fact that the heated season was at its height at that time, he named it *La Baie des Chaleurs* (the Bay of Heats). On the earliest maps it is also called *La Baie des Espagnols*, indicating that it was frequented by Spanish vessels, probably for the purposes of fishing.

In these waters is located the scene of the old legend of the Massachusetts coast, relative to Skipper Ireson's misdeed, which, with the record of its punishment, has been commemorated in the poetry of Whittier: —

" Small pity for him! — He sailed away
From a leaking ship in Chaleur Bay, —
Sailed away from a sinking wreck,
With his own town's-people on her deck !
' Lay by ! lay by !' they called to him ;
Back he answered, ' Sink or swim !
Brag of your catch of fish again !'
And off he sailed through the fog and rain.
Old Floyd Ireson, for his hard heart,
Tarred and feathered and carried in a cart
By the women of Marblehead.

" Fathoms deep in dark Chaleur
That wreck shall lie forevermore.
Mother and sister, wife and maid,
Looked from the rocks of Marblehead
Over the moaning and rainy sea, —
Looked for the coming that might not be !
What did the winds and the sea-birds say
Of the cruel captain who sailed away ? —
Old Floyd Ireson, for his hard heart,
Tarred and feathered and carried in a cart
By the women of Marblehead."

When well within the bay the steamer assumes a course nearly S. W., leaving Miscou and Shippigan Islands astern. The broad *Caraquette Bay* is on the S., and the New-Bandon shores (see page 66) are followed into Nepisiguit Bay. The harbor of Bathurst is entered by a strait two cables wide, between Alston Point and Carron Point, on the former of which there are red and white beacon-lights.

Bathurst (*Bay View Hotel*) is the capital of Gloucester County, and has about 600 inhabitants. It is favorably situated on a peninsula in the harbor, 2½ M. from the bay, and is connected by a bridge with the village of St. Peter's. Large quantities of fish are sent hence to the American cities during the summer; and the exportation of frozen salmon has become an important business. The Intercolonial Railway has a station near Bathurst, which will probably be one of its chief ports on this bay. The beautiful Basin of Bathurst receives the waters of four rivers, and its shores are already well populated by farmers.

The Basin of Bathurst was called by the Indians *Winkapiguwick*, or *Nepisiguit*, signifying the "Foaming Waters." It was occupied in 1638 by M. Enaud, a wealthy Basque gentleman, and his retainers, forming a town called St. Pierre. Enaud married a Mohawk princess, founded mills, and established an extensive fur-trade, erecting a commodious mansion at Abshaboo (Coal Point), at the mouth of the Nepisiguit. But some family troubles ensued, and Madame Enaud's brother slew her husband, after which the French settlements were plundered by the Indians, and such of the inhabitants as could not escape by way of the sea were massacred.

By 1670 the Chaleur shores were again studded with French hamlets, and occupied by an industrious farming population. In 1692 the Micmacs confederated against them, and, under the command of the sagamore Halion, completely devastated the whole district and compelled the settlers to fly to Canada, Thenceforward for 74 years this country was unvisited by Europeans. In 1764 a Scotch trading-post and fort was erected at Alston Point, on the N. shore of Bathurst harbor, and thence were exported great quantities of furs, moose-skins, walrus hides and tusks, and salmon. In 1776 this flourishing settlement was destroyed by American privateers, which also devastated the other shores of Chaleur. The present town was founded in 1818 by Sir Howard Douglas, and was named in honor of the Earl of Bathurst.

The **Nepisiguit River** empties into Bathurst harbor, and is famous for its fine fishing (it is now leased). A road ascends the W. bank for about

E

85 M., passing the Rough Waters, the brilliant rapids of the Pabineau Falls (9 M. up), the dark pools of the Betaboc reach, the Chain of Rocks, and the Narrows. The * **Grand Falls** of the Nepisiguit are 20 M. above Bathurst, and consist of 4 distinct and step-like cliffs, with a total height of 140 ft. They are at the head of the Narrows, where the river flows for 3-4 M. through a cañon between high cliffs of slaty rock. The river boldly takes the leap over this Titanic stairway, and the ensuing roar is deafening, while the base of the cliff is shrouded in white spray. From the profound depths at the foot the river whirls away in a black and foam-flecked course for 2 M. The descent of timber over the Falls affords an exciting spectacle, and the logs are sometimes shot out clear beyond the lower terraces and alight in the pool below.

"Good by, lovely Nepisiguit, stream of the beautiful pools, the fisherman's elysium; farewell to thy merry, noisy current, thy long quiet stretches, thy high bluffs, thy wooded and thy rocky shores. Long may thy music lull the innocent angler into day-dreams of happiness. Long may thy romantic scenery charm the eye and gladden the heart of the artist, and welcome the angler to a happy sylvan home." (ROOSEVELT.)

The * *Grand Falls* of the Tete-à-gouche River are about 8 M. W. of Bathurst, and may be visited by carriage. The river here falls about 90 ft., amid a wild confusion of rocks and cliffs.

Tri-weekly stages run E. from Bathurst to Salmon Beach, 8 M.; Jamesville, 12; Clifton, 15; New Bandon, 20; Pockshaw, 23; Grand Anse, 28; Upper Caraquette, 36; Lower Caraquette, 43; Shippigan, 60. Fare to Caraquette, $3.50. This road follows the shores of the Nepisiguit Bay and the Bay of Chaleur for nearly 80 M. The hamlets of *Clifton* (small inn) and *New Bandon* were settled by Irish immigrants, and are now engaged in making grindstones. Pockshaw has an inn and about 600 inhabitants. *Grand Anse* is an Acadian settlement, and has 700 inhabitants, who are engaged in farming and fishing. Thence the road runs 8 M. S. E. to *Upper Caraquette*, where there are about 600 Acadians. *Lower Caraquette* (two inns) is a French village of 1,500 inhabitants, and is famous for its strong, swift boats and skilful mariners.

Caraquette was founded in 1768 by a colony of Bretons, and owed a part of its early growth to intermarriages with the Micmacs. It is a long street of farms in the old Acadian style, and is situated in a fruitful and well-cultivated country. The view from the hills over the village, and especially from the still venerated spot where the old chapel stood, is very pleasant, and includes Miscou and Shippigan, the Gaspé ports, and the bold Quebec shores. The Jersey house of Robin & Co. has one of its fishing-establishments here, and does a large business.

Caraquette is one of the chief stations of the N. shore fisheries. In the year 1873 the fish product of the three lower Maritime Provinces amounted to the value of $9,060,342. Nova Scotia caught $6,577,086 worth of fish; and New Brunswick caught $2,285,660 worth, of which $527,312 were of salmon, $500,806 of herring, $346,925 of lobsters, $388,099 of codfish, $108,514 of alewives, $90,065 of hake, $64,396 of pollock, $45,480 of oysters, $41,851 of smelt, and $85,477 of mackerel.

Daily stages run S. from Bathurst to Chatham (see page 61). Tri-weekly stages follow the coast of the Bay of Chaleur to the N. W. to Medisco: Rochette, 12 M.; Belledune, 20; Belledune River, 24; Armstrong's Brook, 28; River Louison, 33; New Mills, 38; River Charlo, 44; and Dalhousie, 52. Medisco and Rochette are French villages; the others are of British origin, and none of them have as many as 500 inhabitants. Many small streams enter the bay from this coast, and the whole district is famous for its fishing and hunting (water-fowl). The line of this shore is followed by the Intercolonial Railway.

Off Bathurst the Bay of Chaleur is over 25 M. wide, and the steamer passes out and takes a course to the N. W., passing the hamlet of Rochette, and soon rounding Belledune Point. The imposing highlands of the Gaspesian peninsula are seen on the N. with the peak of Tracadiegash. The passage between Tracadiegash Point and Heron Island is about 7 M. wide; and 6–8 M. beyond the steamer passes Maguacha Point (*Maguacha*, Indian for "Always Red") on the r., and enters the Restigouche Harbor.

"To the person approaching by steamer from the sea, is presented one of the most superb and fascinating panoramic views in Canada. The whole region is mountainous, and almost precipitous enough to be alpine; but its grandeur is derived less from cliffs, chasms, and peaks, than from far-reaching sweeps of outline, and continually rising domes that mingle with the clouds. On the Gaspé side precipitous cliffs of brick-red sandstone flank the shore, so lofty that they seem to cast their gloomy shadows half-way across the Bay, and yawning with rifts and gullies, through which fretful torrents tumble into the sea. Behind them the mountains rise and fall in long undulations of ultramarine, and, towering above them all, is the famous peak of Tracadiegash flashing in the sunlight like a pale blue amethyst." (HALLOCK.)

Dalhousie (*Fraser's Hotel*) is a village of 600 inhabitants, situated at the mouth of the long estuary of the Restigouche, and is the capital of Restigouche County. It faces on the harbor from three sides, and has great facilities for commerce and for handling lumber. The manufacture and exportation of lumber are here carried on on a large scale; and the town is also famous for its shipments of lobsters and salmon. The salmon fisheries in this vicinity are of great value and productiveness. The line of the Intercolonial Railway is about 4 M. S. of Dalhousie. The site of this port was called *Sickadomec* by the Indians. 50 years ago there were but two log-houses here, but the district was soon occupied by hardy Highlanders from Arran, whose new port and metropolis was "located in an alpine wilderness." Directly back of the village is *Mt. Dalhousie*, and the harbor is protected by the high shores of Dalhousie Island. Bonami Point is at the entrance of the harbor, and has a fixed white light; and Fleurant Point is opposite the town, across the estuary.

"The Bay of Chaleur preserves a river-like character for some distance from the point where the river may strictly be said to terminate, and certainly offers the most beautiful scenery to be seen in the Province. . . . . From Mr. Fraser's to the sea, a distance of some 20 M. by water, or 14 by land, the course of the river is really beautiful. Swollen to dimensions of majestic breadth, it flows calmly on, among picturesque and lofty hills, undisturbed by rapids, and studded with innumerable islands covered with the richest growth of elm and maple. . . . . The whole of the distance from Campbellton to Dalhousie, a drive of 20 M. along the coast of the Bay of Chaleur, on an excellent high-road, presents a succession of beautiful views across the narrow bay, in which Tracadiegash, one of the highest of the Gaspé mountains, always forms a conspicuous object, jutting forward as it does into the sea opposite Dalhousie." (HON. ARTHUR GORDON.)

"Nothing can exceed the grandeur and beauty of the approach to the estuary of the Restigouche. The pointed hills in the background, the deep green forest with its patches of cultivation, and the clear blue of the distant mountains, form a picture of the most exquisite kind." (SIR R. BONNYCASTLE.)

"The expanse of three miles across the mouth of the Restigouche, the dreamy alpine land beyond, and the broad plain of the Bay of Chaleur, present one of the most splendid and fascinating panoramic prospects to be found on the continent of America, and has alone rewarded us for the pilgrimage we have made." (CHARLES LANMAN.)

The estuary of the Restigouche is 2–4 M. wide, and extends from Dalhousie to Campbellton, about 16 M. *Point à la Garde* is 9 M. above Dalhousie on the N. shore, and is a bold perpendicular promontory overlooking the harbor. On this and *Battery Point* (the next to the W.) were the extensive French fortifications which were destroyed by Admiral Byron's British squadron in 1780. Several pieces of artillery and other relics have been obtained from the water off these points. Battery Point is a rocky promontory 80 ft. high, with a plain on the top, and a deep channel around its shores. *Point Pleasant* is 4 M. distant, and 1 M. back is a spiral mass of granite 700 ft. high, which is accessible by natural steps on the E. 1½ M. from this peak is a pretty forest-lake, in which red trout are abundant. 5 M. N. of Point à la Garde is the main peak of the Scaumenac Mts., which attains an altitude of 1,745 ft.

Campbellton (three hotels) is situated in a diversified region of hills at the head of deep-water navigation on the Restigouche, which is here 1 M. wide. It has about 600 inhabitants and deals chiefly in the exportation of lumber and fish. One of the chief stations of the Intercolonial Railway is located here. The adjacent country is highly picturesque, and is studded with conical hills, the chief of which is Sugar Loaf, 800 ft. high.

Mission Point is nearly opposite Campbellton, and is surrounded by fine hill-scenery, which has been likened to that of Wales. The river is rapid off these shores, and abounds in salmon. This place is also known as Point-à-la-Croix, and is one of the chief villages and reservations of the Micmac Indians. It has about 500 inhabitants, with a Catholic church.

The Micmac language is said to be a dialect of the Huron tongue; while the Millicetes, on the St. John River, speak a dialect of Delaware origin. These two tribes have an annual council at Mission Point, at which delegates from the Penobscot Indians are in attendance. The Micmac nation occupies the waste places of the Maritime Provinces, from Newfoundland to Gaspé, and numbers over 6,000 souls. These Indians are daring and tireless hunters and fishermen, and lead a life of constant roving, gathering annually at the local capitals,—Chapel Island, in Cape Breton; Ponhook Lake, in Nova Scotia; and Mission Point, in Quebec. They are increasing steadily in numbers, and are becoming more valuable members of the Canadian nation. They have hardly yet recovered from the terrible defeat which was inflicted on them by an invading army of Mohawks, in 1689. The flower of the Maritime tribes hastened to the border to repel the enemy, but they were met by the Mohawks in the Restigouche country, and were annihilated on the field of battle.

The chief of the Micmacs at Mission Point visited Queen Victoria in 1850, and was kindly welcomed and received many presents. When Lord Aylmer, Governor-General of Canada, visited Gaspé, he was waited on by 500 Indians, whose chief made him a long harangue. But the tribe had recently recovered from a wreck (among other things) a box of decanter-labels, marked RUM, BRANDY, GIN, etc., and the noble chief, not knowing their purport, had adorned his ears and nose with them, and surrounded his head with a crown of the same materials. When the British officers recognized the familiar names, they burst into such a peal of laughter as drove the astonished and incensed chief from their presence forever.

3 M. above Mission Point is *Point au Bourdo*, the ancient site of La Petite Rochelle, deriving its present name from Capt. Bourdo, of the French frigate *Marchault*, who was killed in the battle off this point and was buried here. Fragments of the French vessels, old artillery, camp equipments, and shells have been found in great numbers in this vicinity.

In 1760 Restigouche was defended by 2 batteries, garrisoned by 250 French regulars, 700 Acadians, and 700 Indians; and in the harbor lay the French war-vessels *Marchault*, 32, *Bienfaisant*, 22, and *Marquis Marloye*, 18, with 19 prize-ships which had been captured from the English. The place was attacked by a powerful British fleet, consisting of the *Fame*, 74, *Dorsetshire*, *Scarborough*, *Achilles*, and *Repulse*, all under the command of Commodore John Byron (grandfather of the poet, Lord Byron). But little resistance was attempted; and the French fleet and batteries surrendered to their formidable antagonist. The captured ships were carried to Louisbourg, and the batteries and the 200 houses of Restigouche were destroyed.

The **Restigouche River** is a stately stream which is navigable for 135 M. above Campbellton. It runs through level lands for several miles above its mouth, and then is enclosed between bold and rugged shores. There are hundreds of low and level islands of a rich and yearly replenished soil; and above the Tomkedgwick are wide belts of intervale. 30 M. from its mouth it receives the waters of the Metapedia River, flowing down from the Metis Mts.; and 35 M. from the mouth is the confluence of the trout-abounding Upsalquitch. 21 M. farther up is the mouth of the Patapedia; and 20 M. beyond this point the Tomkedgwick comes in from the N. W. This system of waters drains over 6,000 square miles of territory, and is connected by portages with the streams which lead into the Bay of Fundy and the River St. Lawrence.

Campbellton to the St. Lawrence River.

The Metapedia Road leaves the N. shore of the Restigouche a few miles above Campbellton, and strikes through the forest to the N. W. for the St. Lawrence River. This is the route of the new Intercolonial Railway, which passes up through the wilderness to St. Flavie. The distance from Campbellton to St. Flavie is 111 M., and the fare by stage is $9. This road leads across the barren highlands of Gaspé, and through one of the most thinly settled portions of Canada.

The French hamlet of *St. Alexis* is near the mouth of the Metapedia River. *Metapedia* is 15 M. above Campbellton, and is situated amid the pretty scenery at the confluence of the Metapedia and Restigouche Rivers. The salmon-fisheries in this vicinity attract a few enthusiastic sportsmen every year. Near the confluence is the old Fraser mansion, famous among the travellers of earlier days. The Intercolonial Railway crosses the Restigouche in this vicinity, and has a station at Metapedia. 60 M. beyond this village is the Metapedia Lake.

The **Metapedia Lake** is 12 M. long by 2 M. wide, and is surrounded by low shores of limestone, above and beyond which are distant ranges of highlands. Its waters abound in tuladi (gray trout), trout, and white-fish, and afford good sporting. The lake contains a large island, which is a favorite breeding-place of loons.

St. Flavie (two inns) is a village of 450 French people, situated on the S. shore of the River St. Lawrence, and is the point where the Intercolonial Railway reaches the river and turns to the S. W. towards Quebec. It is distant from Campbellton, 111 M.; from Father Point, 15 M.; from Rivière du Loup, 76 M.; and from Quebec, 201 M.

16. St. John to Amherst and Halifax.

The Intercolonial Railway.

This route traverses the S. E. counties of New Brunswick, passes the isthmus at the head of the Bay of Fundy, and after crossing the Cobequid Mts. and rounding the head of Cobequid Bay, runs S. W. to the city of Halifax. It traverses some interesting districts and has a few glimpses of attractive scenery, but the views are generally monotonous and without any striking beauties. During calm and pleasant weather the traveller will find the Annapolis route (see Route 18) much the pleasanter way to go from St. John to Halifax.

There is no change of cars between St. John and Halifax, and baggage is checked through. During the summer there is a day express-train, leaving St. John at 7 A. M., and due at Halifax at 7.40 P. M.; and a night express, leaving St. John at 8.30 P. M., and due at Halifax at 9 A. M. Pullman-cars have recently been introduced on this line.

Stations.— St. John; Moosepath, 3 M.; Brookville, 5; Torryburn, 6; Riverside, 7; Rothesay, 9; Quispamsis, 12; Nauwigewauk, 17; Hampton, 22; Passekeag, 26; Bloomfield, 27; Norton, 38; Apohaqui, 39; Sussex, 44; Plumweseep, 47; Penobsquis, 51; Anagance, 60; Petitcodiac, 66; Pollet River, 71; Salisbury, 76; Boundary Creek, 79; Moncton, 89; Humphrey, 91; Painsec Junction, 97 (Dorchester Road, 102; Shediac, 106; Point du Chêne, 108); Meadow Brook, 101; Memramcook, 108; Dorchester, 116; Sackville, 127; Aulac, 131; Amherst, 138; Nappan, 144; Maccan, 147; Athol, 151; Spring Hill, 156; Salt Springs, 164; River Philip, 167; Thompson, 174; Greenville, 181; Wentworth, 187; Folly Lake, 191; Londonderry, 199; Debert, 204; Ishgonish, 208; Truro, 216; Johnson, 220; Brookfield, 224; Polly Bog, 229; Stewiacke, 233; Shubenacadie, 238; Milford, 242; Elmsdale, 247; Enfield, 249; Grand Lake, 254; Wellington, 256; Windsor Junction, 264; Rocky Lake, 266; Bedford, 269; Four-Mile House, 273; Halifax, 276.

Fares from St. John.— To Sussex, 1st class, $ 1.32,— 2d class, 88c.; to Moncton, 1st class, $ 2 67,— 2d class, $ 1.78; to Shediac, 1st class, $ 3,— 2d class, $ 2; to Amherst, 1st class, $ 3 78,— 2d class, $ 2.52; to Truro, 1st class, $ 5.06,— 2d class, $ 3.37; to Halifax, 1st class, $ 6,— 2d class, $ 4.

Fares from Halifax.— To Truro, 1st class, $ 1.83,— 2d class, $ 1.22; to Pictou, 1st class, $ 3.15,— 2d class, $ 2.10; to Amherst, 1st class, $ 3.78,— 2d class, $ 2.52; to Shediac, 1st class, $ 4.55,— 2d class, $ 3.03; to Sussex, 1st class, $ 5.31,— 2d class, $ 3.54; to St. John, 1st class, $ 6,— 2d class, $ 4.

Way-passengers can estimate their expenses easily on the basis of 3c. per mile for 1st class, and 2c. per mile for 2d class tickets, which is the tariff fixed by the Canadian Government for all distances of less than 100 M. on its national railways.

On leaving the Valley station, in the city of St. John (see page 19), the train passes out into the Marsh Valley, which is ascended for several miles (see page 22). A short distance beyond Moosepath Park the line crosses *Lawlor's Lake* on an embankment which cost heavily, on account of the great depth to which the ballasting sunk. The **Kennebecasis Bay** is soon seen, on the l., and is skirted for 5 M., passing the villas of Rothesay (see page 22), and giving pleasant views over the broad waters. Quispamsis station is 3 M. S. of Gondola Point, whence a ferry crosses the Kennebecasis to the pretty hamlet of *Clifton.* The narrowing valley is now followed to the N. E., with occasional glimpses of the river on the l. *Hampton* (two hotels) is the shire-town of Kings County, whose new public buildings are seen to the r. of the track. It is a thriving village of recent origin, and is visited in summer by the people of St. John, on account of the hill-scenery in the vicinity.

St. Martin's, or **Quaco**, is about 20 M. S. E., on the Bay of Fundy, and is to be connected with Hampton by a new railway. (It is now visited by tri-weekly stage from St. John in 32 M., fare $1.50; a rugged road.) This is one of the chief ship-building towns in the province, and has over 1,000 inhabitants, with several churches and other public buildings. It was originally settled by the King's Orange Rangers, and has recently become a favorite point for summer excursions from St. John. The hotel accommodation is inferior. S. of the village is the tall lighthouse on Quaco Head, sustaining a revolving white light. The name *Quaco* is a contraction of the Indian words *Gulwahgahgee,* meaning " the Home of the Sea-cow."

The shores about Quaco are bold and picturesque, fronting the Bay with lofty iron-bound cliffs, among which are small strips of stony beaches. The strata are highly inclined and in some cases are strangely contorted, while their shelves and crevices are adorned with pine-trees. *Quaco Head* is 2 M. from St. Martin's, and is 350 ft. high, surrounded by cliffs of red sandstone 250 ft. in height. This bold promontory rises directly from the sea, and is crowned by forests. The harbor of Quaco is rather pretty, whence it has been likened to the Bay of Naples. *Tracy's Lake* is about 5 M. from Quaco, on the Loch Lomond road, and is noted for an abundance of trout. 10-12 M. N. of the village is the Mount Theobald Lake, a small round forest-pool in which trout are found in great numbers.

Hampton station is 1 M. from the village of Hampton Ferry, and beyond Bloomfield the train reaches *Norton,* whence a road runs 7 M. N. W. to Springfield, at the head of Belleisle Bay. *Apohaqui* (Apohaqui Hotel) is a village of 300 inhabitants, on the upper Kennebecasis, and at the mouth of the Mill-stream Valley.

The train now reaches **Sussex** (*Exchange Hotel*), a pleasant little village of 400 inhabitants, whence the famous farm-lands of the *Sussex Vale* stretch off to the S. E. along the course of Trout Brook. There are several hamlets (with inns) amid the pleasant rural scenery of the Vale, and good trout-fishing is found on the smaller streams. 8 M. up is the prosperous settlement of Seeley's Mills, with 650 inhabitants.

The Sussex Vale was settled by the military corps of the New Jersey Loyalists (most of whom were Germans), soon after the Revolutionary War, and it is now occupied, for the most part, by their descendants. "Good roads, well-executed bridges, cleared land, excellent crops, comfortable houses, high-bred cattle and horses, good conveyances public and private, commodious churches, well-taught schools, well-provided inns, and an intelligent, industrious people, all in the midst of scenery lofty, soft, rounded, beautifully varied with hill and valley, mountain and meadow, forest and flood, have taken the place of the pathless wilderness, the endless trees, the untaught Indian, and the savage moose." (PROF. JOHNSTON.)

Beyond Plumweseep occasional glimpses of the long low ridge of Piccadilly Mt. are obtained on the r., and Mt. Pisgah is just N. of *Penobsquis* station (small inn), which is the seat of the New Brunswick Paper Manufacturing Co. and of several salt-works. Tri-weekly stages run hence 32 M. S. E. to the maritime village of Salmon River, on Chignecto Bay, 4 M. N. W. of the obscure shipping-port of *Point Wolfe* (Stevens's Hotel).

Petitcodiac (*Mansard House; Central Hotel*) is 15 M. beyond Penobsquis, and is a busy village of 400 inhabitants, many of whom are connected with the lumber-trade. 5 M. S. E. is the Pollett River village, near which there is good trouting. In this vicinity are the *Pollett Falls*, where the river, after flowing through a narrow defile between lofty and rugged hills, falls over a line of sandstone ledges, and then whirls away down a dark gorge below. The caverns, crags, and eroded fronts of the sandstone cliffs form picturesque bits of scenery.

15-18 M. N. of Petitcodiac are the famous fishing-grounds of the *Canaan River*. The railway now descends the valley of the Petitcodiac River, which was settled after the Revolutionary War by Germans from Pennsylvania who remained loyal to Great Britain. *Salisbury* (two inns) is a pleasant village of 300 inhabitants.

Stages run from Salisbury, or Moncton, to **Hillsborough** (two hotels), a busy village of 900 inhabitants, whence are shipped the abundant products of the mines of Albert County. The Albert Coal-mines are connected with Hillsborough by a railway 5½ M. long, and produce large quantities of valuable bituminous coal, much of which is sent by sea to Portland and Boston. 2¼ M. from the village are extensive plaster-quarries, whose products are shipped to the American ports. S. E. of Hillsborough, down the Petitcodiac River, are the villages of the parish of Hopewell, of which *Hopewell Cape* is the capital of the county. W. of Hopewell Corner is *Harvey Corner*, whence a pleasant road leads to Rocher. To the S. are the Shepody Lakes and River, beyond which (and 8 M. from Harvey Corner) is Little Rocher, near Cape Enragé on Chignecto Bay (with a fixed light, visible for 15 M.). Off these bold shores are the Albert Quarries and the rocky cliffs of Grindstone Island. The mines and villages of Albert County are being joined with the Intercolonial Railway system by a line called the *Albert Railway*, which intersects the former road and runs down through the lower parishes, meeting with fine scenery in its passage between Shepody Mt. (1,050 ft. high) and the Bay.

Beyond Salisbury station the Halifax train runs 13 M. N. E. to **Moncton** (*King's Hotel*), the headquarters of the Intercolonial Railway and the site of its extensive machine-shops. It is well laid out, and has 4 churches, a weekly paper, and some manufacturing works. Its situation at the head of navigation on the Petitcodiac gives certain commercial advantages, and affords opportunity for the visitor to see the great "Bore," or tide-wave, of the Bay of Fundy. At the beginning of the flood-tide a wall of water 4-6 ft. high sweeps up the river, and within 6 hours the stream rises over 70 ft. On account of the sharp curve in the river at this point, Moncton was known only as "the Bend" for over a century, when it was named in honor of an early English officer of the Acadian wars. This bend also gave rise to the name of the river, which was hence called by the French *Petit Coude* ("Little Elbow").

The new division of the Intercolonial Railway runs N. from Moncton, and is designed to meet the Canadian railway system at Rivière du Loup. It passes through or near the chief towns of the North Shore, and follows the Bay of Chaleur for many miles. A considerable portion of the line will probably be open to travel in the summer of 1875, but the officers of the road cannot yet give precise information. The towns on this line are described in Route 15.

The Halifax train runs out to the N. E. from Moncton, and after passing *Painsec Junction* (see page 59) deflects to the S. E. into the Memramcook

Valley. It soon reaches the connected villages of *Memramcook* and *St. Joseph* (three inns), occupying the centre of a prosperous farming district which is inhabited by over 1,000 Acadians, — a pious and simple-hearted Catholic peasantry, — a large portion of whom belong to the prolific families of Leblanc, Cormier, Gaudet, and Bouque. On the opposite shore is the College of St. Joseph de Memramcook, where about 100 students (mostly from Canada and the United States) are conducted through a high-school curriculum by 12 friars and ecclesiastics. Near the college is the handsome stone building of the Church of St. Joseph de Memramcook.

The scenery is of a bold character as the train descends the r. bank of the Memramcook River, and crosses to **Dorchester** (*Dorchester Hotel*), a prosperous village of 800 inhabitants, situated near the mouth of the river and among the finest wheat-lands in New Brunswick. In this vicinity (and at *Rockland*, 4 M. W.) are large quarries of olive-colored sandstone, most of which is sent to Boston and New York. Dorchester has 3 churches, the public buildings of Westmoreland County, and numerous pleasant residences. Shipbuilding is carried on to some extent.

A ferry crosses Shepody Bay to Hopewell Cape (see page 72); and 6-8 M. W. of Dorchester is Belliveau village, nine tenths of whose inhabitants belong to the families of Belliveau, Gautreault, and Melançon. This settlement was named in honor of the venerable M. Belliveau, whose long life extended from 1730 to 1840. In 1776 many of the Acadians of this vicinity joined the New England forces under Col. Eddy, who occupied Sackville and attacked Fort Cumberland (see page 78).

The train now runs E. 12 M. from Dorchester to **Sackville** (*Brunswick House*), a rising and prosperous village of about 1,500 inhabitants, situated on a red sandstone slope at the mouth of the Tantramar[1] River, near the head of the Bay of Fundy. It has ship-yards, a stove foundry, a newspaper, and 8 churches. Sackville is the seat of the Mount Allison Wesleyan College, an institution which was founded by Mr. C. F. Allison, and is conducted by the Wesleyan Conference of Eastern British America. It includes a small college, a theological hall, and academies for boys and girls. A road leads from Sackville S. E. down the rugged headland between Cumberland Basin and Shepody Bay, passing the marine hamlets of Woodpoint (5 M.), Rockport (12 M.), and N. Joggins, 14 M. from Sackville, and near the highlands of Cape Marangouin.

Sackville is the point established for the outlet of the projected **Baie Verte Canal**, a useful work 18 M. long, which would allow vessels to pass from the Bay of Fundy to the Gulf of St. Lawrence without having to round the iron-bound peninsula of Nova Scotia. This canal has been planned and desired for over a century, but nothing has yet been done, except the surveying of the isthmus. Tri-weekly stages run N. E. along the telegraph-road from Sackville to Jolicœur (10 M.), Baie Verte Road (14 M.), Baie Verte (18 M., small inn), and Port Elgin (20 M. ; inn). About 16 M. N. E. of Port Elgin is **Cape Tormentine,** "the great headland which forms the E. extremity of New Brunswick within the Gulf. Indian Point may be said to form the southern, and Cape Jourimain the northern points of this headland, which is a place of importance in a nautical point of view, not only from

[1] *Tantramar*, from the French word *Tintamarre*, meaning "a thundering noise."

its position, but from its dangerous and extensive shoals." The submarine telegraph to Prince Edward Island crosses from Cape Jourimain ; and it is from this point that the winter mail-service is conducted, when the mails, passengers, and baggage are subjected to an exciting and perilous transit in ice-boats to Cape Traverse. Baie Verte is 9 M. wide and 11 M. deep, but affords no good shelter. It receives the Tignish and Gaspereau Rivers, and at the mouth of the latter are the ancient ruins of Fort Monctou.

At Sackville the Halifax train crosses the Tantramar River, and runs out over the wide **Tantramar Marsh** to Aulac, or Cole's Island (stage to Cape Tormentine), near which it crosses the Aulac River. Trains are sometimes blocked in on these plains during the snow-storms of winter, and the passengers are subjected to great hardships. The Missiguash River is next crossed, with the ruins of Fort Beausejour (Cumberland) on the N., and of Fort Beanbassin (Lawrence) on the S. These forts are best visited from Amherst, which is 4-5 M. distant, and is reached after traversing the *Missiguash Marsh*. The Missiguash River is the boundary between New Brunswick and Nova Scotia, and **Amherst** is the first town reached in the latter Province.

Amherst to Halifax, see Route 17.

NOVA SCOTIA.

THE Province of Nova Scotia is peninsular in location, and is connected with the mainland by an isthmus 8 M. wide. It is bounded on the N. by the Bay of Fundy, the Strait of Northumberland, and the Gulf of St. Lawrence; on the E. and S. by the Atlantic Ocean; and on the W. by the ocean, the Bay of Fundy, and the Province of New Brunswick. Its length, from Cape Canso to Cape St. Mary, is 383 M., and its breadth varies from 50 M. to 104 M. The area of the peninsular portion of the Province is about 16,000 square miles. (The island of Cape Breton is connected with this Province, politically, but its description is reserved for another section of this book.)

"Acadie is much warmer in summer and much colder in winter than the countries in Europe lying under the same parallels of latitude" (Southern France, Sardinia, Lombardy, Genoa, Venice, Northern Turkey, the Crimea, and Circassia). "The spring season is colder and the autumn more agreeable than those on the opposite side of the Atlantic. Its climate is favorable to agriculture, its soil generally fertile. The land is well watered by rivers, brooks, and lakes. The supply of timber for use and for exportation may be considered as inexhaustible. The fisheries on the coasts are abundant. The harbors are numerous and excellent. Wild animals are abundant, among which are remarkable the moose, caribou, and red deer. Wild fowl also are plenty. Extensive tracts of alluvial land of great value are found on the Bay of Fundy. These lands have a natural richness that dispenses with all manuring; all that is wanted to keep them in order is spade-work. As to cereals, — wheat, rye, oats, buckwheat, maize, all prosper. The potato, the hop, flax, and hemp are everywhere prolific. The vegetables of the kitchen garden are successfully raised. Of fruit there are many wild kinds, and the apple, pear, plum, and cherry seem almost indigenous. The vine thrives; good grapes are often raised in the open air. It was said by a French writer that Acadie produced readily everything that grew in Old France, except the olive.

"In the peninsula, or Acadie proper, there is an abundance of mineral wealth. Coal is found in Cumberland and Pictou; iron ore, in Colchester and Annapolis Counties; gypsum, in Hants; marble and limestone, in different localities; freestone, for building, at Remsheg (Port Wallace) and

Pictou; granite, near Halifax, Shelburne, etc.; brick clay, in the counties of Halifax and Annapolis. The amethysts of Parrsborough and its vicinity have been long celebrated, and pearls have been found lately in the Annapolis River. The discovery of gold along the whole Atlantic shore of the peninsula of Nova Scotia has taken place since 1860, and it now gives steady remunerative employment to about 800 or 1,000 laborers, with every expectation of its expansion." (BEAMISH MURDOCH.) The production of gold from the Nova-Scotia mines now amounts to about $400,000 a year.

In 1873 the Nova-Scotians caught $6,577,086 worth of fish, of which $2,531,159 worth were of codfish, $1,411,676 of mackerel, $717,861 of herring, and $865,574 of lobsters.

The territory now occupied by the Maritime Provinces was known for nearly two centuries by the name of *Acadie*,[1] and was the scene of frequent wars between Britain and France. Its first discoverers were the Northmen, about the year 1000 A. D., and Sebastian Cabot rediscovered it in 1498. In 1518 and 1598 futile attempts were made by French nobles to found colonies here, and French fishermen, fur-traders, and explorers frequented these shores for over a century. In 1605 a settlement was founded at Port Royal, after the discoveries of De Monts and Champlain, but it was broken up in 1618 by the Virginians, who claimed that Acadie belonged to Britain by virtue of Cabot's discovery. In 1621 James I. of England granted to Sir William Alexander the domain called NOVA SCOTIA, including all the lands E. of a line drawn from Passamoquoddy Bay N. to the St. Lawrence; but this claim was renounced in 1632, and the rival French nobles, La Tour and D'Aulnay, commenced their fratricidal wars, each striving to be sole lord of Acadie. In 1654 the Province was captured by a force sent out by Cromwell, but the French interest soon regained its former position.

The order of the Baronets of Nova Scotia was founded by King Charles I., in 1625, and consisted of 150 well-born gentlemen of Scotland, who received, with their titles and insignia, grants of 18 square miles each, in the wide domains of Acadia. These manors were to be settled by the baronets at their own expense, and were expected in time to yield handsome revenues. But little was ever accomplished by this order. Meantime Cardinal Richelieu founded and became grand master of a more powerful French association called the Company of New France (1627). It con-

---

[1] *Acadia* is the Anglicized (or Latinized) form of *Acadie*, an Indian word signifying "the place," or "the region." It is a part of the compound words *Segdeben-acadie* (Shubenacadie), meaning "place of wild potatoes"; *Tulluk-cadie* (Tracadie), meaning "dwelling-place"; *Sun-acadie*, or "place of cranberries"; *Kitpoo-acadie*, or "place of eagles," and others of similar form. The Milicete tribes pronounced this word "Quoddy," whence *Pestumoo-quoddy* (Passamoquoddy), meaning "place of pollocks"; *Noodi-quoddy*, or "place of seals," etc. When a British officer was descending the Shubenacadie with a Micmac guide, he inquired how the name originated; the Indian answered, "Because plenty wild potatoes — *segdeben* — once grew here." "Well, '*acadie*,' Paul, what does that mean?" "Means — where you find 'em," rejoined the Micmac.

sisted of 100 members, who received Acadia, Quebec, Florida, and Newfoundland "in simple homage," and had power to erect duchies, marquisates, and seigniories, subject to the royal approval. They allowed French Catholics only to settle on these lands, and were protected by national frigates. This order continued for 40 years, and was instrumental in founding numerous villages along the Nova-Scotian coast.

In 1690 the New-Englanders overran the Province and seized the fortresses, but it was restored to France in 1697. In 1703 and 1707 unsuccessful expeditions were sent from Massachusetts against the Acadian strongholds, but they were finally captured in 1710; and in 1713 Nova Scotia was ceded to Great Britain by the Treaty of Utrecht. The Province was kept in a condition of disorder for the next 40 years, by the disaffection of its French population and the lawlessness of the Indians, and the British fortresses were often menaced and attacked. After the foundation of Halifax, in 1749, a slow tide of immigration set in and strengthened the government. In 1755 the French people in the Province (7,000 in number) were suddenly seized and transported to the remote American colonies, and the French forts on the Baie-Verte frontier were captured.

In 1758 the first House of Assembly met at Halifax, and in 1763 the French power in America was finally and totally crushed. At the close of the Revolution, 20,000 self-exiled Americans settled in Nova Scotia; and in 1784 New Brunswick and Cape Breton were withdrawn and made into separate provinces (Cape Breton was reunited to Nova Scotia in 1820). During the Revolution and the War of 1812 Halifax was the chief station of the British navy, and the shores of the Province were continually harassed by American privateers.

In 1864 a convention was held at Charlottetown, P. E. I., to consider measures for forming a federal union of the Maritime Provinces. During the session Canadian delegates were admitted, on the request of the St. Lawrence Provinces; and a subsequent congress of all the Provinces was held at Quebec, at which the plan of the Dominion of Canada was elaborated. It is now thought that this quasi-national government does not fulfil all the original wishes of the seaboard regions, and that it may be well to unite (or reunite) the Maritime Provinces into one powerful province called ACADIA, by which the expense of three local legislatures and cabinets could be saved, their homogeneous commercial interests could be favored by uniform laws, and the populous and wealthy Provinces of Quebec and Ontario could be balanced in the Dominion Parliament.

"There are perhaps no Provinces in the world possessing finer harbors, or furnishing in greater abundance all the conveniences of life. The climate is quite mild and very healthy, and no lands have been found that are not of surpassing fertility. . . . . Finally, nowhere are there to be seen forests more beautiful or with wood better fitted for buildings and masts. There are in some places copper mines, and in others of coal. . . . . The fish most commonly caught on the coast are the cod, salmon, mackerel, herring, sardine, shad, trout, gotte, gaparot, barbel, sturgeon, goberge, — all fish that can be salted and exported. Seals, walruses, and whales are found in great numbers. . . . . The rivers, too, are full of fresh-water fish, and the banks teem with countless game." (FATHER CHARLEVOIX, 1765.)

"Herewith I enter the lists as the champion of Nova Scotia. . . . . Were I to give a first-class certificate of its general character, I would affirm that it yields a greater variety of products for export than any territory on the globe of the same superficial area. This is saying a great deal. Let us see : she has ice, lumber, ships, salt-fish, salmon and lobsters, coal, iron, gold, copper, plaster, slate, grindstones, fat cattle, wool, potatoes, apples, large game, and furs." (CHARLES HALLOCK, 1873.)

17. St. John to Amherst and Halifax.

St. John to Amherst, see preceding route.

Amherst (*Acadia Hotel*; *Amherst Hotel*) is a flourishing town midway between St. John and Halifax (138 M. from each). It is the capital of Cumberland County, Nova Scotia, and is pleasantly situated at the head of the Cumberland Basin, one of the great arms of the Bay of Fundy. It has 3,606 inhabitants, and is engaged in the lumber trade; while the immense area of fertile meadows about the town furnishes profitable employment for a large rural population. Bi-weekly stages run N. E. up the valley of the La Planche to *Tidnish* (two inns), a village of 300 inhabitants on Baie Verte. Tri-weekly stages run N. E. to *Shinimicas* and the large farming district called the Head of Amherst, which has over 2,000 inhabitants.

The present domain of Nova Scotia was ceded to Great Britain by the Treaty of Utrecht, in 1713, but its boundaries were not defined, and the French determined to limit it on the N. to the Missiguash River. To this end Gov. La Jonquiére sent M. La Corne, with 600 soldiers, to erect forts on the line of the Missiguash. The warrior-priest, the Abbé Laloutre (Vicar-General of Acadie), led many Acadians to this vicinity, where the flourishing settlement of *Beaubassin* was founded. At the same time La Corne established a chain of military posts from the Bay of Fundy to Baie Verte, the chief fort being located on the present site of Fort Cumberland, and bearing the name of *Beausejour*. The governor of Nova Scotia sent out a British force under Major Lawrence, who captured and destroyed Beaubassin, and erected Fort Lawrence near its site. The Acadians were industriously laboring in the peaceful pursuits of agriculture about Beausejour; and the King of France had granted 80,000 livres for the great *aboideau* across the Aulac River. The British complained, however, that the priests were endeavoring to array the Acadians against them, and to entice them away from the Nova-Scotian shores. It was resolved that the French forces should be driven from their position, and a powerful expedition was fitted out at Boston. Three frigates and a number of transports conveying the New-England levies sailed up the Bay of Fundy in May, 1755, and debarked a strong land force at Fort Lawrence. Meantime 1,200 – 1,500 Acadians had been gathered about Beausejour, by the influence of the Abbé Laloutre, and a sharp skirmish was fought on L'Isle de la Valliére. On the 4th of June the Anglo-American forces left their camps on the glacis of Fort Lawrence, routed the Acadians at the fords of the Missiguash, and advanced by parallels and siege-lines against the hostile works. When the British batteries reached Butte-a-Charles the fort was vigorously shelled, and with such disastrous effect that it capitulated on June 16th, the garrison marching out with arms, baggage, and banners. The French troops were paroled and sent to Louisbourg, and the Acadians were suffered to remain. Laloutre, escaping to Quebec, there received an ecclesiastical censure, and was afterwards remanded to France.

In November, 1776, Col. Eddy led a force of Massachusetts troops, men of Maugerville, Acadians, and Indians, against Fort Cumberland. He first cut out a store-vessel from under the guns of the fort, and captured several detachments of the garrison (the Royal Fencibles). The commandant refused to surrender, and repulsed the Americans in a night-attack, by means of a furious cannonade. Eddy then blockaded the fort for several days, but was finally driven off by the arrival of a man-of-war from Halifax, bringing a reinforcement of 400 men. The Massachusetts camp was broken up by a sortie, and all its stores were destroyed. The Americans fled to the forest, and fell back on the St. John River. A large proportion of the men of Cumberland County went to Maine after this campaign, despairing of the success of Republicanism in the Maritime Provinces. Among them were a considerable number of Acadians.

The ruins of Fort Cumberland are a few miles N. W. of Amherst, beyond the Aulac River, and on a high bluff at the S. end of the Point de Bute range of hills. It was kept in repair by the Imperial Government for many years after its capture, and still presents an appearance of strength and solidity, though it has been long deserted. The remains of the besiegers' parallels are also visible near the works. On a bold bluff within cannon-shot, on the farther bank of the Missiguash River, are the scanty remains of the British Fort Lawrence. Numerous relics of the old Acadians may still be traced in this vicinity. 5 M. above the fort, on the Baie Verte road, is *Bloody Bridge*, where a British foraging party under Col. Dixon was surprised and massacred by the Indians (under French officers).

The *view from the bastions of Fort Cumberland is famous for its extent and beauty. It includes Sackville and its colleges on the N. W., Amherst and the Nova-Scotian shores on the S. E., and the bluff and hamlet of Fort Lawrence. The wide and blooming expanse of the Tantramar and Missiguash Marshes is overlooked, — the view including over 50,000 acres of rich marine intervale, — and on the S. the eye travels for many leagues down the blue sheet of the Bay of Fundy (Cumberland Basin).

The great **Tantramar Marsh** is S. of Sackville, and is 9 M. long by 4 M. wide, being also traversed by the Tantramar and Aulac Rivers. It is composed of fine silicious matter deposited as marine alluvium, and is called "red marsh," in distinction from the "blue marsh" of the uplands. The low shores around the head of the Bay of Fundy for a distance of 20 M. have been reclaimed by the erection of dikes, with *aboideaux* at the mouths of the rivers to exclude the flow of the tides. The land thus gained is very rich, and produces fine crops of English hay, averaging from 1½ to 2 tons to the acre. The land seems inexhaustible, having been cultivated now for nearly a century without rotation or fertilization.

The Chignecto Peninsula.

Minudie is 8 M. S. W. of Amherst, with which it is connected by a ferry across the estuaries of the Maccan and Hebert Rivers. It has 600 inhabitants, and is near the rich meadows called the Elysian Fields. In the vicinity are profitable quarries of grindstones, and there are shad-fisheries to the S. W. 6 – 8 M. S. are the Joggins Mines, pertaining to the General Mining Association of London; and the Victoria Mines, on the river Hebert. Coal has been obtained thence for 25 years. This district is reached by stages from Maccan station. About the year 1730 the coal-mines at Chignecto were leased to a Boston company, which was to pay a quit-rent of one penny an acre (on 4,000 acres), and a royalty of 18 pence per chaldron on the coal raised. But this enterprise was broken up in 1732, when the warehouses and machinery were destroyed by the Indians (probably incited by the French at Louisbourg).

The **Joggins Shore** extends to the S.W. along the Chignecto Channel, and is remarkable for its geological peculiarities, which have been visited and studied by European savans. The local explanation of the name is that the cliffs here "jog in" and out in an unexampled manner. The height of the cliffs is from 130 to 400 ft.; and the width of the Chignecto Basin is from 5 to 8 M. 85–40 M. from Amherst is *Apple River*, a sequestered hamlet on the estuary of the Apple River, amidst fine marine scenery. Apple Head is just W. of this place, and is 413 ft. high, overlooking the Chignecto Channel and the New-Brunswick shores. There is a fixed white light on its outer point. To the E., Apple River traverses the Caribou Plains, and on its upper waters affords the best of trout-fishing, with an abundance of salmon between February and July. 15–20 M. S. W. of Apple River, by a road which crosses the Cobequid Mts. E. of Cape Chignecto, is *Advocate Harbor* (see Route 21).

"The road from Amherst to Parrsboro' is tedious and uninteresting. In places it is made so straight that you can see several miles of it before you, which produces an appearance of interminable length, while the stunted growth of the spruce and birch trees bespeaks a cold, thin soil, and invests the scene with a melancholy and sterile aspect." (JUDGE HALIBURTON.) This road is 86 M. long, ascending the valley of the Maccan River, and passing the hamlet of Canuan, near the Cobequid Mts.

The Halifax train runs S. from Amherst to *Maccan* (stages to Minudie and Joggins), in the great coal-field of Cumberland County. Daily stages run from *Athol* station to Parrsboro'. From Athol the line passes to *Spring Hill*, a coal-mining district, whence a railway is being constructed to Parrsboro' (see Route 21). 11 M. beyond is the station at *River Philip* (small hotel), a pleasant stream in which good fishing is found. The salmon are especially abundant during the springtime. *Oxford* station (two inns) has two small woollen factories, and is 14 M. S. W. of Pugwash, on the Northumberland Strait.

The train now passes through extensive forests, in which many sugar-maples are seen, and begins the ascent of the **Cobequid Mts.**, with the Wallace Valley below on the l. The Cobequid range runs almost due E. and W. from Truro, and is 100 M. long, with an average breadth of 10–12 M. It consists of a succession of rounded hills, 800–1,000 feet high, covered with tall and luxuriant forests of beech and sugar-maple. From Thomson, Greenville, and Wentworth stations stages run to Wallace and Pugwash (see page 81), also to Westchester. The railway traverses the hill-country by the *Folly Pass*, and has its heaviest grades between Folly Lake and Londonderry; where are also 2–3 M. of snow-sheds, to protect the deep cuttings from the drifting in of snow from the hills. Fine views of the Wallace Valley are afforded from the open levels of the line. From *Londonderry* a branch-railway runs to the Londonderry Iron-mines, which have been worked for nearly 40 years. The ores are magnetic, specular, and hematite, and occur in a wedge-shaped vein 7 M. long and 120 ft. thick. The iron is of fine quality, but is difficult to work.

The train descends from the Pass along the line of the Folly River, which it crosses on a bridge 200 feet above the water. Beyond the farming settlement of *Debert* (stages to Economy and Five Islands) the descent is continued, and occasional views of the Cobequid Bay are given as the train passes across Onslow to Truro. The landscape now becomes more pleasing and thickly settled.

Truro (*Somerset House; Prince of Wales Hotel; Victoria*) is a wealthy and prosperous town of over 4,000 inhabitants, and occupies a pleasant situation 2 M. from the head of Cobequid Bay (an arm of the Basin of Minas). The level site of the town is nearly surrounded by an amphitheatre of gracefully rounded hills, and on the W. are the old diked meadows of the Acadian era. Truro is the capital of Colchester County and the seat of the Provincial Normal School. Fishing and shipbuilding are carried on here, and there are large and growing manufactures, including boots and shoes, woollens, and iron-wares. The neighboring county has valuable farming-lands, and contains several iron-mines.

Truro was settled at an early date by the Acadian French, and after their expulsion from Nova Scotia was occupied by Scotch-Irish from New Hampshire. In 1761 a large number of disbanded Irish troops settled here, and engaged in the peaceful pursuits of agriculture.

A road runs W. from Truro between the Cobequid Mts. and the Basin of Minas, passing Masstown (10 M.); Folly Village (14 M.), at the mouth of the Folly River; Great Village (18 M.), a place of 600 inhabitants; Highland Village (21 M.); Port au Pique (23 M.); Bass River (27 M.); Upper Economy (28 M.); and Five Islands (45 M.). (See Route 22.) The stages run from Debert station.

Stages also run S. W. to Old Barns, on the S. shore of Cobequid Bay, and S. E. 15 M. to Middle Stewiacke, on the Stewiacke River.

Truro is the point of departure for the Pictou Branch of the Intercolonial Railway (see Route 31).

The North Shore of Nova Scotia.

Blair's express-stages leave Truro on the arrival of the morning trains from Halifax, on Tuesday, Thursday, and Saturday, returning on the alternate days. Truro to Tatamagouche, 29 M.; to Wallace, 42; to Pugwash, 52. Stages also run from Wentworth, Greenville, and Thomson to the N. Shore (according to the Intercolonial Railway circular for 1874), and a tri-weekly line runs between Pictou and Amherst by way of the N. shore.

In passing from Truro to Tatamagouche the road crosses the Cobequid Mts. and descends through a thinly settled region to the N. *Tatamagouche* (two inns) is situated at the head of a large harbor which opens on the Northumberland Strait, and has about 1,500 inhabitants. Some shipbuilding is done here, and there are freestone quarries in the vicinity. 6 M. to the E. is the large village of *Brule Harbor*, and 6 M. farther E., also on the Tatamagouche Bay, and at the mouth of the River John, is the shipbuilding settlement of *River John*, which was founded by Swiss Protestants in 1763. It is 20 M. from this point to Pictou, and the intervening coast is occupied by colonists from the Hebrides.

Blair's stage runs W. from Tatamagouche to *Wallace* (two inns), a town of 2,600 inhabitants, situated on the deep waters of Wallace Harbor (formerly called Remsheg). Plaster, lime, and freestone are found here in large quantities, and the latter is being quarried by several companies. The Provincial Building at Halifax was made of Wallace stone. To the N. E., beyond the lighthouse on Mullin Point, is the marine hamlet of *Fox Harbor*, whose original settlers came from the Hebrides. *Pugwash* (small inn) is 10 M. beyond Wallace, and is a flourishing port with about 3,300 inhabitants. The harbor, though difficult of access, is deep and well sheltered, and has several ship-yards on its shores. The chief exports of Pugwash are deals and lumber, freestone, lime, and plaster.

The Halifax train runs S. from Truro to *Brookfield*, whence hay and lumber are exported, and then to *Stewiacke*, which is 3 M. from the pretty farming village of the same name, on the Stewiacke River. The next station is *Shubenacadie* (International Hotel), a busy little manufacturing village on the river of the same name.

Daily stages descend the valley of the Shubenacadie for 18 M. to the N. to the town of *Maitland* (two inns), at the mouth of the river (see Route 22). Stages also run S. E. (Tuesday and Thursday) to Gay's River (7 M.), Gay's River Road (14 M), Middle Musquodoboit (21 M.), Upper Musquodoboit (25 M.), Melrose, Guysborough, and Port Mulgrave, on the Strait of Canso. Gold was discovered near Gay's River in 1862, in the conglomerate rock of the great ridge called the Boar's Back, which extends for 60 M. through the inland towns. It nearly resembles the alluvial deposits found in the placer-diggings of California, and the stream-washings have yielded as high as an ounce per man daily. Scientific mining was begun in 1863, but has given only light returns. *Middle Musquodoboit* is a farming-town with about 1,000 inhabitants, situated on the S. of the Boar's Back ridge, 42 M. from Halifax. Upper Musquodoboit is about the same size, and beyond that point the stages traverse a dreary and thinly settled district for several leagues, to Melrose.

The Halifax train runs S. W. to *Elmsdale*, a village near the Shubenacadie River, engaged in making leather and carriages. Enfield is the seat of a large pottery. 7 M. N. W. are the *Renfrew Gold-Mines*, where gold-bearing quartz was discovered in 1861. Much money and labor were at first wasted by inexperienced miners, but of late years the lodes have been worked systematically, and are considered among the most valuable in Nova Scotia. The average yield is 16 pennyweights of gold to a ton of quartz, and in 1869 these mines yielded 3,097 ounces of the precious metal, valued at $61,490. The *Oldham Mines* are 3½ M. S. of Enfield, and are in a deep narrow valley, along whose bottom shafts have been sunk to reach the auriferous quartz. Between 1861 and 1869, 9,254 ounces of gold were sent from the Oldham diggings, and it is thought that yet richer lodes may be found at a greater depth.

Soon after leaving Enfield the train passes along the S. E. shore of Grand Lake, which is 8 M. long by 1-2 M. wide. It crosses the outlet stream, runs around Long Lake, and intersects the Windsor Branch Railway at *Windsor Junction*. Station, *Rocky Lake*, on the lake of the same name, where large quantities of ice are cut by the Nova-Scotia Ice Company, for exportation to the United States. 3 M. N. E. of this station are the *Waverley Gold-Mines*, where the gold is found in barrel-quartz, so named because it appears in cylindrical masses like barrels laid side by side, or like a corduroy-road. At its first discovery all the floating population of Halifax flocked out here, but they failed to better their condition, and the total yield between 1861 and 1869 was only about 1,600 ounces. Waverley village is picturesquely situated in a narrow valley between two lakes, and has about 600 inhabitants.

After crossing Rocky Lake the train soon reaches the shores of the beautiful *Bedford Basin*, and follows their graceful curves for several miles. On the l. are fine views of the villas and hills beyond the blue water.

Halifax, see page 93.

18. St. John to Halifax, by the Annapolis Valley.

This is the pleasantest route, during calm weather, between the chief cities of the Maritime Provinces. After a passage of about 5 hours in the steamer, across the Bay of Fundy, the pretty scenery of the Annapolis Basin is traversed, and at Annapolis the passenger takes the train of the Windsor & Annapolis Railway, which runs through to Halifax. The line traverses a comparatively rich and picturesque country, abounding in historic and poetic associations of the deepest interest.

The distance between St. John and Halifax by this route is 86 M. less than by the Intercolonial Railway; but the time on both routes is about the same, on account of the delay in crossing the Bay of Fundy. The Annapolis-Halifax line is only practicable 4 times a week. The steamer leaves St. John at 8 A. M., on Monday, Wednesday, Friday, and Saturday, connecting with the express trains which leave Annapolis at 2 P. M. and arrive at Halifax at about 8 P. M. Express trains leave Halifax at 8.15 A. M. on Tuesday, Thursday, Friday, and Saturday, connecting with the steamer which leaves Annapolis at 2.35 P. M. and arrives at St. John at 8 P. M. (Time-table of 1874.)

Fares.— St. John to Halifax, 1st class, $5; 2d class, $3.50; to Digby, $1.50; to Annapolis $2. Passengers for Halifax dine on the steamer and take tea at Kentville (10 minutes); those for St. John dine at Kentville (18 minutes) and take tea on the boat. There are two through trains each way daily between Halifax and Annapolis. Special rates are made for excursions (limited time) by the agents of this route, Small & Hathaway, 39 Dock St., St. John.

Distances. — St. John to Digby, 43 M.; Annapolis, 61; Round Hill, 68; Bridgetown, 75; Paradise, 80; Lawrencetown, 83; Middleton, 89; Wilmot, 92; Kingston, 96; Morden Road, 101; Aylesford, 103; Berwick, 108; Waterville, 111; Cambridge, 113; Coldbrook, 115; Kentville, 120; Port Williams, 125; Wolfville, 127; Grand Pré, 130; Horton Landing, 131; Avonport, 133; Hantsport, 138; Mount Denson, 140; Falmouth, 143; Windsor, 145; Three-Mile Plains, 148; Newport, 151; Ellershouse, 154; Stillwater, 157; Mount Uniacke, 164; Beaver Bank, 174; Windsor Junction, 177; Rocky Lake, 179; Bedford, 182; Four-Mile House, 186; Halifax, 190.

The steamer *Empress* leaves her wharf at Reed's Point, St. John, and soon passes the heights and spires of Carleton on the r. and the lighthouse on Partridge Island on the l., beyond which Mispeck Point is seen. *Cape Spencer* is then opened to the E., on the New Brunswick coast, and the steamer sweeps out into the open bay. Travellers who are subject to seasickness would do well to avoid this passage during or immediately after a breeze from the N. E. or S. W., or during a gale from any direction. At such times very rough water is found on the bay. It will be remembered that the ocean-steamships *Pactolus* and *Connaught* were lost in these waters. But in ordinary summer weather the bay is quiet, except for a light tidal swell, and will not affect the traveller whose mind is properly fixed on something outside of himself.

Soon after passing Partridge Island, the dark ridge of the North Mt. is seen in advance, cleft by the gap called the *** Digby Gut,** which, in the earlier days, was known as St. George's Channel. The course is laid straight for this pass, and the steamer runs in by *Prim Point*, with its fog-whistle and fixed light (visible 13 M.), and enters the tide-swept defile, with bold and mountainous bluffs rising on either side. The shores on the l. are 610 ft. high, and on the r. 400-560 ft., between which the tide rushes with a velocity of 5 knots an hour, making broad and powerful swirls and eddies over 12-25 fathoms of water. After running for about 2 M. through this passage, the steamer enters the Annapolis Basin, and runs S. by E. 3 M. to Digby.

"The white houses of Digby, scattered over the downs like a flock of washed sheep, had a somewhat chilly aspect, it is true, and made us long for the sun on them. But as I think of it now, I prefer to have the town and the pretty hillsides that stand about the basin in the light we saw them; and especially do I like to recall the high wooden pier at Digby, deserted by the tide and so blown by the wind that the passengers who came out on it, with their tossing drapery, brought to mind the windy Dutch harbors that Backhüysen painted." (WARNER's *Baddeck.*)

Digby (*Daley's Hotel*) is a maritime village of about 1,000 inhabitants, situated on the S. W. shore of the Annapolis Basin, and engaged in ship-building and the fisheries of haddock, mackerel, and herring. The Digby herring are famous for their delicacy, and are known in the Provinces as "Digby chickens." Porpoises, also, are caught in the swift currents of the Digby Gut. The village is visited by summer voyagers on account of its picturesque environs and the opportunities for fishing and sporting in the vicinity; and attempts have been made to erect a large hotel. There was a French fort here in the early days; and in 1783 the township was granted to the ex-American Loyalists, who founded the village of Conway on these shores. Stages run between Digby and Annapolis, and also from Digby to Yarmouth (see Route 25).

It is called 18 M. from Digby to Annapolis (though this distance seems over-estimated when compared with the charts and the course run by the steamer). The * **Annapolis Basin** gradually decreases from a width of nearly 5 M. to 1 M., and is hemmed in between the converging ridges of the *North Mt.* and the *South Mt.* The former range has a height of 6-700 ft., and is bold and mountainous in its outlines. The South Mt. is from 300 to 500 ft. high, and its lines of ascent are more gradual. The North Mt. is composed of trap, resting upon red sandstone; and the South Mt. is of granite and metamorphic slates. The geologic theory is that the North Mt. was once completely insulated, and that the tides flowed through the whole valley, until a shoal at the confluence of the Blomidon and Digby currents became a bar, and this in time became dry land and a water-shed.

Between the head of Argyle Bay and the slopes of the Annapolis Basin are the rarely visited and sequestered hill-ranges called the *Blue Mountains.* "The Indians are said to have formerly resorted periodically to groves among these wilds, which they considered as consecrated places, in order to offer sacrifices to their gods."

"We were sailing along the gracefully moulded and tree-covered hills of the Annapolis Basin, and up the mildly picturesque river of that name, and we were about to enter what the provincials all enthusiastically call the Garden of Nova Scotia. .... It is, — this valley of Annapolis, — in the belief of provincials, the most beautiful and blooming place in the world, with a soil and climate kind to the husbandman, a land of fair meadows, orchards, and vines..... It was not until we had travelled over the rest of the country that we saw the appropriateness of the designation. The explanation is, that not so much is required of a garden here as in some other parts of the world."

Soon after leaving Digby, Bear Island is seen in-shore on the r., in front of the little port of *Bear River* (inn), which has a foundry, tanneries, and saw-mills. Iron and gold are found in the vicinity, and lumber and cordwood are exported hence to the United States and the West Indies. A few miles beyond, and also on the S. shore, is the hamlet of *Clementsport* (two inns), where large iron-works were formerly established, in connection with the ore-beds to the S. Roads lead thence to the S. W. in 10-12 M. to the romantic districts of the Blue Mts. and the upper Liverpool Lakes (see Route 27), at whose entrance is the rural village of *Clementsvale.*

8-10 M. beyond Digby the steamer passes **Goat Island**, of which Lescarbot writes, in *Les Muses de la Nouvelle France* (1609) :

"Adieu mon doux plaisir fonteines et ruisseaux,
Qui les vaux et les monts arrousez de vos eaux.
Pourray-je t'oublier, belle ile forétière
Riche honneur de ce lieu et de cette rivière ?"

In 1707 the British frigate *Annibal* and two brigantines were sailing up the Basin to attack Annapolis, when they met such a sharp volley from the Ile aux Chèvres that they were forced to retire in confusion. The French name of the island was Anglicized by translation. On the point near this island was the first settlement of the French in Nova Scotia. A fort was erected here by the Scottish pioneers, and was restored to France by the Treaty of St. Germain, after which it was garrisoned by French troops. In 1827 a stone block was found on the point, inscribed with a square and compass and the date "1606." In May, 1782, there was a naval combat off Goat Island, in which an American war-brig of 8 guns was captured by H. M. S. *Buckram.*

Above the island the Basin is about 1 M. wide, and is bordered by farm-streets. To the N. E., across a low alluvial point, are seen the spires and ramparts of Annapolis Royal, where the steamer soon reaches her wharf, after passing under the massive walls of the old fortress. There are several inns here, of which the American House is perhaps the best. *The Grange* is about 1 M. from the pier, and is an old country mansion, in broad and shady grounds, now used as a summer hotel. There is also a restaurant near the railway-station. *Stages* run from Annapolis to Digby (Annapolis to Clementsport, 8½ M.; Victoria Bridge, 13¼; Smith's Cove, 16 ; Digby, 20½ ;— Yarmouth, 87½). Stages also run S. E. 78 M. (semi-weekly) to Liverpool (see Route 27).

Annapolis Royal, the capital of Annapolis County, is a maritime and agricultural village, situated at the head of the Annapolis Basin, and contains 5-600 inhabitants. It is frequented by summer visitors on account of its pleasant environs and tempered sea-air, and the opportunities for salt-water fishing in the Basin, and trouting among the hills to the S. The chief object of interest to the passing traveller is the *old fortress which fronts the Basin and covers 28 acres with its ramparts and outworks. It is entered by the way of the fields opposite Perkins's Hotel. The works are disarmed, and have remained unoccupied for many years. One of the last occupations was that of the Rifle Brigade, in 1850; but the post was abandoned soon after, on account of the numerous and successful desertions which thinned the ranks of the garrison. But when Canada passed into a state of semi-independence in 1867, this fortress was one of the few domains reserved to the British Crown. The inner fort is entered by an ancient archway which fronts towards the Basin, giving passage to the parade-ground, on which are the quaint old English barracks, with steep roofs and great chimneys. In the S. E. bastion is the magazine, with a vaulted roof of masonry, near which are the foundations of the French barracks. From the parapet on this side are overlooked the landward outworks and the lines of the old Hessian and Waldecker settlements towards Clementsport. On the hillside beyond the marsh is seen an ancient house of the era of the French occupation, the only one now standing in the valley. In the bastion towards the river is a vaulted room, whence a passage leads down to the French garrison-wharf; but the arched way has fallen in, and the wharf is now but a shapeless pile of stones. The * view from this angle of the works is very beautiful, including the villages of Annapolis Royal and Granville, the sombre heights of the North and South Mts., and the Basin for many miles, with Goat Island in the distance.

The road which leads by the fortress passes the old garrison cemetery, St. Luke's Church, the court-house and county academy, and many quaint and antiquated mansions. A ferry crosses the Annapolis River to *Granville* (two inns), a busy little shipbuilding village, with 3-400 inhabitants and three churches. A road leads hence across the North Mt. in 4-5 M., to the hamlets of *Hillsburn* and *Leitchfield*, on the Bay of Fundy.

"Without the historic light of French adventure upon this town and basin of Annapolis.... I confess that I should have no longing to stay here for a week; notwithstanding the guide-book distinctly says that this harbor has 'a striking resemblance to the beautiful Bay of Naples.' I am not offended at this remark, for it is the one always made about a harbor, and I am sure the passing traveller can stand it, if the Bay of Naples can." (WARNER'S *Baddeck*.)

The Basin of Annapolis was first entered in 1604 by De Monts's fleet, exploring the shores of Acadie; and the beauty of the scene so impressed the Baron de Poutrincourt that he secured a grant here, and named it Port Royal. After the failure of the colony at St. Croix Island, the people moved to this point, bringing all their stores and supplies, and settled on the N. side of the river. In July, 1606, Lescarbot and another company of Frenchmen joined the new settlement, and conducted improvements of the land, while Poutrincourt and Champlain explored the Massachusetts coast. 400 Indians had been gathered by the sagamore Membertou in a stockaded village near the fort, and all went on well and favorably until De Monts's grant was annulled by the King of France, and then the colony was abandoned. Lescarbot says of this expedition, and of Port Royal itself: "I must needs be so bold as to tell in this occurrence, that if ever that country be inhabited with Christians and civil people, the first praise thereof must of right be due to the authors of this voyage.... Finally, being in the port, it was unto us a thing marvellous to see the fair distance and the largeness of it, and the mountains and hills that environed it, and I wondered how so fair a place did remain desert, being all filled with woods, seeing that so many pine away in the world which might make good of this land, if only they had a chief governor to conduct them thither."

ANNAPOLIS ROYAL.   *Route 18.*   87

Four years later the brave Baron de Poutrincourt left his estates in Champagne, with a deep cargo of supplies, descended the rivers Aube and Seine, and sailed out from Dieppe (Feb. 26, 1610) On arriving at Port Royal, everything was found as when left; and the work of proselyting the Indians was at once entered on. Membertou and his tribe were converted, baptized, and feasted, amid salutes from the cannon and the chanting of the *Te Deum;* and numerous other forest-clans soon followed the same course.

Poutrincourt was a Gallican Catholic, and hated the Jesuits, but was forced to take out two of them to his new domain. They assumed a high authority there, but were sternly rebuked by the Baron, who said, "It is my part to rule you on earth, and yours only to guide me to heaven." They threatened to lay Port Royal under interdict; and Poutrincourt's son and successor so greatly resented this that they left the colony on a mission ship sent out by the Marchioness de Guercheville, and founded St. Sauveur, on the island of Mount Desert. In 1613, after the Virginians under Capt. Argall had destroyed St. Sauveur, the vengeful Jesuits piloted their fleet to Port Royal, which was completely demolished. Poutrincourt came out in 1614 only to find his colony in ruins, and the remnant of the people wandering in the forest; and was so disheartened that he returned to France, where he was killed, the next year, in the battle of Méry-sur-Seine.

It is a memorable fact that these attacks of the Virginians on Mount Desert and Port Royal were the very commencement of the wars between Great Britain and France in North America, "which scarcely ever entirely ceased until, at the cost of infinite blood and treasure, France was stripped of all her possessions in America by the peace of 1763."

Between 1620 and 1630 an ephemeral Scottish colony was located at Port Royal, and was succeeded by the French. In 1628 the place was captured by Sir David Kirk, with an English fleet, and was left in ruins. In 1634 it was granted to Claude de Razilly, "Seigneur de Razilly, des Eaux Mesles et Cuon, en Anjou," who afterwards became commandant of Oleron and vice-admiral of France. He was a bold naval officer, related to Cardinal Richelieu; and his brother Isaac commanded at Lahave (see Route 25). His lieutenants were D'Aulnay Charnisay and Charles de la Tour, and he transferred all his Acadian estates to the former, in 1642, after which began the feudal wars between those two nobles (see page 19). Several fleets sailed from Port Royal to attack La Tour, at St. John; and a Boston fleet, in alliance with La Tour, assailed Port Royal.

In 1654 the town was under the rule of Emmanuel le Borgne, a merchant of La Rochelle, who had succeeded to D'Aulnay's estates, by the aid of César, Duke of Vendôme, on account of debts due to him from the Acadian lord. Later in the same year the fortress was taken by a fleet sent out by Oliver Cromwell, but the inhabitants of the valley were not disturbed.

By the census of 1671 there were 361 souls at Port Royal, with over 1,000 head of live-stock and 364 acres of cultivated land. In 1684 the fishing-fleet of the port was captured by English "corsairs"; and in 1686 there were 622 souls in the town. In 1690 the fort contained 18 cannon and 86 soldiers, and was taken and pillaged by Sir William Phipps, who sailed from Boston with 3 war-vessels and 700 men. A few months later it was plundered by corsairs from the West Indies, and in 1691 the Chevalier de Villebon took the fort in the name of France. Baron La Hontan wrote: "Port Royal, the capital, or the only city of Acadia, is in effect no more than a little paltry town that is somewhat enlarged since the war broke out in 1689 by the accession of the inhabitants that lived near Boston, the metropolitan of New England. It subsists upon the traffic of the skins which the savages bring thither to truck for European goods." In the summer of 1707 the fortress was attacked by 2 regiments and a small fleet, from Boston, and siege operations were commenced. An attempt at storming the works by night was frustrated by M. de Subercase's vigilance and the brisk fire of the French artillery, and the besiegers were finally forced to retire with severe loss. A few weeks later a second expedition from Massachusetts attacked the works, but after a siege of 15 days their camps were stormed by the Baron de St. Castin and the Chevalier de la Boularderie, and the feebly led Americans were driven on board their ships. Subercase then enlarged the fortress, made arrangements to run off slaves from Boston, and planned to capture Rhode Island, "which is inhabited by rich Quakers, and is the resort of rascals and even pirates."

In the autumn of 1710 the frigates *Dragon, Chester, Falmouth, Leostaffe, Feversham, Star,* and *Province,* with 20 transports, left Boston and sailed to Port Royal.

88   Route 18.   THE ANNAPOLIS VALLEY.

There were 2 regiments from Massachusetts, 2 from the rest of New England, and 1 of Royal Marines. After the erection of mortar-batteries, several days were spent in bombarding the fort from the fleet and the siege-lines, but the fire from the ramparts was kept up steadily until the garrison were on the verge of starvation; Subercase then surrendered his forces (258 men), who were shipped off to France, and Gen. Nicholson changed the name of Port Royal to ANNAPOLIS ROYAL, in honor of Queen Anne, then sovereign of Great Britain.
In 1711, 80 New-Englanders from the garrison were cut to pieces at Bloody Brook, 12 M. up the river, and the fortress was then invested by the Acadians and Micmacs. For nearly 40 years afterwards Annapolis was almost always in a state of siege, being menaced from time to time by the disaffected Acadians and their savage allies. In 1744 the non-combatants were sent to Boston for safety, and in July of that year the fort was beleaguered by a force of fanatic Catholics under the Abbé Laloutre. Five companies of Massachusetts troops soon joined the garrison, and the besiegers were reinforced by French regulars from Louisbourg. The siege was continued for nearly three months, but Gov. Mascarene showed a bold front, and provisions and men came in from Boston. The town was destroyed by the artillery of the fort and by incendiary sorties, since it served to shelter the hostile riflemen. Soon after Duvivier and Laloutre had retired, two French frigates entered the Basin and captured some ships of Massachusetts, but left four days before Tyng's Boston squadron arrived. A year later, De Ramezay menaced the fort with 700 men, but was easily beaten off by the garrison, aided by the frigates *Chester*, 50, and *Shirley*, 20, which were lying in the Basin. After the deportation of the Acadians, Annapolis remained in peace until 1781, when two American war-vessels ascended the Basin by night, surprised and captured the fortress and spiked its guns, and plundered every house in the town, after locking the citizens up in the old block-house.

The Annapolis Valley.

This pretty district has suffered, like the St. John River, from the absurdly extravagant descriptions of its local admirers, and its depreciation by Mr. Warner (see page 84) expresses the natural reaction which must be felt by travellers (unless they are from Newfoundland or Labrador) after comparing the actual valley with these high-flown panegyrics. A recent Provincial writer says: "The route of the Windsor & Annapolis Railway lies through a magnificent farming-country whose beauty is so great that we exhaust the English language of its adjectives, and are compelled to revert to the quaint old French which was spoken by the early settlers of this Garden of Canada, in our efforts to describe it." In point of fact the Annapolis region is far inferior either in beauty or fertility to the valleys of the Nashua, the Schuylkill, the Shenandoah, and scores of other familiar streams which have been described without effusion and without impressing the service of alien languages. The Editor walked through a considerable portion of this valley, in the process of a closer analysis of its features, and found a tranquil and commonplace farming-district, devoid of salient points of interest, and occupied by an insufficient population, among whose hamlets he found unvarying and honest hospitality and kindness. It is a peaceful rural land, hemmed in between high and monotonous ridges, blooming during its brief summer, and will afford a series of pretty views and pleasing suggestions to the traveller whose expectations have not been raised beyond bounds by the exaggerated praises of well-meaning, but injudicious authors.

It is claimed that the apples of the Annapolis Valley are the best in America, and 50,000 barrels are exported yearly, — many of which are sold in the cities of Great Britain. The chief productions of the district are hay, cheese, and live-stock, a large proportion of which is exported.

The Halifax train runs out from Annapolis over the lowlands, and takes a course to the N. E., near the old highway. **Bridgetown** (*Granville House*) is the first important station, and is 14 M. from Annapolis, at the head of navigation on the river. It has about 1,000 inhabitants, 4 churches, and a weekly newspaper, and is situated in a district of apple orchards and rich pastures. Some manufacturing is done on the water-power of the Annapolis River; and the surrounding country is well populated, and is reputed to be one of the healthiest districts in Nova Scotia. To the S. is *Bloody Brook*, where a detachment of New-England troops was massacred by the French and Indians; and roads lead up over the South Mt. into the howling wilderness of the interior. 5 M. from Bridgetown, over the North Mt., is the obscure marine hamlet of *Hampton.*

Paradise (small inn) is a pleasantly situated village of about 400 inhabitants, with several saw and grist mills and tanneries. The principal exports are lumber and cheese, though there are also large deposits of merchantable granite in the vicinity. A road crosses the North Mt. to *Port Williams*, 7 M. distant, a fishing-village of about 300 inhabitants, situated on the Bay of Fundy. The coast is illuminated here, at night, by two white lights. Farther down the shore is the hamlet of *St. Croix Cove.*

Lawrencetown is a prosperous village of about 600 inhabitants, whence much lumber is exported. In 1754, 20,000 acres in this vicinity were granted to 20 gentlemen, who named their new domain in honor of Gov. Lawrence. 8 M. distant, on the summit of the North Mt., is the hamlet of *Havelock*, beyond which is the farming settlement of *Mt. Hawley*, near the Bay of Fundy. *New Albany* (small inn) is a forest-village 8-10 M. S. E. of Lawrencetown; and about 10 M. farther into the great central wilderness is the farming district of *Springfield*, beyond the South Mt.

Middleton (Middleton Hotel) is a small village near the old iron-mines on the South Mt. A few miles S. of Middleton are the *Nictau Falls*, a pretty cascade on a small mountain-stream. 1½ M. from Middleton is the hamlet of *Lower Middleton*, surrounded by orchards, with an Anglican church, and a seminary for young ladies. *Wilmot* station is ½ M. from *Farmington* (two inns), a pleasant little Presbyterian village. *Margaretsville* (Harris's Hotel) is 7 M. distant, across the North Mt., and is a shipbuilding and fishing settlement of 300 inhabitants, situated on the Bay of Fundy. Fruit and lumber are exported hence to the United States. Near this point is a fixed red light of high power.

The **Wilmot Springs** are about 3 M. from Farmington, and have, for many years, enjoyed a local celebrity for their efficacy in healing cutaneous diseases and wounds. They were formerly much resorted to, but are now nearly abandoned, though bathing-houses and other accommodations are kept here. The springs are situated in a grove of tall trees near the road, filling two large basins; and the water is cold, clear, and nearly tasteless. The principal ingredients are, in each gallon: 78 grains of sulphuric acid, 54½ grains of lime, 6 grains of soda and potash, and 3 grains of magnesia. A few visitors pass the summer at Wilmot every year, on account of the benefits resulting from the use of these waters.

Kingston station is 1½ M. from *Kingston*, 2 M. from *Melvern Square*, 2½ M. from *Tremont*, and 4 M. from *Prince William Street*, rural hamlets in the valley. From Morden Road station a highway runs N. W. 7 M. across the North Mt. to the little port of *Morden*, or *French Cross* (Balcomb's Hotel), on the Bay of Fundy. Station Aylesford (Patterson's Hotel), a small hamlet from which a road runs S. E. to *Factory Dale* (4 M.), a manufacturing hamlet whence the valley is overlooked; and the farming towns of *Jacksonville* and *Morristown* are 5-7 M. away, on the top of the South Mt.

Lake George (*Hall's inn*) is 12 M. distant, whence the great forest-bound chain of the **Aylesford Lakes** may be visited. The chief of these is *Kempt Lake*, which is about 7 M. long. A road runs S. from the Lake George settlement by Lake Paul and Owl Lake to *Falkland* (82 M. from Aylesford), which is on the great Lake Sherbrooke, in Lunenburg County, near the head-waters of the Gold River.
"The great Aylesford sand-plain folks call it, in a ginral way, the Devil's Goose Pasture. It is 13 M. long and 7 M. wide; it ain't just drifting sands, but it's all but that, it's so barren. It's uneven, or wavy, like the swell of the sea in a calm, and is covered with short, thin, dry, coarse grass, and dotted here and there with a half-starved birch and a stunted misshapen spruce. It is jest about as silent and lonesome and desolate a place as you would wish to see. . . . . All that country thereabouts, as I have heard tell when I was a boy, was once owned by the Lord, the king, and the devil. The glebe-lands belonged to the first, the ungranted wilderness-lands to the second, and the sand-plain fell to the share of the last (and people do say the old gentleman was rather done in the division, but that is neither here nor there), and so it is called to this day the *Devil's Goose Pasture*."

Station, *Berwick* (two inns), a prosperous village of 400 inhabitants, where the manufacture of shoes is carried on. A road leads to the N. W. 7 M. across Pleasant Valley and the Black Rock Mt. to *Harborville*, a shipbuilding village on the Bay of Fundy, whence large quantities of cordwood and potatoes are shipped to the United States. Several miles farther up the bay-shore is the village of *Canada Creek*, near which is a lighthouse.

At Berwick the line enters the * **Cornwallis Valley**, which is shorter but much more picturesque than that of Annapolis. Following the course of the Cornwallis River, the line approaches the base of the South Mt., while the North Mt. trends away to the N. E. at an ever-increasing angle. Beyond the rural stations of Waterville, Cambridge, and Coldbrook, the train reaches **Kentville** (*Webster House*; restaurant in the station), the headquarters of the railway and the capital of Kings County. This town has 1,000 inhabitants, 4 churches, and a weekly newspaper; and there are several mills and quarries in the vicinity. Raw umber and manganese have been found here. The roads to the N. across the mountain lead to the maritime hamlets of Hall's Harbor (10 M.), Chipman's Brook (14 M.), and Baxter's Harbor (12 M.); also to Sheffield Mills (7 M.), Canning (8 M.), Steam Mills (2 M.), and Billtown (6 M.).

Kentville to Chester.

The Royal mail-stages leave Kentville at 6 A. M. on Monday and Thursday, reaching Chester in the afternoon. The return trip is made on Tuesday and Friday. The distance between Kentville and Chester is 46 M., and the intervening country is wild and picturesque. After passing the South Mt. by the Mill-Brook Valley, at 8-10 M. from Kentville, the road runs near the *Gaspereaux Lake*, a beautiful forest-loch about 5 M. long, with many islands and highly diversified shores. This water is connected by short straits with the island-studded Two-Mile Lake and the Four-Mile Lake, near which are the romantic Aylesford Lakes. E. and S. E. of the Gaspereaux Lake are the trackless solitudes of the far-spreading *Blue Mts.*, amid whose recesses are the lakelets where the Gold River takes its rise. At 20 M. from Kentville the stage enters the Episcopal village of *New Ross* (Turner's Hotel), at the crossing of the Dalhousie Road from Halifax to Annapolis. From this point the stage descends the valley of the Gold River to Chester (see Route 24).

The Halifax train runs E. from Kentville down the Cornwallis Valley to *Port Williams*, which is 1½ M. from the village of that name, whence daily stages run to Canning. The next station is Wolfville, from which the Land of Evangeline may most easily be visited (see Route 22). The buildings of Acadia College are seen on the hill to the r. of the track.

The Halifax train runs out from Wolfville with the wide expanse of the reclaimed meadows on the l., beyond which is *Cape Blomidon*, looming leagues away. In a few minutes the train reaches **Grand Pré**, and as it slows up before stopping, the tree is seen (on the l. about 300 ft. from the track) which marks the site of the ancient Acadian chapel. Beyond *Horton Landing* the Gaspereaux River is crossed, and the line begins to swing around toward the S. E. At *Avonport* the line reaches the broad Avon River, and runs along its l. bank to *Hantsport* (two inns). This is a large manufacturing and shipbuilding village, where numerous vessels are owned. In the vicinity are productive quarries of freestone. *Mount Denson* station is near the hill whose off-look Judge Haliburton so highly extols:—

"I have seen at different periods of my life a good deal of Europe and much of America; but I have seldom seen anything to be compared with the view of the Basin of Minas and its adjacent landscape, as it presents itself to you on your ascent of Mount Denson. .... He who travels on this continent, and does not spend a few days on the shores of this beautiful and extraordinary basin, may be said to have missed one of the greatest attractions on this side of the water."

The next station is *Falmouth*, in a region which abounds in gypsum. Back toward Central Falmouth there are prolific orchards of apples. The line now crosses the Avon River on the most costly bridge in the Maritime Provinces, over the singular tides of this system of waters.

The traveller who passes from Annapolis to Windsor at the hours of low-tide will sympathize with the author of "Baddeck," who says that the Avon "would have been a charming river if there had been a drop of water in it. I never knew before how much water adds to a river. Its slimy bottom was quite a ghastly spectacle, an ugly rent in the land that nothing could heal but the friendly returning tide. I should think it would be confusing to dwell by a river that runs first one way and then the other and then vanishes altogether."

The remarkable tides of this river are also described by Mr. Noble, as follows: The tide was out, "leaving miles of black" (red) "river-bottom entirely bare, with only a small stream coursing through in a serpentine manner. A line of blue water was visible on the northern horizon. After an absence of an hour or so, I loitered back, when, to my surprise, there was a river like the Hudson at Catskill, running up with a powerful current. The high wharf, upon which but a short time before I had stood and surveyed the black, unsightly fields of mud, was now up to its middle in the turbid and whirling stream."

Windsor (*Clifton House*, large and comfortable; *Avon House*) is a cultured and prosperous village of 2,715 inhabitants, occupying the promontory at the intersection of the Avon and St. Croix Rivers. The adjacent districts of Falmouth and St. Croix have about 3,300 inhabitants. There are in Windsor 7 churches, a bank, and several manufactories; there are also several busy shipyards. The chief exportation of Windsor is plaster of Paris and gypsum, large quantities of which are used in the United

States for fertilizing the soil. Near the end of the railway bridge, on a projecting hill, is the Clifton mansion, formerly the home of the genial and witty Thomas C. Haliburton (born at Windsor in 1797, 13 years a Judge in Nova Scotia, 6 years an M. P. at London, and died in 1865), the author of the "Sam Slick" books.

On the knoll over the village are the crumbling block-houses and earthworks of *Fort Edward*, whence is obtained a pretty view down the widening Avon and out over the distant Basin of Minas. About 1 M. from the station, on a hill which overlooks the fine valley of the Avon and its uncleared mountain-rim, are the plain buildings of **King's College**, the oldest college now existing in Canada.

It was founded in 1787, and chartered by King George III. in 1802. It is under the control and patronage of the Anglican Church, and is well endowed with scholarships, honors, etc., but has only 5 professors and a limited number of students. The Nova-Scotians have not hitherto sought to qualify themselves by culture and study for public honors and preferments, because they knew that all the offices in the Province would be filled by British carpet-baggers. King's College has also a divinity school for Episcopalian students.

The Province of Nova Scotia is occupied by 36 Christian sects. Of its inhabitants, 55,124 belong to the Anglican Church, and are ministered to by a lord bishop, 4 canons, 8 rural deans, and 68 clergymen. There are 102,001 Catholics, 103,539 Presbyterians, 73,430 Baptists, 41,751 Methodists, and 4,958 Lutherans (census of 1871).

The site of Windsor was called by the Indians *Pisiquid*, "the Junction of the Waters," and the adjacent lowlands were settled at an early day by the French, who raised large quantities of wheat and exported it to Boston. The French settled in this vicinity about the middle of the 17th century, but retired far into the interior at the time of the British conquest. Gov. Lawrence issued a proclamation inviting settlers to come in from New England, stating that "100,000 acres of land had been cultivated and had borne wheat, rye, barley, oats, hemp, flax, etc., for the last century without failure." The deserted French hamlets were occupied in 1759-60 by families from Massachusetts and Rhode Island, and their descendants still possess the land. The Rhode-Islanders erected the township of Newport, Massachusetts formed Falmouth, and Windsor was granted to British officers and was fortified in 1759. The broad rich marshes near Windsor had attracted a large Acadian population, and here was their principal church, whose site is still venerated by the Micmac Indians.

"I cannot recall a prettier village than this. If you doubt my word, come and see it. Yonder we discern a portion of the Basin of Minas; around us are the rich meadows of Nova Scotia. Intellect has here placed a crowning college upon a hill; opulence has surrounded it with picturesque villas." (COZZENS.) Another writer has spoken with enthusiasm of Windsor's "wide and beautiful environing meadows and the hanging-gardens of mountain-forests on the S. and W."

The Halifax train sweeps along the St. Croix River around Windsor, passing (on the r.) the dark buildings of King's College, on a hilltop, with the new chapel in front of their line. The character of the landscape begins to change, and to present a striking contrast with the agricultural regions just traversed.

"Indeed, if a man can live on rocks, like a goat, he may settle anywhere between Windsor and Halifax. With the exception of a wild pond or two, we saw nothing but rocks and stunted firs for forty-five miles, a monotony unrelieved by one picturesque feature. Then we longed for the 'Garden of Nova Scotia,' and understood what is meant by the name." (WARNER's *Baddeck*.)

Beyond Three-Mile Plains the train reaches *Newport*, near which large

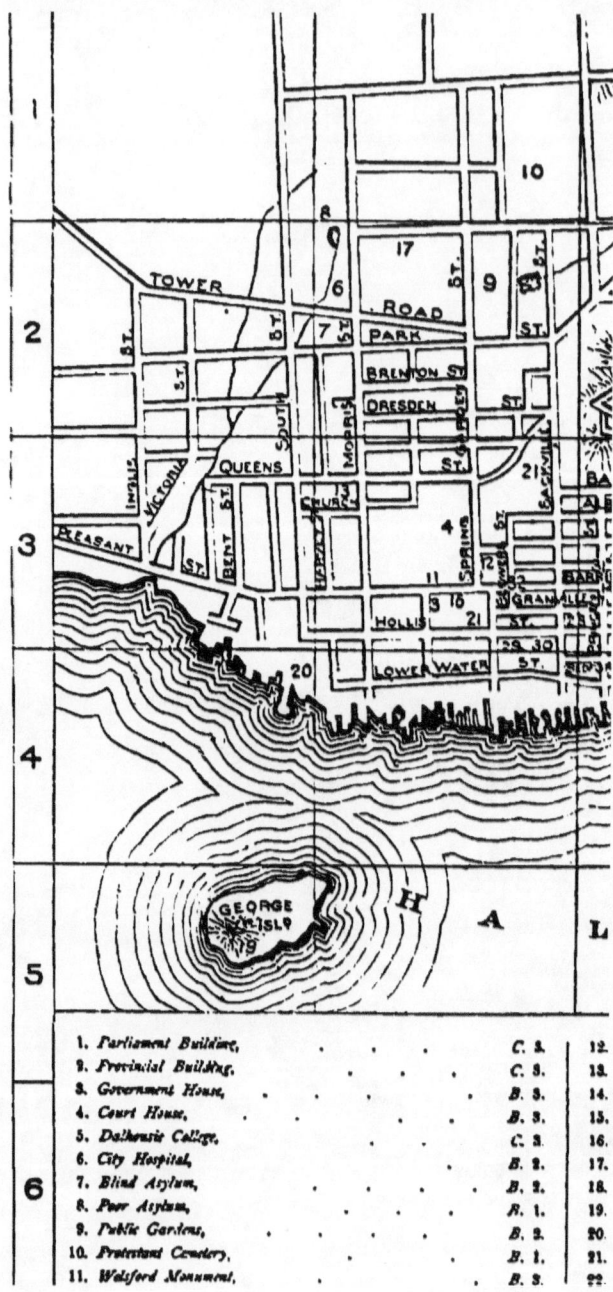

1. Parliament Building,
2. Provincial Building,
3. Government House,
4. Court House,
5. Dalhousie College,
6. City Hospital,
7. Blind Asylum,
8. Poor Asylum,
9. Public Gardens,
10. Protestant Cemetery,
11. Welsford Monument,

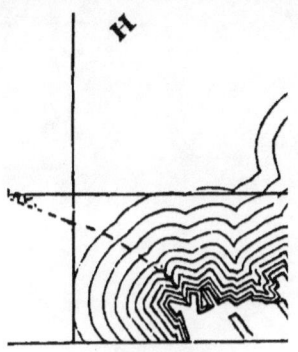

B. 3.	23. Officers' Quarters,
B. 3.	24. Military Hospital,
C. 3.	25. Queen's Dock Yard,
D. 2.	26. Admiralty House,
B. 3.	27. Y. M. C. A.,
R. 2.	28. Halifax Club,
C. 2.	29. Halifax Hotel,
A. 5.	30. International Hotel,
A. 4.	31. Carlton Hotel,
B. 3.	32. Waverley Hotel,
E. 1.	83. Railway Station,

quantities of gypsum are quarried from the veins in the soft marly sandstone. Nearly 3,000 tons of this fine fibrous mineral are shipped yearly from Newport to the United States. To the N. are the villages of Brooklyn (5 M.), devoted to manufacturing; Scotch Village (9 M.), a farming settlement; and Burlington, on the Kennetcook River (10 M.). Chivirie and Walton, 20-22 M. N., on the Basin of Minas, are accessible from Newport by a tri-weekly conveyance. The train passes on to *Ellershouse* (small inn), a hamlet clustered around a furniture-factory and lumber-mills. 2½ M. distant is the settlement at the foot of the *Ardoise Mt.*, which is the highest point of land in the Province, and overlooks Falmouth, Windsor, and the Basin of Minas. The train now crosses the Five-Island Lake, skirts Uniacke Lake, with Mt. Uniacke on the N., and stops at the *Mt. Uniacke* station (small inn). The Mt. Uniacke estate and mansion were founded more than 50 years ago by Richard John Uniacke, then Attorney-General of Nova Scotia. The house occupies a picturesque position between two rock-bound lakes, and the domain has a hardworking tenantry. The *Mt. Uniacke Gold-Mines* are 3 M. from the station, and were opened in 1865. In 1869 the mines yielded $37,340, or $345 to each workman, being 6 ounces and 4 pennyweights from each ton of ore. For the next 10 M. the line traverses an irredeemable wilderness, and then reaches *Beaver Bank*, whence lumber and slate are exported. At **Windsor Junction** the train runs on to the rails of the Intercolonial Railway (see page 82), which it follows to Halifax.

19. Halifax.

Arrival from the Sea. — Cape Sambro is usually seen first by the passenger on the transatlantic steamers, and Halifax Harbor is soon entered between the lighthouses on Chebucto Head and Devil Island. These lights are 7½ M. apart, Chebucto (on the l.) having a revolving light visible for 18 M., and Devil Island a fixed red light on a brown tower. On the W. shore the fishing-hamlets of Portuguese Cove, Bear Cove, and Herring Cove are passed in succession. 4 M. S. E. of Herring Cove is the dangerous Thrumcap Shoal, where H. B. M. frigate *La Tribune*, 44, was wrecked in 1797, and nearly all her people were lost, partly by reason of an absurd stretch of naval punctilio. Between this shoal and McNab's Island on one side, and the mainland on the other, is the long and narrow strait called the Eastern Passage. In 1862 the Confederate cruiser *Tallahassee* was blockaded in Halifax Harbor by a squadron of United-States frigates. The shallow and tortuous Eastern Passage was not watched, since nothing but small fishing-craft had ever traversed it, and it was considered impassable for a steamer like the *Tallahassee*. But Capt. Wood took advantage of the high tide, on a dark night, and crept cautiously out behind McNab's Island. By daylight he was far out of sight of the outwitted blockading fleet.
2 M. from Herring Cove the steamer passes Salisbury Head, and runs between the Martello Tower and lighthouse on Maugher Beach (r. side) and the York Redoubt (1¼ M. apart) Near the Redoubt is a Catholic church, and a little above is the hamlet of Falkland, with its Episcopal church, beyond which the N. W. Arm opens on the l. Passing between the batteries on McNab's Island and Fort Ogilvie, on Point Pleasant, the steamship soon runs by Fort Clarence and the fortress on George Island, and reaches her wharf at Halifax, with the town of Dartmouth and the great Insane Asylum on the opposite shore.
Arrival by Railway. — The station is at Richmond, some distance from the city, but passengers can go in either by carriage, hotel-omnibus, or horse-car. The railway is now being prolonged by a system of costly works, and will soon reach a terminus within the city.

Route 19.   HALIFAX.

Hotels.—The *Halifax, 107 Hollis St., $2 a day; the *International, on Hollis St., $1.75-2 a day; Carlton House, 57 Argyle St., small but aristocratic; Mansion House, 149 Barrington St.; Waverley, 8 Barrington St.; and numerous small second-class houses, of which the Arlington and the Cambridge, nearly opposite the International, are the best situated ($1-1.25 a day). An attempt is now being made to provide for Halifax a first-class modern hotel, like the Victoria at St. John.

Restaurants.—One of the best is that connected with the Acadian Hotel, 64 Granville St. Ices, pastry, and confectionery may be obtained at the shops on Hollis St. *American beverages* are compounded at the bar in the Halifax House.

Reading-Rooms.—The Young Men's Christian Association, corner of Granville and Prince Sts.; the Provincial Library, in the Parliament Buildings; and in the two chief hotels. The Halifax Library is at 197 Hollis St.; and the Citizens' Free Library (founded by Chief Justice Sir William Young) is at 265 Barrington St., and is open from 3 to 6 P. M. The Merchants' Exchange and Reading-Room is at 158 Hollis St.

Clubs.—The Halifax Club has an elegant house at 155 Hollis St.; the Albion is at 87 Hollis St.; the Catholic Young Men's Club, 1 Grafton St. (open from 2 to 10 P. M.); the Highland, North British, St. George's, Charitable Irish, and Germania Societies. The Royal Halifax Yacht Club has a house at Richmond, with billiard and reading rooms, and a line of piers and boat-houses for the vessels of their fleet.

Amusements of various kinds are afforded, at different times, in the Temperance Hall, on Starr St. During the winter some fine skating is enjoyed at the Rink, in the Public Gardens. Good games of cricket and indifferent base-ball playing may be seen on the Garrison Cricket-ground. But Halifax is chiefly famous for the interest it takes in trials of skill between yachtsmen and oarsmen, and exciting aquatic contests occur frequently during the summer.

Horse-cars run every 15 minutes, from 6 A. M. to 10 P. M., from the Richmond Station to the Fresh-Water Bridge, traversing the Campbell Road, Upper Water St., Granville St., Hollis, Morris, and Pleasant Sts. Also on Barrington St. and the Spring Garden Road to the Poor Asylum.

Railways.—The Intercolonial, running to St. John, N. B., in 276 M. (see Routes 16 and 17), and to Pictou in 113 M. (see Route 31); the Windsor & Annapolis, prolonged by a steamship connection to St. John (see Route 18).

Steamships.—The Allan Line, fortnightly, for St. John's, N. F., Queenstown, and Liverpool; Norfolk, and Baltimore. Fares, Halifax to Liverpool, $75 and $25; to Norfolk or Baltimore, $20 and $12. The Anchor Line, for St. John's, N. F., and Glasgow. The Royal Mail Steamers *Alpha* and *Delta* (Cunard Line) leave Halifax for Bermuda and St. Thomas every fourth Monday, connecting at St. Thomas with steamships for all parts of the West Indies, Panama, and the Spanish Main.

The *Carroll* and *Alhambra* leave Esson's Wharf for Boston on alternate Saturdays. Fare, $8; with state-room, $9. The *Falmouth* leaves Dominion Wharf for Portland every Tuesday at 4 P. M. This vessel is nearly new, and is handsomely fitted up for passenger-traffic. Fares, Halifax to Portland, $7 and $5; to Boston, $8 and $6.50; to New York (by the Sound boats), $12 and $10.50.

The *Carroll* or the *Alhambra* leaves Esson's Wharf every Monday noon for the Strait of Canso and Charlottetown, P. E. I. Fares to Charlottetown, cabin, $4; cabin state-room, $5; saloon state-room, $6. The *George Shattuck* leaves Boak's Wharf, fortnightly, for N. Sydney, C. B., and St. Pierre Miq. (see Route 50). The steamship *Virgo* leaves for Sydney, C. B., and St. John's, N. F., every alternate Tuesday (see Routes 36 and 51). Fares, to Sydney, $8; to St. John's, $15; steerage to either port, $5.

The *Micmac* cruises in the harbor during the summer, running from the South Ferry Wharf to McNab's Island and up the N. W. Arm (fare, 25c.). The steam-ferry from Dartmouth has its point of departure near the foot of George St. The *Goliah* makes frequent trips up the Bedford Basin.

Stages leave Halifax daily for Chester, Lunenburg, Liverpool, Shelburne, and Yarmouth (see Route 24), departing at 6 A. M. Stages leave at 6 A. M. on Monday, Wednesday, and Friday for Musquodoboit Harbor, Jeddore, Ship Harbor, Tangier, Sheet Harbor, Beaver Harbor, and Salmon River (see Route 29).

HALIFAX, the capital of the Province of Nova Scotia, and the chief naval station of the British Empire in the Western Hemisphere, occupies a commanding position on one of the finest harbors of the Atlantic coast. It has 29,582 inhabitants (census of 1871), with 7 banks, 4 daily papers and several tri-weeklies and weeklies, and 24 churches (7 Anglican, 5 Presbyterian, 3 each of Catholic, Wesleyan, and Baptist). The city occupies a picturesque position on the E. slope of the peninsula (of 8,000 acres), between the bay, the N. W. Arm, and the Bedford Basin; and looks out upon a noble harbor, deep, completely sheltered, easily accessible, and large enough "to contain all the navies of Europe." In 1869 the imports amounted to $7,202,504, and the exports to $3,169,548; and in 1870 the assessed valuation of the city was $16,753,812. The city has a copious supply of water, which is drawn from the Chain Lakes, about 12 M. distant, and so high above Halifax that it can force jets over the highest houses by its own pressure. There is a fire-alarm telegraph, and an efficient fire department, with several steam-engines.

The city lies along the shore of the harbor for 2½ M., and is about ¾ M. wide. Its plan is regular, and some of the business streets are well built; but the general character of the houses is that of poor construction and dingy colors. It has, however, been much bettered of late years, owing to the improvements after two great fires, and to the wealth which flowed in during the American civil war, and hardly deserves the severe criticism of a recent traveller: "Probably there is not anywhere a more rusty, forlorn town, and this in spite of its magnificent situation."

Hollis and Granville Streets, in the vicinity of the Parliament Buildings, contain the most attractive shops and the headquarters of the great importing houses. Many of the buildings in this section are of solid and elegant construction, though the prevalence of dark colors gives a sombre hue to the street lines.

The **Parliament Building** occupies the square between Hollis, George, Granville, and Prince Streets, and is surrounded with trees. In 1830 this plain structure of gray stone was called the finest building in North America, but American architecture has advanced very far since that time. Opposite the Granville-St. entrance is the *Library*, occupying a very cosey little hall, and supplied with British and Canadian works on law, history, and science. In the N. part of the building is the plain and commodious hall of the House of Assembly; and on the S. is the chamber of the Legislative Council, in which are some fine portraits. On the r. and l. of the vice-regal throne are full-length * portraits of King George III. and Queen Charlotte; on the N. wall are Chief Justice Blowers, King William IV., Judge Haliburton (see page 92), * Sir Thomas Strange (by *Benjamin West*), and Sir Brenton Haliburton. Opposite the throne are Nova Scotia's military heroes, Sir John Inglis (the defender of Lucknow) and Sir W. Fenwick Williams of Kars. On the S. wall are full-length portraits of King George II. and Queen Caroline.

The new **Provincial Building** is E. of the Parliament Building, on Hollis St., and is 140 by 70 ft in area. It is built of brown freestone, in an ornate style of architecture, and cost $120,000. The lower story is occupied by the Post-Office; and the third floor contains the * **Provincial Museum**, which exhibits preserved birds, animals, reptiles, fossils, minerals, shells, coins, and specimens of the stones, minerals, coals, and gold ores of Nova Scotia. There are also numerous Indian relics, curiosities from Japan and China, naval models, and old portraits. Opposite the entrance is a gilt pyramid, which represents the amount of gold produced in the Province between 1862 and 1870, — 5 tons, 8 cwt., valued at $8,373,431. Most of this gold has been coined at the U. S. Mint in Philadelphia, and is purer and finer than that of California and Montana.

On the corner of Granville and Prince Streets, near the Parliament Building, is the new and stately stone building of the Young Men's Christian Association, with its reading-rooms and other departments. The massive brownstone house of the Halifax Club is to the S., on Hollis St.

The * **Citadel** covers the summit of the hill upon whose slopes the city is built, and is 250 ft. above the level of the sea. Visitors are admitted and allowed to pass around the ramparts under escort of a soldier, after registering their names at the gate. The attendant soldier will point out all the objects of interest, and (if he be an artillerist) will give instructive discourse on the armament, though his language may sometimes become hopelessly technical. The Citadel is a fortress of the first class, according to the standards of the old school; though of late years the government has bestowed much attention on the works at George's Island, which are more important in a naval point of view.

The works were commenced by Prince Edward, the Duke of Kent, and the father of Queen Victoria, who was then Commander of the Forces on this station. He employed in the service a large number of the Maroons, who had been conquered by the British, and were banished from Jamaica, and subsequently deported to Sierra Leone. Changes and additions have been made nearly every year since, until the present immense stronghold has been completed. It is separated from the glacis by a deep moat, over which are the guns on the numerous bastions. The massive masonry of the walls seems to defy assault, and the extensive barracks within are said to be bomb-proof. During the years 1873-74 the artillery has been changed, and the previous mixed armament has been to a great degree replaced by muzzle-loading Woolwich guns of heavy calibre, adapted for firing the conical Palliser shot with points of chilled iron. The visitor is allowed to walk around the circuit of the ramparts, and this elevated station affords a broad view on either side. Perhaps the best prospect is that from the S. E. bastion, overlooking the crowded city on the slopes below; the narrow harbor with its shipping; Dartmouth, sweeping up toward Bedford Basin; Fort Clarence, below Dartmouth, with its dark casemates; McNab's Island, crowned with batteries and shutting in the Eastern Passage; the outer harbor, with its fortified points, and the ocean beyond.

Near the portal of the citadel is an outer battery of antiquated guns; and at the S. end of the glacis are the extensive barracks of the Royal Artillery. Other military quarters are seen on the opposite side of the Citadel.

"But if you cast your eyes over yonder magnificent bay, where vessels bearing flags of all nations are at anchor, and then let your vision sweep past and over the islands to the outlets beyond, where the quiet ocean lies, bordered with fog-banks that loom ominously at the boundary-line of the horizon, you will see a picture of marvellous beauty: for the coast scenery here transcends our own sea-shores, both in color and outline. And behind us again stretch large green plains, dotted with cottages, and bounded with undulating hills, with now and then glimpses of blue water; and as we walk down Citadel Hill, we feel half reconciled to Halifax, its quaint mouldy old gables, its soldiers and sailors, its fogs, cabs, penny and half-penny tokens, and all its little, odd, outlandish peculiarities." (COZZENS.)

Lower Water St. borders the harbor-front, and gives access to the wharves of the various steamship and packet lines. It runs from the Ordnance Yard, at the foot of Buckingham St., to the Government reservation near George Island, and presents a remarkably dingy and dilapidated appearance throughout its entire length.

The **Queen's Dockyard** occupies ½ M. of the shore of the upper harbor, and is surrounded on the landward side by a high stone-wall. It contains the usual paraphernalia of a first-class navy-yard, — storehouses, machine-shops, docks, arsenals, a hospital, and a line of officers' quarters. It is much used by the frigates of the British navy, both to repair and to refit, and the visitor may generally see here two or three vessels of Her Britannic Majesty.

The Dockyard was founded in 1758, and received great additions (including the present wall) in 1770. During the two great wars with the United States it was invaluable as a station for the royal navy, whose fleets thence descended upon the American coast. Many trophies of the war of 1812 were kept here (as similar marine mementos of another nation are kept in the Brooklyn and Washington Navy-Yards), including the figure-head of the unfortunate American frigate, the *Chesapeake*, which was captured in 1813, off Boston Harbor, by the British frigate *Shannon*, and was brought into Halifax with great rejoicing. It is, perhaps, in kindly recognition of the new fraternity of the Anglo-American nations, that the Imperial Government has lately caused these invidious emblems of strife to be removed.

The Dockyard is not open to the public, but the superintendent will generally admit visitors upon presentation of their cards.

In the N. W. part of the city, near the foot of Citadel Hill, is the **Military Hospital**, before which is the *Garrison Chapel*, a plain wooden building on whose inner walls are many mural tablets in memory of officers who have died on this station. Beyond this point, Brunswick St. runs N. W. by the Church of the Redeemer to *St. George's Church*, a singular wooden building of a circular form. At the corner of Brunswick and Gerrish Sts. is a cemetery, in which stands a quaint little church dating from 1761, having been erected by one of the first companies of German immigrants.

On Gottingen St. is the Church of St. Joseph, where the Catholic seamen of the fleet attend mass on Sunday at 9½ A. M. Near this building is the Orphan Asylum of the Sisters of Charity.

Farther N. on Gottingen St. is the Deaf and Dumb Asylum, beyond which, on North St., is the Roman Catholic *College of St. Mary*, at Belle Air. This institution is under the charge of the Christian Brothers, and has the same line of studies as an American high-school. Farther out on Gottingen St. is the **Admiralty House**, the official residence of the commander-in-chief of the North-American and West-Indian Squadrons, beyond which are the *Wellington Barracks*, over the Richmond railway-station. From the plateau on which the secluded Admiralty House is located, the visitor can look down on the Queen's Dockyard, the fleet, and the inner harbor.

The Roman Catholic **Cathedral of St. Mary** is on the Spring Garden Road, near its intersection with Pleasant St. It has recently been much enlarged and improved by the addition of an elegant granite façade and spire, in florid Gothic architecture. The Cathedral fronts on an old and honored cemetery, on whose E. side is a finely conceived * monument to Welsford and Parker, the Nova-Scotian heroes of the Crimean War. (Major Welsford was killed in the storming of the Redan.) It consists of a small but massive arch of brownstone, standing on a broad granite base, and supporting a statue of the British lion. Opposite the cemetery, on Pleasant St., is the Presbyterian Church of St. Matthew (under the care of Rev. George M. Grant). Above the Cathedral, on the Spring Garden Road, is the handsome building of the *Court House*, well situated amid open grounds, near the jail and the capacious drill-sheds.

The *Horticultural Gardens* are on the Spring Garden Road, and are well arranged and cared for. They were purchased by the city in September, 1874, and were then united with the Public Gardens, which are just S. of Citadel Hill. Military music is given here by the garrison bands during the summer. Near the Gardens is the Convent of the Sacred Heart, a stately building situated in pleasant grounds. The *Protestant Cemetery* adjoins the Horticultural Gardens on the W., and contains a great number of monuments. In the same quarter of the city, below Morris St., are the new Blind Asylum, the City Hospital, and the immense and stately building of the Poor Asylum, lately completed at a cost of $260,000.

The **Government House** is a short distance beyond St. Matthew's Church, on Pleasant St., and is the official residence of the Lieutenant-Governor of Nova Scotia. It is a plain and massive old stone building, with projecting wings, and is nearly surrounded by trees. Farther S , on Morris St., is the Anglican Cathedral of St. Luke, a plain and homely wooden building. Beyond this point are the pretty wooden churches and villas which extend toward Point Pleasant.

At the foot of South St. are the *Ordnance Grounds*, from whose wharf the lower harbor is overlooked. About 1,800 ft. distant is **George's Island**, on which is a powerful modern fortress, bearing a heavy armament from which immense chilled-iron or steel-pointed shot could be hurled against a hostile fleet. This position is the key to the harbor, and converges its fire with that of *Fort Clarence*, a low but massive casemated work, 1 M. S. E. on the Dartmouth shore, whose guns could sweep the Eastern Passage and the inner harbor. The passage from the outer harbor is defended by the York Redoubt, near Sandwich Point, by a new line of batteries on the N. W. shore of McNab's Island, and by the forts on Point Pleasant.

At the corner of Prince and Barrington Sts. is St. Paul's Episcopal Church, a plain and spacious old building (built in 1750), with numerous mural tablets on the inner walls. *Dalhousie College and University* is at the corner of Duke and Barrington Sts., and was founded by the Earl of Dalhousie while he was Governor-General of Canada. Its design was to provide means for the liberal education of young men who did not wish to go (or were debarred from going) to King's College, at Windsor. There are 7 professors in the academic department, and the medical school has 13 professors.

In the summer of 1746 the great French Armada sailed from Brest to conquer the British North-American coast from Virginia to Newfoundland. It was commanded by the Duke d'Anville, and was composed of the line-of-battle ships *Trident, Ardent, Mars*, and *Alcide*, 64 guns each; the *Northumberland, Carillon, Tigre, Leopard*, and *Renommée*, 60 guns each; the *Diamant*, 50; *Megére*, 30; *Argonaute*, 26; *Prince d'Orange*, 26; the *Parfait, Mercure, Palme, Girous, Perle*, and 22 other frigates, with 30 transports, carrying an army of 3,150 soldiers. D'Anville's orders were to "occupy Louisbourg, to reduce Nova Scotia, to destroy Boston, and ravage the coast of New England." The Armada was dispersed, however, by a succession of unparalleled and disastrous storms, and D'Anville reached Chebucto Bay (Halifax) on Sept. 10, with only 2 ships of the line and a few transports. Six days later the unfortunate Duke died of apoplexy, induced by grief and distress on account of the disasters which his enterprise had suffered. The Vice-Admiral D'Estournelle committed suicide a few days later. Some other vessels now arrived here, and immense barracks were erected along the Bedford Basin. 1,200 men had died from scurvy on the outward voyage, and the camps were soon turned into hospitals. Over 1,000 French soldiers and 2-300 Micmac Indians died around the Basin and were buried near its quiet waters. Oct. 13, the French fleet, numbering 5 ships of the line and 25 frigates and transports, sailed from Halifax, intending to attack Annapolis Royal; but another terrible storm arose, while the vessels were off Cape Sable, and scattered the remains of the Armada in such wide confusion that they were obliged to retire from the American waters.

The Indians called Halifax harbor *Chebucto*, meaning "the chief haven," and the French named it *La Baie Saine*, "on account of the salubrity of the air."

In the year 1748 the British Lords of Trade, incited by the people of Massachusetts, determined to found a city on the coast of Nova Scotia partly in prospect of commercial advantages, and partly to keep the Acadians in check. Parliament voted £40,000 for this purpose; and on June 21, 1749, a fleet of 13 transports and the sloop-of-war *Sphinx* arrived in the designated harbor, bearing 2,376 colonists (of whom over 1,500 were men). The city was laid out in July, and was named in honor of George Montagu, Earl of Halifax, the head of the Lords of Trade. The Acadians and the Indians soon sent in their submission; but in 1751 the suburb of Dartmouth was attacked at night by the latter, and many of its citizens were massacred. 500 Germans settled here in 1751-52, but it was found difficult to preserve the colony, since so many of its citizens passed over to the New-England Provinces. The great fleets and armies of Loudon and Wolfe concentrated here before advancing against Louisbourg and Quebec; and the city afterwards grew in importance as a naval station. Representative government was established in 1758, and the Parliament of 1770 remained in session for 14 years, while Halifax was made one of the chief stations whence the royal forces were directed upon the insurgent American colonies. After the close of the Revolutionary War, many thousands of exiled Loyalists took refuge here; and the wooden walls and towers with which the city had been fortified were replaced with more formidable defences by Prince Edward.

The ancient palisade-wall included the space between the present Salter, Barrington, and Jacob Streets, and the harbor; and its citadel was the small Government House, on the site of the present Parliament Building, which was surrounded with hogsheads filled with sand, over which light cannons were displayed.

The growth of Halifax during the present century has been very slow, in view of its great commercial advantages and possibilities. The presence of large bodies of troops, and the semi-military régime of a garrison-town, have had a certain effect in deadening the energy of the citizens. Great sums of money were, however, made here during the American civil war, when the sympathies of the Haligonians were warmly enlisted in favor of the revolted States, and many blockade-runners sailed hence to reap rich harvests in the Southern ports. The cessation of the war put a stop to this lucrative trade; but it is now hoped that the completion of the Intercolonial Railway to St. John and Quebec will greatly benefit Halifax. There is a rivalry between St. John and Halifax which resembles that between Chicago and St. Louis, and leads to similar journalistic tournaments. St. John claims that she has a first-class hotel and a theatre, which Halifax has not: and the Nova-Scotian city answers, in return, that she has the best cricket-club and the champion oarsman of America.

Sir William Fenwick Williams, of Kars, Bart., K. C. B., D. C. L., was born at Halifax in 1800. After serving in Ceylon, Turkey, and Persia, he instructed the Moslem artillery, and fortified the city of Kars. Here he was besieged by the Russians, under Gen. Mouravieff. He defeated the enemy near the city, but was forced to surrender after a heroic defence of six months, being a sacrifice to British diplomacy. He was afterwards Commander of the Forces in Canada.

Admiral Sir Provo Wallis was born at Halifax in 1791, and was early engaged in the great battle between the *Cleopatra*, 82, and the French *Ville de Milan*, 46. He afterwards served on the *Curieux*, the *Gloire*, and the *Shannon*, to whose command he succeeded after the battle with the *Chesapeake*.

20. The Environs of Halifax.

The favorite drive from Halifax is to the *Four-Mile House*, and along the shores of the *Bedford Basin. This noble sheet of water is 5 M. long and 1-3 M. wide, with from 8 to 86 fathoms of depth. It is entered by way of the *Narrows*, a passage 2½-3 M. long and ½ M. wide, leading from Halifax Harbor. It is bordered on all sides by bold hills 200-830 ft. in height, between which are 10 square miles of secure anchoring-ground. The village of Bedford is on the W. shore, and has several summer hotels (Bellevue, Bedford, etc.). The steamer *Goliah* leaves Halifax for Bedford at 11 A. M. and 2 P. M. daily. During the summer the light vessels of the Royal Halifax Yacht Club are seen in the Basin daily; and exciting rowing-matches sometimes come off near the Four-Mile House.

Along the shores of the Bedford Basin were the mournful camps and hospitals of the French Armada, in 1746, and 1,300 men were buried there. Their remains were found by subsequent settlers. The first permanent colonies along these shores were made by Massachusetts Loyalists in 1784.

Hammond's Plains are 7 M. W. of Bedford, and were settled in 1815 by slaves brought away from the shores of Maryland and Virginia by the British fleets. This is, like the other villages of freed blacks throughout the Province, dirty and dilapidated to the last degree. To the N. W. is the *Pockwock Lake*, 4 M. long, with diversified shores, and abounding in trout.

"The road to Point Pleasant is a favorite promenade in the long Acadian twilights. Midway between the city and the Point lies 'Kissing Bridge,' which the Halifax maidens sometimes pass over. Who gathers toll nobody knows, but—"

Point Pleasant projects between the harbor and the N. W. Arm, and is covered with pretty groves of evergreen trees, threaded by narrow roads, and now being laid out for a public park. The principal fortification is *Fort Ogilvie*, a garrisoned post, whose artillery commands the channel. A short distance to the W. is the antiquated structure called the *Prince of Wales's Tower*, from which fine views are afforded. The *Point Pleasant Battery* is near the water's edge, and is intended to sweep the outer passage.

The Northwest Arm is 4 M. long and ½ M. wide, and is a river-like inlet, which runs N. W. from the harbor to within 2 M. of the Bedford Basin.

DARTMOUTH. *Route 21.* 101

Its shores are high and picturesque, and on the Halifax side are several fine mansions, surrounded by ornamental grounds. In the upper part of the Arm is *Melville Island*, where American prisoners were kept during the War of 1812. *Ferguson's Cove* is a picturesque village on the N. W. Arm, inhabited chie.'y by fishermen and pilots.

The steamer *Micmac* makes regular trips during the summer up the N. W. Arm, and to *McNab's Island*, which is 3 M. long, and has a summer hotel and some heavy military works. The *Micmac* leaves the South Ferry Wharf at 10 A. M. and 12, and 2 and 3 P. M.

Dartmouth (*Acadian House*) is situated on the harbor, opposite the city of Halifax, to which a steam ferry-boat makes frequent trips. It has several pretty villas belonging to Halifax merchants; and at about ½ M. from the village is the spacious and imposing building of the *Mount Hope Asylum for the Insane*, a long, castellated granite building which overlooks the harbor. Dartmouth has 4,358 inhabitants and 5 churches, and derives prosperity from the working of several foundries and steam-tanneries. It is also the seat of the Chebucto Marine Railway. This town was founded in 1750, but was soon afterwards destroyed, with some of its people, by the Indians. In 1784 it was reoccupied by men of Nantucket who preferred royalism to republicanism. The *Montague Gold-Mines* are 4 M. from Dartmouth, and have yielded in paying quantities. *Cow Bay* is a few miles S. E. of Dartmouth, and is much visited in summer, on account of its fine marine scenery and the facilities for bathing. The *Dartmouth Lakes* commence within 1 M. of the town, and were formerly a favorite resort of sportsmen, but are now nearly fished out.

21. The Basin of Minas. — Halifax to St. John.

Halifax to Windsor, see Route 18 (in reverse).
The steamer leaves Windsor every Wednesday at high water, touching at Parrsboro', and thence running down the Bay to St. John.
The steamer leaves St. John (Reed's Point) every Tuesday evening at high water, for Parrsboro' and Windsor. Fares, St. John to Parrsboro' or Windsor, $3; to Londonderry, Maitland, or Halifax, $4.

As the steamer moves out from her wharf at Windsor, a pleasant view is afforded of the old college town astern, with the farming village of Falmouth on the l., and shipbuilding Newport on the r., beyond the mouth of the St. Croix River. The shores are high and ridgy, and the mouth of the Kennetcook River is passed (on the r.) about 5 M. below Windsor. 2-3 M. below is *Hantsport* (l. bank), a thriving marine village opposite the mouth of the Cockmigon River. On Horton Bluff (l. bank) is a lighthouse which sustains a powerful fixed white light, visible for 20 M., and beyond this point the steamer enters the *Basin of Minas*. On the l. are the low ridges of Long Island and Boot Island, rising on the margin of a wide and verdant meadow. The meadow is **Grand Pré**, the land of Evangeline (see Route 22). Mile after mile the fertile plains of Cornwallis open on the l., bounded by the Horton hills and the dark line of the North Mt. In advance is the bold and clear-cut outline of Cape Blomidon, brooding over the water, and on the r. are the low but well-defined bluffs of *Chivirie*, rich in gypsum and limestone. It is about 22 M. from the mouth of the Avon to Parrsboro', and the course of the steamer continually approaches Blomidon.

Cape Blomidon is a vast precipice of red sandstone of the Triassic era, with strong marks of volcanic action. "The dark basaltic wall, covered with thick woods, the terrace of amygdaloid, with a luxuriant growth of light-green shrubs and young trees that rapidly spring up on its rich and moist surface, the precipice of bright red sandstone, always clean and fresh, and contrasting strongly with the trap above, ... constitute a combination of forms and colors equally striking, if seen in the distance from the hills of Horton or Parrsboro', or more nearly from the sea or the stony beach at its base. Blomidon is a scene never to be forgotten by a traveller who has wandered around its shores or clambered on its giddy precipices." The cape is about 570 ft. high, and presents an interesting sight when its dark-red summit is peering above the white sea-fogs. Sir William Lyell, the eminent British geologist, made a careful study of the phenomena of this vicinity.

The Indian legend says that Blomidon was made by the divine Glooscap, who broke the great beaver-dam off this shore and swung its end around into its present position. Afterwards he crossed to the new-made cape and strewed its slopes with the gems that are found there to-day, carrying thence a set of rare ornaments for his ancient and mysterious female companion. The beneficent chief broke away the beaver-dam because it was flooding all the Cornwallis Valley, and in his conflict with the Great Beaver he threw at him huge masses of rock and earth, which are the present Five Islands. W. of *Utkogwnrheech* (Blomidon) the end of the dam swept around and became *Pleegun* (Cape Split).

As Blomidon is left on the port beam, the steamer hurries across the rapid currents of the outlet of the Basin. In front is seen the white village of Parrsboro', backed by the dark undulations of the Cobequid Mts. Just before reaching Parrsboro' the vessel approaches and passes *Partridge Island* (on the l.), a singular insulated hill 250 ft. high, and connected with the mainland at low tide by a narrow beach.

Partridge Island was the *Pulowech Munegoo* of the Micmacs, and was a favorite location for legends of Glooscap. On his last great journey from Newfoundland by Pictou through Acadia and into the unknown West, he built a grand road from Fort Cumberland to this shore for the use of his weary companions. This miraculously formed ridge is now occupied by the post-road to the N. W., and is called by the Indians *Owwokun* (the causeway). At Partridge Island Glooscap had his celebrated revel with the supernatural Kit-poos-e-ag-unow, the deliverer of all oppressed, who was taken out alive from his mother (slain by a giant), was thrown into a well, and, being miraculously preserved there, came forth in due time to fulfil his high duty to men. These marvellous friends went out on the Basin in a stone canoe to fish by torchlight, and, after cruising over the dark waters for some time, speared a monstrous whale. They tossed him into the canoe "as though he were a trout," and made for the shore, where, in their brotherly feast, the whale was entirely devoured.

Parrsboro' (two inns) is prettily situated at the mouth of a small river, and under the shelter of Partridge Island. It has about 900 inhabitants, with three churches, and is engaged in the lumber-trade. The beauty of the situation and the views, together with the sporting facilities in the back-country, have made Parrsboro' a pleasure resort of considerable repute, and the neat hotel called the *Summer House* is well patronized. This is one of the best points from which to enter the fine hunting and fishing districts of Cumberland County, and guides and outfits may be secured here. Amherst (see page 78) is 36 M. distant, by highways following the valleys of the Parrsboro' and Maccau Rivers.

"Parrsboro' enjoys more than its share of broad, gravelly beach, overhung with clifted and woody bluffs. One fresh from the dead walls of a great city would be delighted with the sylvan shores of Parrsboro'. The beach, with all its breadth, a miracle of pebbly beauty, slants steeply to the surf, which is now rolling up in curling clouds of green and white. Here we turn westward into the great bay itself, going with a tide that rushes like a mighty river toward a cataract, whirling, boiling, breaking in half-moons of crispy foam." (L. L. NOBLE.)
"Pleasant Parrsboro', with its green hills, neat cottages, and sloping shores laved by the sea when the tide is full, but wearing quite a different aspect when the tide goes out; for then it is left perched thirty feet high upon a red clay bluff, and the fishing-boats which were afloat before are careened upon their beam ends, high and dry out of water. The long massive pier at which the steamboat lately landed, lifts up its naked bulk of tree-nailed logs, reeking with green ooze and sea-weed; and a high conical island which constitutes the chief feature of the landscape is transformed into a bold promontory, connected with the mainland by a huge ridge of brick-red clay." (HALLOCK.)

Gentlemen who are interested in geological studies will have a rare chance to make collections about Parrsboro' and the shores of Minas. The most favorable time is when the bluffs have been cracked and scaled by recent frosts; or just after the close of the winter, when much fresh *débris* is found at the foot of the cliffs. Among the minerals on Partridge Island are: analcime, apophyllite, amethyst, agate, apatite, calcite (abundant, in yellow crystals), chabazite, chalcedony, cat's-eye, gypsum, hematite, heulandite, magnetite, stilbite (very abundant), jasper, cacholong, opal, semi-opal, and gold-bearing quartz. About Cape Blomidon are found analcime, agate, amethyst, apophyllite. calcite, chalcedony, chabazite-gmelinite, faröelite, hematite, magnetite, heulandite, laumonite, fibrous gypsum, malachite, mesolite, native copper, natrolite, stilbite, psilomelane, and quartz. Obsidian, malachite, gold, and copper are found at Cape d'Or; jasper and five quartz crystals, on Spencer's Island; augite, aminnthus, pyrites, and wad, at Parrsboro'; and both at Five Islands and Scotsman's Bay there are beautiful specimens of moss agate. At Cornwallis is found the rare mineral called Wichtisite (resembling obsidian, in gray and deep blue colors), which is only known in one other place on earth, at Wichtis, in Finland. The purple and violet quartz, or amethyst, of the Minas shores, is of great beauty and value. A Blomidon amethyst is in the crown of France, and it is now 270 years since the Sieur de Monts carried several large amethysts from Partridge Island to Henri IV. of France. These gems are generally found in geodes, or after fresh falls of trap-rock.

Advocate Harbor and Cape d'Or.

A bi-weekly stage runs W. from Parrsboro' through grand coast scenery, for 28 M., passing the hamlets of Fox Harbor and Port Greville, and stopping at **Advocate Harbor**. This is a sequestered marine hamlet, devoted to shipbuilding and the deep-sea fisheries, and has about 600 inhabitants. It is about 60 M. from Amherst, by a road leading across the Cobequid Mts. and through Apple River (see page 80). Some of the finest marine scenery in the Provinces is in this vicinity. 3-4 M. S. is the immense rocky peninsula of * **Cape d'Or,** almost cut off from the mainland by a deep ravine, in whose bottom the salt tides flow. Cape d'Or is 500 ft. high, and has recently become noted for its rich copper deposits. Off this point there is a heavy rip on the flood-tide, which flows with a velocity of 6 knots an hour, and rises 33-39 ft. 8 M. W. of Advocate Harbor, and visible across the open bay, is *Cape Chignecto, a wonderful headland of rock, 730 - 800 ft. high, running down sheer into the deep waters. This mountain-promontory marks the division of the currents of the Minas and Chignecto Channels.

Cape d'Or is sometimes called *Cap Doré* on the ancient maps, and received its name on account of the copper ore which was found here by the early French explorers, and was supposed to be gold. The Acadians afterwards opened mines here, and the name. *Les Mines*, originally applied to a part of this shore, was given to the noble salt-water lake to the E. *Minas* is either an English modification or the Spanish equivalent thereof. Cape d'Or was granted to the Duke of Chandos many years ago, but he did not continue the mining operations.

After leaving Parrsboro' the steamer runs W. through the passage between Cape Blomidon and Cape Sharp, which is 3½ M. wide, and is swept by the tide at the rate of 6-8 knots an hour. On the r. the ravines of Diligent River and Fox River break the iron-bound coasts of Cumberland County; and on the l. is a remarkable promontory, 7 M. long and 1 M. wide, with an altitude of 400 feet, running W. from Blomidon between the channel and the semicircular bight of Scotsman's Bay. *Cape Split* is the end of this sea-dividing mountain, beyond which the S. shores fall suddenly away, and the steamer enters the Minas Channel. 12 M. beyond Cape Split, *Spencer's Island* and Cape Spencer are passed on the N., beyond which are the massive cliffs of **Cape d'Or**. On the l. are the unvarying ridges of the North Mt., with obscure fishing-hamlets along the shore. To the N. the frowning mass of *Cape Chignecto* is seen; and the course passes within sight of the lofty and lonely rock of *Isle Haute*, which is 7 M. from the nearest shore. It is 1½ M. long and 350 ft. high, and is exactly intersected by the parallel of 65° W. from Greenwich.

The steamer now passes down over the open waters of the Bay of Fundy. St. John is about 62 nautical miles from Isle Haute, in a straight line, and is a little N. of W. from that point, but the exigencies of navigation require a course considerably longer and more southerly. This portion of the route is usually traversed at night, and soon after passing the powerful first-class red revolving-light on *Cape Spencer* (New Brunswick), the steamer runs in by the Partridge-Island light, and enters the harbor of St. John about the break of day.

St. John, see page 15.

The Basin of Minas.

The steamer *William Stroud* leaves Parrsboro' several times weekly, for the villages on the N. and E. shores of the Basin of Minas. As the times of her departure are very irregular, owing to the necessity of following the tide, and her landings vary according to circumstances, the following account relates to the line of the coast rather than to her route. She is announced to call at Parrsboro', Londonderry, Maitland, Kingsport, Summerville, and Windsor.

Soon after leaving Parrsboro', *Frazer's Head* is passed on the l., with its cliffs elevated nearly 400 feet above the water. About 15 M. E. of

Parrsboro' are the remarkable insulated peaks of the *Five Islands, the chief of which is 350 ft. high, rising from the waters of the Basin. On the adjacent shore is the village of Five Islands, occupying a very picturesque position, and containing 600 inhabitants. In this vicinity are found iron, copper, and plumbago, and white-lead is extracted in considerable quantities from minerals mined among the hills. Marble was formerly produced here, but the quarries are now abandoned. The massive ridge variously known as Mt. Gerrish, St. Peter's Mt., and Red Head, looms over the village to a height of 500 ft., having a singularly bold and alpine character for so small an elevation. On its lower slopes are found pockets containing fine barytes, of which large quantities are sent to the United States. A mass of over 150 pounds' weight was sent from this place to the Paris Exposition of 1867. A few miles W. of the village are the falls on the North River, which are 90 ft. high; and to the N. is the wild and picturesque scenery of the Cobequid Mts. Five Islands may be visited by the road from Parrsboro' (16-18 M.), which also passes near the North River Falls. The most direct route to the village is by the mail-stage from Debert station, on the Intercolonial Railway (see page 80).

"Before them lay the outlines of Five Islands, rising beautifully out of the water between them and the mainland. . . . . The two more distant were rounded and well wooded; the third, which was midway among the group, had lofty, precipitous sides, and the summit was dome-shaped; the fourth was like a table, rising with perpendicular sides to the height of 200 ft., with a flat, level surface above, which was all overgrown with forest trees. The last, and nearest of the group, was by far the most singular. It was a bare rock which rose irregularly from the sea, terminating at one end in a peak which rose about 200 ft. in the air. . . . . It resembled, more than anything else, a vast cathedral rising out of the sea, the chief mass of the rock corresponding with the main part of the cathedral, while the tower and spire were there in all their majesty. For this cause the rock has received the name of Pinnacle Island. . . . . At its base they saw the white foam of breaking surf; while far on high around its lofty, tempest-beaten summit, they saw myriads of sea-gulls. Gathering in great white clouds about this place, they sported and chased one another; they screamed and uttered their shrill yells, which sounded afar over the sea." (DE MILLE.)

10 M. beyond these islands the steamer passes the lofty and far-projecting peninsula of *Economy*[1] *Point*, and enters the Cobequid Bay (which ascends to Truro, a distance of 36 M.). After touching at *Londonderry*, on the N. shore, the steamer crosses the bay to *Maitland* (two inns), a busy and prosperous shipbuilding village at the mouth of the Shubenacadie River (see page 82).

The S. shore of the Basin of Minas is lined with bluffs 100-180 ft. high, but is far less imposing than the N. shore. *Noel* is about 15 M. W. of Maitland, and is situated on a pretty little bay between Noel Head and Burnt-Coat Head. It has 800 inhabitants, and produces the mineral called *terra alba*, used in bleaching cottons. It is not found elsewhere in America. After leaving Noel Bay and passing the lighthouse on Burnt-Coat

[1] *Economy* is derived from the Indian name *Kenomee*, which was applied to the same place, and means " Sandy Point."

Head, the trend of the coast is followed to the S. W. for about 20 M. to *Walton*, a village of 600 inhabitants, at the mouth of the La Tête River. Many thousand tons of gypsum and plaster of Paris (calcined gypsum) are annually shipped from this port to the United States. Immense quantities are exported also from the coasts of *Chiririe*, which extend from Walton S. W. to the mouth of the Avon River. The whole back country is composed of limestone soil and gypsum-beds, whose mining and shipment form an industry of increasing importance. Beyond the Chivirie coast the steamer ascends the Avon River to Windsor.

The Basin of Minas was the favorite home of GLOOSCAP, the Hiawatha of the Micmacs, whose traditions describe him as an envoy from the Great Spirit, who had the form and habits of humanity, but was exalted above all peril and sickness and death. He dwelt apart and above, in a great wigwam, and was attended by an old woman and a beautiful youth, and "was never very far from any one of them," who received his counsels. His power was unbounded and supernatural, and was wielded against the enchantments of the magicians, while his wisdom taught the Indians how to hunt and fish, to heal diseases, and to build wigwams and canoes. He named the constellations in the heavens, and many of the chief points on the Acadian shores. The Basin of Minas was his beaver-pond; Cape Split was the bulwark of the dam; and Spencer's Island is his overturned kettle. He controlled the elements, and by his magic wand led the caribou and the bear to his throne. The allied powers of evil advanced with immense hosts to overthrow his great wigwam and break his power; but he extinguished their camp-fires by night and summoned the spirits of the frost, by whose endeavors the land was visited by an intense cold, and the hostile armies were frozen in the forest. On the approach of the English he turned his huge hunting-dogs into stone and then passed away; but will return again, right Spencer's Island, call the dogs to life, and once more dispense his royal hospitality on the Minas shores.

"Now the ways of beasts and men waxed evil, and they greatly vexed Glooscap, and at length he could no longer endure them; and he made a rich feast by the shore of the great lake (Minas). All the beasts came to it; and when the feast was over, he got into a big canoe, he and his uncle, the Great Turtle, and they went away over the big lake, and the beasts looked after them till they saw them no more. And after they ceased to see them, they still heard their voices as they sang, but the sounds grew fainter and fainter in the distance, and at last they wholly died away; and then deep silence fell on them all, and a great marvel came to pass, and the beasts who had till now spoken but one language no longer were able to understand each other, and they all fled away, each his own way, and never again have they met together in council. Until the day when Glooscap shall return to restore the Golden Age, and make men and animals dwell once more together in amity and peace, all Nature mourns. The tradition states that on his departure from Acadia the great snowy owl retired to the deep forests to return no more until he could come to welcome Glooscap; and in those sylvan depths the owls, even yet, repeat to the night, 'Koo koo skoos! Koo koo skoos!' which is to say, in the Indian tongue, 'O, I am sorry! O, I am sorry!' And the loons, who had been the huntsmen of Glooscap, go restlessly up and down through the world, seeking vainly for their master, whom they cannot find, and wailing sadly because they find him not."

THE BASIN OF MINAS

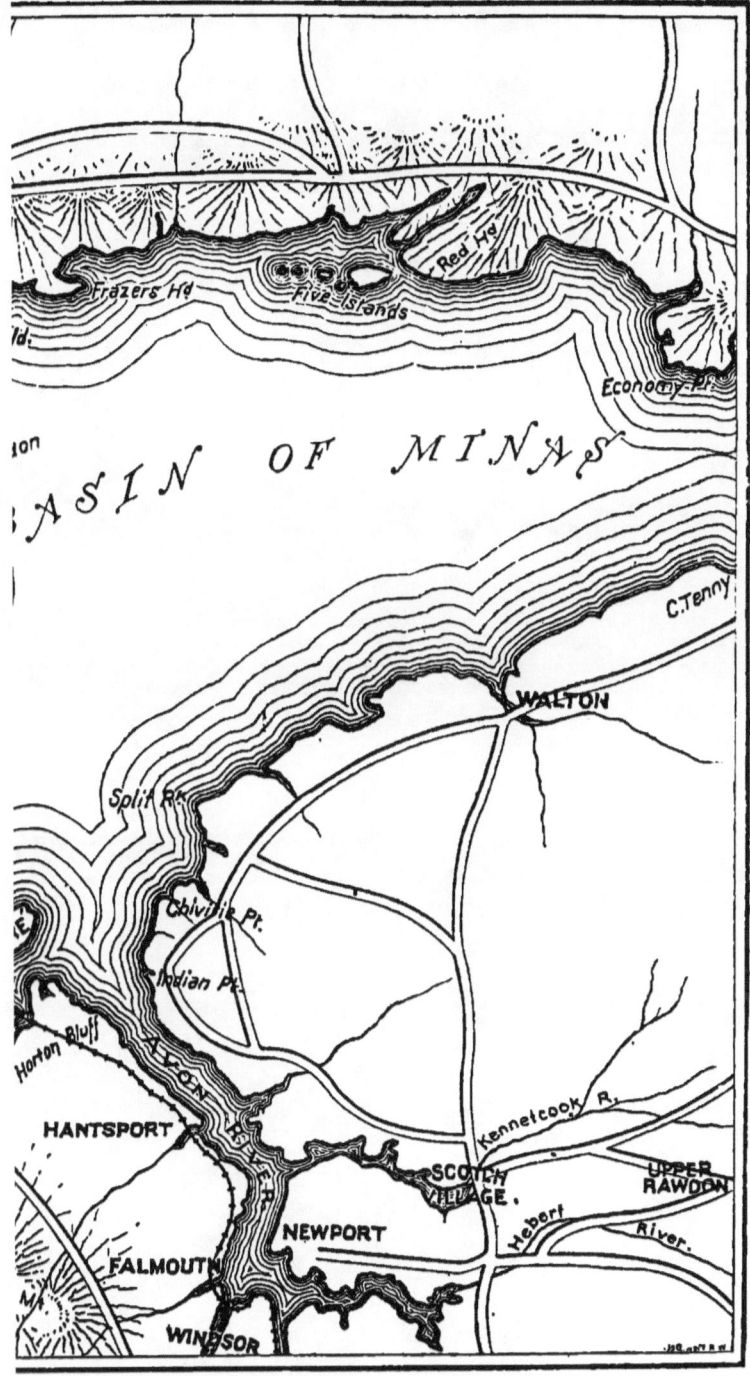

22. The Land of Evangeline.

This beautiful and deeply interesting district is visited with the greatest ease from the academic town of **Wolfville** (*Village Hotel; Acadia Hotel*), which is 127 M. from St. John and 63 M. from Halifax (by Route 18). This quiet settlement is situated on the Cornwallis River, and is engaged in shipbuilding and farming. It has 800 inhabitants, four churches, a ladies' seminary, and the Horton Academy (4 teachers, 60 students). The *Acadia College* is a Baptist institution, with 5 professors, 40 students, and 150 alumni (in 18 years of existence). The college buildings occupy a fine situation on a hill which overlooks "those meadows on the Basin of Minas which Mr. Longfellow has made more sadly poetical than any other spot on the Western Continent." The * view from the belfry of the college is the most beautiful in this vicinity, or even, perhaps, in the Maritime Provinces. Far across the Cornwallis Valley to the N. is the North Mt., which terminates, 15 M. away (21 M. by road), in the majestic bluff of Cape Blomidon, dropping into the Basin of Minas, whose bright waters occupy a broad section of the field of vision. (See Route 21, for Cape Blomidon and the Indian traditions of the Basin of Minas.) To the N. E. is the "great meadow" which gave name and site to the village of Grand Pré. .

A good road leads E. (in 3 M.) from Wolfville to *Lower Horton*, a scattered hamlet among the hills. By passing down from this point to the meadows just beyond the railway-station of **Grand Pré**, the traveller reaches the site of the ancient village. Standing on the platform of the station, he sees a large tree at the corner of the field on the left front. Near that point are the faint remains of the foundations of the Acadian church. The tradition of the country-side claims that the aged willow-tree near by grows on the site of the shop of Basil the Blacksmith, and that cinders have been dug up at its foot. The destruction effected by the British troops was complete, and there are now no relics of the ancient settlement, except the gnarled and knotty trees of the orchards, the lines of willows along the old roads, and the sunken hollows which indicate the sites of former cellars. Near the shore is shown the place where the exiles were put on shipboard. A road leads across the rich diked marsh in 2-3 M. to *Long Island*, a slight elevation fronting on the Basin of Minas, and on which dwells a farming population of about 120 persons. To the N. E. is the mouth of the Gaspereaux River, and on the W. the Cornwallis River is discharged. The early Acadians reclaimed these rich meadows from the sweep of the tides by building light dikes to turn the water. There were 2,100 acres of this gained land in their Grand Pré, but the successive advancing of other lines of aggression has driven back the sea from a much larger area, all of which is very productive and valuable. In 1810 the broad meadow between Grand Pré and Wolfville was enclosed by new dikes and added to the reclaimed domain.

open on the l., bounded by the Horton hills and the dark line of the North Mt. In advance is the bold and clear-cut outline of Cape Blomidon, brooding over the water, and on the r. are the low but well-defined bluffs of *Chivirie*, rich in gypsum and limestone. It is about 22 M. from the mouth of the Avon to Parrsboro', and the course of the steamer continually approaches Blomidon.

Cape Blomidon is a vast precipice of red sandstone of the Triassic era, with strong marks of volcanic action. "The dark basaltic wall, covered with thick woods, the terrace of amygdaloid, with a luxuriant growth of light-green shrubs and young trees that rapidly spring up on its rich and moist surface, the precipice of bright red sandstone, always clean and fresh, and contrasting strongly with the trap above, ... constitute a combination of forms and colors equally striking, if seen in the distance from the hills of Horton or Parrsboro', or more nearly from the sea or the stony beach at its base. Blomidon is a scene never to be forgotten by a traveller who has wandered around its shores or clambered on its giddy precipices." The cape is about 570 ft. high, and presents an interesting sight when its dark-red summit is peering above the white sea-fogs. Sir William Lyell, the eminent British geologist, made a careful study of the phenomena of this vicinity.

The Indian legend says that Blomidon was made by the divine Glooscap, who broke the great beaver-dam off this shore and swung its end around into its present position. Afterwards he crossed to the new-made cape and strewed its slopes with the gems that are found there to-day, carrying thence a set of rare ornaments for his ancient and mysterious female companion. The beneficent chief broke away the beaver-dam because it was flooding all the Cornwallis Valley, and in his conflict with the Great Beaver he threw at him huge masses of rock and earth, which are the present Five Islands. W. of *Utkoguncheech* (Blomidon) the end of the dam swept around and became *Pleegun* (Cape Split).

As Blomidon is left on the port beam, the steamer hurries across the rapid currents of the outlet of the Basin. In front is seen the white village of Parrsboro', backed by the dark undulations of the Cobequid Mts. Just before reaching Parrsboro' the vessel approaches and passes *Partridge Island* (on the l.), a singular insulated hill 250 ft. high, and connected with the mainland at low tide by a narrow beach.

Partridge Island was the *Pulowech Munegoo* of the Micmacs, and was a favorite location for legends of Glooscap. On his last great journey from Newfoundland by Pictou through Acadia and into the unknown West, he built a grand road from Fort Cumberland to this shore for the use of his weary companions. This miraculously formed ridge is now occupied by the post-road to the N. W., and is called by the Indians *Owrokum* (the causeway). At Partridge Island Glooscap had his celebrated revel with the supernatural Kit-poos-e-ag-unow, the deliverer of all oppressed, who was taken out alive from his mother (slain by a giant), was thrown into a well, and, being miraculously preserved there, came forth in due time to fulfil his high duty to men. These marvellous friends went out on the Basin in a stone canoe to fish by torchlight, and, after cruising over the dark waters for some time, speared a monstrous whale. They tossed him into the canoe "as though he were a trout," and made for the shore, where, in their brotherly feast, the whale was entirely devoured.

Parrsboro' (two inns) is prettily situated at the mouth of a small river, and under the shelter of Partridge Island. It has about 900 inhabitants, with three churches, and is engaged in the lumber-trade. The beauty of the situation and the views, together with the sporting facilities in the back-country, have made Parrsboro' a pleasure resort of considerable repute, and the neat hotel called the *Summer House* is well patronized. This is one of the best points from which to enter the fine hunting and fishing districts of Cumberland County, and guides and outfits may be secured here. Amherst (see page 78) is 36 M. distant, by highways following the valleys of the Parrsboro' and Maccau Rivers.

"Parrsboro' enjoys more than its share of broad, gravelly beach, overhung with clifted and woody bluffs. One fresh from the dead walls of a great city would be delighted with the sylvan shores of Parrsboro'. The beach, with all its breadth, a miracle of pebbly beauty, slants steeply to the surf, which is now rolling up in curling clouds of green and white. Here we turn westward into the great bay itself, going with a tide that rushes like a mighty river toward a cataract, whirling, boiling, breaking in half-moons of crispy foam." (L. L. NOBLE.)

"Pleasant Parrsboro', with its green hills, neat cottages, and sloping shores laved by the sea when the tide is full, but wearing quite a different aspect when the tide goes out; for then it is left perched thirty feet high upon a red clay bluff, and the fishing-boats which were afloat before are careened upon their beam ends, high and dry out of water. The long massive pier at which the steamboat lately landed, lifts up its naked bulk of tree-nailed logs, reeking with green ooze and sea-weed; and a high conical island which constitutes the chief feature of the landscape is transformed into a bold promontory, connected with the mainland by a huge ridge of brick-red clay." (HALLOCK.)

Gentlemen who are interested in geological studies will have a rare chance to make collections about Parrsboro' and the shores of Minas. The most favorable time is when the bluffs have been cracked and scaled by recent frosts; or just after the close of the winter, when much fresh *débris* is found at the foot of the cliffs. Among the minerals on Partridge Island are: analcime, apophyllite, amethyst, agate, apatite, calcite (abundant, in yellow crystals), chabazite, chalcedony, cat's-eye, gypsum, hematite, heulandite, magnetite, stilbite (very abundant), jasper, cacholong, opal, semi-opal, and gold-bearing quartz. About Cape Blomidon are found analcime, agate, amethyst, apophyllite. calcite, chalcedony, chabazite-gmelinite, faröelite, hematite, magnetite, heulandite, humonite, fibrous gypsum, malachite, mesolite, native copper, natrolite, stilbite, psilomelane, and quartz. Obsidian, malachite, gold, and copper are found at Cape d'Or; jasper and fine quartz crystals, on Spencer's Island; augite, amianthus, pyrites, and wad, at Parrsboro'; and both at Five Islands and Scotsman's Bay there are beautiful specimens of moss agate. At Cornwallis is found the rare mineral called Wichtisite (resembling obsidian, in gray and deep blue colors), which is only known in one other place on earth, at Wichtis, in Finland. The purple and violet quartz, or amethyst, of the Minas shores, is of great beauty and value. A Blomidon amethyst is in the crown of France, and it is now 270 years since the Sieur de Monts carried several large amethysts from Partridge Island to Henri IV. of France. These gems are generally found in geodes, or after fresh falls of trap-rock.

Advocate Harbor and Cape d'Or.

A bi-weekly stage runs W. from Parrsboro' through grand coast scenery, for 28 M., passing the hamlets of Fox Harbor and Port Greville, and stopping at **Advocate Harbor**. This is a sequestered marine hamlet, devoted to shipbuilding and the deep-sea fisheries, and has about 600 inhabitants. It is about 60 M. from Amherst, by a road leading across the Cobequid Mts. and through Apple River (see page 80). Some of the finest marine scenery in the Provinces is in this vicinity. 3-4 M. S. is the immense rocky peninsula of * **Cape d'Or**, almost cut off from the mainland by a deep ravine, in whose bottom the salt tides flow. Cape d'Or is 500 ft. high, and has recently become noted for its rich copper deposits. Off this point there is a heavy rip on the flood-tide, which flows with a velocity of 6 knots an hour, and rises 33-39 ft. 8 M. W. of Advocate Harbor, and visible across

> Solemnly down the street came the parish priest, and the children
> Paused in their play to kiss the hand he extended to bless them.
> Reverend walked he among them: and up rose matrons and maidens,
> Hailing his slow approach with words of affectionate welcome.
> Then came the laborers home from the field. and serenely the sun sank
> Down to his rest, and twilight prevailed. Anon from the belfry
> Softly the Angelus sounded, and over the roofs of the village
> Columns of pale blue smoke, like clouds of incense ascending.
> Rose from a hundred hearths, the homes of peace and contentment.
> Thus dwelt together in love these simple Acadian farmers, —
> Dwelt in the love of God and of man. Alike were they free from
> Fear, that reigns with the tyrant, and envy, the vice of republics.
> Neither locks had they to their doors, nor bars to their windows;
> But their dwellings were open as day and the hearts of the owners;
> There the richest was poor, and the poorest lived in abundance."

The poet then describes "the gentle Evangeline, the pride of the village."

> "Fair was she to behold, that maiden of seventeen summers,
> Black were her eyes as the berry that grows on the thorn by the wayside,
> Black, yet how softly they gleamed beneath the brown shade of her tresses!
> Sweet was her breath as the breath of kine that feed in the meadows.
> When in the harvest heat she bore to the reapers at noontide
> Flagons of home-brewed ale, ah! fair in sooth was the maiden.
> Fairer was she when, on Sunday morn, while the bell from its turret
> Sprinkled with holy sounds the air, as the priest with his hyssop
> Sprinkles the congregation, and scatters blessings upon them.
> Down the long street she passed, with her chaplet of beads and her missal,
> Wearing her Norman cap, and her kirtle of blue, and the ear-rings, —
> Brought in the olden time from France, and since, as an heirloom,
> Handed down from mother to child, through long generations.
> But a celestial brightness — a more ethereal beauty —
> Shone on her face and encircled her form. when, after confession,
> Homeward serenely she walked, with God's benediction upon her.
> When she had passed, it seemed like the ceasing of exquisite music."

After a beautiful description of the peaceful social life of the Acadians, and the betrothal of Evangeline, the poet tells of the arrival of the English fleet, the convocation of the people, the royal mandate, the destruction of Grand Pré, and the weary exile of the villagers.

> "So passed the morning away. And lo! with a summons sonorous
> Sounded the bell from its tower, and over the meadow a drum beat.
> Thronged erelong was the church with men. Without, in the churchyard,
> Waited the women. They stood by the graves, and hung on the headstones
> Garlands of autumn-leaves and evergreens fresh from the forest.
> Then came the guard from the ships, and marching proudly among them
> Entered the sacred portal. With loud and dissonant clangor
> Echoed the sound of their brazen drums from ceiling and casement, —
> Echoed a moment only, and slowly the ponderous portal
> Closed, and in silence the crowd awaited the will of the soldiers.
> Then uprose their commander, and spake from the steps of the altar,
> Holding aloft in his hands, with its seals, the royal commission.
> 'Ye are convened this day,' he said, 'by his Majesty's orders.
> Clement and kind has he been: but how have you answered his kindness,
> Let your own hearts reply! To my natural make and my temper
> Painful the task is I do, which to you I know must be grievous.
> Yet must I bow and obey, and deliver the will of our monarch:
> Namely, that all your lands, and dwellings, and cattle of all kinds
> Forfeited be to the crown; and that you yourselves from this province
> Be transported to other lands. God grant you may dwell there
> Ever as faithful subjects, a happy and peaceable people!
> Prisoners now I declare you; for such is his Majesty's pleasure.'

There disorder prevailed, and the tumult and stir of embarking.
Busily plied the freighted boats ; and in the confusion
Wives were torn from their husbands, and mothers, too late, saw their children
Left on the land, extending their arms, with wildest entreaties.

Suddenly rose from the south a light, as in autumn the blood-red
Moon climbs the crystal walls of heaven, and o'er the horizon
Titan-like stretches its hundred hands upon mountain and meadow,
Seizing the rocks and the rivers, and piling huge shadows together.
Broader and ever broader it gleamed on the roofs of the village,
Gleamed on the sky and the sea, and the ships that lay in the roadstead.
Columns of shining smoke uprose, and flashes of flame were
Thrust through their folds and withdrawn, like the quivering hands of a martyr.
Then as the wind seized the gleeds and the burning thatch, and uplifting,
Whirled them aloft through the air, at once from a hundred house-tops
Started the sheeted smoke, with flashes of flame intermingled.

Many a weary year had passed since the burning of Grand Pré,
When on the falling tide the freighted vessels departed,
Bearing a nation, with all its household gods, into exile,
Exile without an end, and without an example in story.
Far asunder, on separate coasts, the Acadians landed ;
Scattered were they, like flakes of snow, when the wind from the northeast
Strikes aslant through the fogs that darken the Banks of Newfoundland.
Friendless, homeless, hopeless, they wandered from city to city,
From the cold lakes of the North to sultry Southern savannas,—
From the bleak shores of the sea to the lands where the Father of Waters
Seizes the hills in his hands, and drags them down to the ocean,
Deep in their sands to bury the scattered bones of the mammoth.
Friends they sought and homes ; and many, despairing, heart-broken,
Asked of the earth but a grave, and no longer a friend or a fireside.
Written their history stands on tablets of stone in the churchyards."
           LONGFELLOW'S *Evangeline*.

"Much as we may admire the various bays and lakes, the inlets, promontories, and straits, the mountains and woodlands of this rarely visited corner of creation, — and, compared with it, we can boast of no coast scenery so beautiful,—'the valley of Grand Pré transcends all the rest in the Province. Only our valley of Wyoming, as an inland picture, may match it, both in beauty and tradition. One had its Gertrude, the other its Evangeline." (COZZENS.)

"Beyond is a lofty and extended chain of hills, presenting a vast chasm, apparently burst out by the waters of 19 rivers that empty into the Basin of Minas, and here escape into the Bay of Fundy. The variety and extent of this prospect, the beautiful verdant vale of the Gaspereaux ; the extended township of Horton interspersed with groves of wood and cultured fields, and the cloud-capped summit of the lofty cape that terminates the chain of the North Mt., form an assemblage of objects rarely united with so striking an effect."

"It would be difficult to point out another landscape at all equal to that which is beheld from the hill that overlooks the site of the ancient village of Minas. On either hand extend undulating hills richly cultivated, and intermingled with farmhouses and orchards. From the base of these highlands extend the alluvial meadows which add so much to the appearance and wealth of Horton. The Grand Prairie is skirted by Boot and Long Islands, whose fertile and well-tilled fields are sheltered from the north by evergreen forests of dark foliage. Beyond are the wide expanse of waters of the Basin of Minas, the lower part of Cornwallis, and the isles and blue highlands of the opposite shores. The charm of this prospect consists in the unusual combination of hill, dale, woods, and cultivated fields ; in the calm beauty of agricultural scenery ; and in the romantic wildness of the distant forests. During the summer and autumnal months immense herds of cattle are seen quietly cropping the herbage of the Grand Prairie ; while numerous vessels plying on the Basin convey a pleasing evidence of the prosperity and resources of this fertile district." (HALIBURTON.)

23. Annapolis Royal to Clare and Yarmouth. — The Tusket Lakes.

From St. John or Halifax to Annapolis Royal, see Route 18.

The *Western-Counties Railway* was begun in September, 1874, and is to be finished from Yarmouth to Meteghan (30 M.) by the summer of 1875. It will not reach Annapolis before the latter part of the year 1876.

The Royal mail-stage leaves Annapolis daily on arrival of the morning train from Halifax, and runs S. W. to Clementsport and Digby (distance, 20½ M.; fare, $1.50). A pleasanter route is to go from Annapolis to Digby by the steamboat (75c.; see page 85), which makes four trips weekly. On boat-days the stage leaves Digby for Yarmouth about one hour after her arrival; on other days it leaves at 6 P. M. Digby to Yarmouth, 70 M.; fare, $4.

Itinerary. — Annapolis Royal; Clementsport, 8¼ M.; Victoria Bridge, 13½; Smith's Cove, 16; Digby, 20½; St. Mary's Bay, 27½; Weymouth Road, 32; Weymouth Bridge, 38; Belliveau Cove, 43; Clare, 50; Meteghan Cove, 59; Cheticamp, 63; Bear River, 74; Yarmouth Lakes, 81; Yarmouth, 90.

The traveller will see from the time-table that this is a night-journey, and the return from Yarmouth to Digby is also effected by night. The ensuing descriptions, therefore, will be useful only to such as stop off at some of the roadside villages, or make the journey in their own carriages, by daylight.

Annapolis Royal to Digby, see pages 84, 85 (reversed).

On leaving Digby the stage follows the highway to the S. W., traversing the farming settlement of *Marshalltown*, and crosses the isthmus between the Annapolis Basin and St. Mary's Bay, a distance of about 7 M. Thenceforward, for over 30 M., the highway lies near the beautiful *St. Mary's Bay, which is about 35 M. long, with a width of from 3 to 10 M. On the opposite shore are the highlands of Digby Neck (see Route 24), a continuation of the North Mt. range. On this shore a wide belt of level land has been left between the receding range of the South Mt. (or Blue Mts.) and the bay, and the water-front is occupied by numerous farms.

In St. Mary's Bay the fleet of the Sieur de Monts lay for two weeks, in 1604, while the shores were being explored by boat's-crews. The mariners were greatly rejoiced in finding what they supposed to be valuable deposits of iron and silver. The Parisian priest Aubry was lost on one of these excursions, and roamed through the woods for 16 days, eating nothing but berries, until another vessel took him off. The name *Baie de Ste. Marie* was given by Champlain.

Brighton is at the head of the bay, and is a pleasant agricultural village with a small inn. The hamlets of *Barton* (or Specht's Cove) and *Gilbert's Cove* are soon passed, and the stage enters the pretty village of *Weymouth* (two inns), a seaport which builds some handsome vessels, and has a snug little trade with the United States and the West Indies. It is at the mouth of the Sissiboo River, on whose opposite shore is the Acadian hamlet of *New Edinburgh*. Across St. Mary's Bay is the maritime village of *Sandy Cove*.

The stage now ascends the r. bank of the Sissiboo River to *Weymouth Bridge* (Jones's Hotel), a maritime village of about the same size as Weymouth. It is 4 M. from the mouth of the river; and 2 – 3 M. to the E. are the *Sissiboo Falls*. The shore of St. Mary's Bay is regained at *Belliveau Cove* (small inn), an Acadian hamlet chiefly devoted to agriculture and shipbuilding. From this point down to Beaver River, and beyond through the Tusket and Pubnico regions, the shore is occupied by a range of hamlets which are inhabited by the descendants of the old Acadian-French.

The **Clare Settlements** were founded about 1763 by the descendants of the Acadians who had been exiled to New England. After the conquest of Canada these unfortunate wanderers were suffered to return to Nova Scotia, but they found their former domains about the Basin of Minas already occupied by the New-Englanders. So they removed to the less fertile but still pleasant shores of Clare, and founded new homes, alternating their farm labors with fishing-voyages on St. Mary's Bay or the outer sea. This little commonwealth of 4-5,000 people was for many years governed and directed by "the amiable and venerated Abbé Segoigne," a patrician priest who had fled from France during the Revolution of 1793. His power and influence were unlimited, and were exerted only for the peace and well-being of his people. Under this benign guidance the colony flourished amain; new hamlets arose along the shores of the beautiful bay; and an Acadian village was founded in the oak-groves of Tusket. M. Segoigne also conciliated the Micmacs, learned their language, and was highly venerated by all their tribe.

" When the traveller enters Clare, the houses, the household utensils, the foreign language, and the uniform costume of the inhabitants excite his surprise; because no parish of Nova Scotia has such a distinctive character. The Acadians are far behind their neighbors in modes of agriculture : they show a great reluctance to enter the forest, and in place of advancing upon the highlands, they subdivide their lands along the shore and keep their children about them. They preserve their language and customs with a singular tenacity, and though commerce places them in constant communication with the English, they never contract marriage with them, nor adopt their manners, nor dwell in their villages. This conduct is not due to dislike of the English government; it must be attributed rather to ancient usage, to the national character, and to their systems of education. But if they are inferior to the English colonists in the arts which strengthen and extend the influence of society, they can proudly challenge comparison in their social and domestic virtues. Without ambition, living with frugality, they regulate their life according to their means; devoted to their ancient worship, they are not divided by religious discord; in fine, contented with their lot and moral in their habits of life, they enjoy perhaps as much of happiness and goodness as is possible in the frailty of human nature." (HALIBURTON.)

" Still stands the forest primeval; but under the shade of its branches
Dwells another race, with other customs and language.
Only along the shore of the mournful and misty Atlantic
Linger a few Acadian peasants, whose fathers from exile
Wandered back to their native laud to die in its bosom.
In the fisherman's cot the wheel and the loom are still busy;
Maidens still wear their Norman caps and their kirtles of homespun,
And by the evening fire repeat Evangeline's story,
While from its rocky caverns the deep-voiced neighboring ocean
Speaks, and in accents disconsolate answers the wail of the forest."
LONGFELLOW's *Evangeline.*

The road runs S. W. from Belliveau Cove to Grosses Coques (300 inhabitants) and Port Acadie, Clare, and Saulnierville, a line of hamlets whose inhabitants are engaged in farming and the fisheries. A road runs 7 M. E. to *New Tusket,* an Anglo-Acadian village in the interior, near the island-studded Lake Wentworth. *Meteghan* (German's Hotel) is a bayside village of 500 inhabitants, nearly all of whom are Acadians and farmers. It is thought that the Western-Counties Railway will be completed from Yarmouth to this point by the summer of 1875. Meteghan is the last village on St. Mary's Bay, and the road now turns to the S. and passes

H

the inland hamlet of Cheticamp. *Cape Cove* is an Acadian settlement, and is finely situated on a headland which faces the Atlantic. The stage next passes *Salmon River* (small inn), and descends thence (by Brookville) to *Beaver River* (inn), the first English settlement. It is a village of 400 inhabitants on the Atlantic coast, near the promontory of High Head. The road now leaves the vicinity of the sea and strikes inland through a region of forests and lakes ; reaching Yarmouth about 13 M. S. of Beaver River.

Yarmouth (*United States Hotel*, $6-8 a week; *American Hotel*) is a wealthy and prosperous seaport on the S. W. coast of Nova Scotia, and is situated on a narrow harbor 3 M. from the Atlantic. It has 5,335 inhabitants, with 9 churches, 2 banks, 4 local marine-insurance companies, and 2 weekly newspapers. It has a public library and a small museum of natural history. The schools are said to be the best in the Province, and occupy conspicuous buildings on the ridge back of the town. The Court-House is in the upper part of the town; near which is the spacious Baptist church, built in Novanglian architecture. The Episcopal church is a new building, and is one of the best in Nova Scotia. 1 M. out is a rural cemetery of 40 acres. Yarmouth is built along a line of low rocky heights, over a harbor which is nearly drained at low tide. It receives a goodly number of summer visitors, most of whom pass into the Tusket Lakes or along the coast to the E., in search of sport.

Yarmouth has been called the most American of all the Provincial towns, and is endowed with the energy and pertinacity of New England. Though occupying a remote situation on an indifferent harbor, with a barren and incapable back country, this town has risen to opulence and distinction by the indomitable industry of its citizens. In 1761 the shipping of the country was confined to one 25-ton fishing-boat; in 1869 it amounted to 284 vessels, measuring 93,896 tons, and is now far in advance even of that figure. It is claimed that Yarmouth, for her population, is the largest ship-owning port in the world. In addition to these great commercial fleets, the town has established a steamship-line to St. John and Boston, and is building, almost alone, the Western-Counties Railway to Annapolis. It is expected that great benefit will accrue from the timber-districts which will be opened by this new line of travel. "Yarmouth's financial success is due largely to the practical judgment and sagacity of her mariners. She has reared an army of shipmasters of whom any country might be proud," and it is claimed that a large proportion of the Cape-Ann fishing-captains are natives of this country. On the adjacent coast, and within 12 M. of Yarmouth, are the marine hamlets of Jegoggin, Sandford (Cranberry Head), Arcadia, Hebron, Hartford, Kelley's Cove, Jebogue, Darling's Lake (Short Beach), and Deerfield. These settlements have over 6,000 inhabitants in the aggregate. The coast was occupied by the French during the 17th century, but was afterwards abandoned. About the middle of the last century these deserted shores were taken possession of by colonies of fishermen from Massachusetts and Connecticut, who wished to be nearer their fishing-grounds ; and the present population is descended from these hardy men and the Loyalists of 1783. The ancient Indian name of Yarmouth was *Keespoogwitk*, which means "Land's End."

The steamship *Linda* leaves Yarmouth for Boston every Saturday, and for St. John, N. B., every Thursday.

TUSKET LAKES. *Route 23.* **115**

The Tusket Lakes and Archipelago.

The township of Yarmouth contains 80 lakes, and to a bird flying overhead it must seem like a patchwork of blue and green, in which the blue predominates. They are nearly all connected with the Tusket River, and are generally small, very irregular, and surrounded by young forests. They rarely attain the width of 1 M., and are strung along the course of the river and its tributaries, joined by narrow aisles of water, and breaking off into bays which the unguided voyager would often ascend in mistake for the main channel. In the lower lakes, where the tide flows, near Argyle Bay, are profitable eel-fisheries. The remoter waters, towards the Blue Mts., afford good trout-fishing.

The westerly line of lakes are visited from Yarmouth by riding 5 M. out on the Digby road and then turning off to *Deerfield*, near the Salmon-River Lakes, or passing over to the settlement at *Lake George* (12-14 M. from Yarmouth), which is 1½ M. wide and 3-4 M. long, and is the largest lake in the township. A little farther N. is the Acadian settlement at *Cedar Lake.*

The best route for the sportsman is to follow the Barrington telegraph-road 10 M. N. E. to **Tusket** (two inns), a large and prosperous shipbuilding village, with three churches, near the head of ship-navigation on the Tusket River. The scenery in this vicinity is picturesque, its chief feature being the many green islands off the shores; and the river has been famous for fisheries of salmon and gaspereaux, now impaired by the lumber-mills above. From this point a chain of lakes ascends to the N. for 20 M., including the central group of the Tuskets, and terminating at the island-strewn Lake Wentworth. The best place is found by following the road which runs N. E. 15-18 M., between Vaughan Lake and Butler's Lake, and by many lesser ponds, to the remote settlement of *Kempt* (small hotel), near the head-waters of the central and western groups. To the N. and E. of this point are the trackless forests and savage ridges of the Blue Mts., and the hunter can traverse these wilds for 40 M. to the N. E. (to the Liverpool Lakes), or for 30 M. to the S. E. (to the Shelburne settlements), without meeting any permanent evidences of civilization.

The ancient Indian tradition tells that squirrels were once very numerous in this region, and grew to an enormous size, endangering the lives of men. But the Great Spirit once appeared to a blameless patriarch of the Micmacs, and offered to reward his virtue by granting his utmost desire. After long meditation the chief asked the Divine Visitor to bless the land by taking the power from the mighty squirrels, upon which the mandate was issued and the dreaded animals shrank to their present insignificant size. And hence it is known that ever since that day the squirrel has been querulous at the sight of man.

This great forest was formerly the paradise of moose-hunters, but is now closed to that sport by the recent Provincial law which forbids the killing of moose for the next three years. Poaching is, of course, quite possible, since the forest cannot be studded with game-keepers; but men of culture and foresight will doubtless approve the action of the government, and will abstain from illegally pursuing this noble game, which must become extinct in a very few years unless carefully protected.

S. of Tusket village are the beautiful groups of the **Tusket Isles**, studding the waters of Argyle Bay and the Abuptic Harbor. Like most other collections of islands on this continent, they are popularly supposed to be

365 in number, though they do not claim to possess an intercalary islet like that on Lake George (New York), which appears only every fourth year. The Tuskets vary in size from Morris Island, which is 3 M. long, down to the smallest tuft-crowned rocks, and afford a great diversity of scenery. The outer fringe of the archipelago is threaded by the Halifax and Yarmouth steamship (see page 125).

"The scenery of Argyle Bay is extremely beautiful of its kind; innumerable islands and peninsulas enclose the water in every direction..... Cottages and cultivated land break the masses of forest, and the masts of small fishing-vessels peeping up from every little cove attest the multiplied resources which Nature has provided for the supply of the inhabitants." (CAPT. MOORSON.)

Among these narrow passes hundreds of Acadians took refuge during the persecutions of 1758-60. A British frigate was sent down to hunt them out, but one of her boats' crews was destroyed by the fugitives among the islands, and they were not dislodged. There are now two or three hamlets of Acadians in the region of the upper lakes.

[The Editor deprecates the meagreness of the foregoing account of the Tusket Lakes. It was too late in the season, when he arrived at Yarmouth, to make the tour of this district, and the landlord of the United States Hotel, the best authority on the sporting facilities of the lake-country, was then attending a party of Boston sportsmen among the Blue Mts. The foregoing statements about the district, though obtained from the best accessible sources of information, are therefore given under reserve; and it would be best for gentlemen who wish to summer among the Tuskets to make inquiries by letter of the proprietor of the United States Hotel, Yarmouth, N. S.]

24. Digby Neck.

Tri-weekly stages leave Digby for this remote corner of Nova Scotia. Fare to Sandy Cove, $1.50; to West Port, $2.

Distances. — Digby to Rossway, 8½ M.; Waterford, 12; Centreville, 15; Lakeside, 17; Sandy Cove, 20; Little River, 25; Petite Passage, 30; Free Port; West Port, 40.

The stage runs S. W. from Digby, leaving the settlements of Marshalltown and Brighton on the l., across the Smelt River. The first hamlet reached is *Rossway*, whence a road crosses to Gulliver's Cove on the Bay of Fundy. For over 20 M. the road descends the remarkable peninsula of Digby Neck, whose average width, from bay to bay, is about 1½ M. On the l. is the continuous range of dark hills which marks the W. end of the North Mt. range, where it is sinking towards the sea. Among these hills are found fine specimens of agate and jasper, and the views from their summits (when not hidden by trees) reveal broad and brilliant stretches of blue water on either side. Fogs are, however, very prevalent here, and are locally supposed to be rather healthy than otherwise. On the l. of the road are the broad waters of St. Mary's Bay, far beyond which are the low and rugged Blue Mts.

Sandy Cove (small inn) is the metropolis of Digby Neck, and has 400 inhabitants and two churches. Its people live by farming and fishing, and support a fortnightly packet-boat to St. John, N. B. 4 M. S. E., across St. Mary's Bay, is the port of Weymouth (see page 112). Beyond Little River village the stage crosses the ridge, and the passenger passes the *Petite Passage*, which separates Digby Neck from Long Island. This strait is quite deep and 1 M. wide, and has a red-and-white flashing light on its N. W. point (Boar's Head). On the opposite shore of the passage is a village of 390 inhabitants (mostly fishermen), and the stage now runs down Long Island on the Bay of Fundy side. If there is no fog the view across the bay is pleasing, and is usually enlivened by the sails of passing vessels. Long Island is about 10 M. long, and 2 M. wide, and its village of Free Port has 700 inhabitants.

Near the end of Long Island another ferry-boat is taken, and the traveller crosses the Grand Passage to **West Port** (*Denton's Hotel*), a village of 600 inhabitants, most of whom are fishermen, shipbuilders, or sea-captains. This town is on Brier Island, the S. E. portal of the Bay of Fundy, and is 5 M. long by 2 M. wide. On its E. side are two fixed white lights, and on the W. are a fog-whistle and a powerful white light visible for 15 M.

25. Halifax to Yarmouth.—The Atlantic Coast of Nova Scotia.

The steamers *M. A. Starr* and *Edgar Stuart*, of Fishwick's Express Line, ply along the coast of Nova Scotia. One of them leaves Halifax for Yarmouth on Tuesday, at 6 A. M.; leaving Yarmouth on Thursday, at 9 A. M. (There is also a possibility that a vessel of this line will ply during the present summer between Halifax, Cape Canso, Guysborough, Port Hastings, Port Mulgrave, and Antigonish.)

Fares.—Halifax to Lunenburg, $2; to Liverpool, $3.50; to Shelburne, $4.50; to Yarmouth, $6. Lunenburg to Liverpool, $3; to Shelburne, $3.50; to Yarmouth, $4.50. Liverpool to Shelburne, $2; to Yarmouth, $3.50. Shelburne to Yarmouth, $2.50. Berths are included in these prices, but the meals are extra.

"The Atlantic coast of Nova Scotia, from Cape Canso to Cape Sable, is pierced with innumerable small bays, harbors, and rivers. The shores are lined with rocks and thousands of islands; and although no part of the country can properly be considered mountainous, and there are but few steep high cliffs, yet the aspect of the whole, if not romantically sublime, is exceedingly picturesque; and the scenery, in many places, is richly beautiful.. The landscape which the head of Mahone Bay, in particular, presents can scarcely be surpassed." (M'GREGOR's *British America*.)

"The jagged outline of this coast, as seen upon the map, reminds us of the equally indented Atlantic shores of Scandinavia; and the character of the coast as he sails along it—the rocky surface, the scanty herbage, and the endless pine forests—recall to the traveller the appearance and natural productions of the same European country." (PROF. JOHNSTON.)

The steamer passes down Halifax Harbor (see page 93), and gains the open sea beyond Chebucto Head and the lighthouse on *Sambro Island*. She usually makes a good offing before turning down the coast, in order to avoid the far-reaching and dangerous Sambro Ledges. W. of the open light of Pennant Bay is Mars Head, on whose fatal rocks the ocean steamship *Atlantic* was wrecked.

This line of coast has been famous for its marine disasters. In 1779 the British war-vessels *North* and *Helena* were wrecked near Sambro, and 170 men were drowned. Mars Head derives its name from the fact that the British line-of-battle ship *Mars*, 70 guns, was wrecked upon its black ledges. In 1779 the American war-vessel *Viper*, 22, attacked H. M. S. *Resolution*, just off Sambro, and captured her after a long and desperate battle, in which both ships were badly cut to pieces. **Cape Sambro** was named by the mariners of St. Malo early in the 17th century; and it is thought that the present form of the name is a corruption of *St. Cendre*, the original designation. The ancient Latin book called the *Novus Orbis* (published by Elzevir; Amsterdam, 1633) says that the islands between Cape Sambro (*Sesambre*) and Mahone Bay were called the Martyrs' Isles, on account of the Frenchmen who had there been massacred by the heathen Indians.

Beyond Cape Prospect the deep indentations of St. Margaret's Bay and Mahone Bay make in on the N., and

"breezy Aspotogon
Lifts high its summit blue."

The roughest water of the voyage is usually found while crossing the openings of these bays. The course is laid for *Cross Island*, where there are two lights, one of which is visible for 14 M. Passing close in by this island, the steamer enters that pretty bay which was formerly known to the Indians as *Malagash*, or "Milky," on account of the whiteness of its stormy surf. At the head of this bay the white and compact town of Lunenburg is seen between two round green hills. The steamer passes around the outermost of these, and enters the snug little harbor.

"The town of Lunenburg is situated at the innermost extremity of a peninsula, and to a military traveller presents a more formidable aspect than any other in Nova Scotia, the upper houses being placed on the crests of steep glacis slopes, so as to bear upon all approaches." (CAPT. MOORSON.)

Lunenburg (*King's Hotel*) is a thriving little seaport, situated on a secure and spacious harbor, and enjoying a lucrative West-India trade. Together with its immediate environs, it has 3,231 inhabitants, of whom over half are in the port itself. The German character of the citizens is still retained, though not so completely as in their rural settlements; and the principal churches are Lutheran. The public buildings of Lunenburg County are located here. A large trade in lumber and fish is carried on, in addition to the southern exports. There are numerous farming communities of Germanic origin in the vicinity; and the shore-roads exhibit attractive phases of marine scenery. 7 M. distant is the beautifully situated village of *Mahone Bay* (see Route 26); 4 M. distant are the remarkable seaside ledges called the *Blue Rocks;* to the S. E. is the rural settlement of Lunenburg Peninsula, off which are the sea-girt farms of Heckman's Island; and 12 M. distant is the gold district of *The Ovens*.

This site was anciently occupied by the Indian village of *Malagash*. In 1745 the British government issued a proclamation inviting German Protestants to emigrate to Nova Scotia and take up its unoccupied lands. In 1753, 200 families of Germans and Swiss settled at Lunenburg, and were provided with farming implements and three years' provisions by the government. They fortified their new domains as well as possible, but many of the people were killed by Indians lurking in the woods. The settlement was thus held in check until after the Conquest of Canada, when the Indians ceased hostilities. In 1777 the town was attacked by two American privateers, who landed detachments of armed men and occupied the principal buildings. After plundering the place and securing a valuable booty, these unwelcome visitors sailed away rejoicing, leaving Lunenburg to put on the robes of war and anxiously yearn for another naval attack, for whose reception spirited provisions were made.

Among the people throughout this county German customs are still preserved, as at weddings and funerals; the German language is spoken; and sermons are delivered oftentimes in the same tongue. The cows are made to do service in ploughing, and the farming implements are of a primitive pattern. A large portion of the outdoor work in the fields is done by the women, who are generally strong and muscular.

The steamer leaves Lunenburg Harbor, passes Battery Point and its lighthouse on the l., and descends between the knob-like hills of the outer harbor. On the r. are the shores of the remarkable peninsula of **The Ovens** (distant from Lunenburg, by road, 10 – 12 M.). The low cliffs along this shore are pierced by numerous caverns, three of which are 70 ft. wide at their mouths and over 200 ft. deep. The sea dashes into these dark recesses during a heavy swell with an amazing roar, broken by deep booming reverberations. Certain features in the formation of these caves have led to the supposition that they were made by human labor, though the theorists do not state the probable object for which they were excavated. In 1861 gold was discovered on the Ovens peninsula, and 2,000 ounces were obtained during that autumn, since which the mining fever has subsided, and no earnest work has been done here. The precious metal was obtained chiefly by washing, and but little was effected in the way of quartz-crushing.

Beyond Ovens Head the pretty circular indentation of *Rose Bay* is seen on the r., on whose shores is a settlement of 250 German farmers. The steamer now passes between Cross Island (l.) and Rose Head, which are about 2 M. apart, and enters the Atlantic. When a sufficient offing has been made, the course is laid S. W. $\frac{1}{2}$ W. for $8\frac{3}{4}$ M. Point Enragé is soon passed, and then the vessel approaches * **Ironbound Island**. This remarkable rock is about $\frac{1}{2}$ M. long, and rises from the sea on all sides in smooth curves of dark and iron-like rock, on which the mighty surges of the Atlantic are broken into great sheets of white and hissing foam. Upon this dangerous outpost of Nova Scotia there is a revolving light, which is visible for 13 M. Beyond Ironbound, on the r., is seen the deep estuary of the Lahave River, which is navigable to Bridgewater, a distance of 13 M., passing for 12 M. through the hamlets of New Dublin, and thence through a valley between high and knob-like hills.

At Fort La Héve in 1636–7, died Isaac de Razilly, " Knight Commander of the Order of St. John of Jerusalem, Lieutenant-General of Acadie, and Captain of the West." He was a relative of Cardinal Richelieu, and had fought in the campaigns of La Rochelle and the coast of Morocco. In 1642 D'Aulnay purchased these domains from Claude de Razilly, but soon evacuated the place, removing the people to Port Royal. By 1654 the colony had recovered itself, having "undoubtedly the best port and the best soil in the whole country." It was then attacked by the Sieur le Borgne, who burned all its houses and the chapel. At a later day the new Fort La Héve was attacked by a strong force of New-England troops, who were beaten off several times with the loss of some of their best men. But the brave Frenchmen were finally forced to surrender, and the place was reduced to ruins. In 1705 the settlement was again destroyed by Boston privateers.

When off Cape Lahave the steamer takes a course W. by S., which is followed for 15⅜ M. The fishing hamlet of Broad Cove is on the shore S. W. of Cape Lahave; and when about 9 M. from the cape, the entrance of Port Medway is seen. This harbor is 4 M. long and 1½ M. wide, and receives the waters of the Port Medway and Pedley Rivers. Port Medway (Dunphy's Hotel) is on its W. shore, and has 600 inhabitants, who are engaged in shipbuilding and lumbering.

The steamer soon rounds the revolving red light (visible 16 M.) on Coffin's Island, and turns to the N. W. up Liverpool Bay. The shores are well inhabited, with the settlement of *Moose Harbor* on the l., and *Brooklyn* (or Herring Cove) on the r. The lighthouse on Fort Point is rounded and the vessel enters the mouth of the Liverpool River, with a line of wharves on the l., and the bridge in advance.

Liverpool (*Village Green Hotel*, a comfortable summer-house; and two other inns) is a flourishing seaport with 3,102 inhabitants, 5 churches, a weekly paper, and a bank. Its principal industries are lumbering, fishing, and shipbuilding. The town occupies the rocky shore at the mouth of the Liverpool River, and its streets are adorned with numerous large shade trees. Many summer visitors come to this place, either on account of its own attractions, or to seek the trout on the adjacent streams and lakes (see Route 27). There are pleasant drives also on the Mill-Village Road, and around the shores of the bay.

Liverpool occupies the site of the ancient Indian domain of *Ogumkegeok*, made classic in the traditions of the Micmacs by the celebrated encounter which took place here between the divine Glooscap (see page 106) and the great sorceress of the Atlantic coast. The struggle of craft and malevolence against superior power are quaintly narrated, though taking forms not pleasing to refined minds, and the contest ends in the defeat of the hag of *Ogumkegeok*, who is rent in pieces by the hunting-dogs of Glooscap.

In May, 1604, the harbor of Liverpool was entered by Pierre du Guast, "Sieur de Monts of Saintouge, Gentleman in Ordinary of the Chamber, and Governor of Pons," who had secured a monopoly of the fur-trade between 40° and 54° N. latitude. He found a ship here trading without authority, and confiscated her, naming the harbor Port Rossignol, after her captain, "as though M. de Monts had wished to make some compensation to the man for the loss he inflicted on him, by immortalizing his name." This designation did not hold to the harbor, but has been transferred to the large and beautiful lake near the head-waters of the Liverpool River. About 1684 a shore-fishery was established here by M. Denys and Gov. Razilly. This enterprise was for a long time successful, but was finally crippled by the capture of its heavily laden freighting-ship by the Portuguese. Soon afterward Denys was forced to leave Port Rossignol on account of the machinations of D'Aulnay Charnisay, and the settlement was broken up. By the year 1760 a thriving village stood on this site, and in the War of 1812 many active privateers were fitted out here. In 1882 the port owned 25,000 tons of shipping.

On leaving Liverpool Bay the steamer rounds Western Head and runs S. W. ¼ S. 14 M. On the r. is the deep embayment of Port Mouton, partly sheltered by Mouton Island, and lighted by a fixed red light on Spectacle Island. At its head is the farming and fishing settlement of **Port Mouton**, with 350 inhabitants. This inlet was visited by the exploring ship of the Sieur de Monts in 1604, and received the name which it still bears because a sheep here leaped from the deck into the bay and was drowned. The shores were settled in 1783 by the disbanded veterans of Tarleton's Legion, who had done such valiant service in the Carolinas.

In July, 1622, Sir William Alexander's pioneer-ship entered Port Mouton, "and discovered three very pleasant harbors and went ashore in one of them, which, after the ship's name, they called Luke's Bay, where they found, a great way up, a very pleasant river, being three fathoms deep at the entry thereof, and on every side of the same they did see very delicate meadows, having Roses white and red growing thereon, with a kind of white Lily, which had a dainty smell." These shores, which were hardly so fair as the old mariner painted them, were soon occupied by a French post, after whose destruction they remained in solitude for over a century.

On Little Hope Island is a revolving red light, beyond which the steamer runs W. S. W. 15 M.; then *Port Joli* opens to the N. W., on which is a fishing-village of 200 inhabitants. About 3 M. beyond is *Port Herbert*, a deep and narrow estuary with another maritime hamlet. Farther W. is the mouth of Sable River; but the steamer holds a course too far out to distinguish much of these low shores. 3½ M. N. is *Ram Island*, W. of which are the ledges off Ragged Island Harbor, at whose head is a village of 350 inhabitants. On the W. side of the harbor is *Locke's Island* (two inns), a prosperous little port of 400 inhabitants, whence the West-India trade and the Bank fisheries are carried on. During the season of 1874 70,000 quintals of fish (valued at $ 250,000) were exported from this point. On Carter's Island is a fixed red light, and the sea-swept ledge of *Gull Rock* lies outside of the harbor, and has a powerful white light. Beyond Western Head the steamer runs across the wide estuaries of Green Harbor and the Jordan River, on whose shores are four maritime hamlets. The course is changed to N. W. ½ N., and Bony's and Government Points are passed on the r. On the l. Cape Roseway is approached, on which are two fixed white lights, visible for 10 and 18 M., standing in a black-and-white striped tower. Passing between Surf Point and Sand Point the vessel turns N. by E., leaving *Birchtown Bay* on the l., and runs up to Shelburne. The last few miles are traversed between the picturesque shores of a bay which an enthusiastic mariner has called "the best in the world, except the harbor of Sydney, in Australia."

Shelburne (*Port Roseway House; English and American Hotel*) is the capital of Shelburne County, and has over 1,000 inhabitants and 5 churches. It is engaged chiefly in fishing and shipbuilding, and excels in the latter branch of business. The harbor is 9 M. long and 1-2 M. wide, and has 5-7 fathoms of water, without any shoals or flats. It is completely land-locked, but can never attain any commercial importance, owing to the fact that it is frozen solid during the winter, there being no river currents or strong tides to agitate the water. There are granite-ledges near the village, and the Roseway River empties into the bay 1 M. distant. *Birchtown* is 5 M. from Shelburne, and is at the head of a branch of the bay. It is inhabited by the descendants of the negro slaves brought from Maryland and Virginia by the Loyalist refugees, in 1783. The country back of Shelburne is unimproved, and the roads soon terminate in the great forests about the Blue Mts. Stages run from this town E. and W. Fares, Shelburne to Liverpool, $2.50; to Barrington, $1.50; to Yarmouth, $4.

"The town of Shelburne is situated at the N. extremity of a beautiful inlet, 10 M. in length and 2-3 M. in breadth, in which the whole royal navy of Great Britain might lie completely landlocked." In 1783 large numbers of American Loyalists settled here, hoping to erect a great city on this unrivalled harbor. They brought their servants and equipages, and established a cultured metropolitan society. Shelburne soon ran ahead of Halifax, and measures were taken to transfer the seat of government here. Within one year the primeval forest was replaced by a city of 12,000 inhabitants (of whom 1,200 were negroes). The obscure hamlet which had been founded here (under the name of New Jerusalem) in 1764 was replaced by a metropolis; and Gov. Parr soon entered the bay on the frigate *La Sophie*, amid the roaring of saluting batteries, and named the new city Shelburne. But the place had no rural back-country to supply and be enriched by; and the colonists, mostly patricians from the Atlantic cities, could not and would not engage in the fisheries. The money which they had brought from their old homes was at last exhausted, and then "Shelburne dwindled into insignificance almost as rapidly as it had risen to notoriety." Many of its people returned contritely to the United States; and the population here soon sank to 400. "It is only the sight of a few large storehouses, with decayed timbers and window-frames, standing near the wharves, that will lead him to conclude that those wharves must once have teemed with shipmasters and sailors. The streets of the town are changed into avenues bounded by stone fences on either side, in which grass plants contest the palm of supremacy with stones." Within two years over $2,500,000 were sunk in the founding of Shelburne.

The steamer leaves Shelburne by the same course on which she entered, with the stunted forests of McNutt's Island on the r. Rounding *Cape Roseway* within 1 M. of the lights, she runs down by Gray's Island, passing Round Bay and the hamlet of Black Point, on the bold headland of the same name. *Negro Island* is then seen on the r., and is occupied by a population of fishermen; while its N. E. point has a powerful red-and-white flashing light. Inside of this island is the broad estuary of the Clyde River, and near by is the large and picturesque fishing-village of *Cape Negro*. Cape Negro was so named by Champlain, in 1604, "on account of a rock which at a distance resembles one." The steamer then passes the Salvage Rocks, off Blanche Island (Point Jeffreys), and opens the broad bay of Port Latour on the N. W. This haven was the scene of stirring events during the 17th century, and the remains of the fort of Claude de la Tour are still visible here.

"Claude Turgis de St. Estienne, Sieur de la Tour, of the province of Champagne, quitted Paris, taking with him his son Charles Amador, then 14 years old, to settle in Acadia, near Poutrincourt, who was then engaged in founding Port Royal." 17 years afterwards, Charles succeeded to the government on the death of Biencourt, Poutrincourt's son, and for 4 years held Fort St. Louis, in the present Port Latour. Meantime Claude had been captured by the English and carried to London, where he was knighted, and then married one of the Queen's maids-of-honor. Being a Huguenot, he was the more easily seduced from his allegiance to France, and he offered to the King to procure the surrender of Fort St. Louis (the only French post then held in Acadia) to the English. So he sailed to Nova Scotia with two frigates, and asked his son to yield up the stronghold, offering him high honors at London and the supreme command in Acadia, on behalf of the English power. "Claude at once told his father that he was mistaken in supposing him capable of giving up the place to the enemies of the state. That he would preserve it for the king his master while he had a breath of life. That he esteemed highly the dignities offered him by the English king, but should not buy them at the price of treason. That the prince he served was able to requite him ; and if not, that fidelity was its own best recompense." The father employed affectionate intercession and bold menace, alike in vain ; and the English naval commander then landed his forces, but was severely repulsed from the fort, and finally gave up the siege. A traitor to France and a cause of disaster to England, the unfortunate La Tour dared not return to Europe, but advised his patrician wife to go back with the fleet, since naught now remained for him but penury and misery. The noble lady replied, " that she had not married him to abandon him. That wherever he should take her, and in whatever condition he might be placed, she would always be his faithful companion, and that all her happiness would consist in softening his grief." He then threw himself on the clemency of his son, who tempered filial affection with military vigilance, and welcomed the elder La Tour, with his family, servants, and equipage, giving him a house and liberal subsistence, but making and enforcing the condition that neither himself nor his wife should ever enter Fort St. Louis. There they lived in happiness and comfort for many years. (See also page 19.)

The hamlet of Port Latour is seen on the inner shore, and the vessel rounds the long low promontory of *Baccaro Point*, on which is a small village and a fixed red light (visible 12 M.). On the W. is **Cape Sable Island**, which is 7 M. long and 2 – 3 M. wide, and has a population of 1,636, with three churches. Its first settlers were the French Acadians, who had prosperous little hamlets on the shores. In August, 1758, 400 soldiers of the 35th British Regiment landed here and destroyed the settlements, and carried priest and people away to Halifax. About 1784 the island was occupied by Loyalists from the New-England coasts, whose descendants are daring and adventurous mariners. **Cape Sable** is on an outer islet at the extreme S. point of the island and of Nova Scotia, and is 8 – 9 M. S. W. of Baccaro Point.

It is supposed that Cape Sable and the adjacent shores were the ancient lands of the Norse discoverers, " flat, and covered with wood, and where white sands were far around where they went, and the shore was low." In the year 994 this point was visited by Leif, the son of Eric the Red, of Brattahlid, in Greenland. He anchored his ship off shore and landed in a boat; and when he returned on board he said: " This land shall be named after its qualities, and called MARKLAND " (woodland). Thence he sailed southward, and discovered Vinland the Good, on the S. shores of Massachusetts and Rhode Island, where for many years the bold Norsemen maintained colonies. In the year 1007 Markland was again visited by Thorfinn Karlsefne, who, with 160 men, was sailing south to Vinland. These events are narrated in the ancient Icelandic epics of the Saga of Eric the Red and the Saga of Thorfinn Karlsefne.
In 1347 a ship arrived at Iceland from the shores of Markland, which is described by the Annales Skalholtini and the Codex Flateyensis as having been smaller than any Icelandic coasting-vessel. In such tiny craft did the fearless Norsemen visit these iron-bound shores.
In the autumn of 1750 there was a sharp naval action off the cape between H. M. S. *Albany* and the French war-vessel *St. Francis*. The engagement lasted four hours, and ended in the surrender of the *St. Francis*, whose convoy, however, escaped and reached its destination.
In July, 1812, the Salem privateer *Polly* was cruising off Cape Sable, when she sighted two strange sail, and bore down on them, supposing them to be merchantmen ; but one was a British sloop-of-war, which opened a hot fire upon the incautious *Polly*, and a sharp chase ensued. A calm commenced, during which the frigate's boats and launch attacked the privateer, but were repulsed by heavy discharges of musketry and langrage. The *Polly* made her escape, and during the chase and action the convoy of the frigate had been captured by the privateer *Madison*, and was sent into Salem.
In the same vicinity (Aug. 1, 1812) the Rhode-Island privateer *Yankee* captured the British ship *Royal Bounty*, 10 guns, after a battle of one hour's duration. The

privateer's broadsides were delivered with great precision, and 150 of her shot struck the enemy, while the fire of the *Royal Bounty*, though rapid and heavy, was nearly ineffective. The shattered Briton became unmanageable, and while in that condition was raked from stem to stern by the *Yankee's* batteries.

Cape Sable has long been dreaded by seamen, and has caught up and destroyed many vessels. It is one of the most dangerous prongs of that iron-bound Province for which Edmund Burke could find no better words than "that hard-visaged, ill-favored brat" Probably the most-destructive wreck on this shore was that of the ocean steamship *Hungarian*.

The steamer is now running to the N. W. up the *Barrington Passage*, between Cape Sable Island and the populous Baccaro peninsula. In about 12 M. it lies to off **Barrington**, a thriving maritime village of 1,000 inhabitants, most of whom are engaged in the fisheries and the coasting trade. Clyde River is about 9 M. N. E., and is a lumbering district originally settled by Welshmen. 10-12 M. N. are the Sabimm and Great Pubnico Lakes. Barrington was settled at an early date by the French, but they were crowded off in 1763 by the arrival of 160 families from Cape Cod, who brought hither their household effects on their own vessels. After the Revolution, a colony of Loyalists from Nantucket settled here with their whilom neighbors.

The course is now to the S. W., through a narrow and tide-swept passage between Clement Point and N. E. Point, and thence out through the Barrington West Passage, passing the Baptist church near Clarke's Harbor, and emerging on the open sea between Bear Point and Newell Head. (It is to be noted that, under certain adverse conditions of wind and tide, the steamer does not call at Barrington, but rounds Cape Sable on the outside.) On the l. is *Green Island*, hiding Cape Sable, and the inlet of Shag Harbor is seen on the r. On Bon Portage Island (whose original French name was *Bon Potage*) is a new lighthouse, to warn vessels from the rugged shores on which the *Viceroy* was wrecked. The course soon changes toward the N. W., and **Seal Island**, "the elbow of the Bay of Fundy," is seen on the l., far out at sea, with the tower of its lighthouse (fixed white light, visible 18 M., and fog-whistle) looming above its low shores. On this island the ocean-steamship *Columbia* was lost. The Blonde Rock is 3½ M. S. by W. from the lighthouse, and marks the point where H. B. M. frigate *Blonde* went to pieces, in 1782. Her crew was rescued from the island and was given liberty by the American privateers *Lively* and *Scammell*, which were prowling about Cape Sable at the time of the wreck.

When the Seal Island lighthouse is just abeam, on the other side is seen Cockerwhit and the Mutton Islands; N. of Seal Island the Noddy, Mud, and Round Islands are seen, lying well out at sea. The early French maps (Chaubert's) gave these lonely islands the significant name of *Les Isles aux Loups Marins*.

From Cape Sable " one goes to the *Isle aux Cormorants*, a league distant, so called on account of the infinite number there of those birds, with whose eggs we filled a cask; and from this bay making W. about 6 leagues, crossing a bay which runs in 2-3 leagues to the N., we meet several islands, 2-3 leagues out to sea, which may contain, some 2, others 3 leagues, and others less, according to my judgment. They are mostly very dangerous for vessels to come close to, on account of the great tides and rocks level with the water. These islands are filled with pine-trees, firs, birches, and aspens. A little further on are 4 others. In one there is so great a quantity of birds called *tangueux* that they may be easily knocked down with a stick. In another there are seals. In two others there is such an abundance of birds of different kinds that, without having seen them, could not be imagined, such as cormorants, ducks of three kinds, geese, *marmettes*, bustards, *perroquets de mer*, snipes, vultures, and other birds of prey, *maunes*, sea-larks of two or three kinds, herons, *goillants*, curlews, sea-gulls, divers, kites, *appoils*, crows, cranes, and other sorts, which make their nests here." (CHAMPLAIN.)

"Here are many islands extending into the sea, 4-5 M. distant from the mainland, and many rocks with breaking seas. Some of these islands, on account of the multitude of birds, are called *Isles aux Tangueux*; others are called *Isles aux Loups Marins* (Seal Islands)." (NOVUS ORBIS.)

N. of St. John's Island (on the r.) is seen the deep inlet of Pubnico Harbor, on whose shores is the great fishing-village of **Pubnico** (*Carland's Hotel*), with 1,900 inhabitants, of whom 136 families are Acadian-French, the greater portion belonging to the families of Amiro and D'Entremont. There are valuable eel-fisheries off this coast, and the Acadians own 65 schooners in the Banks fisheries. 5 M. N. is *Argyle*, a settlement of 800 inhabitants, near the island-strewn Abuptic Harbor.

The steamer now crosses the mouth of Argyle Bay and the estuary of the Tusket River (see page 116), and enters the archipelago of the * **Tusket Islands**. In favorable conditions of wind and tide she traverses the *Ellenwood Passage*, passing the Bald Tuskets, Ellenwood, Allen, and Murder Islands, and a multitude of others. The islands are of great variety of size and shape, and are usually thickly covered with low and sturdy trees; and the channels between them are narrow and very deep. The frequent kaleidoscopic changes in the views on either side, and the fascinating commingling and contrast of forest, rock, and water, recall the scenery of the Thousand Islands or the Narrows of Lake George. But the Tuskets are not even embayed; they stand off one of the sharpest angles of the continent, and the deep lanes between them are traversed by the strongest tides of the ocean.

Soon after passing the last Tusket the steamer runs in near the white village on *Jebogue Point*, and enters Yarmouth Sound. On the l. is *Cape Fourchu*, with its fog-whistle and a lofty revolving light which is visible for 18 M. The narrow channel is ascended, with a plain of mud on either side, if the tide is out; and the vessel reaches the end of her journey at the wharves of Yarmouth.

Yarmouth, see page 114.

26. Halifax to Yarmouth, by the Shore Route.—Chester and Mahone Bay.

The easiest route to the chief ports on this coast is by the steamship line (see Route 25); and the new Western-Counties Railway, from Yarmouth to Annapolis, will, when completed, furnish a still more expeditious line of travel. But many points on the Atlantic front of the Province are, and will be, accessible only by stages. This mode of travel is fully as arduous here as in other remote districts, and the accommodations for wayfarers are indifferent.

Distances. — Halifax to St. Margaret's Bay, 21 M.; Hubbard's Cove (McLean's), 82; Chester, 45; Mahone Bay, 62 (branch to Lunenburg in 7 M.); Bridgewater, 70; Mill Village, 88; Liverpool, 97; Port Mouton, 107; Port Joli, 112; Sable River, 122; Jordan River, 130; Shelburne, 137; Barrington, 157; Pubnico, 175; Tusket, 191; Yarmouth, 201. (Certain facts ascertained while travelling over this route have led the Editor to state the distance between Bridgewater and Chester as 4 M. less than that given in the official itinerary.)

Fares. — Halifax to Chester, $2.50; Mahone Bay, $3.50 (Lunenburg, $4); Bridgewater, $4; Liverpool, $6; Shelburne, $8.50; Barrington, $10; Yarmouth, $12.

The stage rattles up the hilly streets of Halifax at early morning, and traverses the wide commons N. of the Citadel, with formal lines of trees on either side. Beyond the ensuing line of suburban villas it descends to the level of the Northwest Arm (see page 100), along whose head it passes. The road then leads along the shores of the lakes whence Halifax draws its water-supply, and enters a dreary and thinly settled region. Dauphiney's Cove is at the head of * **St. Margaret's Bay**, one of the most beautiful bays on all this remarkable coast. It is 12 M. long by 6 M. wide, and is entered by a passage 2 M. wide; and is supposed to have been named (*Baie de Ste. Marguerite*) by Champlain, who visited it in May, 1603. There are several small maritime villages on its shores, and the dark blue waters, bounded by rugged hills, are deep enough for the passage of large ships. The stage runs S. W. along the shore for 11 M., sometimes rolling alongside of beaches of dazzling white sand, then by shingly and stony strands on which the embayed surf breaks lightly, and then by the huts of fishermen's hamlets, with their boats, nets, and kettles by the roadside. *Hubbard's Cove* has a small inn, where passengers get their midday meals.

There was an ancient water-route from this point to the Basin of Minas. 2 M. from the Cove is *Dauphiney's Lake*, which is 4 M. long, whence a carry of 1½ M. leads into the *Ponhook Lake*, a river-like expanse 8 M. long, and nowhere so much as 1 M. wide. A short outlet leads to the Blind Lake, which winds for 7 M. through the forests W. of the Ardoise Mt., and is drained by the St. Croix River, emptying into the Avon at Windsor.

7 M. S. W. of Hubbard's Cove the stage crosses the *East River*, "a glorious runway for salmon, with splendid falls and cold brooks tumbling into it at intervals, at the mouth of which large trout can be caught two at a time, if the angler be skilful enough to land them when hooked." Frequent and beautiful views of Mahone Bay are now gained (on the l.), as the stage sweeps around its head and descends to

Chester (two good inns), a village of about 900 inhabitants, finely situated on a hill-slope which overlooks the Chester Basin and Mahone Bay. It has three churches, and a pleasant summer society. This town was settled about the year 1760 by 144 New-Englanders, who brought an outfit of cattle and farming-tools. In 1784 they were joined by a large number of Loyalist refugees, but these were from the American cities, and soon wearied of farming and returned out of exile. In the woods near the village is a thermal spring 8 ft. around, whence a soft alkaline water is discharged; and on the shores of Sabbatee Lake are found deposits of kaolin, or white pipe-clay.

Mr. Hallock is an enthusiastic admirer of this town, and says: " Three pleasant seasons have I spent at Chester. I idolize its very name. Just below my window a lawn slopes down to a little bay with a jetty, where an occasional schooner lands some stores. There is a large tree, under which I have placed some seats; and off the end of the pier the ladies can catch flounders, tomcods, and cunners, in any quantity. There are beautiful drives in the vicinity, and innumerable islands in the bay, where one can bathe and picnic to heart's-content. There are sailing-boats for lobster-spearing and deep-sea fishing, and row-boats too. From the top of a neighboring hill is a wonderful panorama of forest, stream, and cultivated shore, of bays and distant sea, filled with islands of every size and shape. And if one will go to Gold River he may perchance see, as I have done, caribou quietly feeding on the natural meadows along the upper stream. Beyond Beech Hill is a trackless forest, filled with moose, with which two old hunters living near oft hold familiar intercourse." (*The Fishing Tourist.*)

One of the pleasantest excursions in this district is to Deep Cove and Blandford, 16 M. from Chester, by a road which follows the shores of Mahone Bay. From Blandford the ascent of **Mt. Aspotogon** is easily accomplished, and rewards the visitor by a superb marine * view, including the great archipelago of Mahone Bay, the deep, calm waters of St. Margaret's Bay on the E., the broken and picturesque shores towards Cape Sambro, and a wide sweep of the blue Atlantic. Visitors at Chester also drive down the Lunenburg and Lahave road, which affords pretty sea-views.

A rugged road leads across the Province to Windsor, about 40 M. N., passing through an almost unbroken wilderness of hills, and following the course of the Avon Lakes and River. Semi-weekly stages run from Chester to Kentville (see page 90).

*** Mahone Bay** opens to the S., E. and W. from Chester, and may be explored by boats or yachts from that village. It is studded with beautiful islands, popularly supposed to be 365 in number, the largest of which are occupied by cosey little farms, while the smaller ones are covered with bits of forest. The mainland shores are nearly all occupied by prosperous farms, which are under the care of the laborious Germans of the county. The fogs prevail in these waters to a far less extent than on the outer deep, and it is not infrequently that vessels round the point in a dense white mist and enter the sunshine on the Bay. Boats and boatmen may be obtained at the villages along the shore, and pleasant excursions may be made among the islands, in pursuit of fish. "The unrivalled beauty of Mahone Bay" has been the theme of praise from all who have visited this district. In June, 1813, the line-of-battle-ship *La Hogue* and the frigate *Orpheus* chased the American privateer *Young Teazer* in among these islands. Though completely overpowered, the Yankee vessel refused to surrender, and she was blown up by one of her officers. The whole crew, 94 in number, was destroyed in this catastrophe.

Oak Island is celebrated as one of the places where it is alleged that Capt. Kidd's treasure is hidden. About 80 years ago 3 New-Englanders claimed to have found here evidences of a buried mystery, coinciding with a tradition to the same effect. Digging down, they passed regular layers of flag-stones and cut logs, and their successors penetrated the earth over 100 ft. farther, finding layers of timber, charcoal, putty, West-Indian grass, sawed planks, and other curious substances, together with a quaintly carved stone. The pit became flooded with water, and was pumped out steadily. Halifax and Truro merchants invested in the enterprise, and great stone drains were discovered leading from the sea into the pit. After much money and labor was spent in the excavation, it was given up about 10 years ago, and the object of the great drains and concealed pit still remains a profound mystery.

Big Tancook is the chief of the islands in this bay, and is about 2 M. long. It contains 500 inhabitants, who are engaged in farming and fishing. Between this point and Mt. Aspotogon is Little Tancook Island, with 60 inhabitants. These islands were devastated, in 1756, by the Indians, who killed several of the settlers.

"This bay, the scenery of which, for picturesque grandeur, is not surpassed by any landscape in America, is about 10 M. broad and 12 deep, and contains within it a multitude of beautiful wooded islands, which were probably never counted, but are said to exceed 200."

Soon after the Yarmouth stage leaves Chester "we come to Chester Basin, island-gemmed and indented with many a little cove; and far out to sea, looming up in solitary grandeur, is Aspotogon, a mountain headland said to be the highest land in Nova Scotia (?). The road follows the shore for many a mile, and then turns abruptly up the beautiful valley of Gold River, the finest of all the salmon streams of this grand locality. In it there are eleven glorious pools, all within 2 M. of each other, and others for several miles above at longer intervals."

Mahone Bay (Victoria Hotel) is a village of 800 inhabitants, situated on a pretty cove about 17 M. from Chester. It has 4 churches, and its inhabitants are mostly engaged in fishing and the lumber-trade. In the vicinity are several other populous German settlements, and 7 M. S. is Lunenburg (see page 118). This point was known to the Indians by the name of *Mushamush*, and was fortified by the British in 1754.

The stage now traverses a dreary inland region, inhabited by Germans, and soon reaches *Bridgewater* (two inns), a thriving village on the Lahave River, 13 M. from the sea. It has 1,000 inhabitants and 4 churches, and is largely engaged in the lumber-trade, exporting staves to the United States and the West Indies. The scenery of the Lahave River is attractive and picturesque, but the saw-mills on its upper waters have proved fatal to the fish (see page 119). The road now traverses a dismal region for 18 M., when it reaches *Mill Village* (small hotel), on the Port Medway River. This place has several large saw-mills and a match-factory, and its population numbers about 400. It is near the Doran and Herringcove Lakes, and is 6 M. from the Third Falls of the Lahave. 9 M. S. W. is Liverpool (see page 120).

From Liverpool to Yarmouth the road runs along the heads of the bays and across the intervening strips of land. The chief stations and their distances are given in the itinerary on page 126; the descriptions of the towns may be found in Route 25.

27. The Liverpool Lakes.

This system of inland waters is most easily reached from Halifax or St. John by passing to Annapolis Royal and there taking the stage which leaves at 6 A. M. daily.

Distances.—Annapolis; Milford, 14 M.; Maitland, 27; Northfield, 30; Kempt, 35; Brookfield, 41; Caledonia Corner; Greenfield (Ponhook), 50; Middlefield, 56; Liverpool, 70.

Soon after leaving Annapolis the stage enters the valley of Allen's River, which is followed toward the long low range of the South Mt. At *Milford* (small inn) the upper reservoirs of the Liverpool River are met, and from this point it is possible to descend in canoes or flat-bottomed boats to the town of Liverpool, 60 M. distant. If a competent guide can be secured at Milford this trip can be made with safety, and will open up rare fishing-grounds. The lakes are nearly all bordered by low and rocky shores, with hill-ranges in the distance; and flow through regions which are as yet but little vexed by the works of man. The trout in these waters are abundant and not too coy; though better fishing is found in proportion to the distance to which the southern forest is entered. Mr. McClelland has been the best guide from Milford, but it is uncertain whether he will be available this summer.

Queen's and Lunenburg Counties form "the lake region of Nova Scotia. All that it lacks is the grand old mountains to make it physically as attractive as the Adirondacks, while as for game and fish it is in every way infinitely superior. Its rivers are short, but they flow with full volume to the sea, and yield abundantly of salmon, trout, and sea-trout. Its lakes swarm with trout, and into many of them the salmon ascend to spawn, and are dipped and speared by the Indians in large numbers." (HALLOCK.)

"In the hollows of the highlands are likewise embosomed lakes of every variety of form, and often quite isolated. Deep and intensely blue, their shores fringed with rock bowlders, and generally containing several islands, they do much to diversify the monotony of the forest by their frequency and picturesque scenery." (CAPT. HARDY.)

The Liverpool road is rugged, and leads through a region of almost unbroken forests. Beyond Milford it runs S. E. down the valleys of the Boot Lake and Fisher's Lake, with dark forests and ragged clearings on either side. *Maitland* is a settlement of about 400 inhabitants, and a few miles beyond is Northfield, whence a forest-road leads S. W. 6 M. to the shore of *Fairy Lake*, or the Frozen Ocean, a beautiful island-strewn sheet of water 4 M. long.

The road now enters *Brookfield*, the centre of the new farming settlements of the North District of Queen's County. Several roads diverge hence, and in the vicinity the lakes and tributaries of the Liverpool and Port Medway Rivers are curiously interlaced. 5-6 M. S. E. is the *Malaga Lake*, which is 5 M. long and has several pretty islands. The road passes on to *Greenfield*, a busy lumbering-village at the outlet of Port Medway Great Lake. This long-drawn-out sheet of water is also skirted by the other road, which runs S. from Brookfield through *Caledonia Corner* (small inn). The Ponhook Road is S. W. of Greenfield and runs down through the forest to the outlet of **Ponhook Lake**, "the headquarters of the Micmacs and of all the salmon of the Liverpool River." This Indian village is the place to get guides who are tireless and are familiar with every rod of the lake-district. From this point a canoe voyage of about 8 M. across the Ponhook Lakes leads the voyager into the great * **Lake Rossignol**, which is 12 M. long by 8 M. wide, and affords one of the most picturesque sights in Nova Scotia.

"A glorious view was unfolded as we left the run and entered the still water of the lake. The breeze fell rapidly with the sun and enabled us to steer towards the centre, from which alone the size of the lake could be appreciated, owing to the number of the islands. These were of every imaginable shape and size, — from the grizzly rock bearing a solitary stunted pine, shaggy with *Usnea*, to those of a mile in length, thickly wooded with maple, beech, and birches. . . . . Here and there a bright spot of white sand formed a beach tempting for a disembarkation; and frequent sylvan scenes of an almost fairy-land character opened up as we coasted along the shores, — little harbors almost closed in from the lake, overgrown with water-lilies, arrow-heads, and other aquatic plants, with mossy banks backed by bosky groves of hemlocks." (CAPT. HARDY.)

At the foot of Lake Rossignol is a wide oak-opening, with a fine greensward under groves of white oaks. Near this point the Liverpool River flows out, passing several islets, and affording good trout-fishing. In and about this oak-opening was the chief village of the ancient Micmacs of this region; and here are their nearly obliterated burying-grounds. The site is now a favorite resort for hunting and fishing parties. The name *Ponhook* means "the first lake in a chain"; and these shores are one of the few districts of the vast domains of *Miggumáhghee*, or "Micmac Land," that remain in the possession of the aborigines. From Ponhook 12 lakes may be entered by canoes without making a single portage.

From Lake Rossignol the sportsman may visit the long chain of the *Segum-Sega Lakes*, entered from a stream on the N. W. shore (several portages), and may thence ascend to the region of the Blue Mts. and into Shelburne County. The *Indian Gardens* may also be visited thence, affording many attractions for riflemen. The Micmacs of Ponhook are the best guides to the remoter parts of the forest. There are several gentlemen in the town of Liverpool who have traversed these pleasant solitudes, and they will aid fellow-sportsmen loyally. The Indian village is only about 15 M. from Liverpool, by a road on the l. bank of the river.

Liverpool, see page 120.

28. Halifax to Tangier.

The Royal mail-stage leaves Halifax at 6 A. M. on Monday, Wednesday, and Friday (returning the alternate days), for the villages along the Atlantic shore to the E. The conveyance is not good, and the roads are sometimes in bad condition, but there is pretty coast-scenery along the route.

Distances. — Halifax; Dartmouth; Porter's Lake (Innis's), $16\frac{1}{2}$ M.; Chezzetcook Road (Ormon's), $18\frac{1}{2}$; Musquodoboit Harbor, $28\frac{1}{2}$; Lakeville (Webber's), 40; Ship Harbor, 48; Tangier, 56; Sheet Harbor, 74; Beaver Harbor, 84.

After leaving Dartmouth, the stage runs E. through a lake-strewn country, and passes near the gold-mines of Montague. Beyond the Little Salmon River it traverses *Preston*, with the gold-bearing district of Lawrencetown on the S. The mines and placer-washings at this point drew large and enthusiastic crowds of adventurers in 1861-62, but they are now nearly abandoned. The road rounds the N. end of Echo Lake and ascends a ridge beyond, after which it crosses the long and river-like expanse of *Porter's Lake*, and runs through the post-village of the same name. 3-4 M. to the S. E. is *Chezzetcook Harbor*, with its long shores lined with settlements of the Acadian French, whereof Cozzens writes:—

"But we are again in the Acadian forest; let us enjoy the scenery. The road we are on is but a few miles from the sea-shore, but the ocean is hidden from view by the thick woods. As we ride along, however, we skirt the edges of coves and inlets that frequently break in upon the landscape. There is a chain of fresh-water lakes also along this road. Sometimes we cross a bridge over a rushing torrent; sometimes a calm expanse of water, doubling the evergreens at its margin, comes into view; anon a gleam of sapphire strikes through the verdure, and an ocean-bay with its shingly beach curves in and out between the piny slopes."

Here "the water of the harbor has an intensity of color rarely seen, except in the pictures of the most ultramarine painters. Here and there a green island or a fishing-boat rested upon the surface of the tranquil blue. For miles and miles the eye followed indented grassy slopes that rolled away on either side of the harbor, and the most delicate pencil could scarcely portray the exquisite line of creamy sand that skirted their edges and melted off in the clear margin of the water. Occasional little cottages nestle among these green banks, — not the Acadian houses of the poem, 'with thatched roofs and dormer-windows projecting,' but comfortable, homely-looking buildings of modern shapes, shingled and un-weathercocked. . . . . The women of Chezzetcook appear at daylight in the city of Halifax, and as soon as the sun is up vanish like the dew. They have usually a basket of fresh eggs, a brace or two of worsted socks, a bottle of fir balsam, to sell. These comprise their simple commerce."

Chezzetcook was founded by the French in 1740, but was abandoned during the long subsequent wars. After the British conquest and pacification of Acadia, many of the old families returned to their former homes, and Chezzetcook was re-occupied by its early settlers. They formed an agricultural community, and grew rapidly in prosperity and in numbers. There are about 250 families now resident about the bay, preserving the names and language and many of the primitive customs of the Acadians of the Basin of Minas. (See pages 108 and 113.)

The road passes near the head of Chezzetcook Harbor, on the r., and then turns N. E. between the blue waters of Chezzetcook Great Lake (l.) and Pepiswick Lake (r.). The deep inlet of *Musquodoboit Harbor* is soon reached, and its head is crossed. This is the harbor where Capt. Hardy made his pen-picture of this romantic coast:—

"Nothing can exceed the beauty of scenery in some of the Atlantic harbors of Nova Scotia, — their innumerable islands and heavily-wooded shores fringed with the golden kelp, the wild undulating hills of maple rising in the background, the patches of meadow, and neat little white shanties of the fishermen's clearings, ... the fir woods of the western shores bathed in the morning sunbeams, the perfect reflection of the islands and of the little fishing-schooners, the wreaths of blue smoke rising from their cabin stoves, and the roar of the distant rapids, where the river joins the harbor, borne in cadence on the ear, mingled with the cheerful sounds of awakening life from the clearings."

Near Musquodoboit are some valuable gold-mines, with two powerful quartz-crushing mills, and several moderately rich lodes of auriferous quartz. The stage soon reaches the W. arm of *Jeddore Harbor*, and then crosses the Le Marchant Bridge. The district of Jeddore has 1,623 inhabitants, most of whom are engaged in the fisheries or the coasting trade, alternating these employments with lumbering and shipbuilding. A long tract of wilderness is now traversed, and *Ship Harbor* is reached. A few miles N. W. is the broad expanse of Ship Harbor Lake, reaching nearly to the Boar's Back Ridge, and having a length of 12-14 M. and a width of 2-4 M. To the N. are the hills whence falls the Tangier River, to which the Indians gave the onomatopoetic name of *Ahmagopakegeek*, which signifies "tumbling over the rocks." The post-road now enters the once famous gold-bearing district of *Tangier*.

These mines were opened in 1860, and speedily became widely renowned, attracting thousands of adventurers from all parts of the Atlantic coast. For miles the ground was honeycombed with pits and shafts, and the excited men worked without intermission. But the gold was not found in masses, and only patience and hard work could extract a limited quantity from the quartz, so the crowd became discontented and went to the new fields. Lucrative shore-washings were engaged in for some time, and a stray nugget of Tangier gold weighing 27 ounces was shown in the Dublin Exposition. This district covers about 80 square miles, and has 12 lodes of auriferous quartz. The South Lode is the most valuable, and appears to grow richer as it descends. The mines are now being worked by two small companies, and their average yield is $ 400-500 per miner each year.

Beyond Tangier and Pope's Bay the post-road passes the head of *Spry Bay*, and then the head of Mushaboon Harbor, and reaches *Sheet Harbor* (Farnal's Hotel). This is a small shipbuilding village, at the head of the long harbor of the same name, and is at the outlets of the Middle and North Rivers, famous for their fine salmon fisheries.

From this point a road follows the shore to the N. E. to Sherbrooke, about 50 M. distant, passing the obscure maritime hamlets of Beaver Harbor, Necum Tench, Ekum Sekum, Marie Joseph, and Liscomb Harbor. The back-country on all this route is yet desolate and unsettled. There are so many islands off the shore that this portion of the Atlantic is called the Bay of Islands (old French, *Baie de Toutes les Isles*), although it is not embayed.

Sherbrooke, see page 133.

29. The Northeast Coast of Nova Scotia.

This district is reached by passing on the Intercolonial Railway (see Routes 16 and 17) from St. John or Halifax to New Glasgow, and thence taking the Royal mail-stage to Antigonish (see Route 32).

From Antigonish a stage departs on Monday, Wednesday, and Friday mornings, running 40 M. S. (fare, $2) to **Sherbrooke** (two inns). This is a village on the l. bank of the St. Mary's River, the largest river in Nova Scotia, and is at the head of navigation on that stream. It is engaged in shipbuilding and in the exportation of deals and lumber. The town derives considerable interest from the fact that in the vicinity is one of the broadest and most prolific gold-fields in the Province. *Goldenville* is 3 M. from Sherbrooke, by a road which crosses the St. Mary's on a long bridge. This district covers 18 square miles, and is the richest in the Province, having yielded as high as $2,000 per man per year, or about three times the average production of the best of the Australian mines. The auriferous lodes are operated at Goldenville only, where there are several quartz-crushers on a large scale. These mines were discovered in 1861, and on the first day over $500 worth of gold was found here. Systematic mining operations were soon commenced, and the yield of the precious metal has since been very satisfactory.

The *Wine-Harbor Gold-field* is several miles S. E. of Sherbrooke, near the mouth of the St. Mary's River. The average yield per ton is small, yet the breadth and continuity of the lodes renders the work easy and certain. This district is seamed with abandoned shafts and tunnels, one of which is 700 ft. long. The first discovery of gold was made in 1860 in the sands of the sea-shore, and the quartz lodes on the N. E. side of the harbor were soon opened. Of later years the Wine-Harbor district has greatly declined in popularity and productiveness.
The *Stormont Gold-fields* are 36 M. N. E. of Sherbrooke, and are most easily reached by direct conveyance from Antigonish. Gold was discovered here by the Indians in 1861, and occurs in thick layers of quartz. Owing to its remoteness, this region has remained undeveloped, and its total yield in 1869 was but 227 ounces ($4,540). The chief village in the district is at the head of *Country Harbor*, a picturesque arm of the sea, 8 M. long and 2-3 M. wide. There are fine opportunities for shooting and fishing among the adjacent bays and highlands. All this shore was settled in 1783-4 by Loyalists from North and South Carolina.

Guysborough and Cape Canso.

Guysborough (*Grant's Hotel*) is reached by daily mail-stages from Antigonish, from which it is 31 M. distant (fare, $2.50). After leaving the valley of the South River, the road passes through a rough and hilly region, and descends through the Intervale Settlement and Manchester to Guysborough, a marine village at the head of Chedabucto Bay. It has about 1,500 inhabitants, with a prosperous academy, and is the capital of Guysborough County (named in honor of Sir Guy Carleton). It is engaged in shipbuilding and the fisheries, and has a good and spacious harbor. The noble anchorage of Milford Haven lies between the town and the bay.

A strong post was established at Chedabucto, on the site of Guysborough in 1636, by M. Denys, who had spacious warehouses and a strong fort here, together with 120 men. Here he received and supported the exiled children of D'Aulnay Charnisay; and here also he was vainly besieged for several days by La Giraudière and 100 men from Canso. In 1690 the works were held by De Montorgueuil, and were bravely defended against the attacks of the New-England army under Sir William Phipps. Finally, when the buildings of the fort were all in flames about him, the gallant Frenchman surrendered, and was sent to Placentia with his soldiers. The ruins of the ancient fort are now to be traced near the mouth of the harbor.

A bold ridge runs 31 M. E. from Guysborough along the S. shore of Chedabucto Bay to **Cape Canso**, the most easterly point of Nova Scotia. A road follows the course of the bay to the fishing-village of Cape Canso, which has over 1,000 inhabitants and enjoys a profitable little export trade. Several islands lie off this extreme point of Nova Scotia, one of which bears two powerful white lights and a fog-whistle. Canso Harbor is marked by a fixed red light which is visible for 12 M.

White Haven is on the S. side of the great peninsula of Wilmot, 30 M. from Guysborough, and is a small fishing settlement situated on one of the finest bays on the American coast. It was originally intended to have the Intercolonial Railway terminate here, and connect with the transatlantic steamships. The harbor is easy of access, of capacious breadth, and free from ice in winter. Its E. point is White Head, usually the first land seen by vessels crossing from Europe in this upper latitude, on which is a fixed white light. Just W of White Haven is the fishermen's hamlet of *Molasses Harbor*, near the broad bight of Tor Bay.

30. Sable Island.

The Editor inserts the following sketch of this remotest outpost of the Maritime Provinces, hoping that its quaint character may make amends for its uselessness to the summer tourist. It may also be of service to voyagers on these coasts who should chance to be cast away on the island, since no one likes to be landed suddenly in a strange country without having some previous knowledge of the reception he may get.

A regular line of communication has recently been established between Sable Island and Halifax. The boats run once a year, and are chartered by the Canadian government to carry provisions and stores to the lighthouse people and patrols, and to bring back the persons who may have been wrecked there during the previous year.

Sable Island is about 90 M. S. E. of Cape Canso. It is a barren expanse of sand, without trees or thickets, and is constantly swept by storms, under whose powerful pressure the whole aspect of the land changes, by the shifting of the low dunes. The only products of this arid shore are cranberries, immense quantities of which are found on the lowlands.

"Should any one be visiting the island now, he might see, about 10 M. distance, looking seaward, half a dozen low dark hummocks on the horizon. As he approaches, they gradually resolve themselves into hills fringed by breakers, and by and by the white sea-beach with its continued surf,—the sand-hills, part naked, part waving in grass of the deepest green, unfold themselves,—a house and a barn dot the western extremity,—here and there along the wild beach lie the ribs of unlucky traders half buried in the shifting sand..... Nearly the first thing the visitor does is to mount the flag-staff, and, climbing into the crow's-nest, scan the scene. The ocean bounds him everywhere. Spread east and west, he views the narrow island in form of a bow, as if the great Atlantic waves had bent it around, nowhere much above 1 M. wide, 26 M. long, including the dry bars, and holding a shallow lake 13 M. long in its centre. There it all lies spread like a map at his feet,—grassy hill and sandy valley fading away into the distance. On the foreground the outpost men galloping their rough ponies into headquarters, recalled by the flag flying over his head; the West-end house of refuge, with bread and matches, firewood and kettle, and directions to find water, and headquarters with flag-staff on the adjoining hill. Every sandy peak or grassy knoll with a dead man's name or date has a tradition,— Baker's Hill, Trott's Cove, Scotchman's Head, French Gardens, — traditionary spot where the poor convicts expiated their social crimes, — the little burial-ground nestling in the long grass of a high hill, and consecrated to the repose of many a sea-tossed limb; and 2-3 M. down the shallow lake, the South-side house and barn, and staff and boats lying on the lake beside the door. 9 M. farther down, by the aid of a glass, he may view the flag-staff at the foot of the lake, and 5 M. farther the East-end lookout, with its staff and watch-house. Herds of wild ponies dot the hills, and black-duck and sheldrakes are heading their young broods on the mirror-like ponds. Seals innumerable are basking on the warm sands, or piled like ledges of rock along the shores. The *Glasgow's* bow, the *Maskonemet's* stern, the *East Boston's* hulk, and the grinning ribs of the well-fastened *Guide*, are spotting the sands, each with its tale of last adventure, hardships passed, and toil endured. The whole picture is set in a silver-frosted frame of rolling surf and sea-ribbed sand."

"Mounted upon his hardy pony, the solitary patrol starts upon his lonely way. He rides up the centre valleys, ever and anon mounting a grassy hill to look seaward, reaches the West-end bar, speculates upon perchance a broken spar, an empty bottle, or a cask of beef struggling in the land-wash, — now fords the shallow lake, looking well for his land-range, to escape the hole where Baker was drowned; and coming on the breeding-ground of the countless birds, his pony's hoof with a reckless smash goes crunching through a dozen eggs or callow young. He fairly puts his pony to her mettle to escape the cloud of angry birds which, arising in countless numbers, dent his weather-beaten tarpaulin with their sharp bills, and snap his pony's ears, and confuse him with their sharp, shrill cries. Ten minutes more, and he is holding hard to count the seals. There they lay, old ocean's flocks, resting their wave-tossed limbs, — great ocean bulls, and cows, and calves." (DR. J. B. GILPIN.)

For over a century Sable Island has been famous for its wild horses. They number perhaps 400, and are divided into gangs which are under the leadership of the old males. They resemble the Mexican or Ukraine wild horses, in their large heads, shaggy necks, sloping quarters, paddling gait, and chestnut or piebald colors. Once a year the droves are all her led by daring horsemen into a large pound, where 20 or 30 of the best are taken out to be sent to Nova Scotia. After the horses chosen for exportation are lassoed and secured, the remainder are turned loose again.

Since Sable Island was first sighted by Cabot, in 1497, it has been an object of terror to mariners. Several vessels of D'Anville's French Armada were lost here; and among the many wrecks in later days, the chief have been those of the ocean steamship *Georgia* and the French frigate *L'Africaine*.

In the year 1583, when Sir Humphrey Gilbert was returning from Newfoundland (of which he had taken possession in the name of the English Crown), his little fleet became entangled among the shoals about Sable Island. On one of these outlying bars the ship *Delight* struck heavily and dashed her stern and quarters to pieces. The officers and over 100 men were lost, and 14 of the crew, after drifting about in a pinnace for many days, were finally rescued. The other vessels, the *Squirrel* and the *Golden Hind*, bore off to sea and set their course for England. But when off the Azores the *Squirrel* was sorely tossed by a tempest (being of only 10 tons' burden), and upon her deck was seen Sir Humphrey Gilbert reading a book. As she swept past the *Golden Hind*, the brave knight cried out to the captain of the latter: "Courage, my lads, we are as near heaven by sea as by land." About midnight the *Squirrel* plunged heavily forward into the trough of the sea, and went down with all on board. Thus perished this "resolute soldier of Jesus Christ, ... one of the noblest and best of men in an age of great men."

In 1598 a futile attempt at colonizing Sable Island was made by "Le Sieur Baron de Leri et de St. Just, Vicomte de Gueu." But he left some live-stock here that afterwards saved many lives.

In the year 1598 the Marquis de la Roche was sent by Henri IV. to America, carrying 200 convicts from the French prisons. He determined to found a settlement on Sable Island, and left 40 of his men there to commence the work. Soon after, De la Roche was forced by stress of storm to return to France, abandoning these unfortunate colonists. Without food, clothing, or wood, they suffered intensely, until partial relief was brought by the wrecking of a French ship on the island. For seven years they dwelt in huts built of wrecked timber, dressed in seal-skins, and living on fish. Then King Henri IV. sent out a ship under Chedotel, and the 12 survivors, gaunt, squalid, and long-bearded, were carried back to France, where they were pardoned and rewarded.

An attempt was made about the middle of the 16th century to colonize Cape Breton in the interests of Spain, but the fleet that was transporting the Spaniards and their property was dashed to pieces on Sable Island.

31. St. John and Halifax to Pictou.

By the Pictou Branch Railway, which diverges from the Intercolonial Railway at Truro.

Stations. — *St. John to Pictou.* St. John to Truro, 215 M.; Valley, 219; Union, 224; Riversdale, 228; West River, 236; Glengarry, 243; Hopewell, 250; Stellarton, 255; New Glasgow, 258; Pictou Landing, 266; Steamboat Wharf, 267.

Stations. — *Halifax to Pictou.* Halifax to Truro, 61 M.; Valley, 65; Union, 70; Riversdale, 74; West River, 82; Glengarry, 89; Hopewell, 96; Stellarton, 101; New Glasgow, 104; Pictou Landing, 112; Steamboat Wharf, 113.

St. John to Truro, see Routes 16 and 17.

Halifax to Truro, see Route 17 (reversed).

The train runs E. from Truro, and soon after leaving the environs, enters a comparatively broken and uninteresting region. On the l. are the rolling foot-hills of the Cobequid Range, and the valley of the Salmon River is followed by several insignificant forest stations. *Riversdale* is surrounded by a pleasant diversity of hill-scenery, and has a spool-factory and a considerable lumber trade. 14 M. to the N. is the thriving Scottish settlement of *Earltown*. Beyond West River the train reaches *Glengarry*, which is the station for the Scottish villages of New Lairg and Gairloch. *Hopewell* (Hopewell Hotel) has small woollen and spool factories; and a short distance beyond the line approaches the banks of the East River.

Stellarton is the station for the great **Albion Mines**, which are controlled (for the most part) by the General Mining Association, of London. There is a populous village here, most of whose inhabitants are connected with the mines. The coal-seams extend over several miles of area, and are of remarkable thickness. They are being worked in several pits, and would doubtless return a great revenue in case of the removal of the restrictive trade regulations of the United States. In the year 1864 over 200,000 tons of coal were raised from these mines.

New Glasgow (three inns) is a town of 2,500 inhabitants, largely engaged in shipbuilding and having other manufactures, including foundries and tanneries. It is favorably situated on the East River, and derives considerable importance from being the point of departure for the Royal mail-stages for Antigonish, the Strait of Canso, and Cape Breton; also for Guysborough, Wine Harbor, and Sherbrooke.

The train now descends by the East River to *Fisher's Grant*, opposite the town of Pictou, to which the passengers are conveyed by a steam ferry-boat. If the traveller is about to take the steamship he must remain on the train, which runs down 1 M. farther to her wharf.

Pictou (*St. Lawrence Hall*) is a wealthy and flourishing town on the Gulf shore of Nova Scotia. It has about 3,500 inhabitants, with several churches, a masonic hall, and a weekly paper. The public buildings of Pictou County are also located here, and the academy is the chief educational establishment. The harbor is the finest on the S. shores of the Gulf of St. Lawrence, and can accommodate ships of any burden, having a depth of 5 – 7 fathoms. The town occupies a commanding position on a hillside over a small cove on the N. side of the harbor, and nearly opposite, the basin is divided into three arms, into which flow the East, Middle, and West Rivers. On the East River are the shipping wharves of the Albion and the International Coal Companies, whence immense quantities of coal were exported in the palmy days before the United States punished Canada for aiding her rebel States, by repealing the Reciprocity Treaty.

Pictou has a large coasting trade; is engaged in shipbuilding; and has a marine-railway. It has also tobacco-factories, carding-mills, several saw and grist mills, a foundry, and three or four tanneries. But the chief business is connected with the adjacent mines and the exportation of coal, and with the large freestone quarries in the vicinity.

Stages leave Pictou several times weekly, for River John, Tatamagouche, Wallace, Pugwash, and Amherst (see page 81). Steamships leave (opposite) Pictou for Charlottetown, Summerside, and Shediac, on Monday, Wednesday, and Friday, on the arrival of the Halifax train (see Route 44); also for the Gulf ports and Quebec, every Tuesday at 7 A. M., and alternate Fridays at 1 P. M. (see Route 66); also for Port Hood and the Magdalen Islands (see Route 49); and for Hawkesbury and the Strait of Canso.

After the divine Glooscap (see page 106) had left Newfoundland, where he conferred upon the loons the power of weirdly crying when they needed his aid, he landed at Pictou (from *Piktook*, an Indian word meaning "Bubbling," or "Gas-exploding," and referred to the ebullitions of the water near the great coal-beds). Here he created the tortoise tribe, in this wise: Great festivals and games were made in his honor by the Indians of Pictook, but he chose to dwell with a homely, lazy, and despised old bachelor named Mikchickh, whom, after clothing in his own robe and giving him victory in the games, he initiated as the progenitor and king of all the tortoises, smoking him till his coat became brown and as hard as bone, and then reducing his size by a rude surgical operation.

The site of Pictou was occupied in ancient times by a populous Indian village, and in 1763 the French made futile preparations to found a colony here. In 1765, 200,000 acres of land in this vicinity were granted to a company in Philadelphia, whence bands of settlers came in 1767 – 71. Meantime the site of the town had been given to an army officer, who in turn sold it for a horse and saddle. The Pennsylvanians were disheartened at the severity of the climate and the infertility of the soil, and no progress was made in the new colony until 1773, when the ship *Hector* arrived with 180 persons from the Scottish Highlands. They were brought over by the Philadelphia company, but when they found that the shore lands were all taken, they refused to settle on the company's territory, and hence the agent cut off their supply of provisions. They subsisted on fish and venison, with a little flour from Truro, until the next spring, when they sent a ship-load of pine-timber to Britain, and planted wheat and potatoes. Soon afterwards they were joined by 15 destitute families from Dumfriesshire; and at the close of the Revolutionary War many disbanded soldiers settled here with their families. In 1786 the Rev. James McGregor came to Pictou and made a home, and as he was a powerful preacher in the Gaelic language, many Highlanders from the other parts of the Province moved here, and new immigrations arrived from Scotland. In 1788 the town was commenced on its present site by Deacon Patterson, and in 1792 it was made a shire-town. Great quantities of lumber were exported to Britain between 1805 and 1820, during the period of European convulsion, when the Baltic ports were closed, and while the British navy was the main hope of the nation. The place was captured in 1777 by an American privateer. Coal was discovered here in 1798, but the exportation was small until 1827, when the General Mining Association of London began operations.

J. W. Dawson, LL. D., F. R. S., was born at Pictou in 1820, and graduated at the University of Edinburgh in 1840. He studied and travelled with Sir Charles Lyell, and has become one of the leaders among the Christian scientists. His greatest work was the "Acadian Geology." For the past 20 years he has been Principal of the McGill College, at Montreal.

32. St. John and Halifax to the Strait of Canso and Cape Breton.

By the way of the land, through Antigonish.

(Compare also page 12.) The Royal mail-stage leaves New Glasgow (see page 136) daily, on arrival of the morning train from Halifax, — at about 12.30 P.M., — and runs E. to the Strait of Canso, connecting with other stages for Sherbrooke, Guysborough, and all parts of Cape Breton. This route is served by a line of stage-coaches which are said to be "second to none on the continent," though these excellent conveyances are exchanged at Antigonish for less comfortable vehicles. Passengers take supper at Antigonish, ride on all night, and reach the Strait of Canso before dawn. Gentlemen who are planning a summer trip will see that this mode of travel is utterly unsuited for an element in a pleasure-tour, and that it is nearly impracticable for ladies, at least, to endure such a night-journey. The attractions and discomforts of this route are admirably drawn in Charles Dudley Warner's "Baddeck; and that Sort of Thing" (Chapter III.).

Fares. — Halifax to Antigonish, $5; Guysborough, $6; Port Hawkesbury, $7.25 (exclusive of ferriage across the Strait of Canso); St. Peter's, $9.25; Sydney, $12.

Distances. — New Glasgow to French River, 15 M.; Marshy Hope, 25; Antigonish, 38; Tracadie, 58; Port Mulgrave, 74; Port Hawkesbury, 75½; St. Peter's, 114; Sydney, 179.

On reaching the open country beyond New Glasgow, the road passes on for several miles through an uninteresting region of small farms and recent clearings. At the crossing of the Sutherland River, a road diverges to the N. E., leading to *Merigomish*, a shipbuilding hamlet on the coast, with a safe and well-sheltered harbor. In this vicinity are iron and coal deposits, the latter of which are worked by the Merigomish Coal Mining Company, with a capital of $400,000. Beyond the hamlet at the crossing of French River, — "which may have seen better days, and will probably see worse," — the road ascends a long ridge which overlooks the Piedmont Valley to the N. E. Thence it descends through a sufficiently dreary country to the relay-house at *Marshy Hope*.

"The sun has set when we come thundering down into the pretty Catholic village of **Antigonish**, the most home-like place we have seen on the island. The twin stone towers of the unfinished cathedral loom up large in the fading light, and the bishop's palace on the hill, the home of the Bishop of Arichat, appears to be an imposing white barn with many staring windows. . . . . People were loitering in the street; the young beaux going up and down with the belles, after the leisurely manner in youth and summer. Perhaps they were students from St. Xavier College, or visiting gallants from Guysborough. They look into the post-office and the fancy store. They stroll and take their little provincial pleasure, and make love, for all we can see, as if Antigonish were a part of the world. How they must look down on Marshy Hope and Addington Forks and Tracadie! What a charming place to live in is this!" (BADDECK.)

Antigonish[1] (two good inns), the capital of the county of the same name, is situated at the head of a long and shoal harbor, near St. George's Bay. Some shipbuilding is done here, and many cargoes of cattle and butter are sent hence to Newfoundland. On the E. shore of the harbor are valuable deposits of gypsum, which are sent away on coasting-vessels. The inhabitants of the village and the adjacent country are of Scottish descent, and their unwavering industry has made Antigonish a prosperous and pleasant town. The *College of St. Francis Xavier* is the Diocesan Seminary of the Franco-Scottish Diocese of Arichat, and is the residence of the Bishop. It is a Catholic institution, and has six teachers. The *Cathedral of St. Ninian* was begun in 1867, and was consecrated September 13, 1874, by a Pontifical High Mass, at which 7 bishops and 30 priests assisted. It is in the Roman Basilica style, 170 by 70 ft. in area, and is built of blue limestone and brick. On the façade, between the tall square towers, is the Gaelic inscription, *Tighe Dhe* ("the House of God"). The arched roof is supported by 14 Corinthian columns, and the interior has numerous windows of stained glass. The costly chancel-window represents Christ, the Virgin Mary, and St. Joseph. There is a large organ, and also a chime of bells named in honor of St. Joseph and the Scottish saints, Ninian, Columba, and Margaret, Queen of Scotland.

Stages run from Antigonish S. to Sherbrooke and S. E. to Guysborough (see Route 29). N. W. of the village are the bold and picturesque highlands known as the **Antigonish Mts.**, projecting from the general line of the coast about 15 M. N. into the Gulf. They are, in some places, 1,000 ft. high, and have a strong and well-marked mountainous character. Semi-weekly stages run N. from Antigonish to *Morristown* and *Cape George*, respectively 10 and 18 M. distant. 8 – 10 M. N. of the latter is the bold promontory of **Cape St. George**, on which, 400 ft. above the sea, is a powerful revolving white light, which is visible for 25 M. at sea. From this point a road runs S. W. to *Malignant Cove*, which is also accessible by a romantic road through the hills from Antigonish. This is a small seaside hamlet, which derives its name from the fact that H. B. M. frigate *Malignant* was once caught in these narrow waters during a heavy storm, and was run ashore here in order to avoid being dashed to pieces on the iron-bound coast beyond. 4 – 5 M. beyond the Cove is **Arisaig**, a romantically situated settlement of Scottish Catholics, who named their new home in memory of Arisaig, in the Western Highlands. It has a long wooden pier, under whose lee is the only harbor and shelter against east-winds between Antigonish and Merigonish.

The Canso stage leaves Antigonish after dark, and after running 9 M. out and crossing the South River, reaches *Pomquet Forks*, a Franco-Scottish Catholic village of 400 inhabitants. 4 – 5 M. N. is another seaside hamlet. The new Catholic Church of the Holy Cross was consecrated at Pomquet in 1874. The next station is **Tracadie**, a French village of 1,800 inhabitants, situated on a small harbor near St. George's Bay. There is a

[1] *Antigonish*, — accent on the last syllable. It is an Indian word, meaning "the River of Fish."

wealthy monastery here, pertaining to the austere order of the Trappists. Most of the monks are from Belgium. There is also a Convent of Sisters of Charity. The people of Tracadie belong to the old Acadian race, whose sad and romantic history is alluded to on pages 108 and 113. "And now we passed through another French settlement, Tracadie, and again the Norman kirtle and petticoat of the pastoral, black-eyed Evangeline appear, and then pass like a day-dream." (COZZENS.)

The road is now narrowed between the hills and St. George's Bay, and it is beyond midnight. But the exhausted traveller cannot sleep on this rugged track, and can only watch the stars or the moon and think how "these splendors burn and this panorama passes night after night down at the end of Nova Scotia, and all for the stage-driver, dozing along on his box, from Antigonish to the Strait."

At Port Mulgrave the Strait of Canso is reached, and passengers bound for Cape Breton are here ferried across.

The Strait of Canso, see page 142.

CAPE BRETON.

The island of Cape Breton is about 100 M. long by 80 M. wide, and has an area of 2,000,000 acres, of which 800,000 acres consist of lakes and swamps. The S. part is low and generally level, but the N. portion is very irregular, and leads off into unexplored highlands. The chief natural peculiarities of the island are the Sydney coal-fields, which cover 250 square miles on the E. coast, and the Bras d'Or, a great lake of salt water, ramifying through the centre of the island, and communicating with the sea by narrow channels. The exterior coast line is 275 M. long, and is provided with good harbors on the E. and S. shores.

The chief exports of Cape Breton are coal and fish, to the United States; timber, to England; and farm-produce and live-stock to Newfoundland. The commanding position of the island makes it the key to the Canadas, and the naval power holding these shores could control or crush the commerce of the Gulf. The upland soils are of good quality, and produce valuable crops of cereals, potatoes, and smaller vegetables.

The Editor trusts that the following extract from Brown's "History of the Island of Cape Breton" (London : 1869) will be of interest to the tourist : " The summers of Cape Breton, say from May to October, may challenge comparison with those of any country within the temperate regions of the world. During all that time there are perhaps not more than ten foggy days in any part of the island, except along the southern coast, between the Gut of Canso and Scatari. Bright sunny days, with balmy westerly winds, follow each other in succession, week after week, while the midday heats are often tempered by cool, refreshing sea-breezes. Of rain there is seldom enough; the growing crops more often suffer from too little than too much."

" To the tourist that loves nature, and who, for the manifold beauties by hill and shore, by woods and waters, is happy to make small sacrifices of personal comfort, I would commend Cape Breton. Your fashionable, whose main object is company, dress, and frivolous pleasure with the gay, and whose only tolerable stopping-place is the grand hotel, had better content himself with reading of this island." (NOBLE.)

The name of the island is derived from that of its E. cape, which was given in honor of its discovery by Breton mariners. In 1713 the French authorities bestowed upon it the new name of *L'Isle Royale,* during the reign of Louis XIV. At this time, after the cession of Acadia to the British Crown, many of its inhabitants emigrated to Cape Breton; and in August, 1714, the fortress of Louisbourg was founded. During the next half-century occurred the terrible wars between France and Great Britain, whose chief incidents were the sieges of Louisbourg and the final demolition of that redoubtable fortress. In 1765 this island was annexed to the Province of Nova Scotia. In 1784 it was erected into a separate Province, and continued as such until 1820, when it was reannexed to Nova Scotia. In 1815 Cape Breton had about 10,000 inhabitants, but in 1871 its population amounted to 75,503, a large proportion of whom were from the Scottish Highlands.

33. The Strait of Canso.

The Gut of Canso, or (as it is now more generally called) the Strait of Canso, is a picturesque passage which connects the Atlantic Ocean with the Gulf of St. Lawrence, and separates the island of Cape Breton from the shores of Nova Scotia. The banks are high and mountainous, covered with spruce and other evergreens, and a succession of small white hamlets lines the coves on either side. This grand avenue of commerce seems worthy of its poetic appellation of "The Golden Gate of the St. Lawrence Gulf." It is claimed that more keels pass through this channel every year than through any other in the world except the Strait of Gibraltar. It is not only the shortest passage between the Atlantic and the Gulf, but has the advantage of anchorage in case of contrary winds and bad weather. The shores are bold-to and free from dangers, and there are several good anchorages, out of the current and in a moderate depth of water. The stream of the tide usually sets from the S., and runs in great swirling eddies, but is much influenced by the winds. The strait is described by Dawson as "a narrow transverse valley, excavated by the currents of the drift period," and portions of its shores are of the carboniferous epoch.

The Strait of Canso is traversed by several thousand sailing-vessels every year, and also by the large steamers of the Boston and Colonial Steamship Company, and (as far as Port Hawkesbury) by the vessels of the P. E. I. Steam Navigation Company.

"So with renewed anticipations we ride on toward the strait 'of unrivalled beauty,' that travellers say ' surpasses anything in America.' And, indeed, Canseau can have my feeble testimony in confirmation. It is a grand marine highway, having steep hills on the Cape Breton Island side, and lofty mountains on the other shore; a full, broad, mile-wide space between them; and reaching, from end to end, fifteen miles, from the Atlantic to the Gulf of St. Lawrence." (COZZENS.)

Vessels from the S., bound for the Strait of Canso, first approach the Nova-Scotian shores near *Cape Canso* (see page 134), whose lights and islands are rounded, and the course lies between N. W. and W. N. W. towards Eddy Point. If a fog prevails, the steam-whistle on Cranberry Island will be heard giving out its notes of warning, sounding for 8 seconds in each minute, and heard for 20 M. with the wind, for 15 M. in calm weather, and 5 - 8 M. in stormy weather and against the wind. On the l. is Chedabucto Bay, stretching in to Guysborough, lined along its S. shore by hills 3 - 700 ft. high; and on the r. the Isle Madame is soon approached. 28 - 30 M. beyond Cape Canso the vessel passes *Eddy Point*, on which are two fixed white lights (visible 8 M.). On the starboard beam is Janvrin Island, beyond which is the broad estuary of *Habitants Bay*. On the Cape-Breton shore is the hamlet of Bear Point, and on the l. are Melford Creek (with its church), Steep Creek, and Pirate's Cove. The hamlets of Port Mulgrave and Port Hawkesbury are now seen, nearly opposite each other, and half-way up the strait.

Port Mulgrave (two inns) is a village of about 400 inhabitants, on the Nova-Scotia side of the strait. It is engaged in the fisheries, and has a harbor which remains open all the year round. Gold-bearing quartz is found in the vicinity; and bold hills tower above the shore for a long distance. A steam ferry-boat plies between this point and Port Hawkesbury, 1½ M. distant.

Port Hawkesbury (*Hawkesbury Hotel*, comfortable; *Acadia Hotel*) is a village of about 700 inhabitants, on the Cape-Breton side of the strait. It is situated on Ship Harbor, a snug haven for vessels of 10-ft. draught, marked by a fixed red light on Stapleton Point. This is the best harbor on the strait, and has very good holding-ground. The village is of a scattered and half-finished appearance, and has two small churches. There are several wharves here, which are visited by the Boston and the Prince Edward Island steamers, and other lines. Stages run hence to Sydney, Arichat, and West Bay, on the Bras d'Or; and a railway has been surveyed to the latter point. The fare from Port Hawkesbury to Charlottetown, by the vessels of the P. E. I. Steam Navigation Company, is $ 3.50.

Port Hastings (more generally known as *Plaster Cove*) is about 4 M. above Port Hawkesbury, on the Cape-Breton shore, and is built on the bluffs over a small harbor in which is a Government wharf. From this point the Cape-Breton mails are distributed through the island by means of the stage-lines. The village is about the size of Port Hawkesbury, and has a lucrative country-trade, besides a large exportation of fish and cattle to Newfoundland and the United States. It derives its chief interest from being the point where the Atlantic-Cable Company transfers its messages, received from all parts of Europe and delivered under the sea, to the Western Union Telegraph Company, by which the tidings are sent away through the Dominion and the United States. The telegraph-office is in a long two-story building near the strait, and 20 - 30 men are employed therein. The hotel at this village has been justly execrated in several books of travel, but occupies a noble situation, overlooking, from a high bluff, the Strait of Canso for several miles to the S. E. Near this building is the consulate of the United States, over which floats the flag of the Republic.

Totiusque Americæ Septentrionalis Domina, Fidei Defensor, etc. In Cujus harum insularum vulgo Cape Breton, Proprietatis et Dominii Testimonium, Hoc Erexit Monumentum, Suæ Majestatis Servus, et Subditus fidelissimus, D. Hovenden Walker, Eques Auratus, Omnium in America Navium Regalium, Præfectus et Thalassiarcha. Monte Septembris, Anno Salutis MDCCXI."

The first civil governor of Cape Breton after its severance from Nova Scotia (1784) was Major Desbarres, a veteran of the campaigns of the Mohawk Valley, Lake George, Ticonderoga, Louisbourg, and Quebec. One of his chief steps was to select a site for the new capital of the island, and the location chosen was the peninsula on the S. arm of the capacious harbor called Spanish River. The seat of government thus established was named Sydney, in honor of Lord Sydney, Secretary of State for the Colonies, who had erected Cape Breton into a separate Province. In the spring of 1785 the Loyalists under Abraham Cuyler (ex-Mayor of Albany, N. Y.) came from Louisbourg to Sydney, cut down the forests, and erected buildings.

In 1781 a sharp naval battle was fought off Sydney Harbor, between the French frigates *L'Astrée* and *L'Hermione* (of 44 guns each) and a British squadron consisting of the *Charlestown*, 28, *Allegiance*, 16, *Vulture*, 16, *Little Jack*, 6, and the armed transport *Vernon*. 16 coal-ships which were under convoy of the British fleet fled into Sydney harbor, while the frigates rapidly overhauled the escort and brought on a general engagement. After a long and stubborn action, the *Little Jack* surrendered, and the remainder of the fleet would have shared the same fate, had it not been for the approach of night, under whose shelter the shattered British vessels bore away to the eastward and escaped. They had lost 18 men killed and 28 wounded. The senior captain of the victorious French vessels was La Perouse, who started in 1788, with two frigates, on a voyage of discovery around the world, but was lost, with all his equipage, on the Isle of Vanikoro.

37. The East Coast of Cape Breton.—The Sydney Coal-Fields.

The Sydney Mines are on the N. side of Sydney Harbor, and are connected with N. Sydney by a coal-railway and also by a daily stage (fare, 75c.). They are on the level land included between the Little Bras d'Or and the harbor of Sydney, and are worked by the General Mining Association of London. Nearly 500 men are employed in the pits, and the village has a population of 2,500.

The International Mines are at Bridgeport, 13 M. N. E. of Sydney, and are connected with that harbor by a railway that cost $500,000. The seashore is here lined with rich coal-deposits, extending from Lingan Harbor to Sydney. It is probable that the submarine mining, which has already been commenced, will follow the carboniferous strata far beneath the sea.

The Victoria Mines are W. of this district, and near Low Point, 9 M. from Sydney. The company has a railway which extends to their freighting station on Sydney Harbor, and is at present doing a prosperous business.

The Lingan Mines are near Bridgeport, and are reached by a tri-weekly stage from Sydney (15 M.; fare, $1.50). Lingan is derived from the French word L'Indienne, applying to the same place. It was occupied and fortified by the British early in the 18th century, and a garrison of 50 men was stationed here to guard the coal-mines. At a later day the French army at Louisbourg was supplied with large quantities of coal from this point, and several cargoes were sent away. During the summer of 1752 the mine was set on fire, and the fort and buildings were all destroyed.

The Little Glace Bay Mines are 18 M. from Sydney, and are reached by a tri-weekly stage (fare, $ 1). They are situated on Glace Bay and Glace Cove, and about Table Head, and are carried on by a Halifax company, which employs 300 miners. The deposits are very rich along this shore, and extend far out beneath the sea.

The Gowrie and Block-House Mines are on Cow Bay, and are among the most extensive on this coast. They are 22 M. from Sydney, and are reached by a tri-weekly stage. They employ over 600 men, and have formed a town of 2,000 inhabitants. Large fleets gather in the bay for the transportation of the coal to the S., and while lying here are in considerable peril during the prevalence of easterly gales, which have a full sweep into the roadstead. Nearly 70 vessels were wrecked here during the Lord's-Day Gale, and the shores were strewn with broken hulks and many yet sadder relics of disaster. The S. portal of the bay is Cape Morien, and on the N. is Cape Perry, off which is the sea-surrounded Flint Island, bearing a revolving white light.

The coal-beds of Cape Breton were first described by Denys, in 1672, and from 1677 to 1690 he had a royalty of 20 sous per ton on all the coal that was exported. Some of it was taken to France, and great quantities were sent into New England. In 1720 a mine was opened at Cow Bay, whence the French army at Louisbourg was supplied, and numerous cargoes were shipped to Boston. Between 1745 and 1749 the British garrison at Louisbourg was abundantly supplied with fuel from mines at Burnt Head and Little Bras d'Or, which were protected against the Indians by fortified outposts. The Abbé Raynal says that there was "a prodigious demand for Cape-Breton coal from New England from the year 1745 to 1749." But this trade was soon stopped by the British government, and only enough mining was done to supply the troops at Louisbourg and Halifax. The "coal-smugglers" still carried on a lucrative business, slipping quietly into the harbors and mining from the great seams in the face of the cliffs. In 1785 the Sydney vein was opened by Gov. Desbarres, but its profitable working was prevented by heavy royalties. The Imperial Government then assumed the control, and its vessels captured many of the light craft of the smugglers. In 1828 the General Mining Association was formed in London, and secured the privilege of the mines and minerals of Nova Scotia and Cape Breton from the Duke of York, to whom they had been granted by King George IV. Under the energetic management of the Association the business increased rapidly, and became profitable. Between 1827 and 1857 (inclusive), 1,931,684 tons of coal were mined in Cape Breton, of which 605,008 tons were sent to the United States. Between 1857 and 1870 there were sold at the mines 3,323,981 tons. By far the greater part of these products came from the Sydney field, but of late years considerable exportations are being made from the mines at Glace Bay, Cow Bay (Block-House), Gowrie, and Lingan. The Caledonia, Glace Bay, and Block-House coals are used for making gas at Boston and Cambridge, and the gas of New York is made from International, Glace Bay, Caledonia, and Block-House coals.

35. The Strait of Canso to Sydney, C. B.

By the way of the land, through St. Peter's.

The Royal mail-stage leaves Port Hawkesbury every morning, some time after the arrival of the Antigonish stage, and runs E. and N. E. to Sydney. Fare, $5. This is one of the most arduous routes by which Sydney can be approached, and leads through a thinly settled and uninteresting country until St. Peter's is reached. Beyond that point there is a series of attractive views of the Great Bras d'Or and St. Andrew's Channel, continuing almost to Sydney.

Distances.—(Port Hastings to Port Hawkesbury, 4-5 M.) Port Hawkesbury to Grand Anse, 21 M.; St. Peter's, 35; Red Island, 52; Irish Cove, 64; Sydney, 100.

There is but little to interest the traveller during the first part of the journey. After leaving Port Hawkesbury, the stage enters a rugged and unpromising country, leaving the populous shores of Canso and pushing E. to the River Inhabitants. Crossing that stream where it begins to narrow, the road continues through a region of low bleak hills, with occasional views, to the r., of the deeper coves of the Lennox Passage. Before noon it reaches the narrow Haulover Isthmus, which separates St. Peter's Bay, on the Atlantic side, from St. Peter's Inlet, on the Bras d'Or side. At this point is situated the village of **St. Peter's** (two inns), a Scottish settlement near the bay. The canal which has been constructed here to open communication between the Atlantic and the Bras d'Or is ½ M. long, 26 ft. wide, and 13 ft. deep, and is expected to be of much benefit to the Bras d'Or villages. It has been finished within a few years, and pertains to the Government, which takes a small toll from the vessels passing through. S. E. of St. Peter's are the bluff heights of Mt. Granville, and to the N. W. are the uninhabited highlands which are called on the maps the Sporting Mts.

St. Peter's was founded by M. Denys, about the year 1636, to command the lower end of the Bras d'Or, as his post at St. Anne's commanded the upper end. He built a portage-road here, opened farm-lands, and erected a fort which mounted several cannon. The Indians residing on the most remote arms of the Bras d'Or were thus enabled to visit and carry their furs and fish to either one of Denys's forts. Denys himself, together with the fort, the ship, and all other property here, was captured soon after by a naval force sent out by M. le Borgne. But in 1656 Denys retook his posts, guarded by a charter from King Louis. A few years later St. Peter's was captured by La Giraudière, but was afterwards restored to Denys, who, however, abandoned the island about 1670, when all his buildings at this post were destroyed by fire. In 1737 St. Peter's was fortified by M. de St. Ovide, the commandant at Louisbourg; but during the New-England crusade against the latter city, in 1745, it was captured and plundered by Col. Moulton's Massachusetts regiment. In 1752 St. Peter's was the chief depot of the fur-trade with the Micmacs, and was surrounded with fruitful farms. It was then called *Port Toulouse*, and was connected with Louisbourg by a military road 18 leagues in length, constructed by the Count de Raymond. Besides the garrison of French troops, there was a civil population of 230 souls; and in 1760 Port Toulouse had grown to be a larger town than even Louisbourg itself. The King of France afterwards reprimanded the Count de Raymond for constructing his military road, saying that it would afford the English an opportunity to attack Louisbourg on the landward side.

From the Strait of Canso to Grand River the coast is occupied by a line of humble and retired villages, inhabited by Acadian-French fishermen. 7-8 M. S. E. of St. Peter's are the *L'Ardoise* settlements (so named because a slate-quarry was once worked here). In 1750 there was a large French village here, with a garrison of troops, and L'Ardoise was the chief depot of the fur-trade with the Indians. At Grand River the character of the population changes, though the names of the settlement would indicate, were history silent, that the towns beyond that point were originally founded by the French. They are now occupied exclusively by the Scotch, whose light vessels put out from the harbors of Grand River, L'Archevêque, St. Esprit, Blancherotte, Framboise, and Fourchu, on which are fishing-villages.

A few miles N. E. of St. Peter's the stage crosses the Indian Reservation near Louis Cove. *Chapel Island* is a little way off shore, and is the largest of the group of islets at the mouth of St. Peter's Inlet. These islands were granted by the government, in 1792, to the Micmac chiefs Bask and Tomma, for the use of their tribe, and have ever since been retained by their descendants. On the largest island is a Catholic chapel where all the Micmacs of Cape Breton gather, on the festival of St. Anne, every year, and pass several days in religious ceremonies and aboriginal games. Beyond this point the road runs N. E. between Soldier's Cove and the bold highlands on the r. and traverses the Red-Island Settlement, off which are the Red Islands.

"The road that skirts the Arm of Gold is about 100 M. in length. After leaving Sydney you ride beside the Spanish River a short distance, until you come to the portage, which separates it from the lake, and then you follow the delicious curve of the great beach until you arrive at St. Peter's. . . . . There is not a lovelier ride by white-pebbled beach and wide stretch of wave. Now we roll along amidst primeval trees, — not the evergreens of the sea-coast, but familiar growths of maple, beech, birch, and larches, juniper, or hackmatack, — imperishable for shipcraft; now we cross bridges, over sparkling brooks alive with trout and salmon. . . . . To hang now in our curricle, upon this wooded hill-top, overlooking the clear surface of the lake, with leafy island, and peninsula dotted in its depths, in all its native grace, without a touch or trace of handiwork, far or near, save and except a single spot of sail in the far-off, is holy and sublime." (COZZENS.)

About 10 M. beyond the Red Island Settlement is the way-office and village at Irish Cove, whence a road runs 10 – 12 M. S. E. across the highlands to the Grand-River Lake, or *Loch Lomond*, a picturesque sheet of water 5 – 6 M. long, studded with islets and abounding in trout. The Scottish hamlets of Loch Lomond and Lochside are on its shores; and on the N., and connected by a narrow strait, is *Loch Uist*. The road crosses the lake and descends to Framboise Harbor, on the Atlantic coast.

N. of Loch Uist, and about 7 M. from the Bras d'Or, is a remarkable saline spring, containing in each gallon 343 grains of chloride of sodium, 308 of chloride of calcium, and 9 of the chlorides of magnesium and potassium. This water is singularly free from sulphurous contamination, and has been found very efficient in cases of asthma, rheumatism, and chronic headache. There are no accommodations for visitors.

About 6 M. N. W. of Irish Cove is seen *Benacadie Point*, at the entrance to the **East Bay**, a picturesque inlet of the Bras d'Or, which ascends for 18 – 20 M. to the N. E., and is bordered by lines of bold heights. Near its N. shore are several groups of islands, and the depth of the bay is from 8 to 32 fathoms. The stage follows its shore to the upper end. Above Irish Cove the road lies between the bay and a mountain 600 ft. high, beyond which is *Cape Rhumore*. 3 – 4 M. farther on is Loch an Fad, beyond which a roadside chapel is seen, and the road passes on to *Edoobekuk*, between the heights and the blue water. The opposite shore (4 M. distant) is occupied by the Indians, whose principal village is called *Escasoni*, and is situated near the group of islands in Crane Cove. The bay now diminishes to 2 M. in width, and is followed to its source in the lagoon of Tweednogie. The aggregate number of inhabitants, Scottish and Indian, along the shores of the East Bay, is a little over 2,000. The stage crosses the narrow isthmus (4-5 M.), and then follows the line of the Forks Lake and the Spanish River, to the town of Sydney.

Sydney, see page 150.

36. Halifax to Sydney, Cape Breton.
By the Sea.

There is an indirect route by the Boston & Colonial steamships to Port Hawkesbury, thence by stage to West Bay, and up the Bras d'Or to Sydney.

The Anglo-French Steamship Company's vessel, the *George Shattuck*, leaves Halifax on alternate Saturdays at 3 P. M., for Sydney and St. Pierre (see Route 50). Fares (including meals), Halifax to Sydney, cabin, $10; steerage, $6.

The Eastern Steamship Company's vessel, the *Virgo*, leaves Halifax on alternate Tuesdays for Sydney and St. John's, N. F. (see also Route 51). Fares to Sydney, $8 and $5 (including state-rooms, but not meals); or for an excursion to Sydney and return, $12. For further particulars, or to make certain of the days of starting, address Wilkinson, Wood, & Co., Halifax, N. S. The *Virgo* is much better adapted for carrying passengers than is the *George Shattuck*.

Halifax Harbor, see page 93.

The course of the steamship is almost always within sight of land, a cold, dark, and rock-bound coast, off which are submerged ledges on which the sea breaks into white foam. This coast is described in Routes 28 and 29; but of its aspect from the sea the Editor can say nothing, as he was obliged to traverse the route as far as Canso by night.

After passing the bold headland of Cape Canso, the deep bight of Chedabucto Bay is seen on the W., running in to Guysborough and the Strait of Canso. Between Cape Canso and Red Point, on Cape Breton, the opening is about 30 M. wide, inside of which are Isle Madame (Route 34) and St. Peter's Bay. The course of the vessel, after crossing this wide opening, converges toward the Breton coast, which is, however, low and without character, and is studded with white fishing-hamlets. *St. Esprit* is visible, with its little harbor indenting the coast.

About the middle of the last century the British frigate *Tilbury*, 64, was caught on this shore during a heavy gale of wind, and was unable to work off, in spite of the utmost exertions of her great crew. The Tilbury Rocks, off St. Esprit, still commemorate the place where she finally struck and went to pieces. 200 sailors were either drowned or killed by being dashed on the sharp rocks, and 200 men and 15 officers were saved from the waves by the French people of St. Esprit, who nourished and sheltered them with tender care. England and France being then at war, the survivors of the *Tilbury's* crew were despatched to France as prisoners, on the French frigate *Hermione*. This vessel was, however, captured in the English Channel, and the sailors were released.

Beyond St. Esprit the coves of Framboise and Fourchu make in from the sea, and above the deep inlet of Gabarus Bay the lighthouse of Louisbourg (see Route 38) may perhaps be seen.

In 1744 the French ships *Notre Dame de la Délivrance, Louis Érasme,* and *Marquis d'Antin* sailed from Callao (Peru), with a vast amount of treasure on board, concealed under a surface-cargo of cocoa. The two latter were captured off the Azores by the British privateers *Prince Frederick* and *Duke,* but during the 3 hours' action the *Notre Dame* escaped. Not daring to approach the French coast while so many hostile privateers were cruising about, she crowded all sail and bore away for Louisbourg. 20 days later she sighted Scatari, and it seemed that her valuable cargo was already safe. But she was met, a short distance to the S., by a British fleet, and became a prize. Among the people captured on the *Notre Dame* was Don Antonio d'Ulloa, the famous Spanish scientist, who was kept here in light captivity for two months, and who afterwards wrote an interesting book about Cape Breton. The lucky vessels that made the capture were the *Sunderland, Boston,* and *Chester,* and their crews had great prize-money, — for over $4,000,000 was found on the *Notre Dame,* in bars and ingots of gold and silver.
In 1756 the French frigate *Arc-en-Ciel,* 50, and the *Amitié* were captured in these waters by H. B. M. ships *Centurion* and *Success.* In July, 1756, the French vessels *Héros,* 74, *Illustre,* 64, and two 36-gun frigates met H. B. M. ships *Grafton,* 70, *Nottingham,* 70, and the *Jamaica* sloop, and fought from mid-afternoon till dark. The action was indecisive, and each fleet claimed that the other stole away at night. The loss of men on both sides was considerable.
In May, 1745, a gallant naval action was fought hereabouts between the French ship-of-the-line *Vigilant* and Com. Warren's fleet, consisting of the *Superb* (60-gun ship), and the *Launceston, Mermaid,* and *Eltham* (40-gun frigates). The *Vigilant* was carrying a supply of military goods from Brest to Louisbourg, and met the *Mermaid,* standing off and on in the fog. The latter made sail and fled toward the squadron, and the *Vigilant* swept on in the fog and ran into the midst of the British fleet. Warren's ships opened fire on every side, but the French captain, the Marquis de Maisonforte, refused to surrender, though his decks were covered with stores and his lower batteries were below the water-line by reason of the heavy cargo. The battle was terrific, and lasted for 7 hours, while Maisonforte kept his colors flying and his cannon roaring until all his rigging was cut away by the British shot, the rudder was broken, the forecastle battered to pieces, and great numbers of the crew wounded or dead.

The steamship now runs out to round Scatari, traversing waters which maintain a uniform depth of over 80 fathoms. On the W. is the promontory of **Cape Breton,** from which the island receives its name. It is a low headland, off which is the dark rock of Porto Nuevo Island.

There is an old French tradition to the effect that Verazzano, the eminent Florentine navigator, landed near Cape Breton on his last voyage, and attempted to found a fortified settlement. But being suddenly attacked and overpowered by the Indians, himself and all his crew were put to death in a cruel manner. It is known to history that this discoverer was never heard from after leaving France on his last voyage (in 1525).
It is believed that Cape Breton was first visited by the *Marigold* (70 tons), in 1593; whereof it is written : " Here diuers of our men went on land vpon the very cape, where, at their arriuall they found the spittes of oke of the Sauages which had roasted meate a little before. And as they viewed the countrey they saw diuers beastes and foules, as blacke foxes, deeres, otters, great foules with redde legges, penguines, and certaine others." Thence the *Marigold* sailed to the site of Louisbourg, where her crew landed to get water, but were driven off shore by the Indians.
The cape probably owes its name to the fact of its being visited by the Breton and Basque fishermen, who in those days frequented these seas. Cape Breton was at that time a prosperous commercial city, near Bayonne, in the South of France. It was frequented by the Huguenots about this time, and had large fleets engaged in the fisheries. By the changing of the course of the Adour River, and the drifting of sand into its harbor, its maritime importance was taken away, and in 1841 it had but 920 inhabitants. (*Dictionnaire Encyclopédique.*)
In 1629 Lord Ochiltree, the son of the Earl of Arran, came out with 60 colonists,

39. The North Shore of Cape Breton. — St. Anne's Bay and St. Paul's Island.

Conveyances may be hired at Baddeck (see page 162) by which to visit St. Anne's. The distance is about 10 M. to the head of the harbor. The first part of the way leads along the shores of Baddeck Bay, with the promontory of Red Head over the water to the r. The road then crosses a cold district of denuded highlands, and descends to the * *Valley of St. Anne.* As the harbor is approached, the traveller can see the amphitheatrical glens in which the great Holy Fairs or annual religious communions of the people are held. These quaint Presbyterian camp-meetings are said to be a relic of the ancient churches in the Scottish Highlands. The shores of the harbor were occupied in 1820 by immigrants from the Highlands, who are now well located on comfortable farms. The road follows the S. Arm, and to the l. is seen the N. Arm, winding away among the tall mountains. Just E. of the N. Arm is St. Anne's Mt., which is 1,070 ft. high, and pushes forward cliffs 960 ft. high nearly to the water's edge.

" There is no ride on the continent, of the kind, so full of picturesque beauty and constant surprises as this around the indentations of St. Anne's harbor. High bluffs, bold shores, exquisite sea-views, mountainous ranges, delicious air," are found here in abundance. About opposite the lighthouse on the bar, at the mouth of the harbor, is *Old Fort Point*, on which the French batteries were established. Near this point is the hamlet of *Englishtown,* chiefly interesting as containing the grave of the once famous " Nova-Scotia Giant." The mountains back of Englishtown are over 1,000 ft. high, and run N. E. to Cape Dauphin, whence they repel the sea. Imray's *Sailing Directions* states that " on the N. side the land is very high, and ships-of-war may lie so near the shore that a water-hose may reach the fresh water." As to the harbor, the ancient description of Charlevoix still holds good: —

" Port Ste. Anne, as already stated, has before it a very sure roadstead between the Cibou Islands. The port is almost completely closed by a tongue of land, leaving passage for only a single ship. This port, thus closed, is nearly two leagues in circuit, and is oval in form. Ships can everywhere approach the land, and scarcely perceive the winds, on account of its high banks and the surrounding mountains. . . . . The fishing is very abundant; great quantities of good wood are found there, such as maple, beech, wild cherry, and especially oaks very suitable for building and masts, being 28 – 38 ft. high ; marble is common ; most of the land good, — in Great and Little Labrador, which are only a league and a half off, the soil is very fertile, and it can contain a very large number of settlers."

In St. Anne's Bay the English ship *Chancewell* was wrecked in 1597, and while she lay aground " there came aboord many shallops with store of French men, who robbed and spoyled all they could lay their hands on, pillaging the poore men euen to their very shirts, and vsing them in sauage manner ; whereas they should rather as Christians haue aided them in that distresse." In 1629 this harbor was occupied by the *Great St. Andrew* and the *Marguerite*, armed vessels of France, whose crews, together with their English prisoners, constructed a fort to command the entrance It was armed with 8 cannon, 1,800 pounds of powder, pikes, and muskets, and was garrisoned by 40 men. The commander of the fleet raised the arms of the King and of Cardinal Richelieu over its walls, and erected a chapel, for whose care he left two

INGONISH. *Route 39.* 159

Jesuits. He then named the harbor St. Anne's. Before the close of that winter more than one third of the troops died of the scurvy, and the commandant assassinated his lieutenant on the parade-ground. In 1634 the Jesuits founded an Indian mission here, but both this and the garrison were afterwards withdrawn. Some years later a new battery and settlement were erected here by Nicholas Denys, Sieur de Fronsac, who traded hence with the Indians of the N. of Cape Breton.
The valley of the N. Arm of St. Anne's was granted, in 1713, to M. de Rouville, a captain in the infantry of France, and brother of that Hertel de Rouville who led the forces that destroyed Schenectady, Deerfield, and Haverhill. The N. Arm was long called Rouville's River. At a later day Costabelle, Beaucourt, Soubras, and other French officers had fishing-stations on the bay. In 1745 a frigate from Com. Warren's fleet (then blockading Louisbourg) entered the harbor, and destroyed all the property on its shores. St. Anne's Bay was afterwards called *Port Dauphin* by the French, and the government long hesitated as to whether the chief fortress of Cape Breton should be located here or at Louisbourg. The perfect security of the harbor afforded a strong argument in favor of St. Anne's, and it seemed capable of being made impregnable at slight expense. After the foundation of Louisbourg 1,000 cords of wood were sent to that place annually from St. Anne's.

The road from the Bras d'Or to the N. shore of Cape Breton diverges from the St. Anne road before reaching the harbor, and bears to the N. E., along the W. Branch. It rounds the North-River Valley by a great curve, and then sweeps up the harbor-shore, under the imposing cliffs of St. Anne's Mt. From St. Anne's to Ingonish the distance is about 40 M., by a remarkably picturesque road between the mountains and the Atlantic, on a narrow plain, which recalls Byron's lines: —

"The mountains look on Marathon,
And Marathon looks on the sea."

"Grand and very beautiful are the rocky gorges and ravines which furrow the hills and precipices between St. Anne's and Ingonish. . . . . Equally grand and picturesque is the red syenitic escarpment of Smoky Cape, capped with the cloud from which it derives its name, with many a lofty headland in the background, and the peak of the Sugar-loaf Mountain just peeping above the far-distant horizon." (BROWN.)

The proud headland of **Cape Smoky** (the *Cap Enfumé* of the French) is 950 ft. high, and runs sheer down into the sea. To the W. there are peaks 1,200 – 1,300 ft. high; and as the road bends around the deep bights to the N., it passes under summits more than 1,400 ft. high. Among these massive hills, and facing Cape Smoky, is the village of **Ingonish**, inhabited by Scottish Catholic fishermen, 800 of whom are found in this district. On the island that shelters the harbor is a fixed white light, 237 ft. above the sea, and visible for 15 M.

Ingonish was one of the early stations of the French. In 1729 a great church was built here, whose foundations only remain now; and in 1849 a church-bell, marked *St. Malo*, 1729, and weighing 200 pounds, was found buried in the sands of the beach. The settlement here was probably ruined by the drawing away of its people to aid in holding Louisbourg against the Anglo-American forces. In 1740 Ingonish was the second town on the island, and its fleet caught 13,560 quintals of fish. It was destroyed, in 1745, by men-of-war from Com. Warren's fleet.
The highland region back of Ingonish has always been famous for its abundance of game, especially of moose and caribou. In the winter of 1789 over 9,000 moose were killed here for the sake of their skins, which brought ten shillings each; and for many years this wholesale slaughter went on, and vessels knew when they were approaching the N. shore of Cape Breton by the odor of decaying carcasses which came from the shore. Finally the outraged laws of the Province were vindicated by the occupation of Ingonish by a body of troops, whose duty it was to restrain the

Totiusque Americæ Septentrionalis Domina, Fidei Defensor, etc. In Cujus harum insularum vulgo Cape Breton, Proprietatis et Dominii Testimonium, Hoc Erexit Monumentum, Suæ Majestatis Servus, et Subditus fidelissimus, D. Hovenden Walker, Eques Auratus, Omnium in America Navium Regalium, Præfectus et Thalassiarcha. Monte Septembris, Anno Salutis MDCCXI."

The first civil governor of Cape Breton after its severance from Nova Scotia (1784) was Major Desbarres, a veteran of the campaigns of the Mohawk Valley, Lake George, Ticonderoga, Louisbourg, and Quebec. One of his chief steps was to select a site for the new capital of the island, and the location chosen was the peninsula on the S. arm of the capacious harbor called Spanish River. The seat of government thus established was named Sydney, in honor of Lord Sydney, Secretary of State for the Colonies, who had erected Cape Breton into a separate Province. In the spring of 1785 the Loyalists under Abraham Cuyler (ex-Mayor of Albany, N. Y.) came from Louisbourg to Sydney, cut down the forests, and erected buildings.

In 1781 a sharp naval battle was fought off Sydney Harbor, between the French frigates *L'Astrée* and *L'Hermione* (of 44 guns each) and a British squadron consisting of the *Charlestown*, 28, *Allegiance*, 16, *Vulture*, 16, *Little Jack*, 6, and the armed transport *Vernon*. 16 coal-ships which were under convoy of the British fleet fled into Sydney harbor, while the frigates rapidly overhauled the escort and brought on a general engagement. After a long and stubborn action, the *Little Jack* surrendered, and the remainder of the fleet would have shared the same fate, had it not been for the approach of night, under whose shelter the shattered British vessels bore away to the eastward and escaped. They had lost 18 men killed and 28 wounded. The senior captain of the victorious French vessels was La Perouse, who started in 1788, with two frigates, on a voyage of discovery around the world, but was lost, with all his equipage, on the Isle of Vanikoro.

37. The East Coast of Cape Breton.—The Sydney Coal-Fields.

The *Sydney Mines* are on the N. side of Sydney Harbor, and are connected with N. Sydney by a coal-railway and also by a daily stage (fare, 75c.). They are on the level land included between the Little Bras d'Or and the harbor of Sydney, and are worked by the General Mining Association of London. Nearly 500 men are employed in the pits, and the village has a population of 2,500.

The *International Mines* are at Bridgeport, 13 M. N. E. of Sydney, and are connected with that harbor by a railway that cost $500,000. The sea-shore is here lined with rich coal-deposits, extending from Lingan Harbor to Sydney. It is probable that the submarine mining, which has already been commenced, will follow the carboniferous strata far beneath the sea.

The *Victoria Mines* are W. of this district, and near Low Point, 9 M. from Sydney. The company has a railway which extends to their freighting station on Sydney Harbor, and is at present doing a prosperous business.

The *Lingan Mines* are near Bridgeport, and are reached by a tri-weekly stage from Sydney (15 M.; fare, $1.50). Lingan is derived from the French word L'Indienne, applying to the same place. It was occupied and fortified by the British early in the 18th century, and a garrison of 50 men was stationed here to guard the coal-mines. At a later day the French army at Louisbourg was supplied with large quantities of coal from this point, and several cargoes were sent away. During the summer of 1752 the mine was set on fire, and the fort and buildings were all destroyed.

The *Little Glace Bay Mines* are 18 M. from Sydney, and are reached by a tri-weekly stage (fare, $1). They are situated on Glace Bay and Glace Cove, and about Table Head, and are carried on by a Halifax company, which employs 300 miners. The deposits are very rich along this shore, and extend far out beneath the sea.

The *Gowrie and Block-House Mines* are on Cow Bay, and are among the most extensive on this coast. They are 22 M. from Sydney, and are reached by a tri-weekly stage. They employ over 600 men, and have formed a town of 2,000 inhabitants. Large fleets gather in the bay for the transportation of the coal to the S., and while lying here are in considerable peril during the prevalence of easterly gales, which have a full sweep into the roadstead. Nearly 70 vessels were wrecked here during the Lord's-Day Gale, and the shores were strewn with broken hulks and many yet sadder relics of disaster. The S. portal of the bay is Cape Morien, and on the N. is Cape Perry, off which is the sea-surrounded Flint Island, bearing a revolving white light.

The coal-beds of Cape Breton were first described by Denys, in 1672, and from 1677 to 1690 he had a royalty of 20 sous per ton on all the coal that was exported. Some of it was taken to France, and great quantities were sent into New England. In 1720 a mine was opened at Cow Bay, whence the French army at Louisbourg was supplied, and numerous cargoes were shipped to Boston. Between 1745 and 1749 the British garrison at Louisbourg was abundantly supplied with fuel from mines at Burnt Head and Little Bras d'Or, which were protected against the Indians by fortified outposts. The Abbé Raynal says that there was " a prodigious demand for Cape-Breton coal from New England from the year 1745 to 1749." But this trade was soon stopped by the British government, and only enough mining was done to supply the troops at Louisbourg and Halifax. The " coal-smugglers " still carried on a lucrative business, slipping quietly into the harbors and mining from the great seams in the face of the cliffs. In 1785 the Sydney vein was opened by Gov. Desbarres, but its profitable working was prevented by heavy royalties. The Imperial Government then assumed the control, and its vessels captured many of the light craft of the smugglers. In 1828 the General Mining Association was formed in London, and secured the privilege of the mines and minerals of Nova Scotia and Cape Breton from the Duke of York, to whom they had been granted by King George IV. Under the energetic management of the Association the business increased rapidly, and became profitable. Between 1827 and 1857 (inclusive), 1,931,634 tons of coal were mined in Cape Breton, of which 605,008 tons were sent to the United States. Between 1857 and 1870 there were sold at the mines 3,323,981 tons. By far the greater part of these products came from the Sydney field, but of late years considerable exportations are being made from the mines at Glace Bay, Cow Bay (Block-House), Gowrie, and Lingan. The Caledonia, Glace Bay, and Block-House coals are used for making gas at Boston and Cambridge, and the gas of New York is made from International, Glace Bay, Caledonia, and Block-House coals.

38. The Fortress of Louisbourg.

Louisbourg is reached (until the railway is finished) by a weekly stage from Sydney, in 24 M. A road runs hence 15 - 18 M. N. E. along an interesting coast, to *Cape Breton* (see page 149), passing the hamlets of Big and Little Loran, "named in honor of the haughty house of Lorraine." Cape Breton itself is nearly insulated by the deep haven of Balcine Cove, and just off its S. point is the rock of Porto Nuevo, rising boldly from the sea. Beyond the cape and the hamlet of Main-à-Dieu the Mira Bay road passes the hamlet of *Catalogne* (18 M. from Sydney), at the outlet of the broad lagoon of the Catalogne Lake, and follows the Mira River from the village of Mira Gut to the drawbridge on the Louisbourg road, where the farming hamlet of *Albert Bridge* has been established (12 M. from Sydney). A road runs hence S. W. 12 - 14 M. to *Marion Bridge*, a Scottish settlement near the long and narrow Mira Lake. The road ascends thence along the valley of the Salmon River to the vicinity of Loch Uist and Loch Lomond (see page 147).

Gabarus Bay is 8 - 10 M. S. W. of Louisbourg, and is a deep and spacious but poorly sheltered roadstead. It has a large and straggling fishing-settlement, near the Gabarus, Belfry, and Mira Lakes.

Louisbourg at present consists of a small hamlet occupied by fishermen, whose vessels sail hence to the stormy Grand Banks. The adjacent country is hilly and unproductive, and contains no settlements. The harbor is entered through a passage 10 fathoms deep, with a powerful white light on the N. E. headland, and is a capacious basin with 5 - 7 fathoms of water, well sheltered from any wind. On Point Rochfort, at the S. W. side of the harbor, are the ruins of the ancient French fortress and city.

"The ruins of the once formidable batteries, with wide broken gaps (blown up by gunpowder), present a melancholy picture of past energy. The strong and capacious magazine, once the deposit of immense quantities of munitions of war, is still nearly entire, but, hidden by the accumulation of earth and turf, now affords a commodious shelter for flocks of peaceful sheep, which feed around the burial-ground where the remains of many a gallant Frenchman and patriotic Briton are deposited; while beneath the clear cold wave may be seen the vast sunken ships of war, whose very bulk indicates the power enjoyed by the Gallic nation ere England became mistress of her colonies on the shores of the Western Atlantic. Desolation now sits with a ghastly smile around the once formidable bastions. All is silent except the loud reverberating ocean, as it rolls its tremendous surges along the rocky beach, or the bleating of the scattered sheep, as with tinkling bells they return in the dusky solitude of eve to their singular folds." (MONTGOMERY MARTIN.)

"If you ever visit Louisbourg, you will observe a patch of dark greensward on Point Rochfort, — the site of the old burying-ground. Beneath it lie the ashes of hundreds of brave New-Englanders. No monument marks the sacred spot, but the waves of the restless ocean, in calm or storm, sing an everlasting requiem over the graves of the departed heroes." (R. BROWN.)

The port of Louisbourg was called from the earliest times *Havre d l'Anglois*, but no important settlements were made here until after the surrender of Newfoundland and Acadia to Great Britain, by the Treaty of Utrecht. Then the French troops and inhabitants evacuated Placentia (N. F.) and came to this place. In 1714 M. de St. Ovide de Brouillan was made Governor of Louisbourg; and the work of building the fortress was begun about 1720.

THE FORTRESS OF LOUISBOURG. *Route 38.* 155

The powerful defences of "the Dunkirk of America" were hurried to completion, and the people of New England "looked with awe upon the sombre walls of Louisbourg, whose towers rose like giants above the northern seas." Over 30,000,000 livres were drawn from the French royal treasury, and were expended on the fortifications of Louisbourg; and numerous cargoes of building-stone were sent hither from France (as if Cape Breton had not enough, and little else). Fleets of New-England vessels bore lumber and bricks to the new fortress; and the Acadians sent in supplies and cattle. For more than 20 years the French government devoted all its energy and resources to one object,— the completion of these fortifications. Inhabitants were drawn to the place by bounties; and Louisbourg soon had a large trade with France, New England, and the West Indies.

The harbor was guarded by a battery of 30 28-pounders, on Goat Island; and by the Grand (or Royal) Battery, which carried 30 heavy guns and raked the entrance. On the landward side was a deep moat and projecting bastions; and the great careening-dock was opposite. The land and harbor sides of the town were defended by lines of ramparts and bastions, on which 80 guns were mounted; and the West Gate was overlooked by a battery of 16 24-pounders. The Citadel was in the gorge of the King's Bastion. In the centre of the city were the stately stone church, nunnery, and hospital of St. Jean de Dieu. The streets crossed each other at right angles, and communicated with the wharves by five gates in the harborward wall. The fortress was in the first system of Vauban, and required a large garrison.

Early in 1745 the Massachusetts Legislature determined to attack Louisbourg with all the forces of the Province; and Gov. Shirley, the originator of the enterprise, gave the military command to Col. Wm. Pepperell. Massachusetts furnished 3,250 men; New Hampshire, 300; and Connecticut, 500; and George Whitefield gave the motto for the army, "*Nil desperandum, Christo duce*," thus making the enterprise a sort of Puritan crusade. The forces were joined at Canso by Commodore Warren's West-India fleet, and a lauding was soon effected in Gabarus Bay. The garrison consisted of 750 French veterans and 1,500 militia, and the assailants were "4,000 undisciplined militia or volunteers, officered by men who had, with one or two exceptions, never seen a shot fired in anger all their lives, encamped in an open country, ...and sadly deficient in suitable artillery." The storehouses up the harbor were set on fire by Vaughan's New-Hampshire men; and the black smoke drove down on the Grand Battery, so greatly alarming its garrison that they spiked their guns and fled. The fort was occupied by the Americans and soon opened on the city. Fascine batteries were erected at 1,550 and 950 yards from the West Gate, and a breaching battery was reared at night within 250 yards of the walls. Amid the roar of a continual bombardment, the garrison made sorties by sea and land; and 1,500 of the Americans were sick or wounded, 600 were kept out in the country watching the hostile Indians, and 200 had been lost in a disastrous attempt at storming the Island Battery. Early in June, the guns of the Circular Battery were all dismounted, the King's Bastion had a breach 24 feet deep, the town had been ruined by a rain of bombs and red-hot balls, and the Island Battery had been rendered untenable by the American cannonade. On the 15th the fleet (consisting of the *Superb, Sunderland, Canterbury,* and *Princess Mary,* 60 guns each; and the *Launceston, Chester, Lark, Mermaid, Hector,* and *Eltham,* of 40 guns each) was drawn up off the harbor; and the army was arrayed "to march with drums beating and colours flying to the assault of the West Gate." But Gov. Duchambon saw these ominous preparations and surrendered the works, to avoid unnecessary carnage. "As the troops, entering the fortress, beheld the strength of the place, their hearts for the first time sank within them. 'God has gone out of his way,' said they, 'in a remarkable and most miraculous manner, to incline the hearts of the French to give up and deliver this strong city into our hand.'" Pepperell attributed his success, not to his artillery or the fleet of line-of-battle ships, but to the prayers of New England, daily arising from every village in behalf of the absent army. "The news of this important victory filled New England with joy and Europe with astonishment." Boston and London and the chief towns of America and England were illuminated; the batteries of London Tower fired salutes; and King George II. made Pepperell a baronet, and Warren a rear-admiral. (For the naval exploits, see page 149.)

4,130 French people were sent home on a fleet of transports; the siege-batteries were levelled, and 266 guns were mounted on the repaired walls; and in the following April the New-England troops were relieved by two regiments from Gibraltar, and went home, having lost nearly 1,000 men. The historian Smollet designated the capture of Louisbourg, "the most important achievement of the war of 1745"; and the authors of the "Universal History" considered it "an equivalent for all the successes of the French upon the Continent." The siege is minutely described (with maps) in Brown's "History of the Island of Cape Breton," pages 168-248.

"That a colony like Massachusetts, at that time far from being rich or populous, should display such remarkable military spirit and enterprise, aided only by the smaller Province of New Hampshire; that they should equip both land and sea forces to attack a redoubtable fortress called by British officers impregnable, and on which the French Crown had expended immense sums; .... that 4,000 rustic militia, whose officers were as inexperienced in war as their men, although supported by naval forces, should conquer the regular troops of the greatest military power of the age, and wrest from their hands a place of unusual strength, all appear little short of miracle." (BEAMISH MURDOCH.)

So keenly did the French government feel the loss of Louisbourg that the great French Armada was sent out in 1746 to retake it and to destroy Boston. After the disastrous failure of this expedition (see page 99), La Jonquière was despatched with 16 men-of-war and 28 other vessels, on the same errand, but was attacked by the fleets of Anson and Warren off Cape Finisterre, and lost 9 ships of war, 4,000 men, and $ 8,000,000 worth of the convoyed cargoes. In 1749 the war was ended, Louisbourg and Cape Breton were restored to France, and "after four years of warfare in all parts of the world, after all the waste of blood and treasure, the war ended just where it began."

When war broke out again between England and France, in 1755, Louisbourg was blockaded by the fleet of Admiral Boscawen. England soon sent 11 line-of-battle ships, a squadron of frigates, and 50 transports, bearing 6,000 soldiers, to reduce the fortress; but France was too prompt to be surprised, and held it with 17 sail of the line and 10,000 men. The vast English fleet got within 2 M. of Louisbourg and then recoiled, sailed to Halifax, and soon broke up, sending the army to New York and the ships to England. France then equipped fleets at Toulon and Rochfort, to reinforce Louisbourg; but the *Foudroyant*, 84, the *Orpheus*, 64, and other vessels were captured. Six men-of-war and sixteen transports reached Louisbourg, with a great amount of military supplies.

Great Britain now fitted out an immense fleet at Spithead, consisting of the *Namur*, 90 guns; *Royal William*, 80; *Princess Amelia*, 80; *Terrible*, 74; the *Northumberland*, *Oxford*, *Burford*, *Vanguard*, *Somerset*, and *Lancaster*, 70 guns each; the *Devonshire*, *Bedford*, *Captain*, and *Prince Frederick*, 64 each; the *Pembroke*, *Kingston*, *York*, *Prince of Orange*, *Defiance*, and *Nottingham*, 60 guns each; the *Centurion* and *Sutherland*, 50 each; the frigates *Juno*, *Grammont*, *Nightingale*, *Hunter*, *Boreas*, *Hind*, *Trent*, *Port Mahon*, *Diana*, *Shannon*, *Kennington*, *Scarborough*, *Squirrel*, *Hawk*, *Beaver*, *Tyloe*, and *Halifax*; and the fire-ships *Etna* and *Lightning*. There were also 118 transports, carrying 13.600 men, in 17 regiments. Boscawen commanded the fleet, Amherst the army, and Wolfe was one of the brigadiers.

This powerful armament soon appeared off Louisbourg, and at dawn on the 8th of June, 1758, the British troops landed at Gabarus Bay, and pushed through the fatal surf of Freshwater Cove, amid the hot fire of the French shore-batteries. After losing 110 men they carried the entrenchments at the point of the bayonet, and the French fell back on Louisbourg. The fortress had been greatly strengthened since the siege of 1745, and was defended by 3,400 men of the Artillery and the regiments of Volontaires Etrangers, Artois, Bourgogne, and Cambisé, besides large bodies of militia and Indians. In the harbor were the ships-of-war, *Prudent*, 74; *Entreprenant*, 74; *Capricieux*, 64; *Célèbre*, 64; *Bienfaisant*, 64; *Apollon*, 50; *Diane*, 36; *Aréthuse*, 36; *Fidèle*, 36; *Echo*, 32; *Biche*, 16; and *Chèvre*, 16.

Wolfe's brigade then occupied the old Lighthouse Battery, and opened fire on the city, the French fleet, and the Island Battery. The latter was soon completely destroyed by Wolfe's tremendous cannonade; and since the harbor was thus left unguarded, Gov. Drucour sank the frigates *Diane*, *Apollon*, *Biche*, *Fidèle*, and *Chèvre* at its entrance. Meantime the main army was erecting works on Green Hill and opposite the Queen's and Princess's Bastions, under the fire of the French ramparts and ships, and annoyed on the rear by the Indians. During a bloody sortie by the French, the Earl of Dundonald and many of the Grenadiers were killed. The heavy siege-batteries were advanced rapidly, and poured in a crushing fire on the doomed city, destroying the Citadel, the West Gate, and the barracks. The magazine of the *Entreprenant*, 74, blew up, and the *Capricieux* and *Célèbre*,

THE FORTRESS OF LOUISBOURG. *Route 38.* 157

catching the fire in their sails, were burned at their moorings. The *Aréthuse* and *Echo* ran out of the harbor in foggy weather, but the latter was captured. Only two French frigates remained, and these were both captured by boats from the fleet, after a daring attack. On the 26th of July the Chevalier de Drucour surrendered the city, with 5,637 men, 236 pieces of artillery, and immense amounts of stores and supplies. The French had lost about 1,000 men, the British nearly 600, during the siege.

All England rang with the tidings of the fall of " the Dunkirk of America," special prayers and thanksgivings were read in all the churches of the kingdom ; and 11 sets of colors from Louisbourg were presented to the King at Kensington Palace, whence they were borne with great pomp to St. Paul's Cathedral. Marine insurance on Anglo-American vessels fell at once from 30 to 12 per cent, because the French privateers were driven from the western seas by the closing of their port of refuge.

In 1759 the great fleet and army of Gen. Wolfe gathered at Louisbourg and sailed away to the Conquest of Canada. Halifax was a fine naval station, and it was deemed inexpedient to maintain a costly garrison at Louisbourg ; so sappers and miners were sent there in the summer of 1760, and " in the short space of six months all the fortifications and public buildings, which had cost France 25 years of labor and a vast amount of money, were utterly demolished, — the walls and glacis levelled into the ditch, — leaving, in fact, nothing to mark their former situation but heaps of stones and rubbish. Nothing was left standing but the private houses, which had been rent and shattered during the siege, the hospital, and a barrack capable of lodging 300 men. .... All the artillery, ammunition, stores, implements, — in short, everything of the slightest value, even the hewn stones which had decorated the public buildings, were transported to Halifax."

The British garrison was withdrawn in 1768, and after the foundation of Sydney " the most splendid town of La Nouvelle France " was completely deserted by its people.

During a year or two past a scheme has been agitated whose fulfilment would restore Louisbourg to more than its former importance. It is proposed to construct a first-class railway from this point to some station on the Pictou Branch of the Intercolonial Railway, crossing the Strait of Canso either by a lofty suspension-bridge or a steam ferry-boat on which the trains would be carried. It is thought that the freight and passenger receipts from the coal-mines and the settlements on the territory traversed would more than defray the cost of construction and maintenance. The projectors then intend to make Louisbourg a port of call for the ocean-steamships, for whose use this safe and accessible harbor is peculiarly adapted. This port is on the 60th parallel of W. longitude, and is 11 degrees E. of Boston and 14 degrees E. of New York, or so much farther advanced on the route to Europe. When the through railway is completed to Boston, Montreal, and New York, it is thought that most of the better class, at least, of transatlantic travellers would prefer to save time and nearly 1,000 M. of ocean-voyaging, by leaving or taking the steamship here. Extensive surveys have already been made in this vicinity, and real estate in Louisbourg has rapidly advanced in value.

39. The North Shore of Cape Breton.—St. Anne's Bay and St. Paul's Island.

Conveyances may be hired at Baddeck (see page 162) by which to visit St. Anne's. The distance is about 10 M. to the head of the harbor. The first part of the way leads along the shores of Baddeck Bay, with the promontory of Red Head over the water to the r. The road then crosses a cold district of denuded highlands, and descends to the * *Valley of St. Anne*. As the harbor is approached, the traveller can see the amphitheatrical glens in which the great Holy Fairs or annual religious communions of the people are held. These quaint Presbyterian camp-meetings are said to be a relic of the ancient churches in the Scottish Highlands. The shores of the harbor were occupied in 1820 by immigrants from the Highlands, who are now well located on comfortable farms. The road follows the S. Arm, and to the l. is seen the N. Arm, winding away among the tall mountains. Just E. of the N. Arm is St. Anne's Mt., which is 1,070 ft. high, and pushes forward cliffs 960 ft. high nearly to the water's edge.

"There is no ride on the continent, of the kind, so full of picturesque beauty and constant surprises as this around the indentations of St. Anne's harbor. High bluffs, bold shores, exquisite sea-views, mountainous ranges, delicious air," are found here in abundance. About opposite the lighthouse on the bar, at the mouth of the harbor, is *Old Fort Point*, on which the French batteries were established. Near this point is the hamlet of *Englishtown*, chiefly interesting as containing the grave of the once famous "Nova-Scotia Giant." The mountains back of Englishtown are over 1,000 ft. high, and run N. E. to Cape Dauphin, whence they repel the sea. Imray's *Sailing Directions* states that "on the N. side the land is very high, and ships-of-war may lie so near the shore that a water-hose may reach the fresh water." As to the harbor, the ancient description of Charlevoix still holds good:—

"Port Ste. Anne, as already stated, has before it a very sure roadstead between the Cibou Islands. The port is almost completely closed by a tongue of land, leaving passage for only a single ship. This port, thus closed, is nearly two leagues in circuit, and is oval in form. Ships can everywhere approach the land, and scarcely perceive the winds, on account of its high banks and the surrounding mountains. .... The fishing is very abundant; great quantities of good wood are found there, such as maple, beech, wild cherry, and especially oaks very suitable for building and masts, being 28-38 ft. high; marble is common; most of the land good,—in Great and Little Labrador, which are only a league and a half off, the soil is very fertile, and it can contain a very large number of settlers."

In St. Anne's Bay the English ship *Chancewell* was wrecked in 1597, and while she lay aground "there came aboord many shallops with store of French men, who robbed and spoyled all they could lay their hands on, pillaging the poore men euen to their very shirts, and vsing them in sauage manner; whereas they should rather as Christians haue aided them in that distresse." In 1629 this harbor was occupied by the *Great St. Andrew* and the *Marguerite*, armed vessels of France, whose crews, together with their English prisoners, constructed a fort to command the entrance It was armed with 8 cannon, 1,800 pounds of powder, pikes, and muskets, and was garrisoned by 40 men. The commander of the fleet raised the arms of the King and of Cardinal Richelieu over its walls, and erected a chapel, for whose care he left two

INGONISH. *Route 39.* 159

Jesuits. He then named the harbor St. Anne's. Before the close of that winter more than one third of the troops died of the scurvy, and the commandant assassinated his lieutenant on the parade-ground. In 1634 the Jesuits founded an Indian mission here, but both this and the garrison were afterwards withdrawn. Some years later a new battery and settlement were erected here by Nicholas Denys, Sieur de Fronsac, who traded hence with the Indians of the N. of Cape Breton.
The valley of the N. Arm of St. Anne's was granted, in 1713, to M. de Rouville, a captain in the infantry of France, and brother of that Hertel de Rouville who led the forces that destroyed Schenectady, Deerfield, and Haverhill. The N. Arm was long called Rouville's River. At a later day Costabello, Beaucourt, Soubras, and other French officers had fishing-stations on the bay. In 1745 a frigate from Com. Warren's fleet (then blockading Louisbourg) entered the harbor, and destroyed all the property on its shores. St. Anne's Bay was afterwards called *Port Dauphin* by the French, and the government long hesitated as to whether the chief fortress of Cape Breton should be located here or at Louisbourg. The perfect security of the harbor afforded a strong argument in favor of St. Anne's, and it seemed capable of being made impregnable at slight expense. After the foundation of Louisbourg 1,000 cords of wood were sent to that place annually from St. Anne's.

The road from the Bras d'Or to the N. shore of Cape Breton diverges from the St. Anne road before reaching the harbor, and bears to the N. E., along the W. Branch. It rounds the North-River Valley by a great curve, and then sweeps up the harbor-shore, under the imposing cliffs of St. Anne's Mt. From St. Anne's to Ingonish the distance is about 40 M., by a remarkably picturesque road between the mountains and the Atlantic, on a narrow plain, which recalls Byron's lines: —

"The mountains look on Marathon,
And Marathon looks on the sea."

"Grand and very beautiful are the rocky gorges and ravines which furrow the hills and precipices between St. Anne's and Ingonish..... Equally grand and picturesque is the red syenitic escarpment of Smoky Cape, capped with the cloud from which it derives its name, with many a lofty headland in the background, and the peak of the Sugar-loaf Mountain just peeping above the far-distant horizon." (BROWN.)

The proud headland of **Cape Smoky** (the *Cap Enfumé* of the French) is 950 ft. high, and runs sheer down into the sea. To the W. there are peaks 1,200 – 1,300 ft. high; and as the road bends around the deep bights to the N., it passes under summits more than 1,400 ft. high. Among these massive hills, and facing Cape Smoky, is the village of **Ingonish**, inhabited by Scottish Catholic fishermen, 800 of whom are found in this district. On the island that shelters the harbor is a fixed white light, 237 ft. above the sea, and visible for 15 M.

Ingonish was one of the early stations of the French. In 1729 a great church was built here, whose foundations only remain now; and in 1849 a church-bell, marked *St. Malo,* 1729, and weighing 200 pounds, was found buried in the sands of the beach. The settlement here was probably ruined by the drawing away of its people to aid in holding Louisbourg against the Anglo-American forces. In 1740 Ingonish was the second town on the island, and its fleet caught 13,560 quintals of fish. It was destroyed, in 1745, by men-of-war from Com. Warren's fleet.
The highland region back of Ingonish has always been famous for its abundance of game, especially of moose and caribou. In the winter of 1789 over 9,000 moose were killed here for the sake of their skins, which brought ten shillings each; and for many years this wholesale slaughter went on, and vessels knew when they were approaching the N. shore of Cape Breton by the odor of decaying carcasses which came from the shore. Finally the outraged laws of the Province were vindicated by the occupation of Ingonish by a body of troops, whose duty it was to restrain the

whence small cargoes of produce are annually shipped to Newfoundland. Near this point is a marble cave, with several chambers 6 – 8 ft. high; and foxes are often seen among the hills. It is claimed that valuable deposits of magnetic and hematitic iron-ore have been found in this vicinity. Stages run 30 M. S. W. from Whycocomagh to Port Hastings, on the tame and uninteresting road known as the Victoria Line.

"What we first saw was an inlet of the Bras d'Or, called by the driver Hogamah Bay. At its entrance were long, wooded islands, beyond which we saw the backs of graceful hills, like the capes of some poetic sea-coast. .... A peaceful place, this Whycocomagh. The lapsing waters of the Bras d'Or made a summer music all along the quiet street; the bay lay smiling with its islands in front, and an amphitheatre of hills rose beyond." (WARNER's *Baddeck*.)

On leaving Whycocomagh the quaint double peaks of Salt Mt. are seen in retrospective views, and the road soon enters the *Skye Glen,* a long, narrow valley, which is occupied by the Highlanders. The wagon soon reaches the picturesque gorge of the *Mabou Valley,* with the mountainous mass of Cape Mabou in front. The Mull River is seen on the l., glittering far below in the valley, and erelong the widenings of the sea are reached, and the traveller arrives at the wretched inn of *Mabou.* The stage for Port Hood (10 M. S.) leaves about midnight, reaching Port Hastings at 9 A. M. (see Route 42).

The steamer *Neptune* ascends St. Patrick's Channel to Whycocomagh every week, on its alternate trips passing around from Sydney to the Channel by way of the Great Bras d'Or (Sydney to Whycocomagh, $2). This route is much easier for the traveller than that by the stage, and reveals as much natural beauty, if made during the hours of daylight. The passage of the Little Narrows and the approach to Whycocomagh are its most striking phases.

42. The West Coast of Cape Breton.—Port Hood and Margaree.

The Royal mail-stage leaves Port Hastings (Plaster Cove) every morning, after the arrival of the Halifax mail. Fare to Port Hood, $3.

Distances.—Port Hastings; Low Point, 7 M.; Creignish, 9; Long Point, 14; Judique, 18; Little Judique, 24; Port Hood, 28; Mabou, 38; Broad Cove Intervale, 56; Margaree Forks, 68; Margaree, 76; Cheticamp, 88.

The first portion of this route is interesting, as it affords frequent pleasant views of the Strait of Canso and its bright maritime processions. The trend of the coast is followed from Port Hastings to the N. W., and a succession of small hamlets is seen along the bases of the highlands. Just beyond *Low Point* is the Catholic village of the same name, looking out over the sea. The road now skirts the wider waters of St. George's Bay, over which the dark Antigonish Mts. are visible. Beyond the settlements of Creignish and Long Point is the populous district of *Judique,* inhabited by Scottish Catholics, who are devoted to the sea and to agriculture. The Judiquers are famous throughout the Province for their great stature, and are well known to the American fishermen on account of their pugnacity. Yankee crews landing on this coast are frequently assailed by these pugilistic Gaels, and the stalwart men of Judique usually come off victorious in the fistic encounters. The district has about 2,000 inhabitants.

Port Hood (two inns) is the capital of Inverness County, and is a picturesque little seaport of about 800 inhabitants. The American fishermen in the Gulf frequently take shelter here during rough weather, and 400 sail have been seen in the port at one time. There are large coal-deposits in the vicinity, which, however, have not yet been developed to any extent. The town was founded by Capt. Smith and a party of New-Englanders, in 1790. "This port affords the only safe anchorage on the W. coast of Cape Breton to the N. of the Gut of Canso," and is marked by a red-and-white light, near the highway, on the S. Off shore is Smith's Island, which is 2 M. long and 210 ft. high, beyond which are the high shores of Henry Island. The Magdalen-Islands steamer touches at Port Hood (see Route 49) and a stage-road runs N. E. to Hillsborough, where it meets the road from Mabou, and thence passes E. to Whycocomagh (see page 167).

Mabou (uncomfortable inn) is 10 M. N. E. of Port Hood, and is reached by a daily stage passing along the shore-road. It is at the mouth of the broad estuary of the Mabou River, amid bold and attractive scenery, and contains about 800 inhabitants. To the N. E. is the highland district of Cape Mabou, averaging 1,000 ft. in height, and thickly wooded. The Gulf-shore road to Margaree runs between this range and the sea, passing the marine hamlets of Cape Mabou and Sight Point. There is an inland road, behind the hills, which is entered by following the Whycocomagh road to the head of the estuary of the Mabou and then diverging to the N. E. This road is traversed by a tri-weekly stage, and leads up by the large farming-settlement at Broad Cove Intervale, to the W. shores of *Lake Ainslie* (see page 167), which has several small Scottish hamlets among the glens.

"The angler who has once driven through Ainslie Glen to the shores of the lake, launched his canoe upon its broad waters, and entered its swiftly running stream, will never be content to return until he has fished its successive pools to its very mouth."

A road leads out from near the W. shore of the lake to the village of *Broad Cove Chapel*, on the Gulf coast, traversing a pass in the highlands. The stage runs N. between the hills and the valley of the Margaree (S. W. Branch), "one of the most romantic and best stocked salmon-rivers in the world." Beyond the settlement of Broad Cove Marsh, a road runs out to the Gulf abreast of *Sea-Wolf Island*, on whose cliffs is a fixed light, 300 ft.

S. W. for nearly 80 M., between the mountains of St. Anne and the highlands of Boularderie.

The *Neptune* soon traverses the narrow channel of the Little Bras d'Or and enters a broader bay. Beyond Grove Point it reaches a beautiful sound which is followed for 25 M., and is 3-4 M. wide. (It is called *St. Andrew's Channel* on the Admiralty charts, but that name is elsewhere applied to the East Bay.) Near George Mt., on the l., are the low shores of Long Island; and the steamer sometimes stops off Beaver Harbor, or Boisdale. The course is now laid towards the W. shore, rounds Kempt Head, the S. extremity of Boularderie Island, and passes Coffin Island on the r., beyond which is seen the long channel of the Great Bras d'Or. The course is nearly N. W., and lies between Red Point (r. side) and Mackay Point (l. side), which are about 3 M. apart. In front is seen the village of Baddeck, while inside of the points Baddeck Bay extends to the r. and St. Patrick's Channel to the l.

Baddeck (*Telegraph House*, comfortable; *Bras d' Or Hotel*) is the capital of Victoria County, and the chief village on the Bras d'Or. It has three churches, a court-house, and a quaint little jail, and is the centre of a group of farming-settlements whose aggregate population is 1,749. The harbor can accommodate vessels of 500 tons, and from this point several cargoes of produce are annually sent to Newfoundland. Gold has been found in the vicinity, and there is a saline spring farther down the shore. This locality was first visited by the French, from whom it received the name *Bedeque*, since Scotticized to *Baddeck* (accent on the last syllable). It was first settled by the disbanded soldiers of the Royal Rangers, and in 1793 there were 10 inhabitants here.

"Although it was Sunday, I could not but notice that Baddeck was a clean-looking village of white wooden houses, of perhaps 7-800 inhabitants; that it stretched along the shore for a mile or more, straggling off into farm-houses at each end, lying for the most part on the sloping curve of the bay. There were a few country-looking stores and shops, and on the shore three or four rather decayed and shaky wharves ran into the water, and a few schooners lay at anchor near them; and the usual decaying warehouses leaned about the docks. A peaceful and perhaps a thriving place, but not a bustling place. . . . .

"Having attributed the quiet of Baddeck on Sunday to religion, we did not know to what to lay the quiet on Monday. But its peacefulness continued. I have no doubt that the farmers began to farm, and the traders to trade, and the sailors to sail; but the tourist felt that he had come into a place of rest. The promise of the red sky the evening before was fulfilled in another royal day. There was an inspiration in the air that one looks for rather in the mountains than on the sea-coast, it seemed like some new and gentle compound of sea-air and land-air, which was the perfection of breathing material. In this atmosphere, which seems to flow over all these Atlantic isles at this season, one endures a great deal of exercise with little fatigue; or he is content to sit still and has no feeling of sluggishness. Mere living is a kind of happiness, and the easy-going traveller is satisfied with little to do and less to see. Let the reader not understand that we are recommending him to go to Baddeck. Far from it. . . . . There are few whom it would pay to go a thousand miles for the sake of sitting on the dock at Baddeck when the sun goes down, and watching the purple lights on the islands and the distant hills, the red flush on the horizon and on the lake, and the creeping on of gray twilight. You can see all this as well elsewhere? I am not so sure. There is a harmony of beauty about the

Bras d'Or at Baddeck which is lacking in many scenes of more pretension."
(CHARLES DUDLEY WARNER'S *Baddeck; and that Sort of Thing.*)

The tourist who stops at Baddeck should visit the Indian village which occupies a grassy point near the town. It pertains to one of the clans of the Micmac tribe, and usually has 12-15 wigwams. Visitors are received with a not unkindly indifference, and may here study Indian domestic life, the curious manner of carrying babies, and the architecture of the wigwam. Some of the people can talk English. The visitor should endeavor to see one of the Micmac Catholic prayer-books, printed (at Vienna) in a singular hieroglyphic, and bought by the Indians at the Trappist monastery in Tracadie. The camp at Baddeck is broken up in the autumn and the people retire to their reservations near the hunting-grounds.

The Micmacs of Nova Scotia and Cape Breton still retain many of their ancient customs, and are of purer blood than any other tribe on the Atlantic coast. They number about 1,600 (and 1,400 in New Brunswick), and occupy several reservations in the Province, where they are cared for and protected by the Dominion government. Under this paternal care (strongly contrasting with the Indian policy of the United States) the aborigines are steadily increasing in numbers and approaching a better standard of civilization, and are loyal and useful subjects of their "great mother," Queen Victoria. The discipline of families is well preserved by the use of corporeal punishment. Warm parental affection is a strongly marked feature, and the subordination of the women is still maintained, though ameliorated by the influences of civilization. The Micmacs have exchanged their former belief in and worship of the hostile principles of good and evil for the creed of the Roman Catholic Church, of which they are devout communicants.

Their language has many curious verbal coincidences with that of the Gaelic race, and is said to be "copious, flexible, and expressive." Philologists have also traced a marked analogy between the Greek and Micmac languages, basing thereon a sharp rebuke to Renan's flippant attack on the aboriginal tongues of America.

Baddeck to Whycocomagh, see Route 41. Baddeck to St. Anne's Bay, see Route 39. A road runs from this point nearly N. for 10 M. to the forks of the Big Baddeck River, where trout are found. To the N. are the Baddeck Mts., an unexplored and savage highland region which extends for 60 M. to the N., as far as Cape North, with a breadth of 15-25 M. This mountain-region has been the favorite hunting-ground for the moose and caribou (none of which can be shot between 1874 and 1877, according to the Provincial game-law), and it also contains bears, wolves and foxes, rabbits and hares, beaver, mink, and muskrats.

The *Margaree River* may be reached from Baddeck (in 28 M.) by a picturesque road, ascending the long valley, and crossing the Hunter's Mt., with fine views over the Bras d'Or. The pleasant rural district of the Middle Valley is then traversed, and the road leads through a remarkable pass of the hills and enters the rich valley of the Margaree, famous for its fishing (see Route 42). Visitors to this district usually board in the farm-houses, where plain and substantial fare is given.

The *Middle River* lies to the W. of Baddeck, and is approached by the Whycocomagh road (Route 41). The valley has over 1,000 inhabitants, of the Gaelic Highland race, many of whom are unacquainted with the English language. Near their settlements are prolific trout-streams, where fine sport may be enjoyed in the early summer. The chief settlements are respectively 12, 13, and 16 M. from Baddeck, and near the head of the river is an undeveloped gold district. A few miles up this

PRINCE EDWARD ISLAND.

PRINCE EDWARD ISLAND is situated in the southern portion of the Gulf of St. Lawrence, and is bounded on the S. by the Northumberland Strait. It is 30 M. from Cape Breton Island, 15 M. from Nova Scotia, and 9 M. from New Brunswick, and is surrounded by deep and navigable waters. The extreme length is 130 M.; the extreme breadth, 34 M.; and the area is 2,133 square miles. The surface is low or gently undulating, with small hills in the central parts, and the soil is mostly derived from red sandstone, and is very fertile. The air is balmy and bracing, less foggy than the adjacent shores, and milder than that of New Brunswick. The most abundant trees are the evergreens, besides which the oak and maple are found. The shores are deeply indented by harbors, of which those toward the Gulf are obstructed by sand, but those on the S. are commodious and accessible.

The island is divided into 3 counties, including 13 districts, or 67 townships and 3 royalties. It has 94,021 inhabitants, of whom 40,765 are Catholics, 29,579 are Presbyterians, 8,361 Methodists, and 7,220 Episcopalians. The majority of the people are Gaelic, and there are 300-400 Micmac Indians. The local government is conducted by the Legislative Council (13 members) and the House of Assembly (28 members), and the political parties which form about the petty questions of the island display a partisan acrimony and employ a caustic journalism such as are not seen even in the United States. The Province is provided with governor and cabinet, supreme and vice-admiralty courts, a public debt and a public domain, on the same plan as those of the great Provinces of Quebec and Ontario. The land is in a high state of cultivation, and nearly all the population is rural. Manufactories can scarcely be said to exist, but the fisheries are carried on to some extent, and shipbuilding receives considerable attention. The roads are good in dry weather, and lead through quiet rural scenery, broken every few miles by the blue expanses of the broad bays and salt-water lagoons. The chief exports consist of oats, wheat, barley, hay, potatoes, fish, live-stock, and lumber.

It has been claimed that Prince Edward Island was discovered by Cabot, in 1497, but there is no certainty on this subject. It was visited by Champlain on St. John's Day, 1608, and received from him the name of *L'Isle St. Jean.* The whole country was then covered with stately forests, abounding in game, and was inhabited by a clan of the Micmac Indians, who called it *Epayguit* ("Anchored on the Wave"). It was included in the broad domain of Acadia, over which France and England waged such disastrous wars, but was not settled for over two centuries after Cabot's voyage. In 1663 this and the Magdalen Islands were granted to M. Doublet, a captain in the French navy, who erected summer fishing-stations here, but abandoned them every autumn. After England had wrested Nova Scotia from France, a few Acadians crossed over to L'Isle St. Jean and became its first settlers. In 1728 there were 60 French families here; in 1745 there were about 800 inhabitants; and during her death-struggles with the Anglo-American armies, the Province of Quebec drew large supplies of grain and cattle from these shores. The capital was at Port la Joie (near Charlottetown), where there was a battery and garrison, dependent on the military commandant of Louisbourg. It is claimed by Haliburton that the island was captured by the New-Englanders in 1745, but it is known only that Gen. Pepperell ordered 400 of his soldiers to sail from Louisbourg and occupy L'Isle St. Jean. It does not appear whether or not this was done. After the expulsion of the Acadians from Nova Scotia, many of them fled to this island, which contained 4,100 inhabitants in 1758. In that year Lord Rollo took possession of it, according to the capitulation of Louisbourg, with a small military force.

In 1763 the island was ceded to Great Britain by the Treaty of Fontainebleau, and became a part of the Province of Nova Scotia. It was surveyed in 1764-6, and was granted to about 100 English and Scottish gentlemen, who were to pay quitrents and to settle their lands with 1 person to every 200 acres, within 10 years, the colonists to be Protestants from the continent of Europe. When the 10 years had elapsed, many of the estates were forfeited or sold to other parties, and only 19 of the 67 townships had any settlers. In 1770 the island was made a separate Province, and in 1773 the first House of Assembly met. In 1775 the Americans captured the capital, and in 1778 four Canadian companies were stationed there. In 1780 the Province was called New Ireland, but the King vetoed this name, and in 1800 it was entitled Prince Edward Island, in honor of His Royal Highness Prince Edward, Duke of Kent, then Commander of the Forces in British North America (afterwards father of Queen Victoria). In 1803 the Earl of Selkirk sent over 800 Highlanders, and other proprietors settled colonies on their domains. The complicated questions arising from the old proprietary estates have engrossed most of the legislation of the island for 70 years, and are being slowly settled by the purchase of the lands by the government. Prince Edward Island long refused to enter the Dominion of Canada, but yielded at last on very favorable terms, one of the conditions being that the Confederacy should build a railway throughout the Province.

Denys Basin by a range of massive highlands on the N. The N. shore hills are 700-770 ft. high, and those on the S. shore are 250-620 ft. high. The shores are thinly inhabited, and the only hamlets are at the head of the channel. (For the rest, the Editor has all these shores minutely outlined on the Admiralty chart now before him; but what shall it profit the traveller to know the precise locality of the Crammond Isles, or Calder Hill, or Ballam Head?)

"The only other thing of note the Bras d'Or offered us before we reached West Bay was the finest show of medusæ or jelly-fish that could be produced. At first there were dozens of these disk-shaped transparent creatures, and then hundreds, starring the water like marguerites sprinkled on a meadow, and of sizes from that of a teacup to a dinner-plate. We soon ran into a school of them, a convention, a herd as extensive as the vast buffalo droves on the plains, a collection as thick as clover-blossoms in a field in June, miles of them apparently: and at length the boat had to push its way through a mass of them which covered the water like the leaves of the pond-lily, and filled the deeps far down with their beautiful contracting and expanding forms. I did not suppose there were so many jelly-fishes in all the world." (WARNER'S *Baddeck*.)

"The scenery of the lakes is exceedingly striking and diversified. Long rocky cliffs and escarpments rise in some places abruptly from the water's edge; in others, undulating or rolling hills predominate, fringed on the shores by low white cliffs of gypsum or red conglomerate; whilst the deep basins and channels, which branch off in all directions from the central expanse of waters, studded with innumerable islets covered with a rich growth of spruce and hemlock, present views the most picturesque and diversified imaginable." (BROWN.)

"The scenery of this vast inlet is in some places beautifully picturesque, and in some others monotonous and uninteresting, but in many parts of a sublime character, which exhibits the sombre gloom of pine forests, the luxuriant verdure of broad valleys and wooded mountains, and the wild features of lofty promontories frowning in stubborn ruggedness over the waters of the rivers and inlets." (M'GREGOR.)

"So wide is it, and so indented by broad bays and deep coves, that a coasting journey around it is equal in extent to a voyage across the Atlantic. Besides the distant mountains that rise proudly from the remote shores, there are many noble islands in its expanse, and forest-covered peninsulas, bordered with beaches of glittering white pebbles. But over all this wide landscape there broods a spirit of primeval solitude..... For, strange as it may seem, the Golden Arm is a very useless piece of water in this part of the world; highly favored as it is by nature, landlocked, deep enough for vessels of all burden, easy of access on the Gulf side, free from fogs, and only separated from the ocean at its southern end by a narrow strip of land, about ¾ M. wide; abounding in timber, coal, and gypsum, and valuable for its fisheries, especially in winter, yet the Bras d'Or is undeveloped for want of that element which seems to be alien to the Colonies, namely, *enterprise*." (COZZENS.)

The Bras d'Or to Halifax.

When the steamer arrives at *West Bay*, a collection of singularly assorted vehicles is seen waiting by the wharf, and the passengers are conveyed on this motley train over 13 M. of uninteresting country to Port Hawkesbury (see page 143). The morning mail-stage may be taken from the opposite side of the Strait of Canso to Antigonish and New Glasgow (see Route 32); thence by railway to Halifax. But a pleasanter route (in calm weather) is to go on board the P. E. Island steamboat, which arrives during the evening, and pass to Pictou, through St. George's Bay and the Northumberland Strait. Pictou to Halifax, see Route 31.

41. Baddeck to Mabou and Port Hood. — St. Patrick's Channel and Whycocomagh.

This route is traversed by the Royal mail-stage on Monday and Wednesday, leaving Baddeck at noon, and reaching Whycocomagh after 4 o'clock, and Mabou at 9 P. M. The distance is about 50 M.; the fare is $ 2.50. The Royal mail-stage on this route is a one-horse wagon with a single seat, so that the accommodations for travel are limited.

Mr. Warner thus describes the road between Whycocomagh and Baddeck: "From the time we first struck the Bras d'Or for thirty miles we rode in constant sight of its magnificent water. Now we were two hundred feet above the water, on the hillside skirting a point or following an indentation; and now we were diving into a narrow valley, crossing a stream, or turning a sharp corner, but always with the Bras d'Or in view, the afternoon sun shining on it, softening the outlines of its embracing hills, casting a shadow from its wooded islands. Sometimes we opened upon a broad water plain bounded by the Watchabaktchkt hills, and again we looked over hill after hill receding into the soft and hazy blue of the land beyond the great mass of the Bras d'Or. The reader can compare the view and the ride to the Bay of Naples and the Cornice Road; we did nothing of the sort; we held on to the seat, prayed that the harness of the pony might not break, and gave constant expression to our wonder and delight."

St. Patrick's Channel is 20 M. long by 1–3 M. wide, and is made highly picturesque by its deep coves, wooded points, and lofty shores. Its general course is followed by the highway, affording rich views from some of the higher grades. After leaving Baddeck the road strikes across the country for about 5 M. to the Baddeck River, in whose upper waters are large trout. Beyond this point the road swings around the blue expanse of Indian Bay, approaching a bold hill-range 650 ft. high, and crosses the Middle River, at whose mouth is an Indian reservation. Frequent glimpses are afforded of St. Patrick's Channel, well to the l. across the green meadows. A range of lofty heights now forces the road nearer to the water, and it passes within 2 M. of the remarkable strait known as the *Little Narrows*, about which there are 150 inhabitants.

A road leads N. W. 5 M. into *Ainslie Glen*, and to the great **Ainslie Lake**, which covers 25 square miles, and is the source of the Margaree River. Its shores are broken and rugged, and are occupied by a hardy population of Highlanders. Petroleum springs have been found in this vicinity (see page 169).

Beyond the Little Narrows is a magnificent basin, 15 M. long and 3–5 M. wide, into whose sequestered and forest-bound waters large ships make their way, and are here laden with timber for Europe. On his second trip up this Basin, the Editor was startled, on rounding a promontory, at seeing a large Liverpool ship lying here, at anchor, with her yard-arms almost among the trees. The road runs around the successive spurs of the *Salt Mt.*, a massive ridge on the N. shore of the Basin, and many very attractive views are gained from its upper reaches. The water is of a rich blue, partly owing to its depth, which is from 3 to 20 fathoms.

Whycocomagh (*Inverness House*) is a Scottish Presbyterian hamlet, situated at the N. W. angle of the Basin, and surrounded by pretty Trosach-like scenery. There are about 400 inhabitants in this neighborhood, whence small cargoes of produce are annually shipped to Newfoundland. Near this point is a marble cave, with several chambers 6 – 8 ft. high; and foxes are often seen among the hills. It is claimed that valuable deposits of magnetic and hematitic iron-ore have been found in this vicinity. Stages run 30 M. S. W. from Whycocomagh to Port Hastings, on the tame and uninteresting road known as the Victoria Line.

"What we first saw was an inlet of the Bras d'Or, called by the driver Hogamah Bay. At its entrance were long, wooded islands, beyond which we saw the backs of graceful hills, like the capes of some poetic sea-coast. . . . . A peaceful place, this Whycocomagh. The lapsing waters of the Bras d'Or made a summer music all along the quiet street ; the bay lay smiling with its islands in front, and an amphitheatre of hills rose beyond." (WARNER's *Baddeck*.)

On leaving Whycocomagh the quaint double peaks of Salt Mt. are seen in retrospective views, and the road soon enters the *Skye Glen*, a long, narrow valley, which is occupied by the Highlanders. The wagon soon reaches the picturesque gorge of the *Mabou Valley*, with the mountainous mass of Cape Mabou in front. The Mull River is seen on the l., glittering far below in the valley, and erelong the widenings of the sea are reached, and the traveller arrives at the wretched inn of *Mabou*. The stage for Port Hood (10 M. S.) leaves about midnight, reaching Port Hastings at 9 A. M. (see Route 42).

The steamer *Neptune* ascends St. Patrick's Channel to Whycocomagh every week, on its alternate trips passing around from Sydney to the Channel by way of the Great Bras d'Or (Sydney to Whycocomagh, $2). This route is much easier for the traveller than that by the stage, and reveals as much natural beauty, if made during the hours of daylight. The passage of the Little Narrows and the approach to Whycocomagh are its most striking phases.

42. The West Coast of Cape Breton.—Port Hood and Margaree.

The Royal mail-stage leaves Port Hastings (Plaster Cove) every morning, after the arrival of the Halifax mail. Fare to Port Hood, $3.

Distances.— Port Hastings ; Low Point, 7 M.; Creignish, 9 ; Long Point, 14 ; Judique, 18 ; Little Judique, 24 ; Port Hood, 28 ; Mabou, 88 ; Broad Cove Intervale, 56 ; Margaree Forks, 68 ; Margaree, 76 ; Cheticamp, 88.

The first portion of this route is interesting, as it affords frequent pleasant views of the Strait of Canso and its bright maritime processions. The trend of the coast is followed from Port Hastings to the N. W., and a succession of small hamlets is seen along the bases of the highlands. Just beyond *Low Point* is the Catholic village of the same name, looking out over the sea. The road now skirts the wider waters of St. George's Bay, over which the dark Antigonish Mts. are visible. Beyond the settlements of Creignish and Long Point is the populous district of *Judique*, inhabited by Scottish Catholics, who are devoted to the sea and to agriculture. The Judiquers are famous throughout the Province for their great stature, and are well known to the American fishermen on account of their pugnacity. Yankee crews landing on this coast are frequently assailed by these pugilistic Gaels, and the stalwart men of Judique usually come off victorious in the fistic encounters. The district has about 2,000 inhabitants.

Port Hood (two inns) is the capital of Inverness County, and is a picturesque little seaport of about 800 inhabitants. The American fishermen in the Gulf frequently take shelter here during rough weather, and 400 sail have been seen in the port at one time. There are large coal-deposits in the vicinity, which, however, have not yet been developed to any extent. The town was founded by Capt. Smith and a party of New-Englanders, in 1790. "This port affords the only safe anchorage on the W. coast of Cape Breton to the N. of the Gut of Canso," and is marked by a red-and-white light, near the highway, on the S. Off shore is Smith's Island, which is 2 M. long and 210 ft. high, beyond which are the high shores of Henry Island. The Magdalen-Islands steamer touches at Port Hood (see Route 49) and a stage-road runs N. E. to Hillsborough, where it meets the road from Mabou, and thence passes E. to Whycocomagh (see page 167).

Mabou (uncomfortable inn) is 10 M. N. E. of Port Hood, and is reached by a daily stage passing along the shore-road. It is at the mouth of the broad estuary of the Mabou River, amid bold and attractive scenery, and contains about 800 inhabitants. To the N. E. is the highland district of Cape Mabou, averaging 1,000 ft. in height, and thickly wooded. The Gulf-shore road to Margaree runs between this range and the sea, passing the marine hamlets of Cape Mabou and Sight Point. There is an inland road, behind the hills, which is entered by following the Whycocomagh road to the head of the estuary of the Mabou and then diverging to the N. E. This road is traversed by a tri-weekly stage, and leads up by the large farming-settlement at Broad Cove Intervale, to the W. shores of *Lake Ainslie* (see page 167), which has several small Scottish hamlets among the glens.

"The angler who has once driven through Ainslie Glen to the shores of the lake, launched his canoe upon its broad waters, and entered its swiftly running stream, will never be content to return until he has fished its successive pools to its very mouth."

A road leads out from near the W. shore of the lake to the village of *Broad Cove Chapel*, on the Gulf coast, traversing a pass in the highlands. The stage runs N. between the hills and the valley of the Margaree (S. W. Branch), "one of the most romantic and best stocked salmon-rivers in the world." Beyond the settlement of Broad Cove Marsh, a road runs out to the Gulf abreast of *Sea-Wolf Island*, on whose cliffs is a fixed light, 300 ft.

high. *Margaree Forks* is a rural village at the junction of the N. E. and S. W. Branches of the famous **Margaree River**, where salmon abound from June 15 until July 15.

"In Cape Breton the beautiful Margaree is one of the most noted streams for sea-trout, and its clear water and picturesque scenery, winding through intervale meadows dotted with groups of witch-elm, and backed by wooded hills over a thousand feet in height, entitle it to pre-eminence amongst the rivers of the Gulf."

There are several small hamlets in this region, with a total population of over 4,000. Margaree is on the harbor of the same name, near the Chimney-Corner coal-mines, 48 M. from Port Hood, and has a small fleet of fishing-vessels. A shore-road runs N. E. 12 M. to *Cheticamp*, a district containing about 2,000 inhabitants, most of whom are of the old Acadian race. It is a fishing station of Robin & Co., an ancient and powerful commercial house on the Isle of Jersey; and was founded by them in 1784, and settled by Acadian refugees from Prince Edward Island. The harbor is suitable for small vessels, and is formed by Cheticamp Island, sheltering the mouth of the Cheticamp River. There is a powerful revolving white light on the S. point of the island, 150 ft. high, and visible for 20 M. at sea.

N. E. and E. of Cheticamp extends the great highland-wilderness of the N. part of Cape Breton (see page 163), an unexplored and trackless land of forests and mountains. There are no roads above Cheticamp, and the most northerly point of the Province, *Cape St. Lawrence* (see page 159), is 30 M. N. E. by E. ½ E. from the N. part of Cheticamp Island.

The terrible storm which swept the Gulf of St. Lawrence in August, 1873, and wrecked hundreds of vessels, attained its greatest force around the island of Cape Breton and in the narrow seas to the W., towards Prince Edward's Island and the Magdalen Island. It lasted only a few hours, but was fearfully destructive in its effects, and strewed all these coasts with drowned mariners The following spirited poem is inserted here, by the kind permission of its author, Mr. Edmund C. Stedman.

The Lord's-Day Gale.

In Gloucester port lie fishing craft,—
 More staunch and trim were never seen :
They are sharp before and sheer abaft,
 And true their lines the masts between.
Along the wharves of Gloucester Town
 Their fares are lightly landed down,
 And the laden flakes to sunward lean.

Well know the men each cruising-ground,
 And where the cod and mackerel be ;
Old Eastern Point the schooners round
 And leave Cape Ann on the larboard lee :
Sound are the planks, the hearts are bold,
 That brave December's surges cold
 On George's shoals in the outer sea.

And some must sail to the banks far north
 And set their trawls for the hungry cod,—
In the ghostly fog creep back and forth
 By shrouded paths no foot hath trod ;
Upon the crews the ice-winds blow,
The bitter sleet, the frozen snow,—
 Their lives are in the hand of God !

New England ! New England!
 Needs sail they must, so brave and poor,
Or June be warm or Winter storm,
 Lest a wolf gnaw through the cottage-door !
Three weeks at home, three long months gone,
While the patient good-wives sleep alone,
 And wake to hear the breakers roar.

The Grand Bank gathers in its dead,—
 The deep sea-sand is their winding-sheet ;
Who does not George's billows dread
 That dash together the drifting fleet ?
Who does not long to hear, in May,
The pleasant wash of Saint Lawrence Bay,
 The fairest ground where fishermen meet ?

There the west wave holds the red sunlight
 Till the bells at home are rung for nine ;
Short, short the watch, and calm the night ;
 The fiery northern streamers shine ;
The eastern sky anon is gold,
And winds from piny forests old,
 Scatter the white mists off the brine.

THE LORD'S-DAY GALE.   *Route 42.*

The Province craft with ours at morn
  Are mingled when the vapors shift;
All day, by breeze and current borne,
  Across the bay the sailors drift;
With toil and seine its wealth they win,—
  The dappled, silvery spoil come in
Fast as their hands can haul and lift.

New England! New England!
  Thou lovest well thine ocean main!
It spreadeth its locks among thy rocks,
  And long against thy heart hath lain;
Thy ships upon its bosom ride
And feel the heaving of its tide;
  To thee its secret speech is plain.

Cape Breton and Edward Isle between,
  In strait and gulf the schooners lay;
The sea was all at peace, I ween,
  The night before that August day;
Was never a Gloucester skipper there,
But thought erelong, with a right good fare,
  To sail for home from Saint Lawrence Bay.

New England! New England!
  Thy giant's love was turned to hate!
The winds control his fickle soul,
  And in his wrath he hath no mate.
Thy shores his angry scourges tear,
And for thy children in his care
  The sudden tempests lie in wait.

The East Wind gathered all unknown,—
  A thick sea-cloud his course before;
He left by night the frozen zone
  And smote the cliffs of Labrador;
He lashed the coasts on either hand,
And betwixt the Cape and Newfoundland
  Into the Bay his armies pour.

He caught our helpless cruisers there
  As a gray wolf harries the huddling fold;
A sleet — a darkness — filled the air,
  A shuddering wave before it rolled:
That Lord's-Day morn it was a breeze,—
At noon, a blast that shook the seas,—
  At night — a wind of Death took hold!

It leaped across the Breton bar,
  A death-wind from the stormy East!
It scarred the land, and whirled afar
  The sheltering thatch of man and beast;
It mingled rick and roof and tree,
And like a besom swept the sea,
  And churned the waters into yeast.

From Saint Paul's Light to Edward's Isle
  A thousand craft it smote amain;
And some against it strove the while,
  And more to make a port were fain:
The mackerel-gulls flew screaming past,
And the stick that bent to the noonday blast
  Was split by the sundown hurricane.

Woe, woe to those whom the islands pen!
  In vain they shun the double capes;
Cruel are the reefs of Magdalen;
  The Wolf's white fang what prey escapes?
The Grin'stone grinds the bones of some,
And Coffin Isle is craped with foam;—
  On Deadman's shore are fearful shapes!

O, what can live on the open sea,
  Or moored in port the gale outride?
The very craft that at anchor be
  Are dragged along by the swollen tide!
The great storm-wave came rolling west,
And tossed the vessels on its crest:
  The ancient bounds its might defied!

The ebb to check it had no power;
  The surf ran up to an untold height;
It rose, nor yielded, hour by hour,
  A night and day, a day and night;
Far up the seething shores it cast
The wreck of hull and spar and mast,
  The strangled crews,— a woful sight!

There were twenty and more of Breton sail
  Fast anchored on one mooring-ground;
Each lay within his neighbor's hail,
  When the thick of the tempest closed them round:
All sank at once in the gaping sea,—
Somewhere on the shoals their corses be,
  The foundered hulks, and the seamen drowned.

On reef and bar our schooners drove
  Before the wind, before the swell;
By the steep sand-cliffs their ribs were stove,—
  Long, long their crews the tale shall tell!
Of the Gloucester fleet are wrecks threescore;
Of the Province sail two hundred more
  Were stranded in that tempest fell.

The bedtime bells in Gloucester Town
  That Sabbath night rang soft and clear;
The sailors' children laid them down,
  Dear Lord! their sweet prayers couldst thou hear?
'T is said that gently blew the winds;
The good-wives, through the seaward blinds,
  Looked down the bay and had no fear.

New England! New England!
  Thy ports their dauntless seamen mourn;
The twin capes yearn for their return
  Who never shall be thither borne;
Their orphans whisper as they meet;
The homes are dark in many a street,
  And women move in weeds forlorn.

And wilt thou fail, and dost thou fear?
  Ah, no! though widows' cheeks are pale,
The lads shall say: "Another year,
  And we shall be of age to sail!"
And the mothers' hearts shall fill with pride,
Though tears drop fast for them who died
  When the fleet was wrecked in the Lord's-Day gale.

PRINCE EDWARD ISLAND.

PRINCE EDWARD ISLAND is situated in the southern portion of the Gulf of St. Lawrence, and is bounded on the S. by the Northumberland Strait. It is 30 M. from Cape Breton Island, 15 M. from Nova Scotia, and 9 M. from New Brunswick, and is surrounded by deep and navigable waters. The extreme length is 130 M.; the extreme breadth, 34 M.; and the area is 2,133 square miles. The surface is low or gently undulating, with small hills in the central parts, and the soil is mostly derived from red sandstone, and is very fertile. The air is balmy and bracing, less foggy than the adjacent shores, and milder than that of New Brunswick. The most abundant trees are the evergreens, besides which the oak and maple are found. The shores are deeply indented by harbors, of which those toward the Gulf are obstructed by sand, but those on the S. are commodious and accessible.

The island is divided into 3 counties, including 13 districts, or 67 townships and 3 royalties. It has 94,021 inhabitants, of whom 40,765 are Catholics, 29,579 are Presbyterians, 8,361 Methodists, and 7,220 Episcopalians. The majority of the people are Gaelic, and there are 300-400 Micmac Indians. The local government is conducted by the Legislative Council (13 members) and the House of Assembly (28 members), and the political parties which form about the petty questions of the island display a partisan acrimony and employ a caustic journalism such as are not seen even in the United States. The Province is provided with governor and cabinet, supreme and vice-admiralty courts, a public debt and a public domain, on the same plan as those of the great Provinces of Quebec and Ontario. The land is in a high state of cultivation, and nearly all the population is rural. Manufactories can scarcely be said to exist, but the fisheries are carried on to some extent, and shipbuilding receives considerable attention. The roads are good in dry weather, and lead through quiet rural scenery, broken every few miles by the blue expanses of the broad bays and salt-water lagoons. The chief exports consist of oats, wheat, barley, hay, potatoes, fish, live-stock, and lumber.

It has been claimed that Prince Edward Island was discovered by Cabot, in 1497, but there is no certainty on this subject. It was visited by Champlain on St. John's Day, 1608, and received from him the name of *L'Isle St. Jean.* The whole country was then covered with stately forests, abounding in game, and was inhabited by a clan of the Micmac Indians, who called it *Epayguit* ("Anchored on the Wave"). It was included in the broad domain of Acadia, over which France and England waged such disastrous wars, but was not settled for over two centuries after Cabot's voyage. In 1663 this and the Magdalen Islands were granted to M. Doublet, a captain in the French navy, who erected summer fishing-stations here, but abandoned them every autumn. After England had wrested Nova Scotia from France, a few Acadians crossed over to L'Isle St. Jean and became its first settlers. In 1728 there were 60 French families here; in 1745 there were about 800 inhabitants; and during her death-struggles with the Anglo-American armies, the Province of Quebec drew large supplies of grain and cattle from these shores. The capital was at Port la Joie (near Charlottetown), where there was a battery and garrison, dependent on the military commandant of Louisbourg. It is claimed by Haliburton that the island was captured by the New-Englanders in 1745, but it is known only that Gen. Pepperell ordered 400 of his soldiers to sail from Louisbourg and occupy L'Isle St. Jean. It does not appear whether or not this was done. After the expulsion of the Acadians, from Nova Scotia, many of them fled to this island, which contained 4,100 inhabitants in 1758. In that year Lord Rollo took possession of it, according to the capitulation of Louisbourg, with a small military force.

In 1763 the island was ceded to Great Britain by the Treaty of Fontainebleau, and became a part of the Province of Nova Scotia. It was surveyed in 1764-6, and was granted to about 100 English and Scottish gentlemen, who were to pay quitrents and to settle their lands with 1 person to every 200 acres, within 10 years, the colonists to be Protestants from the continent of Europe. When the 10 years had elapsed, many of the estates were forfeited or sold to other parties, and only 19 of the 67 townships had any settlers. In 1770 the island was made a separate Province, and in 1773 the first House of Assembly met. In 1775 the Americans captured the capital, and in 1778 four Canadian companies were stationed there. In 1780 the Province was called New Ireland, but the King vetoed this name, and in 1800 it was entitled Prince Edward Island, in honor of His Royal Highness Prince Edward, Duke of Kent, then Commander of the Forces in British North America (afterwards father of Queen Victoria). In 1803 the Earl of Selkirk sent over 800 Highlanders, and other proprietors settled colonies on their domains. The complicated questions arising from the old proprietary estates have engrossed most of the legislation of the island for 70 years, and are being slowly settled by the purchase of the lands by the government. Prince Edward Island long refused to enter the Dominion of Canada, but yielded at last on very favorable terms, one of the conditions being that the Confederacy should build a railway throughout the Province.

43. Shediac to Summerside and Charlottetown. — The Northumberland Strait.

St. John to Shediac, see Routes 14 and 16.

It is probable that steamers of the P. E. I. Steam Navigation Company will leave Shediac (Point du Chêne) every day during the summer season, on arrival of the morning train from St. John. The fare from Shediac to Summerside is $1.50 ; and from Summerside to Charlottetown, $ 1.50.

The distance from Shediac to Summerside is 35 M. Soon after leaving the wharf at Point du Chêne the steamer passes out through Shediac Bay, and enters the Northumberland Strait. The course is a little N. of E., and the first point of the island to come into sight is **Cape Egmont**, with its lines of low sandstone cliffs. The traveller now sees the significance of the ancient Indian name of this sea-girt land, *Epayguit*, signifying " Anchored on the Wave."

After passing Cape Egmont on the l., the steamer enters Bedeque, or Halifax, Bay, and runs in toward the low shores on the N. E. After passing Indian Point and Island it enters the harbor of Summerside, with the estuary of the Dunk River on the r.

Summerside, see page 179.

Upon leaving Summerside the steamer passes Indian Point on the l., and, after running by *Salutation Point*, enters the Northumberland Strait. The course is nearly S. E. 9 M. from Salutation Point is **Cape Traverse**, and on the S. shore is Cape Tormentine. At this, the narrowest part of the strait, the mails are carried across by ice-boats in winter, and passengers are transported by the same perilous route. A submarine cable underlies the strait at this point. It is 20 M. from Cape Traverse to St. Peter's Island, and along the island shores are the villages of Tryon, Crapaud, De Sable, and Bonshaw. On passing St. Peter's Island, the steamer enters *Hillsborough Bay* and runs N., with Orwell and Pownal Bays opening on the E.

" Charlottetown Harbor, at its entrance between the cliffs of Blockhouse and Sea-Trout Point, is 450 fathoms wide, and, in sailing in, York River running northward, the Hillsborough River eastwardly, and the Elliot to the westward, surround the visitor with beautiful effects, and as he glides smoothly over their confluence, or what is called the Three Tides, he will feel, perhaps, that he has seen for the first time, should a setting sun gild the horizon, a combination of color and effect which no artist could adequately represent."

Charlottetown, see page 175.

44. Pictou to Prince Edward Island.

To Charlottetown.

The steamships of the P. E. I. Steam Navigation Company leave Pictou for Charlottetown every Wednesday and Saturday (hours not yet regulated). Fare, $2. The distance is a little over 50 M.

Soon after leaving the safe and pleasant harbor of Pictou, the steamer approaches *Pictou Island*, a hilly and well-wooded land 4 M. long, with a lighthouse and some farms. On the W. is Caribou Island, consisting of several islets united by sand-bars, and guarded by a lighthouse. There are pleasant views of the receding highlands of Nova Scotia; and the vessel moves easily through the quiet waters of the Northumberland Strait. "Prince Edward Island, as we approached it, had a pleasing aspect, and none of that remote friendlessness which its appearance on the map conveys to one; a warm and sandy land, in a genial climate, without fogs, we are informed."

After passing (on the r.) the long low *Point Prim*, the steamer sweeps around to the N. into Hillsborough Bay, and enters the harbor of Charlottetown.

Pictou to Georgetown.

The P. E. I. Steam Navigation Company's steamships leave Pictou for Georgetown every Tuesday and Friday; leaving Georgetown for Pictou on the same days. Fare from port to port, $2. The distance is nearly 70 M.

The chief incidents of this short voyage are the views of Pictou Island; the approach to Cape Bear, the S. E. point of P. E. Island, backed by hills 200 ft. high; and the ascent of the noble sheet of Cardigan Bay, between Boughton and Panmure Islands.

Georgetown, see page 181.

45. Charlottetown.

Arrival. — The steamer passes between St. Peter's Island (l.) and Governor's Island (r.) and ascends Hillsborough Bay for about 6 M. It then passes between Blockhouse Point (on the l., with a lighthouse) and Sea-Trout Point, and enters the harbor of Charlottetown, where there are 7-10 fathoms of water. Powerful currents are formed here by the tides of the Hillsborough, York, and Elliot Rivers (or East, North, and West Rivers), which empty into this basin.

Hotels. — St. Lawrence Hotel, Water St.; Revere House, near the steamboat wharf; City Hotel. The hotels of Charlottetown are only boarding-houses of average grade, and will hardly satisfy American gentlemen. Attempts are being made to erect a large summer-hotel here, though there seems to be but little to warrant such an enterprise.

Steamships. — The *Alhambra* and the *Carroll* leave Charlottetown every Thursday for the Strait of Canso, Halifax, and Boston. Fares to Halifax, saloon state-room, $6; cabin state-room, $5; cabin, $4; Halifax to Boston, $9, $7.50, and § 5.50. The P. E. I. Steam Navigation Company's vessels *St. Lawrence* and *Princess of Wales* run between Charlottetown, Shediac, and Pictou (see Routes 43 and 44). The *Heather Belle* plies about the bay and up the Hillsborough River, making also trips to Crapaud and Orwell. She runs up the Hillsborough River to Mount Stewart on Monday, Tuesday, Friday and Saturday; to Crapaud on Wednesday; and to Orwell on Wednesday, Thursday and Friday (time-table of 1874).

CHARLOTTETOWN, the capital of Prince Edward Island, is situated on gently rising ground on the N. side of the Hillsborough River, and fronts on a good harbor. It has about 8,000 inhabitants, with 6 weekly newspapers, 2 banks, and 10 churches. The plan of the city is very regular, and consists of 6 streets, each 100 ft. wide, running E. and W., intersecting 9 streets running from N. to S. There are four large public squares.

The **Colonial Building** is the only fine structure in the city. It stands on Queen's Square, at the head of Great George St., and is built of Nova-Scotia freestone (at a cost of $85,000). The halls of the Legislative Council and House of Assembly are on the second floor, and are handsomely furnished and adorned with portraits of the statesmen of Prince Edward Island. On the same floor is the *Colonial Library*, containing a good collection of books relating to the history, laws, and physical characteristics of Canada and the British Empire. A pleasant view of the city and the rivers may be obtained from the cupola of the building. The *Post Office* is also on Queen's Square, and is a new and handsome stone building. Just beyond is the Market House, a great wooden structure covered with shingles. The principal shops of Charlottetown are about Queen's Square, and offer but little to be desired. The Roman Catholic Cathedral of St. Dunstan is a spacious wooden edifice on Great George St., near the Square. The extensive Convent of Notre Dame is on Hillsborough Square, and occupies a modern brick building. The Prince of Wales College and the Normal School are on Weymouth St., in this vicinity.

The old barracks and drill-shed are W. of Queen's Square, between Pownal and Sydney Sts., and are fronted by a parade-ground. The *Government House* is on a point of land W. of the city, and overlooks the harbor.

In 1748 the government of the island was vested in civil and military officers, whose residence was established at the W. entrance to the harbor of Port la Joie (Charlottetown), where they had a battery and a small garrison. It is said that the first French sailors who entered the inner harbor were so pleased with its tranquil beauty that they named it *Port la Joie*. There were no houses on the site of the city in 1752. The harbor was held by three British frigates in 1746, but was ravaged by 200 Micmacs under the French Ensign Montesson. All the English found on the shore were captured, but the Indians refused to attack the war-vessels.

In 1768 Morris and Deschamps arrived here with a small colony, and erected huts. They laid out the streets of Charlottetown, which was soon established as the capital of the island. In 1775 it was captured by two American war-vessels, which had been cruising in the Gulf to carry off the Quebec storeships. The sailors plundered the town, and led away several local dignitaries as prisoners, but Washington liberated the captives, and reprimanded the predatory cruisers.

Charlottetown "has the appearance of a place from which something has departed; a wooden town, with wide and vacant streets, and the air of waiting for something..... That the productive island, with its system of free schools, is about to enter upon a prosperous career, and that Charlottetown is soon to become a place of great activity, no one who converses with the natives can doubt, and I think that even now no traveller will regret spending an hour or two there; but it is necessary to say that the rosy inducements for tourists to spend the summer there exist only in the guide-books."

Environs of Charlottetown.

The *Wesleyan College* is on an eminence back of the city, and overlooks the harbor and the rivers. It has 10 instructors and about 300 students. *St. Dunstan's College* is a Catholic institution, which occupies the crest of a hill 1 M. from the city, and has 4 professors. There are several pretty villas in the vicinity of Charlottetown; and the roads are very good during dry weather. Some travellers have greatly admired the rural scenery of these suburban roads, but others have reported them as tame and uninteresting. The same conflict of opinion exists with regard to the scenery of the whole island.

Southport is a village opposite Charlottetown, in a pretty situation on the S. shore of the Hillsborough River. It is reached by a steam ferry-boat, which crosses every hour. 3 M. from this place is the eminence called *Tea Hill*, whence a pleasing view of the parish and the bay may be obtained. A few miles beyond is the village of *Pownal*, at the head of Pownal Bay, and in a region prolific in oats and potatoes.

46. Charlottetown to Summerside and Tignish. — The Western Shores of Prince Edward Island.

This region is traversed by the Prince Edward Island Railway, a narrow-gauge road which has recently been built by the Canadian government. This line was opened late in 1874, and its stations are not yet fully established, many of them being merely platforms, at which the trains do not stop unless there are passengers to be put down or taken up. During the winter of 1874-5 this line ran three trains a week, " weather permitting."

Stations. — Charlottetown; Royalty Junction, 5 M.; N. Wiltshire, 17; Hunter River, 21; Kensington, 41; Summerside, 49; Wellington, 61; Tyne, or Port Hill, 71; O'Leary Road, 89; Alberton, 104; Tignish, 117.

After leaving the commodious station-building, in the E. part of Charlottetown, the train sweeps around the city, turning to the N. from the bank of the Hillsborough River. The suburban villas are soon passed, and the line traverses a level country to *Royalty Junction*, where the tracks to Souris and Georgetown (see Route 47) diverge to the N. E. The train now enters the main line, and runs W. through a fertile farming country, — "a sort of Arcadia, in which Shenstone would have delighted." The hamlets are small and the dwellings are very plain, but it is expected that the stations of the new railway will become the nuclei of future villages. The train soon crosses the head-waters of the York River, and reaches *N. Wiltshire*, beyond which is a line of low hills, extending across the island. 4 M. beyond this point is the station of *Hunter River*, whence a much-

8 *   L

travelled road leads to the N. to New Glasgow and Rustico, locally famous for pleasant marine scenery.

Rustico is a quiet marine settlement, with two churches and a bank, and about 300 inhabitants. It is near Grand Rustico Harbor, and is one of the chief fishing stations of the N. shore. The original settlers were Acadians (in the year 1710), many of whose descendants remain in the township, and are peaceful and unprogressive citizens. The *Ocean House* (40 guests) is a small summer hotel near the sand-hills of the beach; and the facilities for boating, bathing, fishing, and gunning are said to be excellent. The great fleets of the Gulf fishermen are sometimes seen off these shores. There is a pleasant drive up the Hunter River to *New Glasgow* (Rockem's inn), which was settled by men of Glasgow, under Alexander Cormack, the Newfoundland explorer, in 1829. The Hunter River affords good trouting.

Grand Rustico Harbor is rendered unsafe by shifting bars of sand, and it was off this port that the Government steamer *Rose* was lost. On the coast to the N. W. are the hamlets of N. Rustico and Cavendish, the latter of which is a Presbyterian farming settlement of 200 inhabitants.

Kensington station is about 41 M. from Charlottetown, and is near the petty hamlet of the same name. To the N. E. is Grenville Harbor, with the estuaries of three rivers, the chief of which is the Stanley. There are several maritime hamlets on these shores, and on the W. is *New London*, a neat Scottish settlement with two churches. A road also leads N. W. from Kensington to *Princetown*, a village of 400 inhabitants, situated on the peninsula between Richmond Bay, March Water, and the Darnley Basin. This town was laid out (in 1766) with broad streets and squares, and was intended for the metropolis of the N. coast, but the expectations of the government were never realized, and "the ploughshare still turns up the sod, where it was intended the busy thoroughfare should be." Malpeque Harbor is the finest and safest on the N. shore of Prince Edward Island. A few miles E. are the lofty sandstone cliffs of Cape Tryon, near New London harbor. Princetown fronts on *Richmond Bay*, a capacious haven which runs in to the S. W. for 10 M., and contains 7 islands. Travellers have praised the beauty of the road from Princetown to Port Hill, which affords many pleasant views over the bay.

Beyond Kensington the train runs S. W. across the rural plains of St. David's Parish, and passes out on the isthmus between Richmond Bay and Bedeque Bay, where the island is only 3-4 M. wide. 9 M. from Kensington it reaches Summerside.

Summerside (two inns) is situated on the N. side of Bedeque Harbor, and is a town of about 2,000 inhabitants, with 8 churches, 5 schools, 2 weekly newspapers, and 2 banks. It is the port whence most of the products of the W. part of the island are sent out, and has grown rapidly of late years. The chief exports in 1870 were 268,000 bushels of oats, 37,393 bushels of potatoes, 10,300 bushels of barley, 86,450 dozen of eggs, and 4,337 barrels of the famous Bedeque oysters. The wharves are long, in order to reach the deep water of the channel; and the houses of the town are mostly small wooden buildings. Considerable shipbuilding is done here.

The * *Island Park Hotel* is a summer resort on an islet off the harbor, and is patronized by American tourists. There are accommodations for fishing and bathing, and a steam ferry-boat plies between the island and the town. The hotel commands a pleasant view of the Bedeque shores and the Strait of Northumberland.

"This little seaport is intended to be attractive, and it would give these travellers great pleasure to describe it if they could at all remember how it looks. But it is a place that, like some faces, makes no sort of impression on the memory. We went ashore there, and tried to take an interest in the shipbuilding, and in the little oysters which the harbor yields; but whether we did take an interest or not has passed out of memory. A small, unpicturesque, wooden town, in the languor of a provincial summer; why should we pretend an interest in it which we did not feel? It did not disturb our reposeful frame of mind, nor much interfere with our enjoyment of the day." (WARNER's *Baddeck*.)

On leaving Summerside, the train runs out to the W., over a level region. To the N. is the hamlet of *St. Eleanors* (Ellison's Hotel), a place of 400 inhabitants, situated in a rich farming country. It enjoys the honor of being the shire-town of Prince County, and is about 2½ M. from Summerside. 3 M. from St. Eleanors is the rural village of *Miscouche*, inhabited by French Acadians. *Wellington* (Western Hotel) is a small hamlet and station 12 M. beyond Summerside, near the head of the Grand River, which flows into Richmond Bay. The Acadian settlements about Cape Egmont are a few miles to the S. W.

The line passes on to *Port Hill*, a prosperous shipbuilding village on Richmond Bay. Near this place is *Lennox Island*, which is reserved for the Micmac Indians, and is inhabited by about 150 persons of that tribe. Between the bay and the Gulf of St. Lawrence is George Island, which is composed of trap-rock and amygdaloid, and is regarded as a curious geological intrusion in the red sandstone formations of the Prince-Edward shores. The train runs N. W. over the isthmus between the Cavendish Inlet and the Percival and Enmore Rivers, and soon enters the North Parish. This region is thinly inhabited by French and British settlers, and is one of the least prosperous portions of the island. The line passes near *Brae*, a settlement of 300 Scotch farmers, near the trout-abounding streams of the Parish of Halifax. To the S. W. is the sequestered marine hamlet of *West Point*, where a town has been laid out and preparations made for a commerce which does not come. The coast trends N. by E. 6 M. from West Point to Cape Wolfe, whence it runs N. E. by E. 27 M. to North Point, in a long unbroken strand of red clay and sandstone cliffs.

Alberton (two inns) is one of the northern termini of the railway, and is a prosperous village of 700 inhabitants, with two churches and an

American consular agency. It is situated on Cascumpec[1] harbor, and is engaged in shipbuilding and the fisheries. The American fishing-schooners often take refuge in this harbor. The neighboring rural districts are fertile and thickly populated, and produce large quantities of oats and potatoes. This town was the birthplace of the Gordons, the heroic missionaries at Eromanga, one of whom was martyred in 1861, the other in 1872. S. of Alberton is Holland Bay, which was named in honor of himself by Major Holland, the English surveyor of the island; and 6 - 8 M. N. is Cape Kildare.

Tignish (*Ryan's Hotel*) is the extreme northern point reached by the railway, and is 117 M. from Charlottetown. It has about 200 inhabitants, and is one of the most important fishing-stations on the island. The inhabitants are mostly French and Scotch, and support a Catholic church and convent. There are several other French villages in this vicinity, concerning which the historian of the island says: "They are all old settlements. The nationality of the people has kept them together, until their farms are subdivided into small portions, and their dwellings are numerous and close together. Few are skilful farmers. Many prefer to obtain a living by fishing rather than farming. They are simple and inoffensive in their manners; quiet and uncomplaining, and easily satisfied. The peculiarities of their race are not yet extinct; and under generous treatment and superior training, the national enterprise and energy, politeness and refinement, would gradually be restored."

North Point is about 8 M. N. of Tignish, and is reached by a sea-viewing road among the sand-dunes. It has a lighthouse, which sustains a powerful light, and is an important point in the navigation of the Gulf.

47. Charlottetown to Georgetown.

By the Prince Edward Island Railway.
Stations. — Charlottetown; Royalty Junction, 5 M.; Mount Stewart, 22; Cardigan, 40; Georgetown, 46.

Beyond Royalty Junction the train diverges to the N. E., and follows the course of the Hillsborough River, though generally at some distance from the shore. The banks of this stream are the most favored part of that prosperous land of which Dr. Cuyler says: "It is one rich, rolling, arable farm, from Cape East clear up to Cape North." As early as 1758 there were 2,000 French colonists about this river. The Hillsborough is 30 M. long, and the tide ascends for 20 M. Much produce is shipped from these shores during the autumnal months. About 8 M. beyond the Junction the line crosses French Fort Creek, on whose banks the French troops erected a fortification to protect the short portage (1½ M.) across the island, from the river to Tracadie Harbor. Here the military domination was surren-

[1] *Cascumpec*, an Indian word, meaning "Flowing through Sand."

dered to the British expeditionary forces. To the N. W. are the Gaelic villages of Covehead and Tracadie, now over a century old; near which is the sandy lagoon of Tracadie Harbor. At the place called Scotch Fort the French built the first church on the island, and in this vicinity the earliest British settlers located. From the French Catholic church on the lofty hill at St. Andrews, a few miles to the N. E., a beautiful view is obtained over a rich rural country.

Mount Stewart (two inns) is a prosperous little shipbuilding village, whence the steamer *Heather Belle* runs to Charlottetown. The train crosses the river at this point, and at Mount Stewart Junction it turns to the S. E., while the Souris Railway diverges to the N. E. The country which is now traversed is thinly settled, and lies about the head-waters of the Morrell and Pisquid Rivers. There are several small lakes in this region, and forests are seen on either hand. At *Cardigan* (small inn) the line reaches the head-waters of the eastern rivers. A road leads hence to the populous settlements on the Vernon River and Pownal Bay.

Georgetown (*Commercial Hotel*) is the capital of King's County, and has about 800 inhabitants. It is situated on the long peninsula between the Cardigan and Brudenelle Rivers, and its harbor is one of the best on the island, being deep and secure, and the last to be closed by ice. The county buildings, academy, and Episcopal church are on Kent Square. The chief business of the town is in the exportation of produce, and ship-building is carried on to some extent. The town is well laid out, but its growth has been very slow. Steamers ply between this port, Pictou, and the Magdalen Islands (see Routes 44 and 49). The harbor is reached by ascending Cardigan Bay and passing the lighthouses on Panmure Head and St. Andrew's Point.

Montague Bridge (Montague House) is reached from Georgetown by a ferry of 6 M. and 11 M. of staging. It has 350 inhabitants and several mills. To the S. E. is St. Mary's Bay. About 20 M. S. of Georgetown is *Murray Harbor*, on which there are several Scottish villages. From Cape Bear the coast trends W. for 27 M. to Point Prim.

> " No land can boast more rich supply,
> That e'er was found beneath the sky ;
> No purer streams have ever flowed,
> Since Heaven that bounteous gift bestowed.
>
> And herring, like a mighty host,
> And cod and mackerel, crowd the coast."

> "In this fine island, long neglected,
> Much, it is thought, might be effected
> By industry and application, —
> Sources of wealth with every nation."

48. Charlottetown to Souris.

By the Prince Edward Island Railway.
Stations.—Charlottetown : Royalty Junction, 5 M. ; Mount Stewart, 22 ; Morrell, 30 ; St. Peter's, 38½ ; Harmony, 55 ; Souris, 60½.

Charlottetown to Mount Stewart, see page 181.

At Mount Stewart Junction the train diverges to the N. E., and soon reaches *Morrell*, a fishing-station on the Morrell River, near St. Peter's Bay.

St. Peter's (*Prairie Hotel*) was from the first the most important port on the N. shore of the island, on account of its rich salmon-fisheries. About the year 1750 the French government endeavored to restrict the fishing of the island, and to stimulate its agriculture, by closing all the ports except St. Peter's and Tracadie. The village is now quite small, though the salmon-fishery is valuable. St. Peter's Bay runs 7 M. into the land, but it is of little use, since there is only 5 ft. of water on its sandy bar. From this inlet to East Point the shore is unbroken, and is formed of a line of red sandstone cliffs, 33 M. long.

"The sea-trout fishing, in the bays and harbors of Prince Edward Island, especially in June, when the fish first rush in from the gulf, is really magnificent. They average from 3 to 5 pounds each. I found the best fishing at St. Peter's Bay, on the N. side of the island, about 28 M. from Charlottetown. I there killed in one morning 16 trout, which weighed 80 pounds. In the bays and along the coasts of the island they are taken with the scarlet fly, from a boat under easy sail, with a 'mackerel breeze,' and sometimes a heavy 'ground swell.' The fly skips from wave to wave at the end of 30 yards of line, and there should be at least 70 yards more on the reel. It is splendid sport, as a strong fish will make sometimes a long run, and give a good chase down the wind." (PERLEY.)

Harmony station is near Rollo Bay, which was named in honor of Lord Rollo, who occupied the island with British troops in 1758. There is a small hamlet on this bay; and to the S. W. are the Gaelic settlements of Dundas, Bridgetown, and Annandale, situated on the Grand River.

Souris (three inns) is a village of Catholic Highlanders, pleasantly situated on the N. side of Colville Bay, and divided into two portions by the Souris River. The harbor is shallow, but is being improved by a breakwater. The shore-fishing is pursued in fleets of dories, and most of the produce of the adjacent country is shipped from Souris to the French Isle of St. Pierre (see page 185). There is a long sandy beach on the W. of the village, and on the S. and E. is a bold headland. Souris was settled by the Acadians in 1748; and now contains about 500 inhabitants.

The East Parish extends for several leagues E. of Souris, and includes the sea-shore hamlets of Red Point, Bothwell, East Point, North Lake, and Fairfield. The East and North Lakes are long and shallow lagoons on the coast. **East Point** is provided with a first-class fixed light, which is 130 ft. above the sea and is visible for 18 M.

49. The Magdalen Islands.

These remote islands are sometimes visited, during the summer, by fishing-parties, who find rare sport in catching the white sea-trout that abound in the vicinity. The accommodations for visitors are of the most primitive kind, but many defects are atoned for by the hospitality of the people.

The mail-steamer *Albert* leaves Pictou for Georgetown (P. E. I.) and the Magdalen Islands every alternate Wednesday. She also leaves Pictou for Port Hood (Cape Breton) every Monday evening, returning on the following morning. (Time-table of 1874.)

Fares. — Halifax to Port Hood, $4.60; to Georgetown, $4.10; to the Magdalen Islands, $8. Further particulars may be obtained by addressing James King, mail-contractor, Halifax.

The **Magdalen Islands** are thirteen in number, and are situated at the entrance to the Gulf of St. Lawrence, 50 M. from East Point (P. E. I.), 60 M. from Cape North (C. B.), 120 M. from Cape Ray (N. F.), and 150 M. from Gaspé. When they are first seen from the sea, they present the appearance of well-detached islets, but on a nearer approach several of them are seen to be connected with each other by double lines of sandy beaches, forming broad and quiet salt-water lagoons. The inhabitants are mostly Acadian fishermen (speaking French only), devoted to the pursuit of the immense schools of cod and mackerel that visit the neighboring waters. At certain seasons of the year the harbors and lagoons are filled with hundreds of sail of fishing-vessels, most of which are American and Provincial. Seal-hunting is carried on here with much success, as extensive fields of ice drift down against the shores, bearing myriads of seals. On one occasion over 6,000 seals were killed here in less than a fortnight by parties going out over the ice from the shore. This is also said to be the best place in America for the lobster fishery, and a Portland company has recently founded a canning establishment here. On account of their abundant returns in these regards the Magdalen Islands have received the fitting title of " The Kingdom of Fish." In order to protect these interests the Dominion armed cutter *La Canadienne* usually spends the summer in these waters, to prevent encroachments by Americans and Frenchmen.

Amherst Island is the chief of the group, and is the seat of the principal village, the custom-house, and the public buildings. On its S. point is a red-and-white revolving light which is visible for 20 M.; and the hills in the interior, 550 ft. high, are seen from a great distance by day. The village has 3 churches and the court-house, and is situated on a small harbor which opens on the S. of *Pleasant Bay*, a broad and secure roadstead where hundreds of vessels sometimes weather heavy storms in safety. 1 M. N. W. of the village is the singular conical hill called the *Demoiselle* (280 ft. high), whence the bay and a great part of the islands may be seen.

Grindstone Island is 5 - 6 M. N. of Amherst, and is connected with it by a double line of sand-beaches, which enclose the wide lagoon called Basque Harbor. It is 5 M. long, and has a central hill 550 ft. high, while on the W. shore is the lofty conical promontory of sandstone which the

Acadians call *Cap de Meule*. On the same side is the thriving hamlet of *L'Étang du Nord*. On the E., and containing 7 square miles, is **Alright Island**, terminated by the grayish-white cliffs of Cape Alright, over 400 ft. high. A sand-beach runs N. E. 10 M. from Grindstone to Wolf Island, a sandstone rock ¾ M. long; and another beach runs thence 9 M. farther to the N. E. to *Grosse Island*, on the Grand Lagoon. This island has another line of lofty cliffs of sandstone. To the E. is *Coffin Island*, and 4 M. N. is *Bryon Island*, beyond which are the Bird Isles.

Entry Island lies to the E. of Amherst Island, off the entrance to Pleasant Bay, and is the most picturesque of the group. Near the centre is a hill 580 ft. high, visible for 25 M., and from whose summit the whole Magdalen group can be overlooked. The wonderful cliffs of red sandstone which line the shores of this island are very picturesque in their effect, and reach a height of 400 ft.

Deadman's Isle is a rugged rock 8 M. W. of Amherst, and derives its name from the fancied resemblance of its contour to that of a corpse laid out for burial. While passing this rock, in 1804, Tom Moore wrote the poem which closes:

" There lieth a wreck on the dismal shore
Of cold and pitiless Labrador,
Where, under the moon, upon mounts of frost,
Full many a mariner's bones are tossed.

"Yon shadowy bark hath been to that wreck,
And the dim blue fire that lights her deck

Doth play on as pale and livid a crew
As ever yet drank the churchyard dew.

" To Deadman's Isle in the eye of the blast,
To Deadman's Isle she speeds her fast;
By skeleton shapes her sails are furled,
And the hand that steers is not of this world."

The **Bird Isles** are two bare rocks of red sandstone, ¾ M. apart, the chief of which is known as Gannet Rock, and is 1,300 ft. long and 100–140 ft. high, lined with vertical cliffs. These isles are haunted by immense numbers of sea-birds, gannets, guillemots, puffins, kittiwakes, and razor-billed auks. "No other breeding-place on our shore is so remarkable at once for the number and variety of the species occupying it." Immense quantities of eggs are carried thence by the islanders, but to a less extent than formerly.

This great natural curiosity was visited in 1632 by the Jesuits (who called the rocks *Les Colombiers*), by Heriot in 1807, by Audubon, and in 1860 by Dr. Bryan. The Dominion has recently erected a lighthouse here at great expense, and to the imminent peril of those engaged in the work, since there is no landing-place, and in breezy weather the surf dashes violently against the cliffs all around. The tower bears a fixed white light of the first class, which is visible for 21 M.

Charlevoix visited these islands in 1720, and wondered how, "in such a Multitude of Nests, every Bird immediately finds her own. We fired a Gun, which gave the Alarm thro' all this flying Commonwealth, and there was formed above the two Islands, a thick Cloud of these Birds, which was at least two or three Leagues around."

The Magdalen Islands were visited by Cartier in 1534, but the first permanent station was founded here in 1663 by a company of Honfleur mariners, to whom the islands were conceded by the Company of New France. In 1720 the Duchess of Orleans granted them to the Count de St. Pierre. In 1763 they were inhabited by 10 Acadian families, and in 1767 a Bostonian named Gridley founded on Amherst

Island an establishment for trading and for the seal and walrus fisheries. During the Revolution American privateers visited the islands, and destroyed everything accessible. Gridley returned after the war, but the walrus soon became extinct, and the islanders turned their attention to the cod and herring fisheries. When Admiral Coffin received his grant there were 100 families here; in 1831 there were 1,000 inhabitants; and the present population is about 3,500. In the mean time three colonies have been founded and populated from these islands, on Labrador and the N. shore. The Lord's-Day Gale (see page 170) wrought sad havoc among the fleets in these waters.

Tradition tells that when Capt. Coffin was conveying Governor-General Lord Dorchester to Canada in his frigate, a furious storm arose in the Gulf, and the skilful mariner saved his vessel by gaining shelter under the lee of these islands. Dorchester, grateful for his preservation, secured for the captain the grant of the islands "in free and common soccage," with the rights of building roads and fortifications reserved to the Crown. The grantee was a native of Boston and a benefactor of Nantucket, and subsequently became Admiral Sir Isaac Coffin. The grant now belongs to his nephew, Admiral Coffin, of Bath, and is an entailed estate of the family. In 1873, 75 years after the grant, the legislature of Quebec (in whose jurisdiction the islands lie) made extensive investigations with a view to buy out the proprietor's claim, since many of the islanders had emigrated to Labrador and the Mingan Isles, dissatisfied with their uncertain tenure of the land.

50. St. Pierre and Miquelon.

The Anglo-French Steamship Company dispatches the steamer *George Shattuck* from Halifax to Sydney and St. Pierre every alternate Saturday during the season of navigation. She leaves St. Pierre every alternate Friday. The voyage to Sydney has recently been made by way of St. Peter's Canal and the Bras d'Or, but it is not likely that that route will be adopted in preference to the outside course.

Fares from Halifax to Sydney, cabin, $10, steerage, $6; to St. Pierre, cabin, $15, steerage, $8; Sydney to St. Pierre, cabin, $9, steerage, $6. The price of meals is included in the cabin-fares. Further information may be obtained by addressing Joseph S. Belcher, Boak's Wharf, Halifax.

St. Pierre may also be visited by the Western Coastal steamer from St. John's, N. F. (see Route 60).

There are several French *cafés* and *pensions* in the village of St. Pierre, at which the traveller can find indifferent accommodations. The best of these is that at which the telegraph-operators stop.

On entering the harbor of St. Pierre, the steamer passes *Galantry Head*, on which is a red-and-white flash-light which is visible for 20 M., and also two fog-guns. Within the harbor are two fixed lights, one white and one red, which are visible for 6 M.; and the *Isle aux Chiens* contains a scattered fishing-village.

The island of St. Pierre is about 12 M. from Point May, on the Newfoundland coast, and is 12 M. in circumference. It is mostly composed of rugged porphyritic ridges, utterly arid and barren, and the scenery is of a striking and singular character. Back of the village is the hill of *Calvaire*, surmounted by a tall cross; and to the S. W., beyond Ravenel Bay, is the lakelet called *L'Étang du Savoyard*. The town is compactly built on the harbor at the E. of the island, and most of its houses are of stone. It is guarded by about 50 French soldiers, whose presence is necessary to keep the multitudes of fearless and pugnacious sailors from incessant rioting. There is a large force of telegraph-operators here, in charge of the two cables from America to Great Britain by way of Newfoundland, and of the Franco-American cable, which runs E. to Brest and S. W. to Duxbury, in Massachusetts.

The only good house in the town is that of the Governor; and the Catholic church and convent rise prominently over the low houses of the fishermen. Near the sea is a battery of ancient guns, which are used only for warning in season of fogs. The buildings are nearly all of wood, and include many shops, where every variety of goods may be obtained. The merchants are connected with French and American firms. There are numerous *cabarets*, or drinking-saloons; and the *auberges*, or small taverns, are thoroughly French. The citizens are famed for their hospitality to properly accredited strangers; and the literary culture of the community is served by a diminutive weekly paper called *La Feuille Officielle*, printed on a sheet of foolscap, and containing its serial Parisian *feuilleton*.

The street of St. Pierre presents a very interesting sight during the spring and fall. It is crowded with many thousands of hardy fishermen, arrayed in the quaint costumes of their native shores, — Normans, Bretons, Basques, Provincials, and New-Englanders, — all active and alert; while the implements of the fisheries are seen on every side. The environs of the town are rocky and utterly unproductive, so that the provisions used here are imported from the Provinces.

The resident population is 3,187 (of whom 24 are Protestant), and the government is conducted by a Commandant, a Police Magistrate, Doctor, Apostolic Prefect, and Engineer, with a few artillerists and gens-d'armes. There is usually one or more French frigates in the harbor, looking after the vast fisheries which employ 15,000 sailors of France, and return 30,000,000 francs' worth of fish.

St. Pierre is the chief rendezvous of the French fishermen, and immense fleets are sometimes gathered here. Over 1,000 sail of square-rigged vessels from France are engaged in these fisheries, and on the 29th of June, 1874, the roadstead near the island contained 350 sail of square-rigged vessels and 300 fore-and-aft vessels. They are here furnished with supplies, which are drawn from the adjacent Provinces, and in return leave many of the luxuries of Old France. It is claimed that the brandy of St. Pierre is the best in America. The fishermen leave their fish here to be cured, and from this point they are sent S. to the United States and the West Indies.

Little Miquelon Island, or Langley Island, lies 3 M. N. W. of St. Pierre, and is about 24 M. around. It is joined to Great Miquelon Island by a long and narrow sandy isthmus. The latter island is 12 M. long, and looks out on Fortune Bay. Near its N. end are the singular hills known as Mt. Chapeau and Mt. Calvaire. On this island, during the summer of 1874, was wrecked H. B. M. frigate *Niobe*, the brave ship that trained her guns on Santiago de Cuba, and prevented a total massacre of the *Virginius* prisoners.

St. Pierre was captured by a British fleet in 1793, and all its inhabitants, 1,502 in number, were carried away to Halifax, whence they were soon afterwards sent to France. In 1796 a French Republican fleet under Admiral Richery visited the deserted island, and completely destroyed its buildings and wharves. It was, however, restored to France in 1814, together with her ancient privileges in these waters. "All the island is only a great laboratory for the preparation, curing, and exportation of codfish. For the rest, not a tree, not a bush, above 25 centimetres."

NEWFOUNDLAND

Is bounded on the W. by the Gulf of St. Lawrence, on the N. by the Strait of Belle Isle, and on the E. and S. by the Atlantic Ocean. From N. to S. it is 350 M. long, and the average breadth is 130 M., giving an estimated area of 40,200 square miles. The coast is steep and bold, and is indented with numerous deep bays and fiords. Mines of lead and copper are being worked with much success, and there are large undeveloped deposits of coal on the W coast.

" Up go the surges on the coast of Newfoundland, and down again into the sea. The huge island .... stands, with its sheer, beetling cliffs, out of the ocean, a monstrous mass of rock and gravel, almost without soil, like a strange thing from the bottom of the great deep, lifted up suddenly into sunshine and storm, but belonging to the watery darkness out of which it has been reared. The eye accustomed to richer and softer scenes finds something of a strange and almost startling beauty in its bold, hard outlines, cut out on every side against the sky. .... Inland, surrounded by a fringe of small forests on the coasts, is a vast wilderness of moss, and rock, and lake, and dwarf firs about breast-high. These little trees are so close and stiff and flat-topped that one can almost walk on them. Of course they are very hard things to make way through and among. . . . . In March or April almost all the men go out in fleets to meet the ice that floats down from the northern regions and to kill the seals that come down on it. In early summer a third part or a half of all the people go, by families, in their schooners, to the coast of Labrador, and spend the summer fishing there ; and in the winter, half of them are living in the woods, in tilts, to have their fuel near them. At home or abroad, during-the season, the men are on the water for seals or cod. The women sow, and plant, and tend the little gardens, and dry the fish ; in short, they do the land-work, and are the better for it." (R. T. S. LOWELL.)

Two of the most remarkable features of the natural history of the island are thus quaintly set forth by Whitbourne (*anno* 1622): "Neither are there any Snakes, Toads, Serpents, or any other venomous Wormes that ever were knowne to hurt any man in that country, but only a very little nimble fly (the least of all other flies), which is called a Miskieto; those flies seem to have a great power and authority upon all loytering and idle people that come to the Newfoundland." Instances have been known where the flies have attacked men with such venom and multitudes that fatal results have followed. In the interior of the island are vast unexplored regions, studded with large lakes and mountain-ranges. Through these solitudes roam countless thousands of deer, which are pursued by the Micmac hunters.

Newfoundland was discovered by the Norsemen in the tenth century, but they merely observed the coast and made no further explorations.

There is good reason for supposing that it was frequented by Breton and Norman fishermen during the fourteenth century. In 1497 the island was formally discovered by John Cabot, who was voyaging under the patronage of Henry VII. of England. The explorations of Cortereal (1501), Verazzano (1524), and Cartier (1534), all touched here, and great fishing-fleets began to visit the surrounding seas. Sir Humphrey Gilbert took possession of Newfoundland in the name of England, in 1583, making this the most ancient colony of the British Empire. The settlements of Guy, Whitbourne, Calvert, and others were soon established on the coast.

The fishermen were terribly persecuted by pirates during the earlier part of the 17th century. Peter Easton alone had 10 sail of corsairs on the coast, claiming that he was "master of the seas," and levying heavy taxes on all the vessels in these waters. Between 1612 and 1660 alone, the pirates captured 180 pieces of ordnance, 1,080 fishermen, and large fleets of vessels.

Between 1692 and 1713 the French made vigorous attempts to conquer the island, and the struggle raged with varying fortunes on the E. and S. shores. By the Treaty of Utrecht the French received permission to catch and cure fish along the W. coast (see Route 61). In 1728 Newfoundland was formed into a Province, and courts were established. The French made determined attacks in 1761 and 1796, and the people were reduced to great extremity by the Non-Intercourse Act passed by the American Congress in 1776 and again in 1812–14. In 1817 there were 80,000 inhabitants, and 800 vessels were engaged in the fisheries, whose product was valued at $10,000,000 a year. In 1832 the first Legislative Assembly was convened; in 1838 a geological survey was made; and in 1858 the Atlantic telegraph-cable was landed on these shores. Newfoundland has refused to enter the Dominion of Canada, and is still governed directly by the British Crown.

51. Halifax to St. John's, Newfoundland.

The ocean steamships between Halifax and Liverpool call at St. John's fortnightly. Their course after leaving Halifax is directly to the N. E. across the open sea, giving Cape Race a wide berth. The fare on these vessels is higher than it is on the *Virgo*, and the accommodations are superior; but the voyager does not get the interesting views of the Canso and Cape-Breton shores.

The Eastern Steamship Company's vessel, the *Virgo*, leaves Halifax on alternate Tuesdays, for Sydney and St. John's, carrying the Royal mails. The fare is $15 (steerage, $5). This steamship is large, and is well arranged for passenger-traffic.

Halifax to Sydney, see page 148.

After leaving the harbor of Sydney, Flint Island is seen on the r., and the blue ranges of the St. Anne Mts. on the l. The course is but little N. of E., and the horizon soon becomes level and landless. Sometimes the dim blue hills of St. Pierre are the first land seen after the Cape-Breton coast

ST. JOHN'S. *Route 52.* 189

sinks below the horizon; but generally the bold mountain-promontory of Cape Chapeau Rouge is the first recognizable shore. Then the deep bight of Placentia Bay opens away on the N. After rounding Cape Race (see Route 22), the steamship stretches away up the Strait Shore past a line of fishing hamlets, deep fiords, and rocky capes.

" When the mists dispersed, the rocky shores of Newfoundland were close upon our left, — lofty cliffs, red and gray, terribly beaten by the waves of the broad ocean. We amused ourselves, as we passed abreast the bays and headlands and rugged islands, with gazing at the wild scene, and searching out the beauty timidly reposing among the bleak and desolate. On the whole, Newfoundland, to the voyager from the States, is a lean and bony land, in thin, ragged clothes, with the smallest amount of adornment. Along the sides of the dull, brown mountains there is a suspicion of verdure, spotted and striped here and there with meagre woods of birch and fir. The glory of this hard region is its coast: a wonderful perplexity of fiords, bays and creeks, islands, peninsulas and capes, endlessly picturesque, and very often magnificently grand. Nothing can well exceed the headlands and precipices, honeycombed, shattered, and hollowed out into vast caverns, and given up to the thunders and the fury of the deep-sea billows. .... The brooks that flow from the highlands, and fall over cliffs of great elevation into the very surf, and that would be counted features of grandeur in some countries, are here the merest trifles, a kind of jewelry on the hem of the landscape." (NOBLE.)

"The first view of the harbor of St. John's is very striking. Lofty precipitous cliffs, of hard dark-red sandstone and conglomerate, range along the coast, with deep water close at their feet. Their beds plunge from a height of 400-700 ft., at an angle of 70°, right into the sea, where they are ceaselessly dashed against by the unbroken swell of the Atlantic waves." (JUKES.)

52. St. John's, Newfoundland.

Arrival from the Sea. — "The harbor of St. John's is certainly one of the most remarkable for bold and effective scenery on the Atlantic shore. .... We were moving spiritedly forward over a bright and lively sea, watching the stern headlands receding in the south, and starting out to view in the north, when we passed Cape Spear, a lofty promontory, crowned with a lighthouse and a signal-staff, upon which was floating the meteor flag of England, and at once found ourselves abreast the bay in front of St. John's. Not a vestige, though, of anything like a city was in sight, except another flag flitting on a distant pinnacle of rock. Like a mighty Coliseum, the sea-wall half encircled the deep water of this outer bay, into which the full power of the ocean let itself under every wind except the westerly. Right towards the coast where it gathered itself up into the greatest massiveness, and tied itself into a very Gordian knot, we cut across, curious to behold when and where the rugged adamant was going to split and let us through. At length it opened, and we looked through, and presently glided through a kind of mountain-pass, with all the lonely grandeur of the Franconia Notch. Above us, and close above, the rugged, brown cliffs rose to a fine height, armed at certain points with cannon, and before us, to all appearance, opened out a most beautiful mountain lake, with a little city looking down from the mountain-side, and a swamp of shipping along its shores. We were in the harbor, and before St. John's." (NOBLE.)

Hotels. — The Union House, 379 Water St. (nearly 1 M. from the Custom House), is the best; Atlantic House, Water St. There are also two or three boarding-houses, which are preferable to the hotels, if a long stay is to be made. Mrs. Simms's, 353 Water St., is one of the best of these; and Knight's Home, 173 Water St., is tolerable. The accommodations for visitors to St. John's are not such as might be desired or expected in a city of so much importance.

Carriages may be engaged at the stands on Water St. (near the Post-Office). The rate per hour is 80c.

Amusements, generally of merely local interest, are prepared in Temperance Hall or the Avalon (Victoria) Rink. Boat-racing is frequently carried on at Quiddy-Viddy Pond. Cricket-matches are also played on the outskirts of the city.

Post-Office, at the Market House, on Water St. *Telegraph*, New York, Newfoundland, and London Co., at the Market House.

Consulates.—American, 149 Water St.; French, Signal-Hill Road; German, 227 Water St.; Spanish, 116 Cochrane St.; Portuguese, 385 Water St.

Mail-wagons leave St. John's for Portugal Cove, daily; for Topsail, Holyrood, Harbor Main, Brigus, Bay Roberts, and Harbor Grace, Mondays and Thursdays (or after arrival of mail from Halifax) in winter, and once weekly in summer; to Bay Bulls and Ferryland, weekly; to Salmonier and Placentia, on the day of arrival of the Halifax mail.

Steamships.— For Bay Verd, Old Perlican, Trinity, Catalina, Bonavista, Greenspond, Fogo, Twillingate, Exploits Island, Tilt Cove, Little Bay Island, Nipper's Harbor, and the Labrador coast, fortnightly, on Monday; to Ferryland, Renewse, Trepassey, Burin, St. Pierre, Harbor Briton, Burgeo, Little Bay (La Poile), and Channel, fortnightly, on Thursday or Friday (passing on to Sydney, C. B., each alternate trip); to the ports on Conception Bay several times a week, from Portugal Cove (see Route 56); to Halifax, fortnightly, by the Eastern Steamship Co.'s vessel, the *Virgo*; to Halifax, fortnightly, by steamships of the Allan Line; to Londonderry and Liverpool, fortnightly, by the Allan Line.

St. John's, the capital of the Province of Newfoundland, is situated in latitude 47° 33′ 6″ N., and longitude 52° 44′ 7″ W., and is built on the slope of a long hill which rises from the shore of a deep and secure harbor. At the time of the census of 1869 there were 22,555 inhabitants in the city (there are now over 25,000); but the population, owing to the peculiar character of its chief industry, is liable at any time to be increased or diminished by several thousand men. The greater part of the citizens are connected with the fisheries, directly or indirectly, and large fleets are despatched from the port throughout the season. Their return, or the arrival of the sealing-steamers, with their great crews, brings new life to the streets, and oftentimes results in such general "rows" as require the attendance of a large police-force. The interests of the city are all with the sea, from which are drawn its revenues, and over which pass the fleets which bring in provisions from the Provinces and States to the S. W. The manufactures of St. John's are insignificant, and consist, for the most part, of biscuit-bakeries and oil-refineries (on the opposite side of the harbor). An immense business is done by the mercantile houses on Water St. in furnishing supplies to the outports (a term applied to all the other ports of Newfoundland except St. John's); and one firm alone has a trade amounting to $12,000,000 a year. For about one month, during the busy season, the streets are absolutely crowded with the people from the N. and W. coasts, selling their fish and oil, and laying in provisions and other supplies for the ensuing year. The commercial interests are served by three banks and a chamber of commerce; and the literary standard of society is maintained by the St. John's Athenæum and the Catholic Institute. The city is supplied with gas, and water is brought in from a lake 4½ M. distant, by works which cost $360,000.

"In trying to describe St. John's, there is some difficulty in applying an adjective to it sufficiently distinctive and appropriate. We find other cities coupled with words which at once give their predominant characteristic: London the richest, Paris the gayest, St. Petersburg the coldest. In one respect the chief town of Newfoundland has, I believe, no rival; we may, therefore, call it the fishiest of modern capitals. Round a great part of the harbor are sheds, acres in extent, roofed with cod split in half, laid on like slates, drying in the sun, or rather the air, for there is

ST. JOHN'S. *Route 52.* 191

not much of the former to depend upon. . . . . The town is irregular and dirty, built chiefly of wood, the dampness of the climate rendering stone unsuitable." (ELIOT WARBURTON.)

The harbor is small, but deep, and is so thoroughly landlocked that the water is always smooth. Here may generally be seen two or three British and French frigates, and at the close of the season these narrow waters are well filled with the vessels of the fishing-fleets and the powerful sealing-steamers. Along the shores are the fish-stages, where immense quantities of cod, herring, and salmon are cured and made ready for exportation. On the S. shore are several wharves right under the cliffs, and also a floating dock which takes up vessels of 800 tons' burden. The entrance to the harbor is called the *Narrows, and is a stupendous cleft in the massive ridge which lines the coast. It is about 1,800 ft. long, and at its narrowest point is but 660 ft. wide. On either side rise precipitous walls of sandstone and conglomerate, of which Signal Hill (on the N. side) reaches an altitude of 520 ft., and the southern ridge is nearly 700 ft. high. Vessels coming in from the ocean are unable to see the Narrows until close upon it, and steer for the lofty block-house on Signal Hill. The points at the entrance were formerly well fortified, and during war-time the harbor was closed by a chain drawn across the Narrows, but the batteries are now in a neglected condition, and are nearly disarmed.

The city occupies the rugged hill on the N. of the harbor, and is built on three parallel streets, connected by steep side-streets. The houses are mostly low and unpainted wooden buildings, crowding out on the sidewalks, and the general appearance is that of poverty and thriftlessness. Even the wealthy merchants generally occupy houses far beneath their station, since they seem to regard Newfoundland as a place to get fortunes in and then retire to England to make their homes. This principle was universally acted on in former years, but latterly pleasant villas are being erected in the suburbs, and a worthier architectural appearance is desired and expected for the ancient capital. *Water Street* is the main business thoroughfare, and follows the curves of the harbor-shore for about 1½ M. Its lower side is occupied by the great mercantile houses which supply "fish-and-fog-land" with provisions, clothing, and household requirements; and the upper side is lined with an alternation of cheap shops and liquor-saloons. In the N. part is the *Custom House*, and near the centre is the spacious building of the Market-Hall and the Post-Office. To the S., Water Street connects with the causeway and bridge of boats which crosses the head of the harbor. Admonished by several disastrous fires, the city has caused Water St. to be built upon in a substantial manner, and the stores, though very plain, are solidly and massively constructed.

The **Anglican Cathedral** stands about midway up the hill, over the old burying-ground. It was planned by Sir Gilbert Scott, the most eminent British architect of the present era, and is in the early English Gothic architecture. Owing to the inability of the Church to raise sufficient funds (for the missions at the outports demand all her revenues), the cathedral is but partly finished, the end of the nave being boarded up and furnished with a cheap temporary altar. The lofty proportions of the interior and the fine Gothic colonnades of stone between the nave and aisles, together with the high lancet-windows, form a pleasant picture.

The * **Roman Catholic Cathedral** is the most stately building in Newfoundland, and occupies the crest of the ridge, commanding a noble * view over the city and harbor and adjacent country, and looking through the Narrows on to the open sea. The prospect from the cathedral terrace on a moonlight night or at the time of a clear sunrise or sunset is especially to be commended. In the front part of the grounds is a colossal statue of St. Peter, and other large statues are seen near the building. The cathedral is an immense stone structure, with twin towers on the front, and is surrounded with a long internal corridor, or cloister. There are no aisles, but the whole building is thrown into a broad nave, from which the transepts diverge to N. and S. The stone of which it is constructed was brought from Conception Bay and from Dunleary, Ireland, and the walls were raised by the free and voluntary labors of the people. Clustered about the cathedral are the *Bishop's Palace*, the convent and its schools, and *St. Bonaventure's College* (5 professors), where the missionaries are disciplined and the Catholic youth are taught in the higher branches of learning.

Catholicism was founded on the island by Sir George Calvert (see Route 54) and by the Bishop of Quebec; suffered persecution from 1762 to 1784, when all priests were banished (though some returned in disguise): and afterwards gained the chief power as a consequence of Irish immigration, upon which the bishops became arrogant and autocratic, and the Province was, practically, governed from Cathedral Hill. The great pile of religious buildings then erected on this commanding height cost over $500,000, and the present revenues of the diocese are princely in amount, being collected by the priests, who board the arriving fishing-vessels and assess their people. The Irish Catholics form a great majority of the citizens of St. John's.

Near the cathedral are the old barracks of the Royal Newfoundland Companies and the garrisons from the British army. The *Military Road* runs along the crest of the heights, and affords pleasant views over the harbor. On this road is the **Colonial Building**, a substantial structure of gray stone, well retired from the carriage-way, and adorned with a massive portico of Doric columns upholding a pediment which is occupied by the Royal Arms of Great Britain and Ireland. The colonial legislature meets in this building, and occupies plain but comfortable halls. The *Government House* is N. of the Colonial Building, and is the official mansion of the governor of the Province (Col. S. J. Hill, C. B.). It was built in 1828-30, and cost $240,000. The surrounding grounds are pleasantly diversified with groves, flower-beds, and walks, and are much visited by the aristocracy of St. John's, during the short but brilliant summer season.

Passing out through the poor suburb called "Maggotty Cove," a walk of about 20 minutes leads to the top of * **Signal Hill.**

" High above, on our r., a ruined monolith, on a mountain-peak (Crow's Nest), marks the site of an old battery, while to the l., sunk in a hollow, a black bog lies sheltered amid the bare bones of mother earth, here mainly composed of dark red sandstones and conglomerate, passing down by regular gradations to the slate below. A sudden turn of the road reveals a deep solitary tarn, some 350 ft. above the sea, in which the guardian rocks reflect their purple faces, and where the ripple of the muskrat, hurrying across, alone disturbs the placid surface. We pass a hideous-looking barrack, and, crossing the soft velvety sward on the crest, reach a little battery, from the parapets of which we look down, almost 500 ft. perpendicularly, right into 'the Narrows,' the strait or creek between the hills connecting the broad Atlantic with the oval harbor within. The great south-side hills, covered with luxuriant wild vegetation, and skeined with twisting torrents, loom across the strait so close that one might fancy it almost possible a stone could fly from the hand to the opposite shore. On our left the vast ocean, with nothing — not a rock — between us and Galway; on our right, at the other end of the narrow neck of water directly beneath, the inner basin, expanding towards the city, with the background of blue hills as a setting to the picture, broken only in their continuous outline by the twin towers of the Catholic cathedral, ever thus from all points performing their mission of conspicuity. Right below us, 400 ft. perpendicular, we lean over the grass parapet and look carefully down into the little battery guarding the narrowest part of the entering-strait, where, in the old wars, heavy chains stretched from shore to shore. . . . . The Narrows are full of fishing-boats returning with the silver spoils of the day glistening in the hold of the smacks, which, to the number of forty or fifty at a time, tack and fill like a fleet of white swans against the western evening breeze. Even as we look down on the decks, they come, and still they come, round the bluff point of Fort Amherst, from the bay outside." (Lt.-Col. McCrea.)
" After dinner we set off for Signal Hill, the grand observatory of the country, both by nature and art. . . . . Little rills rattled by; paths wound among rocky notches and grassy chasms, and led out to dizzy 'over-looks' and 'short-offs.' The town with its thousand smokes sat in a kind of amphitheatre, and seemed to enjoy the spectacle of sails and colors in the harbor. . . . . We struck into a fine military road, and passed spacious stone barracks, soldiers and soldiers' families, goats and little gardens. From the observatory, situated on the craggy peninsula, both the rugged interior and the expanse of ocean were before us." (NOBLE.)

" *Britones et Normani anno a Christo nato MCCCCCIIII. has terras invenere* "; and in August, 1527, 14 sail of Norman, Breton, and Portuguese vessels were sheltered in the harbor of St. John's. In 1542 the Sieur de Roberval, Viceroy of New France, entered here with 3 ships and 200 colonists bound for Quebec. He found 17 vessels at anchor in the harbor, and soon afterward there arrived Jacques Cartier and the Quebec colonists, discouraged, and returning to France. Roberval ordered him back, but he stole out of the harbor during the darkness of night and returned to France. A few years later the harbor was visited by the exploring ship *Mary of Guilford*, and the reverend Canon of St Paul, who had undertaken the unpriestly function of a discoverer, sent hence a chronicle of the voyage to Cardinal Wolsey.
In August, 1583, Sir Humphrey Gilbert (see page 135) entered the harbor of St. John's, with a fleet consisting of the *Delight*, *Golden Hind*, *Swallow*, and *Squirrel*. He took formal possession of the port and of the island of Newfoundland, receiving the obedience of 36 ship-masters then in the harbor. But the adventurous mariners were discontented with the rudeness of the country, and the learned Parmenius wrote back to Hakluyt: " My good Hakluyt, of the manner of this country what shall I say, when I see nothing but a very wildernesse." In view of the date of Gilbert's occupation, Newfoundland claims the proud distinction of being the most ancient colony of the British Empire. In 1584 St. John's was visited by the fleet of Sir Francis Drake, which had swept the adjacent seas and left a line of burning wrecks behind.
In 1696 the town was so strongly guarded that it easily repulsed the Chevalier Nesmond, who attacked it with ten French men-of-war. The expedition of the daring Iberville was more successful, and occupied the place. In November, 1704,

a fleet from Quebec landed a French and Indian force at Placentia, whence they advanced about the middle of January. They were about 400 strong, and crossed the Peninsula of Avalon on snow-shoes. The town of Bay Bulls (*Béboulle*) surrendered on their approach, and a long and painful midwinter march ensued, over the mountains and through the deep snows. The French militia of Placentia were sent in at dawn to surprise the fort at St. John's, but could not enter the works for lack of scaling-ladders; so they contented themselves with occupying the town and Quiddy Viddy. The fort was now besieged for 33 days, in a season of intense cold, when even the harbor was frozen over; but the English held out valiantly, and showered balls and bombs upon the town, finally succeeding in dislodging the enemy and putting them in full retreat.

In June, 1762, the Count d'Hausonville entered the Bay Bulls with a powerful French fleet, consisting of the *Robuste*, 74; *L'Eveillé*, 64; *La Garonne*, 44; and *La Licorne*, 30. He escorted several transports, whence 1,500 soldiers were landed. This force marched on St. John's, which surrendered on summons, together with the English frigate *Grammont*. Lord Colville's fleet hastened up from Halifax and blockaded Admiral De Ternay in the harbor of St. John, while land forces were debarked at Torbay and Quiddy Viddy. The last-named detachment (Royal Americans and Highlanders) proceeded to storm the works on Signal Hill, but the French fought desperately, and held them at bay until the English forces from Torbay came in and succeeded in carrying the entire line of heights. In the mean time, a dense fog had settled over the coast, under whose protection De Ternay led his squadron through the British line of blockade, and gained the open sea. In 1796 a formidable French fleet, under Admiral Richery (consisting of 7 line-of-battle ships and several frigates), menaced St. John's, then commanded by Admiral Sir James Wallace. Strong batteries were erected along the Narrows; fire-ships were drawn up in the harbor; a chain was stretched across the entrance; and the entire body of the people was called under arms. The hostile fleet blockaded the port for many days, but was kept at bay by the batteries on Signal Hill; and after an ineffectual attempt at attack, sailed away to the S. Feb. 12, 1816, a disastrous fire occurred at St. John's, by which 1,500 persons were left homeless; and great suffering would have ensued had it not been for the citizens of Boston, who despatched a ship loaded with provisions and clothing for gratuitous distribution among the impoverished people. Nov. 7, 1817, another terrible fire occurred here, by which $2,000,000 worth of property was destroyed; and this was followed, within 2 weeks, by a third disastrous conflagration. This succession of calamities came near resulting in the abandonment of the colony, and the people were goaded by hunger to a succession of deeds of crime and to organized violations of the laws. In 1825 the first highway was built (from St. John's to Portugal Cove); in 1833 the first session of the Colonial Parliament was held; and the first steamship in the Newfoundland waters arrived here in 1840.

In 1860 the city was convulsed by a terrible riot, arising from politico-religious causes, and threatening wide ruin. An immense mob of armed Irishmen attacked and pillaged the stores on Water St., and filled the lower town with rapine and robbery. The ancient organization called the Royal Newfoundland Companies was ordered out and posted near the Market House, where the troops suffered for hours the gibes of the plunderers, until they were fired upon in the twilight, when they returned a point-blank volley, which caused a sad carnage in the insurgent crowd. Then the great Cathedral bells rang out wildly, and summoned all the rioters to that building, where the Bishop exhorted them to peace and forbearance, under pain of excommunication. After a remarkable interview, the next day, between the Bishop and Gov. Sir Alexander Bannerman, this tragical revolt was ended.

In 1870 St. John's had 21 sailing-vessels and 6 steamers engaged in the sealing business, and their crews amounted to 1.584 men. In 1869 (the latest accessible statistics) 688 vessels, with a tonnage of 109,043 tons, and employing 5,466 men, entered this port; and in the same year there were cleared hence 577 vessels, with 4,937 men.

53. The Environs of St. John's.

"On either side of the city of St. John's, stretching in a semicircle along the rugged coast, at an average radius from the centre of 7 or 8 M., a number of little fishing-coves or bays attract, during the sweet and enjoyable summer, all persons who can command the use of a horse to revel in their beauties. Each little bay is but a slice of the high cliffs scooped out by the friction of the mighty pressure of the Atlantic waves; and leading down to its shingled beach, each boasts of a lovely green valley through which infallibly a tumbling noisy trout-burn pours back the waters evaporated from the parent surface." (LT.-COL. MCCREA.)

The country about the capital is not naturally productive, but has been made to bring forth fruit and vegetables by careful labor, and now supports a considerable farming population. The roads are fine, being for the most part macadamized and free from mud. 3 M. beyond the city is the Lunatic Asylum, pleasantly situated in a small forest.

Quiddy-Viddy Lake is frequently visited by the people of St. John's. The favorite drive is to **Portugal Cove,** over a road that has been described as possessing a "sad and desolate beauty." This road passes the *Windsor Lake*, or Twenty-Mile Pond, "a large picturesque sheet of water, with some pretty, lonely-looking islands." The inn at Portugal Cove looks out on a handsome cascade, and is a favorite goal for wedding-tours from St. John's.

"The scenery about Portugal Cove well repays the ride of nearly 10 M. on a good road from St. John's. It is wildly romantic, and just before entering the village is very beautiful. A succession of lofty hills on each side tower over the road, and shut out everything but their conical or mammillated peaks, covered with wild stunted forest and bold masses of rock, breaking through with a tiny waterfall from the highest, which in winter hangs down in perpendicular ridges of yellow ice. Turning suddenly out of one of the wildest scenes, you cross a little bridge, and the romantic scattered village is hanging over the abrupt rocky shore, with its fish-flakes and busy little anchorage open to the sight, closed in the distance by the shores of Conception Bay, lofty and blue, part of which are concealed by the picturesque Belle Isle." (SIR R. BONNYCASTLE.)

"On approaching Portugal Cove, the eye is struck by the serrated and picturesque outline of the hills which run along the coast from it towards Cape St. Francis, and presently delighted with the wild beauty of the little valley or glen at the mouth of which the cove is situated. The road winds with several turns down the side of the valley, into which some small brooks hurry their waters, flashing in the sunshine as they leap over the rocks and down the ledges, through the dark green of the woods. On turning the shoulder of one of the hill-slopes, the view opens upon Conception Bay, with the rocky points of the cove immediately below." (PROF. JUKES.)

Another favorite excursion is to *Virginia Water*, the former summer residence of the governors of Newfoundland. It is reached by way of the King's Bridge and the pretty little Quiddy-Viddy Lake, beyond which the Ballyhaly Bog is crossed, and the carriage reaches the secluded domain of Virginia Water. It is situated on a beautiful lake of deep water, 3 M. in circumference, "indented with little grass-edged bays, fringed and feathered to the limpid edge with dark dense woods." Beyond this point the drive may be protracted to *Logie Bay*, a small cove between projecting cliffs, with bold and striking shore scenery. Logie Bay is 4 M., and Torbay is 8-9 M. from St. John's, by a fine road which crosses the high and mossy barrens, and affords broad sea-views from the cliffs. The country is thinly settled, and is crossed by several trout-brooks.

Logie Bay is remarkable for the wildness of its rock and cliff scenery. "Nothing like a beach is to be found anywhere on this coast, the descent to the sea being always difficult and generally impracticable. In Logie Bay the thick-bedded dark sandstones and conglomerates stand bold and bare in round-topped hills and precipices 3-400 ft. in height, with occasional fissures traversing their jagged cliffs, and the boiling waves of the Atlantic curling around their feet in white eddies or leaping against their sides with huge spouts of foam and spray." (PROF. JUKES.)

"Torbay is an arm of the sea, — a short, strong arm with a slim hand and finger, reaching into the rocky land and touching the waterfalls and rapids of a pretty brook. Here is a little village, with Romish and Protestant steeples, and the dwellings of fishermen, with the universal appendages of fishing-houses, boats, and flakes. One seldom looks upon a hamlet so picturesque and wild." On the N. shore of the bay is a long line of cliffs, 3-400 ft. high, surf-beaten and majestic, and finely observed by taking a boat out from Torbay and coasting to the N. "At one point, where the rocks recede from the main front and form a kind of headland, the strata, 6-8 ft. thick, assume the form of a pyramid, from a broad base of a hundred yards or more running up to meet in a point. The heart of this vast cave has partly fallen out, and left the resemblance of an enormous tent with cavernous recesses and halls, in which the shades of evening were already lurking, and the surf was sounding mournfully. Occasionally it was musical, pealing forth like the low tones of a great organ with awful solemnity. Now and then, the gloomy silence of a minute was broken by the crash of a billow far within, when the reverberations were like the slamming of great doors."

"After passing this grand specimen of the architecture of the sea, there appeared long rocky reaches, like Egyptian temples, old dead cliffs of yellowish gray checked off by lines and seams into squares, and having the resemblance, where they have fallen out into the ocean, of doors and windows opening in upon the fresher stone." (NOBLE.)

54. The Strait Shore of Avalon.—St. John's to Cape Race.

That portion of the Peninsula of Avalon which fronts to the eastward on the Atlantic has been termed the *Strait Shore*, on account of its generally undeviating line of direction. Its outports may be visited either by the Friday mail-conveyance, through Petty Harbor, Bay Bulls, Ferryland, and Renewse, or by the Western Coastal steamer (see Route 60).

Distances by Road.— St. John's to Blackhead, 4 M.; Petty Harbor, 10; Bay Bulls, 19; Witless Bay, 22; Mobile, 24; Toad Cove, 26; La Manche, 32; Brigus, 34; Cape Broyle, 38; Caplin Cove, 42; Ferryland, 44; Aquafort, 48; Fermeuse, 51; Renewse, 54; Cape Race, 64.

"The road, one of the finest I ever saw,— an old-fashioned English gravel-road, smooth and hard almost as iron, a very luxury for the wheels of a springless wagon, — keeps up the bed of a small river, a good-sized trout-stream, flowing from the inland valley into the harbor of St John's. Contrasted with the bold regions that front the ocean, these valleys are soft and fertile. We passed smooth meadows, and sloping plough-lands, and green pastures, and houses peeping out of pretty groves. One might have called it a Canadian or New-Hampshire vale." The road passes several lakelets and trout-streams, and gives fine views of the ocean on the l., being also one of the most smooth and firmly built of highways. "No nation makes such roads as these, in a land bristling with rugged difficulties, that has not wound its way up to the summit of power and cultivation." The hills along the coast closely resemble the Cordillera peaks; and from the bald summits on the W., Trinity Bay may be seen.

The mail-road running S. from St. John's passes Waterford Bridge and soon approaches *Blackhead*, a Catholic village near an iron-bound shore whose great cliffs have been worn into fantastic shapes by the crash and attrition of the Atlantic surges. Near this place is **Cape Spear**, the most easterly point of North America, 1,656 M. from Valentia Bay, in Ireland. On the summit of the cape, 264 ft. above the sea, is a red-and-white striped tower sustaining a revolving light which is visible for 22 M.

The road now passes between "woody banks running through an undulating country but half reclaimed on the r., while on the l. the slopes stretch up to the breezy headlands, beyond which there is nothing but sea and cloud from this to Europe." *Petty Harbor* is 4 M. S. W. of Cape Spear and 10 M. from St. John's, and is a village of 900 inhabitants, with a refinery of cod-liver oil and long lines of evergreen fish-flakes. Off this point H. B. M. frigate *Tweed* was wrecked in 1814, and 60 men were drowned. The houses of Petty Harbor are situated in a narrow glen at the foot of frowning and barren ridges. The harbor at the foot of this ravine is small and insecure. The dark hills to the W. attain a height of 700 ft. along the unbroken shore which leads S. to Bay Bulls; and at about 4 M. from Petty Harbor is the *Spout, a deep cavern in the seaward cliffs, in whose top is a hole, through which, at high tide and in a heavy sea, the water shoots up every half-minute in a roaring fountain which is seen 3 M. off at sea. The road now approaches Ionclay Hill (810 ft. high), the chief elevation on this coast, and reaches **Bay Bulls**, a village of 700 inhabitants. This is one of the most important of the outports, and affords a refuge to vessels that are unable, on account of storms or ice, to make the harbor of St. John's. There are several farms near the bay, but most of the inhabitants are engaged in the cod-fishery, which is carried on from large open boats. This ancient settlement was exposed to great vicissitudes during the conflicts between the French and the English for the possession of Newfoundland, and was totally destroyed by Admiral Richery (French) in 1796. Fine sporting is found in this vicinity, all along shore, and shooting-parties leave St. John's during the season for several days' adventure hereabouts.

In 1696 the French frigates *Pelican, Diamant, Count de Toulouse, Vendange, Philippe*, and *Harcourt* met the British man-of-war *Sapphire* off Cape Spear, and chased it into Bay Bulls. A naval battle of several hours' duration was closed by the complete discomfiture of the British, who set fire to the shattered *Sapphire* and abandoned her. The French sailors boarded her immediately, but were destroyed by the explosion of the magazine.

Witless Bay is the next village, and has nearly 1,000 inhabitants, with a large and prominent Catholic church. Cod-fishing is carried on to a great extent off this shore, also off Mobile, the next settlement to the S. Beyond the rock-bound hamlets of Toad Cove, La Manche, and Brigus, the road reaches Cape Broyle.

In 1628 Cape Broyle was captured by Admiral de la Rade, with three French war-vessels, who also took the fishing-fleet then in the harbor. But Sir George Calvert sent from the capital of Avalon two frigates (one of which carried 24 guns) and several hundred men, on whose approach "the French let slip their cables, and made to sea as fast as they could." Calvert's men retaliated by harrying the French stations at Trepassey, where they captured six ships of Bayonne and St. Jean de Luz.

Cape Broyle is a prosperous fishing-settlement on Broyle Harbor, near the mountainous headland of Cape Broyle (552 ft. high). There is good salmon-fishing on the river which runs S. E. to the harbor from the foot of Hell Hill.

FERRYLAND.

Ferryland is 2 M. beyond the Caplin-Cove settlement, and is the capital of the district of Ferryland. It has about 700 inhabitants, and is well located on level ground near the head of the harbor. In the immediate vicinity are several prosperous farms, and picturesque scenery surrounds the harbor on all sides. To the S. E. is Ferryland Head, on which is a fixed white light, 200 ft. above the sea, and visible for 16 M. Off this point are the slender spires of rock called the *Hare's Ears*, projecting from the sea to the height of 50 ft.

In 1614 (1622) King James I. granted the great peninsula between Trinity and Placentia Bays to Sir George Calvert, then Secretary of State. The grantee named his new domain Avalon, in honor of the district where Christian tradition claims that the Gospel was first preached in Britain (the present Glastonbury). It was designed to found here a Christian colony, with the broadest principles of toleration and charity. Calvert sent out a considerable company of settlers, under the government of Capt. Wynne, and a colony was planted at Ferryland. The reports sent back to England concerning the soil and productions of the new country were so favorable that Sir George Calvert and his family soon joined the colonists. Under his administration an equitable government was established, fortifications were erected, and other improvements instituted. Lord Baltimore had but little pleasure of his settlement in Avalon. He found that he had been greatly deceived about the climate and the nature of the soil. The Puritans also began to harass him; and Erasmus Stourton, one of their ministers, not only preached dissent under his eyes at Ferryland, but went to England and reported to the Privy Council that Baltimore's priests said mass and had "all the other ceremonies of the Church of Rome, in the ample manner as 't is used in Spain." Finally, after trials by storm and by schismatics, Lord Baltimore died (in 1632), leaving to his son Cecil, 2d Lord Baltimore, the honor of founding Maryland, on the grant already secured from the king. In that more favored southern clime afterwards arose the great city which commemorates and honors the name of BALTIMORE.

In 1637 Sir David Kirke was appointed Count Palatine of Newfoundland, and established himself at Ferryland. He hoisted the royal standard on the forts, and maintained a strong (and sometimes harsh) rule over the island. At the outbreak of the English Revolution (1642), Kirke's brothers joined King Charles's forces and fought bravely through the war, while Sir David strengthened his Newfoundland forts and established a powerful and well-armed fleet. He offered the King a safe asylum in his domain; and the fiery Prince Rupert, with the royal Channel fleet, was sailing to Newfoundland to join Kirke's forces, when he was headed off by the fleet of the Commonwealth, under Sir George Ayscue. After the fall of the Stuarts, Sir David was carried to England in a vessel of the Republic (in 1651), to be tried on various charges; but he bribed Cromwell's son in-law, and was released, returning to Ferryland, where he died in 1656, after having governed the island for over 20 years. At a later day this town became a port of some importance, and was the scene of repeated naval attacks during the French wars. In 1673 it was taken and plundered by 4 Dutch frigates.

In 1694 Ferryland was attacked by 2 large French frigates, carrying 90 guns, which opened a furious cannonade on the town. But the *William and Mary*, 16, was lying in the harbor, with 9 merchant-ships, and their crews built batteries at the harbor-mouth, whence, with the guns of the privateer, they inflicted such damage on the enemy that they withdrew, after a 5 hours' cannonade, having lost about 90 men. In 1762 the powerful French fleet of Admiral de Ternay was driven off by a battery on Bois Island.

Aquafort lies S. W. of Ferryland, and is a small hamlet situated on a long, deep, and narrow harbor embosomed in lofty hills. The next settlement is *Fermeuse*, with 600 inhabitants and a Catholic church and convent. It is on the shore of Admiral's Cove, in the deep and secure harbor of Fermeuse, and the people are engaged in the cod and salmon fisheries. *Reneuse* is an ancient and decadent port 16 M. S. of Ferryland, situated on an indifferent harbor which lies between Burnt Point and Renewse Head. 3 - 4 M. inland are the rugged hummocks called the Red Hills, whence the eastern hill range runs 30 M. N. across Avalon to Holyrood.

6 - 8 M. from Renewse are the tall and shaggy hills called the **Butterpots**, which command broad views over Avalon, and from Bay Bulls to the W. shore of Trepassey Bay. The Butterpots of Holyrood are also seen from this point; and Prof. Jukes counted 80 lakes in sight from the main peak (which is 955 ft. high).

S. of this point extends a fatal iron-bound coast, on which scores of vessels, veiled in impenetrable fog or swept inward by resistless storms, have been dashed in pieces. A very slight error in reckoning will throw vessels bound S. of Cape Race upon this shore, and then, if the Cape Race and Ferryland lights are wrapped in the dense black fog peculiar to these waters, the chances of disaster are great. The erection of a fog-whistle on the cape has greatly lessened the perils of navigation here. The ocean steamships *Anglo-Saxon, Argo*, and *City of Philadelphia* were lost on Cape Race.

Cape Race is the S. E. point of Newfoundland, and is a rugged headland of black slaty rock thrown up in vertical strata. It is provided with a powerful light, 180 ft. above the sea, and visible for 19 M. The great polar current sweeps in close by the cape and turns around it to the W. N. W., forming, together with the ordinary tides and the bay-currents, a complexity of streams that causes many wrecks.

Icebergs are to be seen off this shore at almost all seasons, and the dense fogs are often illumined by the peculiar white glare which precedes them. Field-ice is also common here during the spring and early summer, but is easily avoided by the warning of the "ice blink." Throughout the summer and autumn the fog broods over this shore almost incessantly, and vessels are navigated by casting the lead and following the soundings which are marked out with such precision on the Admiralty charts. 6 M. E. of Cape Race is the *Ballard Bank*, which is 18 M. long and 2 - 12 M. wide, with a depth of water of 15 - 26 fathoms.

Cape Race is distant, by great-circle sailing, from New York, 1,010 M.; Boston, 820; Portland, 779; St. John, N. B., 715; Halifax, 463; Miramichi, 492; Quebec, 836; Cape Clear, 1,713; Galway, 1,721; Liverpool, 1,970.

The **Grand Banks of Newfoundland** are about 50 M. E. of Cape Race. They extend for 4 degrees N. and S. and 5 degrees E. and W. (at 45° N. latitude) running S. to a point. They consist of vast submerged sandbanks, on which the water is from 30 to 60 fathoms deep, and are strewn with shells. Here are found innumerable codfish, generally occupying the shallower waters over the sandy bottoms, and feeding on the shoals of smaller fish below. They pass out into the deeper waters late in November, but return to the Banks in February, and fatten rapidly. Immense fleets are engaged in the fisheries here, and it is estimated that over 100,000 men are dependent on this industry.

Throughout a great part of the spring, summer, and fall, the Grand Banks are covered by rarely broken fogs, through which falls an almost incessant slow rain. Sometimes these fogs are so dense that objects within 60 ft. are totally invisible, at which times the fishing-vessels at anchor are liable to be run down by the great Atlantic steamers. The dangerous proximity of icebergs (which drift across and ground on the Banks) is indicated by the sudden and intense coldness which they send through even a midsummer day, by the peculiar white glare in the air about them, and by the roaring of the breakers on their sides.

It was on the Grand Banks, not far from Cape Race, that the first battle of the Seven Years' War was fought. June 8, 1755, the British 60-gun frigates *Dunkirk* and *Defiance* were cruising about in a dense fog, when they met the French men-of-war *Alcide* and *Lys*. For five hours the battle continued, and a continual cannonade was kept up between the hostile ships. The French were overmatched, but fought valiantly, inflicting heavy losses on the assailants (the *Dunkirk* alone lost 90 men). When they finally surrendered, the *Lys* was found to contain $400,000 in specie and 8 companies of infantry.

The vicinity of Cape Race was for some time the cruising-ground of the U. S. frigate *Constitution*, in 1812, and in these waters she captured the *Adiona*, the *Adeline*, and other vessels.

Near the edge of the Grand Bank (in lat. 41° 41′ N., long. 55° 18′ W.) occurred the famous sea-fight between the *Constitution* and the *Guerrière*, whose result filled the United States with rejoicing, and impaired the prestige of the British navy. On the afternoon of Aug. 19, 1812, the *Constitution* sighted the *Guerrière*, and bore down upon her with double-shotted batteries. The British ship was somewhat inferior in force, but attacked the American with the confidence of victory. The *Constitution* received several broadsides in silence, but when within half pistol-shot discharged her tremendous batteries, and followed with such a fire of deadly precision that the *Guerrière* was soon left a dismasted and shattered wreck. The British ship then surrendered, having lost 101 men in the action, while her antagonist lost but 14. The *Guerrière* had 38 guns, and the *Constitution* had 44.

Among the American privateers that cruised about the Grand Banks in 1812-14, none was more successful than the *Mammoth*, of Baltimore. She captured the ships *Ann and Eliza*, *Urania*, *Anisby*, *Dobson*, *Sallust*, *Uniza*, *Sarah*, *Sir Home Popham*, *Champion*, *Mentor*, and many other rich prizes.

" Far off by stormy Labrador —
Far off the Banks of Newfoundland,
Where angry seas incessant roar,
And foggy mists their wings expand,
The fishing-schooners, black and low,
For weary mouths sail to and fro."

55. St. John's to Labrador. — The Northern Coast of Newfoundland.

The mail-steamer *Leopard* leaves St. John's, N. F., every alternate Monday during the season of navigation, and visits the chief outports on the N. coast (so called). The fares are as follows: St. John's to Old Perlican or Bay Verd, $2, — steerage, $1.50; to Trinity, $4.40, — steerage, $2; to Bonavista, $5.50, — steerage, $2.80; to Greenspond, $6, — steerage, $3; to Fogo, $6.50; to Twillingate, $7; to Exploits Island, $7.50; to Tilt Cove, Little Bay Island, or Nipper's Harbor, $8, — steerage, $4.50; to Battle Harbor, $12. At Battle Harbor the *Leopard* meets the *Hercules*, the Labrador mail-steamer.

The fare on the Labrador steamer is $2 a day, which includes both passage and meals. The northern boats are powerful and seaworthy, but the fare at their tables is necessarily of the plainest kind. The time which will be required for the Labrador trip is nearly four weeks (from St. John's back to St. John's again). The expense is about $50. The journey should be begun before the middle of July, in order to avail of the short summer in these high latitudes. It would be prudent for gentlemen who desire to make this tour to write early in the season to the agents of the steamship lines, to assure themselves of due connections and to learn other particulars. Mr. J. Taylor Wood is the agent at Halifax for the steamer from that port to St. John's; and Walter Grieve & Co., St. John's, N. F., are the agents for the Northern Coastal Line.

Passing out between the stern and frowning portals of the harbor of St. John's, the steamer soon takes a northerly course, and opens the indentation of *Logie Bay* on the W. (see page 196). After running by the tall cliffs of Sugar Loaf and Red Head (700 ft. high), *Torbay* is seen opening to the W., within which is the village of the same name.

About 8 M. beyond Torbay, the white shore of **Cape St. Francis** is seen on the port bow, and, if the water is rough, the great breakers may be seen whitening over the rocks which are called the Brandies. The course is now laid across the mouth of Conception Bay, which is seen extending to the S. W. for 30 M. 18 M. from Cape St. Francis, and about 40 M. from St. John's, the steamer passes between Bay Verd Head and Split Point, and stops off *Bay Verd*, a village of about 600 inhabitants, situated on a broad and unsheltered bight of the sea. The fishing-grounds in this vicinity are among the best on the American coast, and attract large fleets of boats and schooners. The attention of the villagers is divided between farming and fishing, the latter industry being by far the most lucrative. Roads lead out from Bay Verd S. to Carbonear and Harbor Grace (see Route 56), and N. W. to the settlements on Trinity Bay. Soon after leaving Bay Verd, the steamer passes *Baccalieu Island*, a high and ridgy land 3½ M. long, and nearly 2 M. from the main. On its N. end is a powerful flashing light, elevated 380 ft. above the sea, and visible for 28 M.

Although Cabot was the first professional discoverer (if the term may be used) to visit and explore the shores of Newfoundland, there is no doubt that these waters had long been the resort of the fishing-fleets of the Normans, Bretons, and Basques. Lescarbot claims that they had fished off these shores "for many centuries," and Cabot applied the name "Baccalaos" to the country because "in the seas thereabout he found so great multitudes of certain bigge fishes, much like unto Tunnies (which the inhabitants call *Baccalaos*), that they sometimes stayed his shippes." Baccalaos is the ancient Basque name for codfish, and its extensive use by the natives in place of their own word *Apegé*, meaning the same thing, is held as conclusive proof that they had been much in communication with Basque fishermen before the arrival of Cabot. Cabot gave this name to the continent as far as he explored it, but in the map of 1640 it is applied only to the islet which now retains it.

On her alternate trips the vessel rounds in about Grates Point, and stops at *Old Perlican* (see Route 57). Otherwise, it runs across the mouth of Trinity Bay for about 20 M., on a N. W. course, and enters the harbor of **Trinity**, 115 M. from St. John's. The entrance is bold and imposing, and the harbor is one of the best on the island, affording a land-locked anchorage for the largest fleets. It is divided into two arms by a high rocky peninsula (380 ft. high), on whose S. side are the wharves and houses of the town. Trinity has about 1,500 inhabitants, and is a port of entry and the capital of the district of Trinity. Considerable farming is done in the coves near the head of the harbor. Roads lead out to the S. shore (see Route 57), and also to Salmon Cove, 5 M; English Harbor, 7; Ragged Harbor, 16; and Catalina, 20.

On leaving Trinity Harbor, the course is S. E until Green Bay Head and the Horse Chops are passed, when it turns to the N. E., and runs along within sight of a high and cliffy shore. Beyond the Ragged Isles is seen *Green Island*, where there is a fixed white light, visible for 15 M., around which (through rough water if the wind is E.) the vessel passes, threading a labyrinth of shoals and rocks, and enters the harbor of **Catalina**, remarkable for its sudden and frequent intermittent tides. The town of Catalina has 1,300 inhabitants, with 2 churches, of which that of the Episcopalians is a fine piece of architecture, though built of wood. The main part of the settlement is on the W. side of the harbor, and has a considerable maritime trade. The adjacent waters abound in salmon, and delicious edible whelks are found on the rocks. Besides the highway to Trinity (20 M.), a rugged road leads N. to Bonavista in 10 M. Catalina was visited in 1534 by Cartier, who named it *St. Catherine*.

On leaving Catalina Harbor, North Head is passed, and after running N. E. by N. 3 M. Flowers Head is left on the port bow. About 2 M. beyond, the Bird Islets are seen on the l., near which is the fishing-settlement of *Bird Island Cove* (670 inhabitants), with its long and handsome beach. A short distance inland is seen the Burnt Ridge, a line of dark bleak hills rising to a height of 500 ft. The Dollarman Bank, famous for codfish, is now crossed, and on the l. is seen Cape Largent and Spiller Point, off which are the precipitous and tower-like * *Spiller Rocks*, surrounded by the sea. The steamer now passes **Cape Bonavista**, on which is a red-and-white flashing-light, 150 ft. above the sea, and visible for 15 M.

The re-discovery of Newfoundland (after the Northmen's voyages 5 centuries before) was effected in June, 1497, by Cabot, a Venetian in the service of Henry VII. of England, sailing in the ship *Matthew*, of Bristol. He gave the name of *Bona Vista* (" Fair View "), or *Prima Vista* (" First View "), to the first point of the coast which he saw, and that name has since been attached to this northerly cape, since it is believed that this was the location of the new-found shore. (The reader of Biddle's "Memoirs of Sebastian Cabot" will, however, be much puzzled to know what point, if any, Cabot actually saw on these coasts.) The rocks and shoals to the N. are prolific in fish, and are visited by great flotillas of boats.

After rounding the light, the steamer enters Bonavista Bay, a great bight of the sea extending between Capes Bonavista and Freels, a distance of 37 M. About 4 M. S. W. of the cape, the steamer enters the harbor of **Bonavista**, an ancient marine town with 2,600 inhabitants and 3 churches. It is the capital of the district of the same name, and is also a port of entry, having a large and increasing commerce. The harbor is not secure, and during long N. W. gales the sea breaks heavily across the entrance. The Episcopal church is a fine building in English Gothic architecture, but the houses of the town are generally mean and small. Considerable farming is done on the comparatively fertile lands in the vicinity, and it is claimed that the climate is much more genial and the air more clear than on the S. shores of the island. The town is 146 M. from St. John's, and is 30 M. by road from Trinity and 10 M. from Catalina. It is one of the most ancient settlements on the coast, and signalized itself in 1696 by beating off the French fleet which had captured St. John's and ravaged the S. coasts.

Bonavista Bay.

A road leads S. W. from Bonavista to Birchy Cove, 9 M.; Amherst Cove, 12; King's Cove, 20; Keels Cove, 26; Tickle Cove, 33; Open Hole, 36; Plate Cove, 38; and Indian Arm, 43.

King's Cove is a village of Labrador fishermen, with 550 inhabitants and 2 churches. It is on a narrow harbor between the lofty cliffs of the coast range, through whose passes a road runs S. to Trinity in 13 M. 3 M. from King's Cove is *Broad Cove* village, under the shadow of the peak of Southern Head. *Keels* is 6 M. from King's Cove, and does a considerable lumber business. Thence the road descends through Tickle Cove (2 M. from the picturesque Red Cliff Island) to the three villages on the S., each of which has 2 - 300 inhabitants. To the W. are the deep estuaries of Sweet Harbor, Clode Sound (20 M. long), and Newman Sound (11 M. long), penetrating the hill-country and exhibiting a succession of views of romantic scenery and total desolation. Boats may be taken from Open Hole to *Barrow Harbor*, a fishing settlement 10 M. N. W., at the mouth of Newman Sound, and to *Salvage*, 16 M. distant, a village of 500 inhabitants. 6 M. N. W., beyond the Bay of Fair and False, is *Bloody Bay*, a deep and narrow inlet with picturesque forest scenery, extending for several miles among the hills. The name was given on account of the frequent conflicts which here ensued between the Red Indians and the fishermen. At the head of the bay is the Terra Nova River, descending from the *Terra Nova Lake*, which is 15 M. distant, and is 12 M. long.

The N. shore of Bonavista Bay is visited most easily from the port of *Greenspond*. The communication is exclusively by boats, which may be engaged at the village. Nearly all the islands in the vicinity and for 10 M. to the S. W. and S. are occupied by small communities of hardy fishermen, and the shores of the mainland are indented with deep and narrow bays and sounds. To the N. are Pool's Island, 3 M.; Pincher's Island, 9; Cobbler's Island, 10; and Middle Bill Cove (near Cape Freels), 15. To the S. and W. are the Fair Island, 7 M.; Deer Island, 11; Cottel's Island (three settlements), 15; the Gooseberry Isles, 12; and Hare Bay, 23. The last-named place is at the entrance of *Freshwater Bay*, which runs in for about 15 M., with deep water and bold shores. The great northern mail-road is being built along the head of this bay; a short distance from which (by the river) are the *Gambo Ponds*, large lakes in the desolate interior, 23 M. long, abounding in fish. One of the best salmon-fisheries on the island is at the head of *Indian Bay*, 12 M. W. of Greenspond.

On leaving Bonavista, the steamer runs N. by W. across Bonavista Bay, passing the Gooseberry Isles on the port bow. After over 3 hours' run, the N. shore is approached, and the harbor of **Greenspond** is entered. This town contains over 1,000 inhabitants, and is situated on an island 1 M. square, so rugged that soil for house-gardens had to be brought from the mainland. A large business is done here in the fisheries and the seal-trade, and most of the inhabitants are connected with either the one or the other. The entrance to the harbor is difficult, and is marked by a fixed red light, visible for 12 M.

The steamer now runs N. E. and N. for about 18 M. to **Cape Freels**, passing great numbers of islands, some of which are inhabited by fishermen, while others are the resort of myriads of sea-birds, who are seen hovering over the rocks in great flocks. Soon after passing the arid highlands of Cape Freels, the course is laid to the N. W. across the opening of Sir Charles Hamilton's Sound, a broad and deep arm of the sea which is studded with many islands. Leaving the Cape Ridge and Windmill Hill astern, the Penguin Islands are seen, 15½ M. from Cape Freels; and 6 M. farther N. W. the *Wadham Isles* are passed, where, on a lonely and surf-beaten rock, is the Offer Wadham lighthouse, a circular brick tower 100 ft. high, exhibiting a fixed white light, which is visible for 12 M. To the N. E., and well out at sea, is *Funk Island*, near which are good sealing-grounds.

Funk Island was visited by Cartier in 1534, who named it (and the adjacent rocks) *Les Isles des Oyseaux*. Here he saw a white bear " as large as a cow," which had swum 14 leagues from Newfoundland. "He then coasted along all the northern part of that great island, and he says that you meet nowhere else better ports or a more wretched country; on every side it is nothing but frightful rocks, sterile lands covered with a scanty moss; no trees, but only some bushes half dried up; that nevertheless he found men there well made, who wore their hair tied on the top of the head." The isles were again visited by Cartier in July, 1535, in the ship *Grand Hermine*. "If the soyle were as good as the harboroughes are, it were a great commoditie; but it is not to be called the new found land, but rather stones and cragges and a place fit for wilde beastes..... In short, I believe this was the land allotted to Caine." Such was the unfavorable description given by Jaques Cartier of the land between Cape Bonavista and the Strait of Belle Isle.
It is supposed that either the Baccalieu or the Penguin Islands were the "Feather Islands," which the *Annales Skalholtini* and *Lögmann's* state were discovered by the Northmen in the year 1285. The Saga of Eric the Red tells that Leif, son of the Earl of Norway, visited the Labrador and Newfoundland shores in 994. "Then sailed they to the land, and cast anchor, and put off boats, and went ashore, and saw there no grass. Great icebergs were over all up the country, but like a plain of flat stones was all from the sea to the mountains, and it appeared to them that this land had no good qualities." Leif named this country Helluland (from *Hella*, a flat stone), distinguishing Labrador as *Helluland it Mikla*. In 1288 King Eric sent the mariner Rolf to Iceland to call out men for a voyage to these shores; and the name *Nyja Land*, or *Nyja Fundu Land*, was then applied to the great island to the S., and was probably adopted by the English (in the Anglicized form of *Newfoundland*) during the commercial intercourse between England and Iceland in the 15th century.

9½ M. N. W. by N., *Cape Fogo* is approached, and is a bold promontory 214 ft. high, terminating Fogo Island on the S. E. The course continues to the N. W. off the rugged shores of the island, and at 6½ M. from Cape Fogo, Round Head is passed, and the steamer assumes a course more to the westward. 6 – 8 M. from Round Head she enters the harbor of **Fogo**, a port of entry and post-town 216 M. from St. John's. The population is 740, with 2 churches; and the town is of great local importance, being the depot of supplies for the fishing-stations of the N. shore. (See also Route 58 for this and other ports in the Bay of Notre Dame.)

"The western headlands of Fogo are exceedingly attractive, lofty, finely broken, of a red and purplish brown, tinted here and there with pale green..... As we pass the bold prominences and deep, narrow bays or fiords, they are continually changing and surprising us with a new scenery. And now the great sea-wall, on our right, opens and discloses the harbor and village of Fogo, the chief place of the island, gleaming in the setting sun, as if there were flames shining through the windows. Looking to the left, all the western region is one fine Ægean, a sea filled with a multitude of isles, of manifold forms and sizes, and of every height, from mountain pyramids and crested ridges down to rounded knolls and tables, rocky ruins split and shattered, giant slabs sliding edgewise into the deep, columns and grotesque masses ruffled with curling surf, — the Cyclades of the west. I climb the shrouds, and behold fields and lanes of water, an endless and beautiful network, a little Switzerland with her vales and gorges filled with the purple sea." (NOBLE.)

In passing out of Fogo Harbor, the bold bluff of Fogo Head (345 ft. high) is seen on the l., back of which is Brimstone Head. The vessel steams in to the W., up the Bay of Notre Dame, soon passing Fogo Head, and opening the Change Island Tickles on the S. Change Island is then seen on the l., and the course is laid across to the lofty and arid hills of Baccalieu Island. At 22 M. from Fogo the steamer enters the harbor of **Twillingate** (the Anglicized form of *Toulinguet*, the ancient French name of the port). The town of Twillingate is the capital of the district of Twillingate and Fogo, the most northerly political and legal division of Newfoundland, and has a population of 2,790, with 3 churches. It is situated on two islands, and the sections are connected by a bridge. Farming is carried on to a considerable extent in the vicinity, but with varying success, owing to the short and uncertain summers. The houses in the town are (as usually in the coast settlements) very inferior in appearance, snugness and warmth being the chief objects sought after in their architecture.

The finest breed of Newfoundland dogs were formerly found about the Twillingate Isles, and were generally distinguished by their deep black color, with a white cross on the breast. They were smaller than the so-called Newfoundland dogs of America and Britain; were almost amphibious; and lived on fish, salted, fresh, or decayed. Like the great mahogany-colored dogs of Labrador, these animals were distinguished for rare intelligence and unbounded affection (especially for children); and were exempt from hydrophobia. A Newfoundland dog of pure blood is now worth from $75 to $100.

The steamer passes out of Twillingate Harbor and runs by Gull Island. The course is to the S. W., off the rugged shores of the Black Islets, and the N. promontory of the great *New World Island*. 14 M. from Twillingate she reaches the post-town of **Exploits Island**, a place of 530 inhabitants, with a large fleet of fishing-boats. (See also Route 58.)

From Exploits Island the Bay of Notre Dame is crossed, and the harbor of **Tilt Cove** is entered. This village has 770 inhabitants, and is prettily situated on the border of a picturesque lake. The vicinity is famous for its copper-mines, which were discovered in 1857 and opened in 1865. Between 1865 and 1870, 45,000 tons of ore, valued at $1,180,810, were extracted and shipped away. It is found in pockets or bunches 3–4 ft. thick, scattered through the heart of the hills, and is secured by level tunnels several thousand feet long, connected with three perpendicular main shafts, 216 ft. deep. There is also a valuable nickel-mine here, with a lode 10 inches thick, worked by costly machinery, and producing ore worth $332 a ton. A superior quality of marble is found in the vicinity, but is too far from a market to make it worth while to quarry. The male inhabitants of Tilt Cove are all miners.

The next stopping-place is at *Nipper's Harbor*, a small fishing-village 10 M. S. W. of Tilt Cove. The harbor is the best on the N. shore of the Bay of Notre Dame, and lies between the Nipper's Isles and the mainland. On alternate trips the mail-steamer calls also at *Little Bay Island*, 6–8 M. S. of Nipper's harbor.

Tilt Cove was the terminus of the Northern Coast Postal Route until the establishment of the mail-service on the Labrador coast, and it is not probable that the steamers will go N. of that point if the Labrador line is discontinued. It is but a short distance from Tilt Cove to the **French Shore** (see Route 61).

In running from Tilt Cove to the Labrador, the steamer first passes out around **Cape St. John**, and then takes a course almost due N., and far out from the land. Belle Isle, Quirpon, and the other points which may be distantly visible from the ship, are described in Route 61, *ad finem*.

At **Battle Harbor** (see Route 62) the new route of the Northern Coastal steamer ends, and the freight, mails, and passengers bound for other ports are transferred to another vessel.

The Labrador Coast, see Routes 62 and 63.

56. St. John's to Conception Bay.

Mail-stages leave St. John's every morning for Portugal Cove, distant $9\frac{1}{2}$ M. At this point the traveller meets the steamer *Lizzie*, whose route was as follows, during the navigable season of 1874: Tuesday, leaves Harbor Grace for Carbonear, Portugal Cove, and Bay Roberts; Wednesday, leaves Bay Roberts for Brigus, Portugal Cove, Carbonear, and Harbor Grace; Thursday, leaves Harbor Grace for Carbonear, Portugal Cove, and Brigus; Friday, leaves Brigus for Portugal Cove and Harbor Grace; Saturday, leaves Harbor Grace for Portugal Cove, Brigus, Carbonear, and Harbor Grace.

Fares.—Portugal Cove to Brigus, 18 M., fare $1.40; to Carbonear, 20 M.; to Bay Roberts, 20 M.; to Harbor Grace, 20 M., fare, $1.50.

There is also a road extending around Conception Bay. It is 20 M. from St. John's to Topsail, by way of Portugal Cove, passing Beachy, Broad, and Horse Coves. The more direct route leads directly across the N. part of Avalon from St. John's to Topsail. The chief villages and the distances on this road are as follows: St. John's to Topsail, 12 M.; Killigrews, 18; Holyrood, 28; Chapel's Cove, 33; Harbor Main, $34\frac{1}{2}$; Salmon Cove, 37; Colliers, 40; Brigus, 46; Port de Grave, 51; Spaniard's Bay, 56; Harbor Grace, 63; Carbonear, $67\frac{1}{2}$; Salmon Cove, 72; Spout Cove, $76\frac{1}{2}$; Western Bay, 82; Northern Bay, 87; Island Cove, $93\frac{1}{2}$; Caplin Cove, 97; Bay Verd, 105.

The stage-road, after leaving St. John's, traverses a singular farming country for several miles, and then enters a rugged region of hills. **Portugal Cove** is soon reached, and is picturesquely situated on the ledges near the foot of a range of highlands. It contains over 700 inhabitants, with 2 churches, and has a few small farms adjacent (see page 195).

Gaspar Cortereal explored this coast in the year 1500, and named Conception Bay. He carried home such a favorable account that a Portuguese colony was established at the Cove, and 50 ships were sent out to the fisheries. In 1578, 400 sail of vessels were seen in the bay at one time, prosecuting the fisheries under all flags. The colony was broken up by the English fleet under Sir Francis Drake, who also drove the French and Portuguese fishermen from the coast.

Belle Isle lies off shore 3 M. from the Cove, whence it may be visited by ferry-boats (also from Topsail). This interesting island is 9 M. long and 3 M. wide, and is traversed by a line of bold hills. It is famous for the richness of its deep black soil, and produces wheat, oats, potatoes, and hay, with the best of butter. The lower Silurian geological formation is here finely displayed in long parallel strata, amid which iron ore is found. The cliffs which front on the shore are very bold, and sometimes overhang the water or else are cut into strange and fantastic shapes by the action of the sea. Two or three brilliant little waterfalls are seen leaping from the upper levels. Belle Isle has 600 inhabitants, located in two villages, Lance Cove, at the W. end, and the Beach, on the S.

HARBOR GRACE.   *Route 56.*   207

The steamer runs out to the S. W. between Belle Isle and the bold heights about Portugal Cove and Broad Cove, and passes up Conception Bay for 18 M., with the lofty Blue Hills on the S. It then enters the narrow harbor of **Brigus** (*Sullivan's Hotel*), a port of entry and the capital of the district of Brigus. It has 2,000 inhabitants, with Wesleyan, Roman, and Anglican churches, and a convent of the Order of Mercy. The town is built on the shores of a small lake between two rugged hills, and presents a picturesque appearance. It has over 800 boats engaged in the cod-fishery, and about 30 larger vessels in trading and fishing. There are a few farms in the vicinity, producing fair crops in return for great labor. The best of these are on the bright meadows near Clark's Beach, 4 M. from the town; and several prosperous villages are found in the vicinity. Near the town is the singular double peak called the Twins, and a short distance S. W. is the sharp and conical *Thumb Peak* (598 ft. high).

The steamer passes out from the rock-bound harbor and runs N. by the bold hill of Brigus Lookout (400 ft. high). Beyond Burnt Head, *Bay de Grave* is seen opening on the l., with several hamlets, aggregating 2,600 inhabitants. Cupids and Bareneed are the chief of these villages, the latter being on the narrow neck of land between Bay de Grave and Bay Roberts, 2½ M. from Blow-me-down Head. Green Point is now rounded, and the course is laid S. W. up Bay Roberts, passing Coldeast Point on the port bow and stopping at the village of **Bay Roberts** (*Moore's Hotel*). This place consists of one long street, with 2 churches and several wharves, and has 1,000 inhabitants, most of whom spend the summer on the Labrador coast.

Passing out from Bay Roberts, Mad Point is soon left abeam, and *Spaniard's Bay* is seen on the l., entering the land for 3½ M., and dotted with fishing-establishments. The bay is surrounded by a line of high hills, on whose promontories are two or three chapels. The hamlet and church of *Bryant's Cove* are next seen, in a narrow glen at the base of the hills, and the steamer passes on around the dangerous and surf-beaten Harbor-Grace Islands (off Feather Point), on one of which is a revolving white-and-red flash light, 151 ft. above the sea, and visible for 18 M.

Harbor Grace (two inferior inns) is the second city of Newfoundland, and is the capital of the district of Harbor Grace. It has 6,770 inhabitants, with several churches, a weekly newspaper, and fire and police departments. The town is built on level land, near the shelter of the Point of Beach, with its wharves well protected by a long sand-strip. The bay is in the form of a wedge, decreasing from 1½ M. in width to ½ M., and is insecure except in the sheltered place before the city. The trade of this port is very large, and about 200 ships enter the harbor yearly. There is a stone court-house and a strong prison, and the Convent of the Presentation is on the Carbonear road. The Roman Catholic cathedral is the finest building in the city, and its high and symmetrical dome is a landmark for vessels entering the port. The interior of the cathedral is profusely ornamented, having been recently enlarged and newly adorned. Most of the houses in the city are mean and unprepossessing, being rudely constructed of wood, and but little improved by painting.

A rugged road runs N. W. 15 M. across the peninsula to **Heart's Content** (see Route 57). A road to the N. reaches (in 1½ M.) the farming village of *Mosquito Cove*, snugly embosomed in a pretty glen near the cultivated meadows. About the year 1610 a colony was planted here by the agents of that English company in which were Sir Francis Bacon, the Earl of Southampton, and other knights and nobles. King James I. granted to this company all the coast between Capes Bonavista and St. Mary, but their enterprise brought no pecuniary returns.

Carbonear is 1½ M. by road from Mosquito Cove (3 M. from Harbor Grace), and is reached by the steamer after passing Old Sow Point and rounding Carbonear Island. This town has 2,000 inhabitants, with 3 churches, and Wesleyan and Catholic schools. Several wharves are built out to furnish winter-quarters for the vessels and to accommodate the large fish-trade of the place. It is 21 M. by boat to Portugal Cove, across Conception Bay. This town was settled by the French early in the 17th century, under the name of Carboniere, but was soon occupied by the British. In 1696 it was one of the two Newfoundland towns that remained in the hands of the English, all the rest having been captured by Iberville's French fleet. Other marauding French squadrons were beaten off by the men of Carbonear in 1705–6, though the adjacent coast was devastated; and in 1762 Carbonear Island was fortified and garrisoned by the citizens.

The mail-road runs N. from Carbonear to Bay Verd, passing the villages of Croker's Cove, 1 M.; Freshwater, 2; Salmon Cove, 5; Perry's Cove, 8; Broad Cove, 15; Western Bay, 17; Northern Bay, 20; Job's Cove, 25; Island Cove, 27; Low Point, 33; Bay Verd, 38. There is no harbor along this shore, the "coves" being mere open bights, swept by sea-winds and affording insecure anchorage. The inhabitants are engaged in the fisheries, and have made some attempts at farming, in defiance of the early and biting frosts of this high latitude. *Salmon Cove* is near the black and frowning cliffs of Salmon Cove Head, and is famous for its great numbers of salmon. Near Ochre Pit Cove are beds of a reddish clay which is used for paint, and it is claimed that the ancient Bœothic tribes obtained their name of "Red Indians" from their custom of staining themselves with this clay.

Bay Verd, see page 201.

57. Trinity Bay.

This district may be visited by taking the Northern Coastal steamer (see Route 55) to Bay Verd, Old Perlican, or Trinity; or by passing from St. John's to Harbor Grace by Route 56, and thence by the road to Heart's Content (15 M.). The latter village is about 80 M. from St. John's by the road around Conception Bay.

Heart's Content is situated on a fine harbor about half-way up Trinity Bay, and has 880 inhabitants, most of whom are engaged in the Labrador fisheries or in shipbuilding. The scenery in the vicinity is very striking, partaking of the boldness and startling contrast which seems peculiar to this sea-girt Province. Just back of the village is a small lake, over which rises the dark mass of *Mizzen Hill*, 604 ft. high. Heart's Content derives its chief importance and a world-wide fame, from the fact that here is the W. terminus of the old Atlantic telegraph-cable. The office of the company is near the Episcopal Church, and is the only good building in the town.

" O lonely Bay of Trinity,
  O dreary shores, give ear!
Lean down into the white-lipped sea,
  The voice of God to hear!

" From world to world His couriers fly,
  Thought-winged and shod with fire ;
The angel of His stormy sky
  Rides down the sunken wire.

" What saith the herald of the Lord ?
  ' The world s long strife is done :
Close wedded by that mystic cord,
  Its continents are one.

"' And one in heart, as one in blood,
  Shall all her peoples be :
The hands of human brotherhood
  Are clasped beneath the sea.'

" Throb on, strong pulse of thunder! beat
  From answering beach to beach ;
Fuse nations in thy kindly heat,
  And melt the chains of each !

" Wild terror of the sky above,
  Glide tamed and dumb below !
Bear gently, Ocean's carrier-dove,
  Thy errands to and fro.

" Weave on, swift shuttle of the Lord,
  Beneath the deep so far,
The bridal robe of earth's accord,
  The funeral shroud of war !

" For lo ! the fall of Ocean's wall
  Space mocked and time outrun ;
And round the world the thought of all
  Is as the thought of one."
  JOHN G. WHITTIER'S *Cable Hymn*.

The road running N. from Heart's Content leads to New Perlican, 3 M. ; Sillee Cove, 6 M. ; Hants Harbor, 12 ; Seal Cove, 19 ; Lance Cove, 24 ; Old Perlican, 28 ; and Grate's Cove, 34.

New Perlican is on the safe harbor of the same name, and has about 420 inhabitants, most of whom are engaged in the cod-fishery and in ship-building. A packet-boat runs from this point across the Bay to Trinity. Near the village is a large table-rock on which several score of names have been inscribed, some of them over two centuries old.

Old Perlican is about the size of Heart's Content, and is scattered along the embayed shores inside of Perlican Island. It is overlooked by a crescent-shaped range of dark and barren hills. The Northern Coastal steamer calls at this port once a month during the season of navigation.

The southern road from Heart's Content leads to Heart's Desire, 6 M. ; Heart's Delight, 9 ; Shoal Bay, 14 ; Witless Bay, 19 ; Green Harbor, 23 ; Hope All, 28 ; New Harbor, 32 ; and Dildo Cove, 35. The villages on this road are all small, and are mostly inhabited by the toilers of the sea. The country about Green Harbor and Hope All is milder and more pastoral than are the cliff-bound regions on either side. From New Harbor a road runs E. by Spaniard's Bay (Conception Bay) to St. John's, in 68 M. To the S. and W. lie the fishing-hamlets on the narrow isthmus of Avalon, which separates Placentia Bay from Trinity Bay by a strip of land 7 M. long, joining the peninsula of Avalon to the main island. The deep estuary called *Bull Arm* runs up amid the mountains to within 2 M. of the Come-by-chance River of Placentia Bay, and here it is proposed to make a canal joining the two bays.

Heart's Ease is 15 M. from Heart's Content (by boat), and is at the S. entrance of Random Sound. It is a fishing-village with 200 inhabitants and a church. To the S. is the grand cliff-scenery around St. Jones Harbor, and the long and river-like *Deer Harbor*, filled with islands, at whose head is Centre Hill, an isolated cone over 1,000 ft. high. From the summit of Centre Hill or of Crown Hill may be seen nearly the whole extent of the Placentia and Trinity Bays, with their capes and islands, villages and harbors. Just above Heart's Ease is Random Island, covering a large area, and separated from the main by the deep and narrow watercourses called Random Sound and Smith's Sound. There is much fine scenery in the sounds and their deep arms, and salmon-fishing is here carried on to a considerable extent. There are immense quantities of slate on the shores, some of which has been quarried (at Wilton Grove). The two sounds are about 30 M. long, forming three sides of a square around Random Island, and have a width of from ¼ M. to 2 M. "The sail up Smith's Sound was very beautiful. It is a fine river-like arm of the sea, 1-2 M. wide, with lofty, and in many places precipitous, rocky banks, covered with wood. .... The character of the scenery of Random Sound is wild and beautiful, and conveying, from its stillness and silence, the feeling of utter solitude and seclusion."

Trinity is the most convenient point from which to visit the N. shore of the Bay (see page 201). The southern road runs to Trouty, 7 M.; New Bonaventure, 12 M.; and Old Bonaventure, 18 M. Beyond these settlements is the N. entrance to Random Sound.

58. The Bay of Notre Dame.

Passengers are landed from the Northern Coastal steamer at Fogo, Twillingate, Little Bay Island, Nipper's Harbor, or Tilt Cove,—all ports on this bay (see pages 204, 205).

Fogo is situated on **Fogo Island**, which lies between Sir Charles Hamilton's Sound and the Bay of Notre Dame. It is 13 M. long from E. to W., and 8 M. wide, and its shores are bold and rugged. There are 10 fishing-villages on the island, with nearly 2,000 inhabitants (exclusive of Fogo), and roads lead across the hills from cove to cove.

It is 9 M. by road from Fogo to Cape Fogo; 7 M. to Shoal Bay; 5 to Joe Batt's Arm (400 inhabitants); 7 to Little Seldom-come-by; and 9 to *Seldom-come-by*, a considerable village on a fine safe harbor, which is often filled with fleets of schooners and brigs. If ice on the coast or contrary winds prevent the fishermen from reaching Labrador in the early summer, hundreds of sail bear away for this harbor, and wait here until the northern voyage is practicable. There is no other secure anchorage for over 50 M. down the coast. *Tilton Harbor* is on the E. coast of the island, and is a Catholic village of about 400 inhabitants. The principal settlements reached by boat from Fogo are Apsey Cove, 14 M.; Indian Islands, 14; Blackhead Cove, 14; Rocky Bay, 25; Barr'd Islands, 4; and Change Islands, 8. 20 M S. W. is Gander Bay, the outlet of the great *Gander-Bay Ponds*, which bathe the slopes of the Blue Hills and the Heart Ridge, a chain of mountains 30 M. long.

From *Exploits Island* (see page 205) boats pass S. 12 M. through a great archipelago to the mouth of the **River of Exploits.** This noble river descends from Red-Indian Pond, about 90 M. to the S. W., and has a strong current with frequent rapids. The *Grand Falls* are 145 ft. high, where the stream breaks through the Chute-Brook Hills. An Indian trail leads from near the mouth of the river S. W. across the vast barrens of the interior, to the Bay of Despair, on the S. coast of Newfoundland. The River of Exploits flows for the greater part of its course through level lowlands, covered with evergreen forests. It may be ascended in steamers for 12 M., to the first rapid, and from thence to the **Red-Indian Pond** by boats (making frequent portages).

The river was first ascended by Lieut. Buchan, R. N., in 1810, under orders to find and conciliate the Red Indians, who had fled to the interior after being nearly exterminated by the whites. He met a party of them, and left hostages in their hands while he carried some of their number to the coast. But his guests decamped, and he returned only to find that the hostages had been cruelly murdered, and the tribe had fled to the remote interior. In 1823 three squaws were captured, taken to St. John, loaded with presents, and released; since which time no Red Indians have been seen, and it is not known whether the tribe is extinct, or has fled to Labrador, or is secluded in some more remote part of the interior. They were very numerous at the time of the advent of the Europeans, and received the new-comers with confidence; but thereafter for two centuries they were hunted down for the sake of the rich furs in their possession, and gradually retired to the distant inland lakes. In 1827 the Bœothic Society of St. John's sent out envoys to find the Red Indians and open friendly intercourse with them. But they were unable to get sight of a single Indian during long weeks of rambling through the interior, and it is concluded that the race is extinct. On the shores of the broad and beautiful Red-Indian Pond Mr. Cormack found several long-deserted villages of wigwams, with canoes, and curious aboriginal cemeteries. This was evidently the favorite seat of the tribe, and from this point their deer-fences were seen for over 30 M. (see also page 218).

Little Bay Island (250 inhabitants), 15 M. from Tilt Cove, is the most favorable point from which to visit **Hall's Bay.** 8 M. S. W. are the settlements at the mouth of Hall's Bay, of which Ward's Harbor is the chief, having 200 inhabitants and a factory for canning salmon. There are valuable salmon-fisheries near the head of the bay. From Hall's Bay to the N. and W., and towards White Bay, are the favorite summer feeding-grounds of the immense herds of deer which range, almost unmolested, over the interior of the island. The hunting-grounds are usually entered from this point, and sportsmen should secure two or three well-certified Micmac guides.

A veteran British sportsman has written of this region: " I know of no country so near England which offers the same amount of inducement to the explorer, naturalist, or sportsman." It is to be hoped, however, that no future visitors will imitate the atrocious conduct of a party of London sportsmen, who recently entered these hunting-grounds and massacred nearly 2,000 deer during the short season, leaving the forests filled with decaying game. Public opinion will sustain the Micmac Indians, who are dependent on the deer for their living, and who have declared that they will prevent a repetition of such carnage, or punish its perpetrators in a summary manner.

The Indians and the half-breed hunters frequently cross the island from Hall's Bay by ascending Iudian Brook in boats for about 25 M., and then making a portage to the chain of ponds emptying into Grand Pond, and descending by Deer Pond and the Humber River (skirting the Long Range) to the Bay of Islands. The transit is both arduous and perilous. 20 M. inland are the mountains called the *Three Towers*, from whose summit may be seen the Grand Pond, the Bay of Exploits, and the Strait of Belle Isle.

The deer migrate to the S. W. in the autumn, and pass the winter near St. George's Bay and Cape Ray The Red Indians constructed many leagues of fence, from the Bay of Notre Dame to Red-Indian Pond, by which they intercepted the herds during their passage to the S., and laid in supplies of provisions for the winter.

Red-Indian Pond is about 30 M. S. W. of Hall's Bay. It is 40 M. long by 5 - 6 M. wide, and contains many islands. To the S. lie the great interior lakes, in an unexplored and trackless region. The chief of these are Croaker's Lake (10 M. distant), filled with islets; Jameson's Lake, 20 M. long, between Serpentine Mt. and Mt. Misery; Lake Bathurst, 17 by 5 M.; and George IV. Lake, 18 by 6 M. 15 M. W. of Red-Indian Pond is Grand Pond, which is 60 M. long. (See page 218.)

From *Nipper's Harbor* the sportsman may pass up Green Bay, to the S. W., and enter the hunting-grounds (having first taken care to secure trusty guides). On the N. side of the bay is a copper-mine that was opened in 1869, and has yielded well.

Tilt Cove is 23 M. from Hall's Bay, 30 M. from New Bay, and 24 M. from Nimrod. 7 M. distant is *Burying Place*, a small fishing-village, near which have been found numerous birch-bark coffins and other memorials of the Red Indians. A road runs N. E from Tilt Cove, passing in 3 M. *Round Harbor*, which is prolific in copper; and in 4 M *Shoe Cove*, famous for trout, and the station of a government boat which here watches the French fisheries. A road runs N. 7 M. from Shoe Cove to *La Scie*, on the French Shore (see Route 61).

59. Placentia Bay

Is included between Cape St. Mary and Cape Chapeau Rouge, and is 48 M. wide. **Placentia** is the capital of the eastern shore, and is a port of entry and post-town, 80 M. from St. John's by road. It is built along a level strand, overshadowed by round detached hills, and maintains a large fleet of fishing-boats. There are remarkable cliffs on Point Verde and Dixon Island, near the town; and the views from Signal Hill and Castle Hill extend far out over the bay. There is much romantic scenery along the narrow channels of the N. E. and S. E. Arms, which extend from the harbor in among the mountains.

In the year 1660 Placentia Bay was entered by two French frigates, which sailed up into the harbor and landed a strong force of soldiers, with heavy artillery and other munitions. Here they erected a strong fort, occupying a point so near the channel that the Baron La Hontan (who was detached for duty here) said that "ships going in graze (so to speak) upon the angle of the bastion." The French held this post until 1713, when it was surrendered, according to the terms of the treaty of Utrecht. The port became famous as the resort of the French privateers which were destroying the English fisheries, and Commodore Warren was sent out (in 1692) with three 60-gun frigates and two smaller vessels to destroy the town. Warren ran in close to Placentia and opened fire, but was warmly received by the batteries at the entrance and by Fort St. Louis. After a heavy cannonade of six hours' duration, the English fleet was forced to draw off. In 1696 Iberville gathered 14 war-vessels at Placentia, and having received 400 men of Quebec, sailed to the E. and overran all the Atlantic coast of Newfoundland, returning with 40-50 prize-ships and 600 prisoners. In 1697 the great French fleet, which (under Iberville) destroyed all the British posts on Hudson's Bay, gathered here. So much did the British dread the batteries of Placentia and the warlike enthusiasm of M. de Costabelle, its commander, that Admiral Walker, anchored at Sydney, with a splendid fleet carrying 4,000 land-soldiers and 900 cannon, refused to obey his orders to reduce this little French fortress, and sailed back to Britain in disgrace. When France surrendered Newfoundland, in 1713, the soldiers and citizens of Placentia migrated to Cape Breton; and in 1744 a French naval expedition under M. de Brotz failed to recapture it from the British. This town afterwards became one of the chief ports of the Province; but has of late years lost much of its relative importance. A road runs hence to St. John's in 80 M.; also through the settlements on the S. to Distress Cove in 26 M.; also S. W. 88 M. to Branch, on St. Mary's Bay.

Little Placentia is on a narrow harbor 5 M. N. of Placentia, and has 383 inhabitants. Near this point is a bold peak of the western range in Avalon, from which 67 ponds are visible. The islands in the bay are visited from this point. Ram's Islands (133 inhabitants) are 10 M. distant; Red Island (227 inhabitants) is 12 M. W.; and about 18 M. distant is Merasheen Island, which is 21 M. long, and has on its W. coast the Ragged Islands, 865 in number. The great lead-mines at *La Manche* are 12 M. N. of Little Placentia, on the Isthmus of Avalon, 7 M. from Trinity Bay. At the head of the bay, 83 M. from Little Placentia, is the village of North Harbor, near the great Powder-Horn Hills, and 7 M. beyond is Black River, famous for its wild-fowl and other game.

Harbor Buffet is 16 M. from Little Placentia, on the lofty and indented Long Island, and has 883 inhabitants. Near the S. W. part of Placentia Bay is the town and port of **Burin**, a station of the Western Coastal steamers (see page 214).

ST. MARY'S BAY. *Route 60.* 213

60. The Western Outports of Newfoundland.—St. John's to Cape Ray.

On alternate Thursdays or Fridays after the arrival of the mails from Europe, the Western Coastal steamer leaves St. John's for the outports on the S. shore of Newfoundland.

Fares.—St. John's to Ferryland, $2; Renewse, $2; Trepassey, $3.50; Burin, $5; St. Pierre, $6.50; Harbor Briton, $7.50; Burgeo, $9; La Poile, $9.50; Rose Blanche, $10; Channel (Port au Basque), $11; Sydney, $14. The fares for steerage passengers are about half the above prices. Meals are included in the price of the tickets. The trip out and back takes 10 to 12 days.

St. John's to Cape Race, see Route 54.

Passing through the rocky portals of the harbor of St. John's, the steamer directs her course to the S. along the iron-bound Strait Shore. After visiting Ferryland and Renewse (see page 198), the Red Hills are seen in the W.; and beyond the lofty bare summit of Cape Ballard, the dreaded cliffs of **Cape Race** (page 199) are rounded well off shore. Off Freshwater Point the course is changed to N. W., and Trepassey Bay is entered. The shores are lofty and bare, and open to the sweep of the sea. 8½ M. from Freshwater Point is Powles Head, on whose W. side the harbor of *Trepassey* is sheltered. The town contains 514 inhabitants, most of whom are engaged in the fisheries, and fronts on a secure harbor which is never closed by ice. Roads lead hence to Salmonier (31 M.) and Renewse.

In 1628 Lord Baltimore's ships of Avalon, the *Benediction* and the *Victory*, entered Trepassey Bay under full sail, bent on attacking the French settlement. The *Benediction* first greeted the fleet with several cannon-shot, after which she sent a terrific broadside among the vessels. The Basque sailors fled to the shore, and the *Victory*, lowering her boats, took possession of all the vessels in the harbor and bore them away as prizes. The town of Trepassey was destroyed by a British naval attack in 1702.

The steamer now runs S. W. to and around **Cape Pine**, on which is a tall circular tower which upholds a fixed light 314 ft. above the sea, visible at a distance of 24 M. 1 M. W. N. W. is Cape Freels, a little beyond which is *St. Shot's Bay*.

This narrow shore between Cape Pine and St. Shot's is said to be the most dangerous and destructive district on the North American coast, and has been the scene of hundreds of shipwrecks. The conflicting and variable currents in these waters set toward the shore with great force, and draw vessels inward upon the ragged ledges. In former years disasters were frequent here, but at present mariners are warned off by the Admiralty charts and the lights and whistles. St. Shot's is as dreaded a name on the N. coast as Cape Hatteras is in the southern sea. In 1816 the transport *Harpooner* was wrecked on Cape Pine, and 200 people were lost.

St. Mary's Bay is bounded by Cape Freels and Lance Point, and extends for 28 M. into the Peninsula of Avalon. On the E. shore is *St. Mary's*, a court-house town and port of entry, situated on a deep land-locked harbor, and largely engaged in fishing. To the S. is the mountainous Cape English, near which a narrow sandy beach separates the bay from *Holyrood Pond*, a remarkable body of fresh water over 12 M. long. It is 65 M. by road from St. Mary's to St. John's; and at 16 M. distance the village of *Salmonier* is reached. This is a fishing and farming town near the outlet of the broad Salmonier River, famous for its great salmon. To the N. W., at the head of the bay, is some striking scenery, near Colinet Bay, where empties the Hodge-Water River, descending from the Quemo-Gospen Ponds, in the interior of Avalon. There are several small hamlets in this vicinity; and *Colinet* is accessible by land from St. John's in 56 M. The W. shore of St. Mary's Bay is mountainous and rugged, and has no settlements of any consequence.

Beyond the bold Cape St. Mary the steamer runs to the N. W. across the wide entrance to Placentia Bay (see page 212). At about 20 M. from Cape St. Mary the sharply defined headland of Cape Chapeau Rouge becomes visible; and the harbor of **Burin** is entered at about 42 M. from Cape St. Mary. This harbor is the finest in Newfoundland, and is sheltered by islands whose cliff-bound shores are nearly 200 ft. high. On Dodding Head is a lighthouse 430 ft. above the sea, bearing a revolving light which is visible for 27 M. Still farther up, and almost entirely land-locked, is the Burin Inlet. The town of Burin has 1,850 inhabitants, and is an important trading-station, supplying a great part of Placentia Bay. The adjacent scenery is of the boldest and most rugged character, the lofty islands vying with the inland mountains.

On leaving Burin the course is laid to the S. W., passing the lofty promontories of Corbin Head, Miller Head, and Red Head. Beyond the tall sugar-loaf on Sculpin Point the deep harbors of Little and Great St. Lawrence are seen opening to the r.; and the sea-resisting rock of *Cape Chapeau Rouge* is next passed. This great landmark resembles in shape the crown of a hat, and is 748 ft. high, with sheer precipices over 300 ft. high. From this point the course is nearly straight for 33 M., to St. Pierre, running well off, but always in sight of a bold and elevated shore.

St. Pierre, see page 185.

On leaving St. Pierre the course is to the N., passing, in 5 M., the low shores of *Green Island*, and then running for a long distance between the Miquelon Islands and May and Dantzic Points (on the mainland), which are about 12 M. apart. When about half-way across Fortune Bay, Brunet Island (5 M. long) is passed, and on its E. point is seen a lighthouse 408 ft. above the sea, showing a flashing light for 25 M. at sea. 6 M. beyond this point is Sagona Island, with its village of fishermen; and 5 M. farther N. the steamer enters Harbor Briton. Here is an Anglican village of about 350 inhabitants, with an extensive local trade along the shores of Fortune Bay. The harbor is very secure and spacious, and runs far into the land. This town was settled in 1616 by Welshmen, and was then named *Cambriol*.

Fortune Bay

is included between Point May and Pass Island, and is 35 M. wide and 66 M. long. **Fortune** is a town of over 800 inhabitants, situated near the entrance of the bay, and on the Lamaline road. Its energies are chiefly devoted to the fisheries and to trading with St. Pierre. 3 M. E. N. E. are the highlands of Cape Grand Bank, from which the shore trends N. E. by the hamlets of Garnish and Frenchman's Cove to Point Enragée. The E. and N. shores are broken by deep estuaries, in which are small fishing-settlements; and in the N. W. corner are the North and East Bays, famous for herring-fisheries, which attract large fleets of American vessels. On the W. shore is the prosperous village of **Belleorem**, engaged in the cod and herring fisheries, and distant 15 M. from Harbor Briton. Roads lead from this point to the villages of Barrow, Blue Pinion, Corbin, English Harbor West, Coombs' Cove, and St. Jaques. The other settlements on the W. shore are mere fishing-stations, closely hemmed in between the mountains and the sea, and are visited by boats from Harbor Briton.

BURGEO. *Route 60.* 215

Hermitage Bay is an extensive bight of the sea to the N. of Pass Island. Its principal town is *Hermitage Cove*, an Anglican settlement 9 M. from Harbor Briton. N. of the bay is Long Island, which is 25 M. around, and shelters the **Bay of Despair**, famous for its prolific salmon-fisheries. From the head of this bay Indian trails lead inland to Loug Pond, Round Pond, and a great cluster of unvisited lakes situated in a land of forests and mountains. From the farther end of these inland waters diverge the great trails to the River of Exploits and Hall's Bay.

After running out to the S. W. between Sagona Island and Connaigre Head, the course is laid along the comparatively straight coast called the *Western Shore*, extending from Fortune Bay to Cape Ray. Crossing the wide estuary of Hermitage Bay, the bold highlands of Cape La Hune are approached, 12 M. N. of the Penguin Islands. About 25 M. W. of Cape La Hune the steamer passes the *Ramea Islands*, of which the isle called Columbe is remarkable for its height and boldness. There is a fishing-community located here; and the August herrings are held as very choice.

The old marine records report of the Ramea Isles: "In which isles are so great abundance of the huge and mightie sea-oxen with great teeth in the monethes of April, May, and June, that there haue been fifteene hundreth killed there by one small barke in the yeere 1591."
In 1597 the English ship *Hopewell* entered the harbor of Ramea and tried to plunder the French vessels there of their stores and powder, but was forced by a shore-battery to leave incontinently.

About 9 M. W. N. W. of Ramea Columbe, the steamer enters the harbor of **Burgeo**, a port of entry and trading-station of 650 inhabitants, situated on one of the Burgeo Isles, which here form several small, snug harbors. This town is the most important on the Western Shore, and is a favorite resort for vessels seeking supplies. 3 M. distant is *Upper Burgeo*, built on the grassy sand-banks of a small islet; and 7 M. N. is the salmon-fishery at Grandy's Brook, on the line of the N. Y., N. F. and London Telegraph.

Beyond the Burgeo Isles the course is laid along the Western Shore, and at about 25 M. the massive heights at the head of Grand Bruit Bay are seen. 5 M. farther on, after passing Ireland Island, the steamer turns into La Poile Bay, a narrow arm of the sea which cleaves the hills for 10 M. The vessel ascends 3 M. to *La Poile* (Little Bay), a small and decadent fishing-village on the W. shore.

The distance from La Poile to Channel, the last port of call, is 30 M., and the coast is studded with small hamlets. *Garia Bay* is 5-6 M. W. of La Poile, and has two or three villages, situated amid picturesque scenery and surrounded by forests. **Rose Blanche** is midway between La Poile and Channel, and is a port of entry with nearly 500 inhabitants, situated on a small and snug harbor among the mountains. It has a considerable trade with the adjacent fishing-settlements. 8 M. beyond Rose Blanche are the *Burnt Islands*, and 3 M. farther on are the *Dead Islands*. At 8-10 M. inland are seen the dark and desolate crests of the Long-Range Mountains, sheltering the Codroy Valley.

The **Dead Islands** (French, *Les Isles aux Morts*) are so named on account of the many fatal wrecks which have occurred on their dark rocks. The name was given after the loss of an emigrant-ship, when the islands were so fringed with human corpses that it took a gang of men five days to bury them. George Harvey formerly lived on one of the islands, and saved hundreds of lives by boldly putting out to the wrecked ships. About 1880 the *Dispatch* struck on one of the isles. She was full of immigrants, and her boats could not live in the heavy gale which was rapidly breaking her up. But Harvey pushed out in his row-boat, attended only by his daughter (17 years old) and a boy 12 years old. He landed every one of the passengers and crew (163 in number) safely, and fed them for three weeks, insomuch that his family had nothing but fish to eat all winter after. In 1888 the Glasgow ship *Rankin* struck a rock off the isles, and went to pieces, the crew clinging to the stern-rail. In spite of the heavy sea, Harvey rescued them all (25 in number), by making four trips in his punt. "The whole coast between La Poile and Cape Ray seems to have been at one time or other strewed with wrecks. Every house is surrounded with old rigging, spars, masts, sails, ships' bells, rudders, wheels, and other matters. The houses too contain telescopes, compasses, and portions of ships' furniture." (PROF. JUKES.)

Channel (or *Port au Basque*) is 3–4 M. W. of the Dead Isles, and 30 M. from La Poile. It is a port of entry and a transfer-station of the N. Y., N. F. and London Telegraph Company, and has nearly 600 inhabitants, with an Anglican church and several mercantile establishments. The fisheries are of much importance, and large quantities of halibut are caught in the vicinity. A few miles to the W. is the great *Table Mt.*, over Cape Ray, beyond which the French Shore turns to the N. A schooner leaves Port au Basque every fortnight, on the arrival of the steamer from St. John's, and carries the mails N. to St. George's Bay, the Bay of Islands, and Bonne Bay (see Route 61).

The steamer, on every alternate trip, runs S. W. from Channel to Sydney, Cape Breton. The course is across the open sea, and no land is seen, after the mountains about Cape Ray sink below the horizon, until the shores of Cape Breton are approached.

Sydney, see page 150.

61. The French Shore of Newfoundland. — Cape Ray to Cape St. John.

It is not likely that any tourists, except, perhaps, a few adventurous yachtsmen, will visit this district. It is destitute of hotels and roads, and has only one short and infrequent mail-packet route. The only settlements are a few widely scattered fishing-villages, inhabited by a rude and hardy class of mariners; and no form of local government has ever been established on any part of the shore. But the Editor is reluctant to pass over such a vast extent of the coast of the Maritime Provinces without some brief notice, especially since this district is in many of its features so unique. The Editor was unable, owing to the lateness of the season, to visit the French Shore in person, but has been aided in the preparation of the following notes, both by gentlemen who have traversed the coast and the inland lakes, and by various statistics of the Province. It is therefore believed that the ensuing itinerary is correct in all its main features. The distances have been verified by comparison with the British Admiralty charts.

The French Shore may be visited by the trading-schooners which run from port to port throughout its whole extent during the summer season. The most interesting parts of it may also be seen by taking the mail-packet which leaves Port au Basque (Channel) fortnightly, and runs N. to Bonne Bay, touching all along the coast.

CAPE RAY. *Route 61.* 217

The French Shore extends from Cape St. John (N. of Notre Dame Bay) around the N. and W. coasts of the island to Cape Ray, including the richest valleys and fairest soil of Newfoundland. It is nearly exempt from fogs, borders on the most prolific fishing-grounds, and is called the "Garden of Newfoundland." By the treaties of 1713, 1763, and 1783, the French received the right to catch and cure fish, and to erect huts and stages along this entire coast, — a concession of which they have availed themselves to the fullest extent. There are several British colonies along the shore, but they live without law or magistrates, since the home government believes that such appointments would be against the spirit of the treaties with France (which practically neutralized the coast). The only authority is that which is given by courtesy to the resident clergymen of the settlements.

It is 9 M. from Channel to Cape Ray, where the French Shore begins. The distances from this point are given as between harbor and harbor, and do not represent the straight course from one outport to another at a great distance.

Cape Ray to Codroy, 13 M.; Cape Anguille, 18 (Crabb's Brook, 45; Middle Branch, 50; Robinson's Point, 55; Flat Bay, 57; Sandy Point, 65; Indian Head, 75); Cape St. George, 54; Port au Port (Long Point), 84; Bay of Islands, 108; Cape Gregory, 125; Bonne Bay, 140; Green Cove, 147; Cow Harbor, 158; Portland Bill, 176; Bay of Ingornachoix (Point Rich), 206; Port au Choix, 208; Point Ferolle, 220; Flower Cove, 245; Savage Cove, 249; Sandy Bay, 250; Green Island. 255; Cape Norman, 285; Pistolet Island, 292; Noddy Harbor, 306; Quirpon (Cape Bauld), 310; Griguet Bay, 321; St. Lunaire, 326; Braha Bay, 330; St. Anthony, 335; Goose Harbor (Hare Bay), 340; Harbor de Veau, 343; St. Julien, 353; Croque, 358; Conche, 373; Canada Bay, 387; Great Harbor Deep, 410; La Fleur de Lis, 432; La Scie, 455; Cape St. John, 460.

* **Cape Ray** is the S. W. point of Newfoundland, and is strikingly picturesque in its outlines. 3 M. from the shore rises a great table-mountain, with sides 1,700 ft. high and an extensive plateau on the summit. Nearer the sea is the *Sugar Loaf*, a symmetrical conical peak 800 ft. high, N. of which is the Tolt Peak, 1,280 ft. high. These heights may be seen for 50 M. at sea, and the flashing light on the cape is visible at night for 20 M. From this point St. Paul's Island bears S. W. 42 M., and Cape North is W. by S. 57 M. (see page 160).

Soon after passing out to the W. of Cape Ray, *Cape Anguille* is seen on the N., — a bold promontory nearly 1,200 ft. high. Between these capes is the valley of the *Great Codroy River*, with a farming population of several hundred souls; and along its course is the mountain-wall called the **Long Range**, stretching obliquely across the island to the shores of White Bay.

St. George's Bay extends for about 50 M. inland, and its shores are said to be very rich and fertile, abounding also in coal. The scenery about the hamlet of *Crabb's Brook* "forms a most lovely and most English picture." There are several small hamlets around the bay, of which *Sandy Point* is the chief, having 400 inhabitants and 2 churches. The people are rude and uncultured, fond of roaming and adventure; but the moral condition of these communities ranks high in excellence, and great deference is paid to the clergy. The Micmac Indians are often seen in this vicinity, and are partially civilized, and devout members of the Catholic Church. The country to the E. is mountainous, merging into wide grassy plains, on which the deer pass the winter season, roaming about the icy levels of the great interior lakes.

10

Grand Pond is usually (and rarely) visited from St. George's Bay. After ascending the broad sound at the head of the bay for about 10 M., a blind forest-path is entered, and the Indian guides lead the way to the N. E. over a vast expanse of moss (very uncomfortable travelling). The Hare-Head Hills are passed, and after about 15 M. of arduous marching, the traveller reaches the Grand Pond. "And a beautiful sight it was. A narrow strip of blue water, widening, as it proceeded, to about 2 M., lay between bold rocky precipices covered with wood, and rising almost directly from the water to a height of 5-600 ft., having bare tops a little farther back at a still greater elevation." The Bay Indians keep canoes on the pond, and there are several wigwams on the shores. Game and fish are abundant in these woods and waters, since it is but once in years that the all-slaying white man reaches the pond, and the prudent Indians kill only enough for their own actual needs. There is a lofty island 20 M. long, on each side of which are the narrow and ravine-like channels of the pond, with an enormous depth of water. The route to **Hall's Bay** (see page 211) leads up the river from the N. E. corner of the pond for about 35 M., passing through four lakes. From the uppermost pond the canoe is carried for ¼ M. and put into the stream which empties into Hall's Bay. 3 M. W. of the inlet of this river into Grand Pond is the outlet of Junction Brook, a rapid stream which leads to the Humber River and Deer Pond in 8-10 M., and is passable by canoes, with frequent portages.

Near the N. end of Grand Pond, about the year 1770, occurred a terrible battle between the Micmacs and the Red Indians, which resulted in the extermination of the latter nation. The Micmacs were a Catholic tribe from Nova Scotia, who had moved over to Newfoundland, and were displacing the aboriginal inhabitants, the Red Indians, or Bœothics. In the great battle on Grand Pond the utmost determination and spirit were shown by the Bœothics, invaded here in their innermost retreats. But they had only bows and arrows, while the Micmacs were armed with guns, and at the close of the battle not a man, woman, or child of the Red Indians of this section was left alive.

This region is densely covered with forests of large trees (chiefly fir and spruce), alternating with "the barrens,"— vast tracts which are covered with thick moss. Gov. Sir John Harvey, after careful inspection, claims that the barrens are underlaid with luxuriant soil, while for the cultivation of grasses, oats, barley, and potatoes there is "no country out of England or Egypt superior to it." The intense and protracted cold of the winter seasons will preclude agriculture on a large scale.

These inland solitudes are adorned, during the short hot summer, with many brilliant flowers. Among these are great numbers of wild roses, violets, irises, pitcher-plants, heather, maiden-hair, and vividly colored lichens; while (says Sir R. Bonnycastle) "in the tribe of lilies, Solomon in all his glory exceeded not the beauty of those produced in this unheeded wilderness." The only white man who ever yet crossed these lonely lands from shore to shore was a Scotchman named Cormack, who walked from Trinity Bay to St. George's Bay, in 1822. He was accompanied by a Micmac Indian, and the trip took several weeks. The maps of Newfoundland cover this vast unexplored region with conjectural mountains and hypothetical lakes. The British Admiralty chart of Newfoundland (Southern Portion) omits most of these, but gives minute and valuable topographical outlines of the lakes and hills N. of the Bay of Despair, the Red-Indian Pond, and River of Exploits, and the region of the Grand Pond and Deer Pond, with their approaches.

Cape St. George thrusts a huge line of precipices into the sea, and 5 M. beyond is *Red Island*, surrounded by dark red cliffs. 25 M. farther to the N. E. is the entrance to *Port au Port*, a great double harbor of noble capacity. It is separated from St. George's Bay by an isthmus but 1 M. wide, at the W. base of the great Table Mt.

The *****Bay of Islands** affords some of the finest scenery in the Province, and is sheltered by several small but lofty islands. The soil along the shores is said to be deep and productive, and adapted to raising grain and produce. Limestone, gypsum, and fine marble are found here in large quantities. There are about 1,000 inhabitants about the bay, most of whom are engaged in the herring-fishery.

At the head of the bay is the mouth of the **Humber River**, the largest river in Newfoundland. In the last 18 M. of its course it is known as the *Humber Sound*, and is 1-2 M. wide and 50-60 fathoms deep, with lofty and rugged hills on either side. Great quantities of timber are found on these shores, and the trout and salmon fisheries are of considerable value. The river flows into the head of the sound in a narrow and swift current, and is ascended by boats to the Deer Pond. Occasional cabins and clearings are seen along the shores, inhabited by bold and hardy pioneers. 3 M. above the head of the sound there is a rapid 1 M. long, up which boats are drawn by lines. Here " the scenery is highly striking and picturesque, — lofty cliffs of pure white limestone rising abruptly out of the woods to a height of 3-400 ft , and being themselves clothed with thick wood round their sides and over their summits." Above the rapids the river traverses a valley 2 M. wide, filled with birch-groves and hemmed in by high hills. The stream is broad and shallow for 6 M. above the rapids, where another series of rapids is met, above which are the broad waters of *Deer Pond*, 2-3 M. wide and 15 M. long. Here is the undisturbed home of deer and smaller game, loons, gulls, and kingfishers. A few Micmac Indians still visit these solitudes, and their wigwams are seen on the low savannas of the shore. (See also pages 211 and 218)

" Beyond the forest-covered hills which surround it are lakes as beautiful, and larger than Lake George, the cold clear waters of which flow to the bay under the name of the river Humber. It has a valley like Wyoming, and more romantic scenery than the Susquehanna. The Bay of Islands is also a bay of streams and inlets, an endless labyrinth of cliffs and woods and waters, where the summer voyager would delight to wander, and which is worth a volume sparkling with pictures."

Bonne Bay is 23 M. N. E. of the Bay of Islands, and is a favorite resort of American and Provincial fishermen. Great quantities of herring are caught in this vicinity. The mountains of the coast-range closely approach the sea, forming a bold and striking prospect; and the rivers which empty into the bay may be followed to the vicinity of the Long Range.

The coast to the N. N. W. for nearly 70 M. is straight, with the slight indentations of the Bay of St. Paul and Cow Bay. The *Bay of Ingornachoix* has comparatively low and level shores, with two excellent harbors. On its N. point (Point Rich) is a lighthouse containing a white flashing-light which is visible for 18 M.; and 2 M. E. is the fishing-station of *Port au Choix*, whence considerable quantities of codfish and herring are exported. The **Bay of St. John** is dotted with islands, and receives the River of Castors, flowing from an unknown point in the interior, and abounding in salmon.

" What a region for romantic excursions ! Yonder are wooded mountains with a sleepy atmosphere, and attractive vales, and a fine river, the River Castor, flowing from a country almost unexplored ; and here are green isles spotting the sea. — the islands of St. John. Behind them is an expanse of water, alive with fish and fowl, the extremes of which are lost in the deep, untroubled wilderness. A month would not suffice to find out and enjoy its manifold and picturesque beauties, through which wind the deserted trails of the Red Indians, now extinct or banished."

The Bay of St. John is separated by a narrow isthmus from St. Margaret's Bay (on the N.), on which are the stations of *New Ferolle* and *Old Ferolle*. Beyond the Bays of St. Genevieve and St. Barbe, with their few score of inhabitants, is *Flower Cove*, containing a small hamlet and an Episcopal church. The great sealing-grounds of the N. shore are next traversed; and the adjacent coast loses its mountainous character, and sinks into wide plains covered with grass and wild grain.

The Strait of Belle Isle.

The Strait of Belle Isle is now entered, and on the N. is the lofty and barren shore of Labrador (or, if it be night, the fixed light on Point Amour). As Green Island is passed, the *Red Cliffs*, on the Labrador shore, are seen at about 10 M. distance. The low limestone cliffs of the Newfoundland shore are now followed to the N. E., and at 30 M. beyond Green Island, **Cape Norman** is reached, with its revolving light upheld on the bleak dreariness of the spray-swept hill. This cape is the most northerly point of Newfoundland.

The *Sacred Islands* are 12 M. S. E. by E. from Cape Norman, and soon after passing them the hamlet of **Quirpon** is approached. This place is situated on Quirpon Island, 4 degrees N. of St. John's, and is devoted to the sealing business. It has an Episcopal church and cemetery. Multitudes of seals are caught off this point, in the great current which sets from the remote N. into the Strait of Belle Isle. Hundreds of icebergs may sometimes be seen hence, moving in stately procession up the strait. In front of Quirpon are the cold highlands of Jaques-Cartier Island. *Cape Bauld* is the N. point of the island of Quirpon, and the most northerly point of the Province.

14 M. N. of Cape Bauld, and midway to the Labrador shore, is **Belle Isle**, in the entrance of the strait. It is 9½ M. long and 3 M. broad, and is utterly barren and unprofitable. On its S. point is a lonely lighthouse, 470 ft. above the sea, sustaining a fixed white light which is visible for 28 M. During the dense and blinding snow-storms that often sweep over the strait, a cannon is fired at regular intervals; and large deposits of provisions are kept here for the use of shipwrecked mariners. Between Dec. 15 and April 1 there is no light exhibited, for these northern seas are then deserted, save by a few daring seal-hunters. There is but one point where the island can be approached, which is 1½ M. from the lighthouse, and here the stores are landed. There is not a tree or even a bush on the island, and coal is imported from Quebec to warm the house of the keeper,—who, though visited but twice a year, is happy and contented. The path from the landing is cut through the moss-covered rock, and leads up a long and steep ascent.

In the year 1527 "a Canon of St. Paul in London, which was a great mathematician, and a man indued with wealth," sailed for the New World with two ships, which were fitted out by King Henry VIII. After they had gone to the westward for many days, and had passed "great Ilands of Ice," they reached "the mayne land, all wildernesse and mountaines and woodes, and no naturall ground but all mosse, and no habitation nor no people in these parts." They entered the Strait of Belle Isle, and then "there arose a great and a maruailous great storme, and much foul weather," during which the ships were separated. The captain of the *Mary of Guilford* wrote home concerning his consort-ship: "I trust in Almightie Jesu to heare good newes of her"; but no tidings ever came, and she was probably lost in the strait, with all on board.

The islands of Belle Isle and Quirpon were called the **Isles of Demons** in the remote past, and the ancient maps represent them as covered with "devils rampant, with wings, horns, and tails." They were said to be fascinating but malicious, and André Thevet exorcised them from a band of stricken Indians by repeating a part of the Gospel of St. John. The mariners feared to land on these haunted shores, and "when they passed this way, they heard in the air, on the tops and about the masts, a great clamor of men's voices, confused and inarticulate, such as you may hear from the crowd at a fair or market-place; whereupon they well knew that the Isle of Demons was not far off." The brave but superstitious Normans dared not land on the Labrador without the crucifix in hand, believing that those gloomy shores were guarded by great and terrible griffins. These quaint legends undoubtedly had a good foundation. In July, 1873, the coasts of the Strait of Belle Isle were ravaged by bands of immense wolves, who devoured several human beings and besieged the settlements for weeks.

An ancient MS. of 1586 relates a curious legend of Belle Isle. Among the company on the fleet which was conducted through the Straits to Quebec in 1542, were the Lady Marguerite, niece of the Viceroy of New France, and her lover. Their conduct was such as to have scandalized the fleet, and when they reached the Isle of Demons, Roberval, enraged at her shamelessness, put her on shore, with her old nurse. The lover leaped from the ship and joined the women, and the fleet sailed away. Then the demons and the hosts of hell began their assaults on the forsaken trio, tearing about their hut at night, menacing them on the shore, and assaulting them in the forest. But the penitent sinners were guarded by invisible bands of saints, and kept from peril. After many months, wearied by these fiendish assaults, the lover died, and was soon followed by the nurse and the child. Long thereafter lived Marguerite alone, until finally a fishing-vessel ran in warily toward the smoke of her fire, and rescued her, after two years of life among demons.

From Cape Bauld the coast runs S. by the French sealing-stations of Griguet, St. Lunaire, Braha, and St. Anthony, to the deep indentation of *Hare Bay*, which is 18 M. long and 6 M. wide. A short distance to the S. is the fine harbor of *Croque*, a favorite resort for the French fleets and a coaling-station for the steamers. The back country is dismal to the last degree.

To the S. E. are the large islands of Groais (7 × 3½ M. in area) and Belle Isle (9 × 6 M.). Running now to the S. W. by Cape Rouge and Botitot, Conche Harbor is seen on the starboard bow, and **Canada Bay** is opened on the W. This great bay is 12 M. long, and is entered through an intricate passage called the Narrows, beyond which it widens into a safe and capacious basin. The shores are solitary and deserted, and far inland are seen the great hill-ranges called The Clouds. 7 M. to the S. W. is the entrance to Hooping Harbor, and 5 M. farther S. is Fourchette, 12 M. beyond which is *Great Harbor Deep*, a long and narrow estuary with such a depth of water that vessels cannot anchor in it. This is at the W. entrance of White Bay, and is 16 M. from Partridge Point, the E. entrance.

White Bay is a fine sheet of water 45 M. long and 10 – 15 M. wide. It is very deep, and has no islands except such as are close in shore. The fisheries are carried on here to a considerable extent, and at Cat Cove, Jackson's Arm, Chouse Brook, Wiseman's Cove, Seal Cove, and Lobster Harbor are small settlements of resident fishermen. *Chouse Brook* is situated amid noble scenery near the head of the bay, 60 M. by boat from La Scie. On the highlands to the W. and S. of White Bay are the haunts of the deer, which are usually entered from Hall's Bay or Green Bay.

3 M. S. E. of Partridge Point is *La Fleur de Lis* harbor, so named from the simulation of the royal flower by a group of three hills near its head. Running thence to the E., the entrances of Little Bay and Ming's Bight open on the starboard side, and on the port bow are the St. Barbe, or Horse Islands. About 20 M. from La Fleur de Lis is *La Scie*, the last settlement on the French Shore, with its three resident families. A road leads S. 7 M. from this point to Shoe Cove, on the Bay of Notre Dame (see page 211); and 5 M. E. of La Scie is *****Cape St. John**, the boundary of the French Shore on the Atlantic.

"The Cape is in full view, a promontory of shaggy precipices, suggestive of all the fiends of Pandemonium, rather than the lovely Apostle whose name has been gibbeted on the black and dismal crags. . . . . As we bear down toward the Cape, we pass Gull Isle, a mere pile of naked rocks delicately wreathed with lace-like mists. Imagine the last hundred feet of Conway Peak, the very finest of the New-Hampshire mountain-tops, pricking above the waves, and you will see this little outpost and breakwater of Cape St. John." (NOBLE.) The Cape presents by far the grandest scenery on the E. coast of Newfoundland, and is an unbroken wall of black rock, 4–500 ft. high and 5 M. long, against whose immediate base the deep sea sweeps.

"OF THE LANDES OF LABRADOR AND BACCALAOS, LYING WEST AND NORTH-WEST FROM ENGLANDE, AND BEINGE PARTE OF THE FIRME LANDE OF THE WEST INDIES.

"Many haue traualyed to search the coast of the lande of Laborador, as well to the intente to knowe howe farre or whyther it reachethe, as also whether there bee any passage by sea throughe the same into the Sea of Sur and the Islandes of Mainca, which are under the Equinoctiall line: thinkynge that the waye thyther shulde greatly bee shortened by this ryage. The Spanyardes, as to whose ryght the sayde islandes of spices perteyne, dyd fyrst seeke to tynde the same by this way. The Portugales also hauynge the trade of spices in theyr handes, dyd trauayle to fynde the same: although hetherto neyther anye suche passage is founde or the ende of that lande. In the yeare a thousande and fine hundredth, Gaspar Cortesreales made a vyage thyther with two caruuelles; but found not the streyght or passage he sought.... He greatly marnayled to beholde the houge quantitie of snowe and ise. For the sea is there frosen exceedyngly. Thinhabitauntes are men of good corporature, although tawny like the Indiess, and laborious. They paynte theyr bodyes, and weare braselettes and hoopes of syluer and copper. Theyr apparel is made of the skynnes of marternes and dyvers other beastes, whiche they weare with the heare inwarde in wynter, and outwarde in soommer. This apparell they gyrde to theyr bodyes with gyrdels made of cotton or the synewes of fysshes and beastes. They eate fysshe more than any other thynge, and especially salmons, althoughe they haue foules and frute. They make theyr houses of timber, whereof they haue great plentie: and in the steade of tyles, couer them with the skynnes of fysshes and beastes. It is said also that there are grifes in this land: and that the beares and many other beastes and foules are white. To this and the islandes aboute the same, the Britons are accustomed to resorte: as men of nature agreeable vnto them, and born vnder the same altitude and temperature. The Norways also sayled thyther with the pylot cauled John Scolno: and the Englyshe men with Sebastian Cabot.

"The coaste of the lande of Baccalaos is a greate tracte, and the altitude thereof is xlviii degrees and a halfe. Sebastian Cabot was the fyrst that browght any knowleage of this land. For being in Englande in the dayes of Kyng Henry the Seuenth, he furnyshed two shippes at his owne charges or (as some say) at the kynges, whom he persuaded that a passage might bee found to Cathay by the North Seas, and that spices myght bee browght from thense soner by that way, then by the vyage the Portugales vse by the Sea of Sur. He went also to knowe what maner of laudes those Indies were to inhabite. He had withe hym 300 men, and directed his course by the tracte of islande vppon the Cape of Laborador at lviii degrees: affirmynge that in the monethe of July there was such could and heapes of ise that he durst passe no further: also that the dayes were very louge, and in maner withowt nyght, and the nyghtes very cleare. Certeyne it is, that at the lx degrees, the longest day is of xviii houres. But consyderynge the coulde and the straungenees of the vnknowne lande, he turned his course from thense to the West, folowynge the coast of the land of Baccalaos vnto the xxxviii degrees, from whense he returned to Englande. To conclude, the Brytons and Danes have sayled to the Baccalaos; and Jacques Cartier, a Frenchman, was there twyse with three galeons.

"Of these lands Jacobus Gastaldus wryteth thus: 'The Newe land of Baccalaos is a coulde region, whose inhabytauntes are idolatours, and praye to the soone and moone and dyvers idoles. They are whyte people, and very rustical. For they eate flesshe and fysshe and all other thynges rawe. Sumtymes also they eate mans flesshe priuilye, so that theyr Caciqui have no knowleage thereof. The apparell of both the men and women is made of beares skynnes, although they have sables and marternes, not greatly esteemed because they are lyttle. Some of them go naked in soomer, and weare apparell only in wynter..... Northwarde from the region of Baccalaos is the land of Laborador, all full of mountaynes and great woodes, in whiche are manye beares and wylde beares. Thinhabitauntes are idolatoures and warlike people, apparelled as are they of Baccalaos. In all this newe lande is neyther citie or castell, but they lyve in companies lyke heardes of beastes.'"

LABRADOR

Is the great peninsular portion of North America which lies to the N. and N. W. of Newfoundland, and is limited by the Gulf of St. Lawrence, the ocean, and Hudson's Bay. It extends from about 50° N. latitude to 60°, and the climate is extremely rigorous, the mean temperature at Nain being 32° 6′. The land is covered with low mountains and barren plateaus, on which are vast plains of moss interspersed with rocks and bowlders. There are no forests, and the inland region is dotted with lakes and swamps. There are reindeer, bears, foxes, wolves, and smaller game; but their number is small and decreasing. The rivers and lakes swarm with fish, and the whole coast is famous for its valuable fisheries of cod and salmon. At least 1,000 decked vessels are engaged in the Labrador fisheries, and other fleets are devoted to the pursuit of seals. The commercial establishments here are connected with the great firms of England and the Channel Islands. The Esquimaux population is steadily dwindling away, and probably consists of 4,000 souls.

"The coast of Labrador is the edge of a vast solitude of rocky hills, split and blasted by the frosts, and beaten by the waves of the Atlantic, for unknown ages. Every form into which rocks can be washed and broken is visible along its almost interminable shores. A grand headland, yellow, brown, and black, in its horrid nakedness, is ever in sight, one to the north of you, one to the south. Here and there upon them are stripes and patches of pale green, — mosses, lean grasses, and dwarf shrubbery. Occasionally, miles of precipice front the sea, in which the fancy may roughly shape all the structures of human art, — castles, palaces, and temples. Imagine an entire side of Broadway piled up solidly, one, two, three hundred feet in height, often more, and exposed to the charge of the great Atlantic rollers, rushing into the churches, halls, and spacious buildings, thundering through the doorways, dashing in at the windows, sweeping up the lofty fronts, twisting the very cornices with silvery spray, falling back in bright green scrolls and cascades of silvery foam; and yet, all this imagined, can never reach the sentiment of these precipices. More frequent than headlands and perpendicular sea-fronts are the sea-slopes, often bald, tame, and wearisome to the eye, now and then the perfection of all that is picturesque and rough, — a precipice gone to pieces, its softer portions dissolved down to its roots, its flinty bones left standing, a savage scene that scares away all thoughts of order and design in nature..... This is the rosy time of Labrador (July). The blue interior hills, and the stony vales that wind up among them from the sea, have a summer-like and pleasant air. I find myself peopling these regions, and dotting their hills, valleys, and wild shores with human habitations. A second thought — and a mournful one it is — tells me that no men toil in the fields away there; no women keep the house off there; there no children play by the brooks or shout around the country school-house; no bees come home to the hive; no smoke curls from the farm-house chimney; no orchard blooms; no bleating sheep fleck the mountain-sides with whiteness, and no heifer lows in the twilight. There is nobody there; there never was but a miserable and scattered few, and there never will be. It is a great and terrible wilderness of a thousand miles, and lonesome to the very wild animals and birds. Left to the still visitation of the light from the sun, moon, and stars, and the auroral fires, it is only fit to look upon and then be given over to its primeval solitariness. But for the living things of its waters, — the cod, the salmon, and the seal, — which bring thousands of adventurous fishermen and traders to its bleak shores, Labrador would be as desolate as Greenland.

" For a few days the woolly flocks of New England would thrive in Labrador. During these few days there are thousands of her fair daughters who would love to tend them. I prophesy the time is coming when the invalid and tourist from the States will be often found spending the brief but lovely summer here, notwithstanding its ruggedness and desolation." (REV. L. L. NOBLE)

" Wild are the waves which lash the reefs along St. George's bank ;
Cold on the coast of Labrador the fog lies white and dank ;
Through storm, and wave, and blinding mist, stout are the hearts which man
The fishing-smacks of Marblehead, the sea-boats of Cape Ann.

" The cold north light and wintry sun glare on their icy forms,
Bent grimly o'er their straining lines, or wrestling with the storms ;
Free as the winds they drive before, rough as the waves they roam,
They laugh to scorn the slaver's threat against their rocky home."
JOHN G. WHITTIER.

62. The Atlantic Coast of Labrador, to the Moravian Missions and Greenland.

The mail-steamer *Hercules* leaves Battle Harbor fortnightly during the summer.

Battle Harbor is a sheltered roadstead between the Battle Islands and Great Caribou Island, ½ M. long and quite narrow. It is a great resort for fishermen, whose vessels crowd the harbor and are moored to the bold rocky shores. Small houses and stages occupy every point along the sides of the roadstead, and the place is very lively during the fishing season. On the W. is Great Caribou Island, which is 9 M. around, and the steep-shored S. E. Battle Island is the easternmost land of the Labrador coast. The water is of great depth in this vicinity, and is noted for its wonderful ground-swell, which sometimes sweeps into St. Lewis Sound in lines of immense waves during the calmest days of autumn, dashing high over the islets and ledges. An Episcopal church and cemetery were consecrated here by Bishop Field in 1850, and the nephew of Wordsworth (the poet) was for some years its rector. The first Esquimaux convert was baptized in 1857.

Fox Harbor is 3-4 hours' sail from Battle Island, across St. Lewis Sound, and is an Esquimaux village with igloes, kayaks, and other curious things pertaining to this unique people There is a wharf, projecting into the narrow harbor (which resembles a mountain-lake); and the houses are clustered about a humble little Episcopal church.

" Caribou Island fronts to the N. on the bay 5-6 M , I should think, and is a rugged mountain-pile of dark gray rock, rounded in its upper masses, and slashed along its shores with abrupt chasms. It drops short off, at its eastern extremity, into a narrow gulf of deep water. This is Battle Harbor. The billowy pile of igneous rock, perhaps 250 ft. high, lying between this quiet water and the broad Atlantic, is Battle Island, and the site of the town..... At this moment (July) the rocky isle,

SANDWICH BAY. *Route 62.* 225

bombarded by the ocean, and flayed by the sword of the blast for months in the year, is a little paradise of beauty. There are fields of mossy carpet that sinks beneath the foot, with beds of such delicate flowers as one seldom sees..... I have never seen such fairy loveliness as I find here upon this bleak islet, where nature seems to have been playing at Switzerland. Green and yellow mosses, ankle-deep and spotted with blood-red stains, carpet the crags and little vales and cradle-like hollows. Wonderful to behold! flowers pink and white, yellow, red, and blue, are countless as dew-drops, and breathe out upon the pure air their odor, so spirit-like. .... Little gorges and chasms, overhung with miniature precipices, wind gracefully from the summits down to meet the waves, and are filled, where the sun can warm them, with all bloom and sweetness, a kind of wild greenhouse."

The course is laid from Battle Harbor N. across *St. Lewis Sound*, which is 4 M. wide and 10 M. deep (to Fly Island, beyond which is the St. Lewis River, which contains myriads of salmon). Passing the dark and rugged hills (500 ft. high) of Cape St. Lewis, the steamer soon reaches the small but secure haven of *Spear Harbor*, where a short stop is made. The next port is at *St. Francis Harbor*, which is on Granby Island, in the estuary of the deep and navigable Alexis River. An Episcopal church is located here. In this vicinity are several precipitous insulated rocks, rising from the deep sea. The harbor is ½ M. W. of Cape St. Francis, and is deep and well protected, being also a favorite resort for the fishing fleets.

Cape St. Michael is next seen on the W., 11 M. above Cape St. Francis, with its mountainous promontory sheltering an island-studded bay. Beyond the dark and rugged Square Island is the mail-port of *Dead Island*. Crossing now the mouth of St. Michael's Bay, and passing Cape Bluff (which may be seen for 50 M. at sea), the steamer next stops between *Venison Island* and the gloomy cliffs beyond. Running next to the N., on the outside of a great archipelago, the highlands of Partridge Bay are slowly passed.

The *Seal Islands* are 24 M. N. E. of Cape St. Michael, and 18 M. beyond is Spotted Island, distinguished by several white spots on its lofty dark cliffs. To the E. is the great Island of Ponds, near which is *Batteau Harbor*, a mail-port at which a call is made. The next station is at Indian Tickle, which is a narrow roadstead between Indian Island and the highlands of Mulgrave Land. Stopping next at S. E. Cove, the course is laid from thence to Indian Harbor, on the W. side of Huntington Island. This island is 7 M. long, and shelters the entrance to **Sandwich Bay** (the Esquimaux *Netsbuctoke*), which is 6-9 M. wide and 54 M. deep, with 13-40 fathoms of water. There are many picturesque islands in this bay, and on the N. shore are the Mealy Mts., reaching an altitude of 1,482 ft. On the W. side are Eagle and West Rivers, filled with salmon; and East River runs into the bottom of the bay, coming from a large lake where immense numbers of salmon, trout, and pike may be found. 4 M. from the mouth of East River is the small settlement of *Paradise*.

At the head of this great bay are *The Narrows*, with Mount Nat and its bold foothills on the S. "On either side hills towered to the height of a thousand feet, wooded with spruce from base to summit, and these twin escarpments abutted ranges of bold bluffs whose shadows seemed almost to meet midway in the narrow channel that separated them. Through this grand gloomy portal there was an unbroken vista for miles, until the channel made an abrupt turn that hid the water from view; but the great gorge continued on beyond till it was lost in blue shadow." On the N. shore of the Narrows is the Hudson's Bay Company's post of **Rigolette**, occupying the site of an older French trading-station. At the head of the Narrows is Melville Lake, a great inland sea, all along whose S. shore are the weird and wonderful volcanic peaks of the lofty Mealy Mountains. 120 M. S. W. of Rigolette, by this route, is the H. B. Company's post of **Norwest**, situated a little way up the N. W. River, near great spruce forests. This is the chief trading-post of the Mountaineers, a tribe of the great Cree nation of the West, and a tall, graceful, and spirited people. In 1840 they first opened communication with the whites. It was this tribe, which, issuing from the interior highlands in resistless forays, nearly exterminated the Esquimaux of the coast. 300 M. from Fort Norwest is *Fort Nascopie*, situated on the Heights of Land, far in the dark and solitary interior. In that vicinity are the **Grand Falls**, which the *voyageurs* claim are 1,000 ft. high, but Factor M'Lean says are 400 ft. high, — and below them the broad river flashes down through a cañon 300 ft. deep, for over 30 M. 300 M. from Fort Nascopie are the shores of Ungava Bay. (The Esquimaux-Bay district is well described in an article by Charles Hallock, Harper's Magazine, Vol. XXII.)

The Moravians state that the Esquimaux are a proud and enterprising people, low in stature, with coarse features, small hands and feet, and black wiry hair. The men are expert in fishing, catching seals, and managing the light and graceful boat called the *kayak*, which outrides the rudest surges of the sea; while the women are skilful in making garments from skins. Agriculture is impossible, because the country is covered with snow and ice for a great part of the year. They call themselves *Innuits* ("men"), the term *Esquimaux* (meaning "eaters of raw flesh") being applied to them by the hostile tribes to the W. On the 500 M. of the Atlantic coast of Labrador there are about 1,000 of these people, most of whom have been converted by the Moravians. They live about the missions in winter, and assemble from the remotest points to celebrate the mysteries of the Passion Week in the churches. They were heathens and demon-worshippers until 1770, when the Moravian Brethren occupied the coast under permission of the British Crown. They were formerly much more numerous, but have been reduced by long wars with the Mountaineers of the interior and by the ravages of the small-pox. The practice of polygamy has ceased among the tribes, and their marriages are celebrated by the Moravian ritual. The missionaries do considerable trading with the Indians, and keep magazines of provisions at their villages, from which the natives are freely fed during seasons of famine. At each station are a church, a store, a mission-house, and shops and warm huts for the converted and civilized Esquimaux, who are fast learning the mechanic arts. The Moravian mission-ship makes a yearly visit to the Labrador station, replenishing the supplies and carrying away cargoes of furs.

Hopedale is 300 M. N. W. of the Strait of Belle Isle, and is one of the chief Moravian missions on the Labrador coast. It was founded in 1782 by the envoys of the church, and has grown to be a centre of civilizing influences on this dreary coast. Its last statistics claim for it 35 houses, with 46 families and 248 persons; 49 boats and 49 kayaks; and a church containing 74 communicants and 85 baptized children. The mean annual temperature here is 27° 82'. The church is a neat plain building, where the men and women occupy opposite sides, and German hymns are sung to the accompaniment of the violin.

Nain is about 80 M. N. W. of Hopedale, and has about 300 inhabitants, of whom 95 are communicants and 94 are baptized children. It was founded by three Moravians in 1771, and occupies a beautiful position, facing the ocean from the bottom of a narrow haven. It is in 57° N. latitude (same latitude as the Hebrides), and the thermometer sometimes marks 75° in summer, while spirits freeze in the intense cold of winter. *Okkak* is about 120 M. N. W. of Nain, towards Hudson Strait, and is a very successful mission which dates from 1776. The station of *Hebron* is still farther up the coast, and has about 800 inhabitants.

Far away to the N. E., across the broad openings of Davis Strait, is **Cape Desolation**, in Greenland, near the settlements of *Julianshaab*.

63. The Labrador Coast of the Strait of Belle Isle.

At Battle Harbor the Northern Coastal steamer connects with the Labrador mail-boat, which proceeds S. W. across the mouth of St. Charles Channel, and touches at Cape Charles, or *St. Charles Harbor*, entering between Fishflake and Blackbill Islands. This harbor is deep and secure (though small), and is a favorite resort for the fishermen. As the steamer passes the Cape, the round hill of St. Charles may be seen about 1 M. inland, and is noticeable as the loftiest highland in this district. Niger Sound and the Camp Islands (250 – 300 ft. high) are next passed, and a landing is made at *Chimney Tickle*. 1½ M. S. W. of the Camp Islands is Torrent Point, beyond which the vessel passes Table Head, a very picturesque headland, well isolated, and with a level top and precipitous sides. It is 200 ft. high, and is chiefly composed of symmetrical columns of basalt. To the S. are the barren rocks of the Peterel Isles and St. Peter's Isles, giving shelter to St. Peter's Bay. In the S. E. may be seen the dim lines of the distant coast of Belle Isle. On the N. is the bold promontory of Sandwich Head. The deep and narrow **Chateau Bay** now opens to the N. W., guarded by the cliffs of York Point (l.) and Chateau Point (on Castle Island, to the r.), and the steamer ascends its tranquil sheet. Within is the noble fiord of *Temple Bay*, 5 M. long, and lined by lofty highlands, approached through the Temple Pass. On the r. is the ridge of the High Beacon (959 ft.). **Chateau** is a small permanent village, with a church and a large area of fish-stages. In the autumn and winter its inhabitants retire into the back country, for the sake of the fuel which is afforded by the distant forests. The port and harbor are named for the remarkable rocks at the entrance. There are fine trouting-streams up Temple Bay; and vast numbers of curlews visit the islands in August.

" This castle is a most remarkable pile of basaltic rock, rising in vertical columns from an insulated bed of granite. Its height from the level of the ocean is upward of 200 ft. It is composed of regular five-sided prisms, and on all sides the ground is strewn with single blocks and clusters that have become detached and fallen from their places. . . . . [It] seemed like some grim fortress of the feudal ages, from whose embrasures big-mouthed cannon were ready to belch forth flame and smoke. On the very verge of the parapet a cross stood out in bold relief in the gleaming moonlight, like a sentinel upon his watch-tower." (HALLOCK, describing Castle Island.)

Chateau was formerly considered the key of the northern fisheries, and its possession was hotly contested by the English and French. At the time of the depopulation of Acadia a number of its people fled hither and established a strong fortress. This work still remains, and consists of a bastioned star-fort in masonry, with gun-platforms, magazines, and block-houses, surrounded by a deep fosse, beyond which were earthworks and lines of stockades. It was abandoned in 1753, and is now overgrown with thickets. In 1763 a British garrison was located at Chateau, in order to protect the fisheries, but the place was captured in 1778 by the American privateer *Minerva*, and 3 vessels and £70,000 worth of property were carried away as prizes. In 1796 the post was again attacked by a French fleet. A long bombardment ensued between the frigates and the shore-batteries, and it was not until their ammunition was exhausted that the British troops retreated into the back country, after having burnt the village. In 1535 the French exploring fleet under the command of Jaques Cartier assembled here.

After emerging from Chateau Bay, the course is laid around York Point, and the **Strait of Belle Isle** is entered (with Belle Isle itself 18 M. E.). The Labrador coast is now followed for about 25 M., with the stern front of its frowning cliffs slightly indented by the insecure havens of Wreck, Barge, and Greenish Bays. Saddle Island is now seen, with its two rounded hills, and the steamer glides into *Red Bay*, an excellent refuge in whose inner harbor vessels sometimes winter. Large forests are seen at the head of the water, and scattering lines of huts and stages show evidences of the occupation of the hardy northern fishermen. Starting once more on the voyage to the S. W., at 7 M. from Red Bay are seen the Little St. Modeste Islands, sheltering Black Bay, beyond which Cape Diable is passed, and Diable Bay (4 M. W. S. W. of Black Bay). 3 M. farther to the W. the steamer enters *Loup Bay*, rounding high red cliffs, and touches at the fishing-establishment and hamlet of *Lance-au-Loup* (which views the Newfoundland coast from Point Ferolle to Cape Norman). Field-ice is sometimes seen off this shore in the month of June. Capt. Bayfield saw 200 icebergs in the strait in August.

The course is now laid to the S. W. for 3-4 M., to round **Point Amour**, which is at the narrowest part of the strait, and has a fixed light, 155 ft. high, and visible for 18 M. From the Red Cliffs, on the E. of Loup Bay, it is but 11 M. S. S. E. to the coast of Newfoundland.

" The Battery, as sailors call it, is a wall of red sandstone, 2-3 M. in extent, with horizontal lines extending from one extreme to the other, and perpendicular fissures resembling embrasures and gateways. Swelling out with grand proportions toward the sea, it has a most military and picturesque appearance. At one point of this huge citadel of solitude there is the resemblance of a giant portal, with stupendous piers 200 ft. or more in elevation. They are much broken by the yearly assaults of the frost, and the eye darts up the ruddy ruins in surprise. If there was anything to defend, here is a Gibraltar at hand, with comparatively small labor, whose guns could nearly cross the strait. Beneath its precipitous cliffs the débris slopes like a glacis to the beach, with both smooth and broken surfaces, and all very handsomely decorated with rank herbage. . . . . The red sandstone shore is exceedingly picturesque. It has a right royal presence along the deep. Lofty semicircular promontories descend in regular terraces nearly down, then sweep out gracefully with an ample lap to the margin. No art could produce better effect. The long terraced galleries are touched with a tender green, and the well-hollowed vales, now and then occurring, and ascending to the distant horizon between ranks of rounded hills, look green and pasture-like. . . . . Among the very pretty and refreshing features of the coast are its brooks, seen occasionally falling over the rocks in white cascades. Harbors are passed now and then, with small fishing-fleets and dwellings." (NOBLE.)

The steamer enters *Forteau Bay*, and runs across to the W. shore, where are the white houses of a prosperous fishing-establishment, with an Episcopal church and rectory. About the village are seen large Esquimaux dogs, homely, powerful, and intelligent. This bay is the best in the strait, and is much frequented by the French fishermen, for whose convenience one of the Jersey companies has established a station here. On the same side of the harbor a fine cascade (100 ft. high) is seen pouring over the cliffs, and the fresh-water stream which empties at the head of the bay contains large numbers of salmon.

7 M. beyond Forteau, Wood Island is passed, and the harbor of *Blanc Sablon* is entered. To the W. are Bradore Bay and Bonne Esperance Bay, with their trading-stations; and a few miles to the N. W. are the Bradore Hills, several rounded summits, of which the chief is 1,264 ft. high.

Blanc Sablon is on the border-line between the sections of Labrador which belong, the one to the Province of Quebec, the other to Newfoundland. It is named from the white sands which are brought down the river at the head of the bay. Several of the great fishing-companies of the Isle of Jersey have stations here, and the harbor is much visited in summer. Blanc Sablon is at the W. entrance to the Strait of Belle Isle, and it is but 21 M. from the Isle-à-Bois (at the mouth of the bay) to the Newfoundland shore. The village is surrounded by a line of remarkable terraced hills. On *Greenly Island*, just outside of the harbor, 32 sail of fishing-vessels were lost on the night of July 2, 1856.

Following the trend of the N. coast of the Gulf of St. Lawrence, Blanc Sablon is distant from Esquimaux Bay 20 M., from Quebec nearly 600 M., and (in a straight line) 218 M. from Anticosti (see Route 65).

From Blanc Sablon the steamer retraces her course through the Strait of Belle Isle to Battle Harbor.

64. The Labrador Coast of the Gulf of St. Lawrence.—The Mingan Islands.

The ports along this coast may be reached by the American fishing-schooners, from Gloucester, although there can be no certainty when or where they will touch. Boats may be hired at Blanc Sablon to convey passengers to the W.

Quebec to the Moisic River.

The steamer *Margaretta Stevenson* leaves Quebec for the Moisic River every week, and may be hired to call at intermediate ports. The passage occupies 30 – 40 hours, and the cabin-fare is $20 (including meals). The round trip to Moisic and back takes nearly a week.

The N. shore of the Gulf of St. Lawrence is a region which is unique in its dreariness and desolation. The scenery is wild and gloomy, and the shore is faced with barren and storm-beaten hills. The climate is rigorous in the extreme. This district is divided into three parts, — the King's Posts, with 270 M. of coast, from Port Neuf to Cape Cormorant; the Seigniory of Mingan, from Cape Cormorant to the River Agwanus (135 M.); and the Labrador, extending from the Agwanus to Blanc Sablon (156 M.). Along this 561 M. of coast there are (census of 1861) but 5,413 inhabitants, of whom 2,612 are French Canadians and 833 are Indians. 1,754 are fishermen, and 1,038 hunters. In the 560 M. there are but 380 houses, 67½ arpents of cultivated land, and 12 horses. There are 3,841 Catholics, 570 Protestants, and 2 Jews.

The wide *Bradore Bay* is near Blanc Sablon, to the W., and has been called "the most picturesque spot on the Labrador." In the back country are seen the sharp peaks of the Bradore Hills, rising from the wilderness (1,264 ft. high). The bay was formerly celebrated for its numerous humpbacked whales. The village is on Point Jones, on the E. side of the bay.

Bradore Bay is of great extent, and is studded with clusters of islets, which make broad divisions of the roadstead. It was known in ancient times as *La Baie des Eettes*, and was granted by France to the Sieur Le Gardeur de Courtemanche (who, according to tradition, married a Princess of France, the daughter of Henri IV.). That nobleman sent out agents and officers, named the new port *Phélypeaux*, and built at its entrance a bulwark called Fort Pontchartrain. From him it descended to Sieur Foucher, who added the title "de Labrador" to his name; and there still exists a semi-noble family in France, bearing the name of *Foucher de Labrador*. On this bay was the town of **Brest**, which, it is claimed, was founded by men of Brittany, in the year 1508. If this statement is correct, Brest was the first European settlement in America, antedating by over thirty years the foundation of St. Augustine, in Florida. In 1535 Jaques Cartier met French vessels searching for this port. About the year 1600 Brest was at the height of its prosperity, and had 1,000 permanent inhabitants, 200 houses, a governor and an almoner, and strong fortifications. After the subjugation of the Esquimaux by the Montaignais, it was no longer dangerous to establish small fishing-stations along the coast, and Brest began to decline rapidly. Ruins of its ancient works may still be found here.

The *Bay of Bonne-Esperance* is one of the most capacious on this coast, and is sheltered from the sea by a double line of islets. The port is called *Bonny* by the American fishermen, who resort here in great numbers during the herring-season. The islands before the harbor were passed by Jaques Cartier, who said that they were "so numerous that it is not possible to count them." They were formerly (and are sometimes now) called Les Isles de la Demoiselle; and Thévet locates here the tragedy of Roberval's niece Marguerite (see page 221).

Esquimaux Bay is N. of Bonne-Esperance, and is 8 M. in circumference. 2 M. above Esquimaux Island is a small trading-post, above which is the mouth of the river, abounding in salmon. There is a great archipelago between the bay and the Gulf of St. Lawrence. On one of these islands an ancient fort was discovered in the year 1840. It was built of stone and turf, and was surrounded by great piles of human bones. It is supposed that the last great battle between the French and Montaignais and the Esquimaux took place here, and that the latter were exterminated in their own fort.

13 M. W. of Whale Island are Mistanoque Island and Shecatica Bay, beyond Lobster and Rocky Harbors. Port St. Augustine is 15 M. W. of Mistanoque, beyond Shag Island and the castellated highlands of Cumberland Harbor. A line of high islands extends hence 21 M. W. by S. to *Great Meccatina Island*, a granite rock 2 × 3 M. in area, and 500 ft. high. The scenery in this vicinity is remarkable for its grandeur and singular features. 58 M. from Great Meccatina Island is *Cape Whittle;* and in the intervening course the Watagheistic Sound and Wapitagun Harbor are passed. A fringe of islands extends for 6-8 M. off this coast, of which the outermost are barren rocks, and the large inner ones are covered with moss-grown hills.

"Now, brothers, for the icebergs
  Of frozen Labrador,
Floating spectral in the moonshine
  Along the low black shore!
Where like snow the gannet's feathers
  On Brador's rocks are shed,
And the noisy murr are flying,
  Like black scuds, overhead;

"Where in mist the rock is hiding,
  And the sharp reef lurks below,
And the white squall lurks in summer,
  And the autumn tempests blow;
Where, through gray and rolling vapor,
  From evening unto morn,
A thousand boats are hailing,
  Horn answering unto horn.

"Hurrah! for the Red Island,
  With the white cross on its crown!
Hurrah! for Meccatina,
  And its mountains bare and brown!
Where the Caribou's tall antlers
  O'er the dwarf-wood freely toss,
And the footstep of the Mickmack
  Has no sound upon the moss.

"Hurrah!—hurrah!—the west-wind
  Comes freshening down the bay,
The rising sails are filling,—
  Give way, my lads, give way!
Leave the coward landsmen clinging
  To the dull earth, like a weed,—
The stars of heaven shall guide us,
  The breath of heaven shall speed!"

JOHN G. WHITTIER'S *Song of the Fishermen.*

From the quantity of wreck found among these islands, no doubt many melancholy shipwrecks have taken place, which have never been heard of; even if the unfortunate crews landed on the barren rocks, they would perish of cold and hunger. The "eggers" carry on their illegal business along these shores, where millions of sea-birds have their breeding-places. They land on the islands and break all the eggs, and when the birds lay fresh ones they gather them up, and load their boats. There are about 20 vessels engaged in this contraband trade, carrying the eggs to Halifax, Quebec, and Boston. "These men combine together, and form a strong company. They suffer no one to interfere with their business, driving away the fishermen or any one else that attempts to collect eggs near where they happen to be. Might makes right with them, if our information be true. They have arms, and are said by the fishermen not to be scrupulous in the use of them. As soon as they have filled one vessel with eggs, they send her to market; others follow in succession, so that the market is always supplied, but never overstocked. One vessel of 25 tons is said to have cleared £ 200 by this 'egging' business in a favorable season." (*Nautical Magazine.*)

To the W. of Cape Whittle are the Wolf, Coacocho, Olomanosheebo, Wash-shecootai, and Musquarro Rivers, on the last three of which are posts of the Hudson's Bay Company. Next come the Kegashka Bay and River, the cliffs of Mont Joli, the cod banks off Natashquan Point, and several obscure rivers.

The **Mingan Islands** are 29 in number, and lie between the mountainous shores of lower Labrador and the island of Anticosti. They abound in geological phenomena, ancient beaches, denuded rocks, etc., and are of very picturesque contours. About their shores of limestone are thick forests of spruce, birch, and poplar; seals and codfish abound in the adjacent waters; and wild fowl are very plentiful in the proper season. Large Island is 11 M. in circumference; and Mingan, Quarry, Niapisca, Esquimaux, and Charles Islands are 2-3 M. in length. They front the Labrador coast for a distance of 45 M.

There are about 600 inhabitants near the islands, most of whom are Indians and French Acadians, for whose spiritual guidance the Oblate Fathers have established a mission. The chief village is at *Mingan Harbor*, on the mainland, back of Harbor Island; and here is a post of the Hudson's Bay Company. The harbor is commodious and easy of access, and has been visited by large frigates. The salmon and trout fisheries of the Seigniory of Mingan are said to be the best in the world. *Long Point* is due N. of the Perroquets, 6 M. from Mingan Harbor, and is a modern fishing-village fronting on a broad beach. The fish caught and cured here are sent to Spain and Brazil, and form an object of lucrative traffic. The fishermen are hardy and industrious men, generally quiet, but turbulent and desperate during their long drinking-bouts.

The Seigniory of the Mingan Islands and the adjacent mainland was granted to the Sieur François Bissot in 1661, and the feudal rights thus conveyed and still maintained by the owners have greatly retarded the progress of this district. The walrus fisheries were formerly of great value here, and their memory is preserved by *Walrus Island*, on whose shores the great sea-cows used to land. "In 1852 there was not a single establishment on the coast, between the Bay of Mingan and the Seven Isles, and not a quintal of codfish was taken, except on the banks of Mingan and at the River St. John, which the American fishermen have frequented for many years. Now, there is not a river, a cove, a creek, which is not occupied, and every year there are taken 20-25,000 quintals of cod, without counting other fish." "The once desolate coasts of Mingan have acquired, by immigration, a vigorous, moral, and truly Catholic population. The men are generally strong and robust, and above all they are hardy seamen."

On the W. edge of the Mingan Islands are the *Perroquets*, a cluster of low rocks where great numbers of puffins burrow and rear their young. On these islets the steamships *Clyde* and *North Briton* were wrecked (in 1857 and 1861).

A beach of white sand extends W. from Long Point to the *St. John River*, a distance of 18-20 M. The river is marked by the tall adjacent peak of Mount St. John (1,416 ft. high); and furnishes very good fishing (see G. C. SCOTT's "Fishing in American Waters").

The *Manitou River* is 34 M. W. of the St. John, and at 1¼ M. from its mouth it makes a grand leap over a cliff 113 ft. high, forming the most magnificent cataract on the N. shore. The coast Indians still repeat the legend of the invasion of this country by the Micmacs (from Acadia), 200 years ago, and its heroic end. The hostile war-party encamped at the falls, intending to attack the Montagnais at the portages, for which purpose forces were stationed above and below. But the local tribes detected their presence, and cut off the guards at the canoes, then surprised the detachment below the falls, and finally attacked the main body above. After the unsparing carnage of a long night-battle, the Micmacs were conquered, all save their great wizard-chief, who stood on the verge of the falls, singing songs of defiance. A Montagnais chief rushed forward to take him, when the bold Micmac seized his opponent and leaped with him into the foaming waters. They were both borne over the precipice, and the falls have ever since been known as the Manitousiu (Conjurer's) Falls.

The **Moisie River** is about 40 M. W. of the Maniton River, and empties into a broad bay which receives also the Trout River. At this point are the Moisie Iron Works, near which there are about 700 inhabitants, most of whom are connected with the mines. This company has its chief office in Montreal, and runs a weekly steamer between Moisie and Quebec (see page 231). There is a hotel here, where visitors can get plain fare at $5 a week (no liquors on the premises). Large quantities of codfish and salmon are exported from Moisie.

The **Seven Islands** are a group of barren "mountain-peaks, starting suddenly from the ocean," and situated several leagues W. of the mouth of the Moisie River. They were visited by Cartier (1535), who reported that he saw sea-horses here; and in 1731 they were included in the *Domaine du Roi*. The trading-post which was established here by the French, 140 years ago, subsequently reverted to the Hudson's Bay Company, and is visited by 3-400 Nasquapee Indians. Since the departure of the H. B. Company, the post itself has lost its importance, but all vessels trading on the N. shore are now obliged to get their clearances here. The Montagnais Indians had a broad trail running thence up a vast and desolate valley to Lake St. John, 300 M. S. W., and the Moisie River was part of the canoe-route to Hudson's Bay. The Montagnais were here secure from the attacks of the dreaded Mohawks on the one side, and the maritime Esquimaux on the other, and here they received the Jesuit missionaries.

The scenery of the Bay of Seven Islands is famed for its wild beauty and weird desolation. The bay is 7 M. long, and is sheltered by the islands and a mountainous promontory on the W. The immediate shore is a fine sandy beach, back of which are broad lowlands, and " the two parallel ranges of mountains, which add so much to the beauty of the distant scenery of this bay, look like huge and impenetrable barriers between the coast and the howling wilderness beyond them." In the spring and autumn this bay is visited by myriads of ducks, geese, brant, and other wild fowl, and the salmon-fishing in the adjacent streams is of great value. The *Great Boule* is the loftiest of the Seven Islands, reaching an altitude of 700 ft. above the sea, and commanding a broad and magnificent view. There are about 300 inhabitants here, a large proportion of whom are Indians who are engaged in the fur-trade. On *Carrousel Island* is a fixed light, 195 ft. above the sea, which is visible for 20 M.

From Carrousel Island to the St. Margaret River it is 8 M.; to the Cawee Islands, 24; to Sproule Point, 28; and still farther W. are the Pentecost River and English Point, off which are the **Egg Islands**, bearing a revolving white light, which warns off mariners from one of the most dangerous points on the coast.

In the spring of 1711 the British government sent against Quebec 15 men-of-war, under Admiral Sir Hovenden Walker, and 40 transports containing 5,000 veteran soldiers. During a terrible August storm, while they were ascending the Gulf of St. Lawrence, the fleet drove down on the Egg Islands. The frigates were saved from the shoals, but 8 transports were wrecked, with 1,383 men on board, and " 884 brave fellows, who had passed scathless through the sanguinary battles of Blenheim, Ramillies, and Oudenarde, perished miserably on the desolate shores of the St. Lawrence." This terrible loss was the cause of the total failure of the expedition. The French vessels which visited the isles after Walker's disaster "found the wrecks of 8 large vessels, from which the cannon and best articles had been removed, and nearly 3,000 persons drowned, and their bodies lying along the shore. They recognized among them two whole companies of the Queen's Guards, distinguished by their red coats, and several Scotch families, intended as settlers in Canada," among them seven women, all clasping each other's hands. The regiments of Kaine, Windresse, Seymour, and Clayton were nearly annihilated in this wreck. " The French colony could not but recognize a Providence which watched singularly over its preservation, and which, not satisfied with rescuing it from the greatest danger it had yet run, had enriched it with the spoils of an enemy whom it had not had the pains to conquer; hence they rendered Him most heart-felt thanks." (CHARLEVOIX.)

Beyond the hamlet on Caribou Point and the deep bight of Trinity Bay is **Point de Monts** (or, as some say, *Point aux Demons*), 280 M. from Quebec. There is a powerful fixed light on this promontory. 8 M. beyond is *Godbout*, with its fur-trading post; and 9 M. farther W. is Cape St. Nicholas. 18 M. from the cape is Manicouagan Point, 20 M. W. of which is the great Indian trading-post at the *Bersimis River*, where 700 Indians have their headquarters; thence to Cape Colombier it is 11½ M.; and to the church and fort at *Port Neuf* it is 12 M. Point Mille Vaches is opposite Biquette, on the S. shore of the St. Lawrence, and is near the *Sault de Mouton*, a fall of 80 ft. There are several settlements of French Catholic farmers along the shore. At *Les Escoumains* there are 500 inhabitants and considerable quantities of grain and lumber are shipped. The coast is of granite, steep and bold, and runs S. W. 16 M. to *Petite Bergeronne*, whence it is 5½ M. to the mouth of the Saguenay River.

65. Anticosti.

The island of Anticosti lies in the mouth of the St. Lawrence River, and is 118 M. long and 31 M. wide. In 1871 it had about 80 inhabitants, in charge of the government lights and stations, and also 50 acres of cleared land and 3 horses. Fox River is 60 M. distant; the Mingan Islands, 30 M.; and Quebec, about 450 M. The island has lately been the scene of the operations of the Anticosti Land Company, which designed to found here a new Prince Edward Island, covering these peat-plains with prosperous farms. The enterprise has as yet met with but a limited success.

Anticosti has some woodlands, but is for the most part covered with black peaty bogs and ponds, with broad lagoons near the sea. The bogs resemble those of Ireland, and the forests are composed of low and stunted trees. The shores are lined with great piles of driftwood and the fragments of wrecks. There are many bears, otters, foxes, and martens; also partridges, geese, brant, teal, and all manner of aquatic fowl. The months of July and August are rendered miserable by the presence of immense swarms of black flies and mosquitoes, bred in the swamps and bogs. Large whales are seen off these shores, and the early codfish are also found here. Fine limestone and marble occur in several places; and marl and peat are found in vast quantities. There are lighthouses at S. W. Point, S. Point (and a fog-whistle), W. Point (and an alarm-gun), and Heath's Point. The government has established supply-huts along the shores since the terrible wreck of the *Granicus*, on the S. E. point, when the crew reached the shore, but could find nothing to eat, and were obliged to devour each other. None were saved.

In 1690 one of Sir William Phipps's troop-ships was wrecked on Anticosti, during the retreat from Quebec, and but 5 of its people survived the winter on the island. When the ice broke up, these brave fellows started in a row-boat for Boston, 900 M. distant; and after a passage of 44 days they reached their old home in safety. The island was granted in 1691 to the Sieur Joliet, who erected a fort here, but was soon plundered and ejected by the English. In 1814 H. B. M. frigate *Leopard*, 50, the same vessel which captured the U. S. frigate *Chesapeake* was lost here.

"The dangerous, desolate shores of Anticosti, rich in wrecks, accursed in human suffering. This hideous wilderness has been the grave of hundreds; by the slowest and ghastliest of deaths they died, — starvation. Washed ashore from maimed and sinking ships, saved to destruction, they drag their chilled and battered limbs up the rough rocks; for a moment, warm with hope, they look around with eager, straining eyes for shelter, — and there is none; the failing sight darkens on hill and forest, forest and hill, and black despair. Hours and days waste out the lamp of life, until at length the withered skeletons have only strength to die." (ELIOT WARBURTON.)

PROVINCE OF QUEBEC.

QUEBEC is bounded on the W. by the Province of Ontario, on the N. by the wilderness towards Hudson's Bay, on the E. by Maine, Labrador, and the Gulf of St. Lawrence, and on the S. by New Brunswick, New England, and New York. It covers 210,020 square miles, and its scenery is highly diversified and often mountainous, contrasting strongly with the immense prairies of Ontario. The stately river St. Lawrence traverses the Province from S. W. to N. E., and receives as tributaries the large rivers Ottawa, Richelieu, St. Maurice, and Saguenay. The Eastern Townships are famed for their fine highland scenery, amid which are beautiful lakes and glens.

The Province of Quebec has 1,191,516 inhabitants (census of 1871), the vast majority of whom are of French descent and language. 1,019,850 of the people are Roman Catholics, and the laws of education are modified to suit the system of parish-schools.

The Dominion of Canada is ruled by a Governor-General (appointed by the British sovereign) and Privy Council, and a Parliament consisting of 81 senators (24 each from Ontario and Quebec, 12 each from Nova Scotia and New Brunswick, and 9 from P. E. Island, Manitoba, and British Columbia) and 207 members of the House of Commons. There is one member for each 17,000 souls, or 88 for Ontario, 65 for Quebec, 21 for Nova Scotia, 16 for New Brunswick, 6 each for Prince Edward Island and British Columbia, and 5 for Manitoba. In 1872 the debt of the Dominion was $ 122,400,179, most of which represents internal improvements. There are 30,144 Canadian militiamen, with a military school at Kingston; and the navy consists of 8 armed screw-steamers (on the lakes and the Gulf). In 1800 Canada had 240,000 inhabitants; in 1825, 581,920; in 1851, 1,842,265; and in 1871, 3,657,887, — a fifteen-fold increase in 70 years. Between 1842 and 1872, 831,168 emigrants from Great Britain entered Canada; and in the same period, 4,338,086 persons, from the same kingdom, emigrated to the United States. In 1871 and 1872 the exports of Canada amounted to $ 156,813,281, and her imports to $ 194,656,594. Her chief trade is with Great Britain and the United States, and the main exports are breadstuffs and timber. In 1872 the Dominion had 2,928 M. of railways, which had cost $ 163,553,000; and there were then 3,943 post-offices.

The first European explorer who visited this country was Jaques Cartier, who landed at Gaspé in 1534, and ascended the St. Lawrence to the site of Montreal during the following year. Seventeen years later the ill-fated Roberval founded an ephemeral colony near Quebec, and thereafter for over half a century Canada was unvisited. In 1603 Champlain ascended to the site of Montreal, and Quebec and Montreal were soon founded; while the labors of explorations, missions, and fighting the Iroquois were carried on without cessation. In 1629 Canada was taken by an English fleet under Sir David Kirke, but it was restored to France in 1632. The Company of the Hundred Associates was founded by Cardinal Richelieu in 1627, to erect settlements in La Nouvelle France, but the daring and merciless incursions of the Iroquois Indians prevented the growth of the colonies, and in 1663 the company was dissolved. Finally, after they had exterminated the unfortunate Huron nation, the Iroquois destroyed a part of Montreal and many of its people (1689). The long and bitter wars between Canada and the Anglo-American colonies had now commenced, and New York and New England were ravaged by the French troops and their allied Indians.

Naval expeditions were sent from Boston against Quebec in 1690 and 1711, but they both ended disastrously. Montreal and its environs were several times assailed by the forces of New York, but most of the fighting was done on the line of Lake Champlain and in the Maritime Provinces. At last these outposts fell, and powerful British armies entered Canada on the E. and W. In 1759 Wolfe's army captured Quebec, after a pitched battle on the Plains of Abraham; and in the following year Montreal was occupied by Gen. Amherst, with 17,000 men. The French troops were sent home; and in 1763, by the Treaty of Paris, France ceded to Great Britain all her immense Canadian domains. There were then 67,000 French people and 8,000 Indians in the Province.

The resident population was conciliated by tolerance to their religion and other liberal measures, and refused to join the American Colonies when they revolted in 1775. The army of Gen. Montgomery took Montreal and the adjacent country, but the Canadians declined either to aid or to oppose the Americans; and when Arnold was defeated in his attempt to storm Quebec, the Continental forces were soon driven back into the United States. In 1791 the Provinces of Upper Canada and Lower Canada were formed, in order to stop the discontent of the French population, who were thus separated from the English and Loyalist settlements to the W. In 1791 representative government was established, and in 1793 slavery was abolished. The War of 1812 was waged beyond the boundaries of Lower Canada, except during the abortive attempt of the Americans to capture Montreal. In 1837 revolutionary uprisings occurred in various parts of Canada, and were only put down after much bloodshed. In 1840 the two Provinces were united, after which the seigniorial tenures were abolished, decimal currency was adopted, the laws were codified, and other

PROVINCE OF QUEBEC. 237

improvements took place. The capital, which had been shifted from Kingston to Montreal, and then to Toronto, was established by the Queen at Ottawa in 1860. The French and English deputies in Parliament were still at odds, and after a long wrangle in 1864, the attention of the country was drawn to the old project of confederation, which was at last realized in 1867, and Canada (then divided into Ontario and Quebec) and the Maritime Provinces were consolidated into the DOMINION OF CANADA. Since that day the councils of the Imperial Government have manifested a desire to give independence to the new State; and the Dominion, endowed with autonomic powers, has made rapid advances, building great railways, bridges, and canals, and forwarding internal improvements. Meantime Ontario has gained a preponderating power in the national councils, and the statesmen of Quebec are now maturing plans for the repatriation of the 500,000 French-Canadians now in the United States, hoping thereby to restore the Province of Quebec to her former pre-eminence and to populate her waste places.

" Like a virgin goddess in a primeval world, Canada still walks in unconscious beauty among her golden woods and along the margin of her trackless streams, catching but broken glances of her radiant majesty, as mirrored on their surface, and scarcely dreams as yet of the glorious future awaiting her in the Olympus of nations." (EARL OF DUFFERIN.)

" The beggared noble of the early time became a sturdy country gentleman; poor, but not wretched; ignorant of books, except possibly a few scraps of rusty Latin picked up in a Jesuit school; hardy as the hardiest woodsman, but never forgetting his quality of *gentilhomme;* scrupulously wearing its badge, the sword, and copying as well as he could the fashions of the court, which glowed on his vision across the sea in all the effulgence of Versailles, and beamed with reflected ray from the chateau of Quebec. He was at home among his tenants, at home among the Indians, and never more at home than when, a gun in his hand and a crucifix on his breast, he took the war-path with a crew of painted savages and Frenchmen almost as wild, and pounced like a lynx from the forest on some lonely farm or outlying hamlet of New England. How New England hated him, let her records tell. The reddest blood-streaks on her old annals mark the track of the Canadian *gentilhomme.*" (PARKMAN.)

" To a traveller from the Old World, Canada East may appear like a new country, and its inhabitants like colonists; but to me, coming from New England, ... it appeared as old as Normandy itself, and realized much that I had heard of Europe and the Middle Ages. Even the names of humble Canadian villages affected me as if they had been those of the renowned cities of antiquity. To be told by a habitant, when I asked the name of a village in sight, that it is *St. Fereole* or *St. Anne,* the *Guardian Angel* or the *Holy Joseph's*; or of a mountain, that it was *Bélange* or *St. Hyacinthe!* As soon as you leave the States, these saintly names begin. *St. John* is the first town you stop at, and thenceforward the names of the mountains and streams and villages reel, if I may so speak, with the intoxication of poetry, — *Chambly, Longueuil, Pointe aux Trembles, Bartholomy,* etc., etc., — as if it needed only a little foreign accent, a few more liquids and vowels perchance in the language, to make us locate our ideals at once. I began to dream of Provence and the Troubadours, and of places and things which have no existence on the earth. They veiled the Indian and the primitive forest, and the woods toward Hudson's Bay were only as the forests of France and Germany. I could not at once bring myself to believe that the inhabitants who pronounced daily those beautiful and, to me, significant names lead as prosaic lives as we of New England.

"One of the tributaries of the St. Anne is named *La Rivière de la Rose*, and farther east are *La Rivière de la Blondelle* and *La Rivière de la Friponne*. Their very *rivière* meanders more than our *river*..... [It is] a more western and wilder Arcadia, methinks, than the world has ever seen; for the Greeks, with all their wood and river gods, were not so qualified to name the natural features of a country as the ancestors of these French Canadians; and if any people had a right to substitute their own for the Indian names, it was they. They have preceded the pioneer on our own frontiers, and named the *prairie* for us." (THOREAU.)

On the question as to whether the Canadians speak good French, Potherie says that "they had no dialect, which, indeed, is generally lost in a colony." Charlevoix observed (about 1720): "The French language is nowhere spoken with greater purity, there being no accent perceptible." Bougainville adds: "They do not know how to write, but they speak with ease and with an accent as good as the Parisian." Prof. Silliman says that they speak as good French as the common Americans speak English.

From the voluminous work of M. Rameau, entitled *La France aux Colonies — Acadiens et Canadiens* (Paris, 1859), we learn that in the year 1920 the valleys of the Saguenay, Ottawa, and Lower St. Lawrence shall be occupied by a Franco-Canadian nation of 5,000,000 souls; that the mournful vices, "impoverishment of intelligence, and corruption of manners," which the Anglo-American race in the United States has suffered, shall be opposed and checked by the fecund genius of the French race, and the "scientific and artistic aptitudes of the Canadians," emanating continent-enlightening radiance from the walls of the Laval University; that the dissolute barbarism of the Americans shall be ameliorated by the sweet influences of the "Greco-Latin idea" of the Franco-Canadians; and that that agricultural and intellectual people, "the general and essential principle of whose material and intellectual power is in their religious faith and in the simplicity of their manners," shall profit by the sad experience of Old France, — and under the conservative influences of a social aristocracy shall erect a New France, to be forever illustrious in its culture "*de l'esprit, la modestie des mœurs, la liberté et la religion.*"

66. Pictou to Quebec. — The Coasts of Gaspé and the Lower St. Lawrence.

This voyage is full of interest to the lover of fine scenery, and leads through some of the most attractive parts of the Provinces. The vessels pass the lofty highlands of Nova Scotia, the Acadian districts on the sandy shores of New Brunswick, the stately mountains about the Bay of Chaleur, and the frowning ridges of Gaspé. Then comes the ascent of the majestic St. Lawrence, with its white French villages, its Alpine shores, and romantic history, terminated by the quaint mediæval towers of Quebec, "the Walled City of the North." The steamers are large and comfortable, and are quite steady in ordinary seasons. The cabin-tables are well supplied, and the attendance is good. There is but little danger from sea-sickness, except in very breezy weather (see also page 3).

This route is served by the vessels of the Quebec and Gulf Ports Steamship Company, the *Secret* and the *Miramichi*. The *Georgia* formerly plied between Pictou, Charlottetown, Shediac, and Quebec, but she was wrecked on the coast of Maine in January, 1875. The times of departure are liable to variations in case that heavy cargoes are to be landed or shipped at any of the ports. The following time-table is that of the Q. & G. P. S. S. Co. for 1874; further particulars, and the details of changes (if there should be any) may be obtained from the company's agents.

Passengers leave Halifax by railway Monday morning, and connect with the steamship, which leaves Pictou at 7 A. M. on Tuesday. By leaving St. John on the Tuesday-morning train, the boat is met at Shediac, which she enters at 5 P. M. on Tuesday, leaving at 7 P. M. She reaches Chatham at 6 A. M. on Wednesday, and leaves at 7 A. M.; Newcastle at 7.30 A. M., Wednesday, leaving at 8 A. M.; Dalhousie at 1 A. M., Thursday, leaving at 4 A. M.; Paspebiac at 9 A. M., Thursday, leaving at 10 A. M.; Percé at 4 P. M., Thursday, leaving at 4.30 P. M.; Gaspé at 7 P. M., Thursday, leaving at 8 P. M.; Father Point at 7 P. M., Friday, leaving at 8 P. M.; and arrives at Quebec at 10 A. M., on Saturday.

CARLETON. *Route 66.* 239

Quebec to Pictou. — The steamship leaves Quebec at 2 P. M., on Tuesday ; Father Point, 6 A. M., Wednesday ; Gaspé, 4 A. M., Thursday ; Percé, 8 A. M., Thursday ; Paspebiac, 3 P. M., Thursday ; Dalhousie, 9 P. M., Thursday ; Chatham, 4 P. M., Friday ; Newcastle, 6 P. M., Friday ; Shediac, 3 A. M., Saturday (morning train to St. John) ; and arrives at Pictou at 1 P. M., Saturday, connecting with the afternoon train to Halifax.
Fares. — (Meals are included in the 1st-class fares, but the state-rooms are extra. The 2d-class fares are without meals.) Halifax to Shediac, $ 5 or $ 3.50 ; to Chatham or Newcastle, $ 8.50 or $ 4.50 ; to Dalhousie, $ 11.50 or $ 6 ; to Paspebiac, $ 12.50 or $ 6.50 ; to Percé or Gaspé, $ 12 or $ 7 ; to Father Point, $ 17 or $ 8 ; to Quebec, $ 17.50 or $ 8.50.
Quebec to Halifax. — Quebec to Father Point, $ 4 or $ 2 ; to Gaspé, $ 10 or $ 4 ; to Percé, $ 11 or $ 4.25 ; to Paspebiac, $ 13 or $ 5 ; to Dalhousie, $ 14 or $ 5.50 ; to Chatham or Newcastle, $ 14 or $ 5 ; to Shediac, $ 15 or $ 7 ; (to St. John by rail, $ 16 or $ 8) ; to Pictou, $ 16 or $ 7.50 ; to Halifax, $ 17.50 or $ 8.50.
Distances. — Pictou to Shediac, 120 M. ; to Chatham, 225 ; Newcastle, 230 ; Dalhousie, 423 ; Paspebiac, 478 ; Percé, 549 ; Gaspé, 578 ; Father Point, 846 ; Quebec, 1,028.

Halifax to Pictou, see Route 31. St. John to Shediac, see Route 14.
After leaving Pictou Harbor, the steamship passes out between Caribou Island and Pictou Island (see also page 175), and enters the Northumberland Strait. On the S. are the dark highlands of Pictou County, among whose glens are scattered settlements of Scottish people. 10 - 12 M. N. are the low hills of Prince Edward Island. The deep bight of Tatamagouche Bay (see page 81) is passed about 35 M. W. of Pictou, and the blue and monotonous line of the Cobequid Mts. may be seen in the S., in very clear weather. Beyond Baie Verte the steamer passes through the narrow part of the Strait between Cape Traverse and Cape Tormentine, and the low red shores of Prince Edward Island are seen on the r. The course is next laid along the level Westmoreland coast (see page 59), and the harbor of Shediac is entered.

The general aspect of the N. Shore of New Brunswick is described in Route 15 (page 60). It is to be remembered, however, that the Gulf-Ports steamships do not stop at Richibucto, Bathurst, or Campbellton. Having, then, described the coast from Shediac to Dalhousie in Route 15, the present route will follow the shores of the great Gaspesian peninsula.

As the steamship leaves the estuary of the Restigouche, the red sandstone cliffs of *Maguacha Point* are passed, on the l., beyond which is the broad lagoon of Carleton Road. The beautiful peak of *** Tracadiegash** is now approached, and after passing the lighthouse on Tracadiegash Point, the white village of **Carleton** is seen on the Quebec shore. This place has about 800 inhabitants and a convent, and is snugly situated under the lee of the mountains, near a bay which is secure during gales from the N. and E. Immense schools of herring visit these shores during the springtime, at the spawning season, and are caught, to be used as food and for fertilizing the ground. The village is enterprising and active, and is inhabited chiefly by Acadians. The steamer stops off the port if there are any passengers or freight to be landed.

"Carleton is a pretty town, to which a little steamer sometimes runs from Dalhousie, rendering the salmon streams in the vicinity quite accessible. When the sun shines, its white cottages, nestling at the foot of the majestic Tracadiegash Mountain, glisten like snow-flakes against the sombre background, and gleam out in lovely contrast with the clouds that cap the summit of this outpost sentinel of the Alleghany range." (HALLOCK.)

The steamer now passes out upon "the undulating and voluptuous Bay of Chaleur, full of long folds, of languishing contours, which the wind caresses with fan-like breath, and whose softened shores receive the flooding of the waves without a murmur." On the N. is *Cascapediac Bay*, on whose shores are the Acadian and Scottish hamlets of Maria and New Richmond, devoted to farming and the fisheries. The rugged peaks of the Tracadiegash range are seen in fine retrospective views.

New Carlisle is near the mouth of the Grand Bonaventure River, and is the capital of Bonaventure County. It has 400 inhabitants, and is engaged in the fisheries, having also a few summer visitors. The churches and court-house occupy a conspicuous position on the high bank which overlooks the bay. This town was founded in 1785 by American Loyalists, who received from the government one year's provisions, lands, seeds, and farming-implements. $400,000 was expended in establishing this settlement and Douglastown.

Paspebiac (*Clarke's Hotel*) is a village of 250 inhabitants, situated on the N. shore of the Bay of Chaleur, 440 M. from Quebec. Its harbor is formed by a fine beach of sand 3 M. long, curving to the S., and forming a natural breakwater against the sea during easterly gales. The church and houses of the village are built above the red cliffs of the shore, and present the neat and orderly appearance of a military post. On the line of the beach are the great white (and red-trimmed) storehouses and shipyards of Charles Robin & Co. and Le Boutillier Brothers, the mercantile establishments which sustain the place.

Robin & Co. is an ancient house which dates from 1768, and has its headquarters at the Isle of Jersey, off the coast of France. Paspebiac was settled in 1766 by Charles Robin, who established here a large fishing-station. In June, 1778, the place was taken by two American privateers, which carried away the vessels *Hope* and *Bee*. The whole fleet was soon afterward captured by H. B. M. frigates *Hunter* and *Piper*, but Robin was forced to pay such heavy salvage that it ruined his business. In 1783 he came back here under French colors, and in 20 years accumulated a great fortune. The firm of Charles Robin & Co. is now the most powerful on all these coasts, and keeps large fleets employed, supporting numerous villages from 7 wealthy establishments. The heads of the firm live in Jersey, and their officers and managers on this coast are forced by rule to lead a life of celibacy. This company employs 750 men, besides 17 vessels and 151 sailors; and the LeBoutilliers have 580 men and 15 vessels. They export vast quantities of fish and oil to the West Indies and the Mediterranean, supplying their Canadian posts, in return, with all needed products of other countries. Paspebiac receives $100,000 worth of goods yearly, and exports $300,000 worth of fish. The best fish is sent to the Mediterranean in bulk, the second grade goes in tubs to Brazil, and the poorest is shipped in casks to the West Indies. The Jersey fleet reaches Paspebiac early in May, spends the summer fishing in the bay and Gulf, and returns in December. The American market is supplied by the Cape-Ann fleet in these waters; and the proceeds of the autumnal months are sold in Upper Canada. The annual yield of the Bay of Chaleur is estimated at 26,000 quintals of dry codfish, 600 quintals of haddock, 3,000 barrels of herring, 300 barrels of salmon, and 15,000 gallons of cod-oil. The fisheries of the bay and Gulf are valued at $ 800,000 a year, and employ 1,500 sail of vessels and 18,000 men.

In January and February the thermometer sometimes sinks to 25° below zero, and the bay is overhung by dark masses of "frost smoke." In this season the Aurora Borealis is seen by night, illuminating the whole northern horizon with steady brilliance. In July and August the thermometer ranges from 65° to 106°, and the air is tempered by fresh sea-breezes.

The name *Paspebiac* means "broken banks," and the inhabitants are called Paspy Jacks or Pospillots. Many of the bits of agate and jasper called " Gaspé pebbles" are found on this shore after the gales of spring and autumn, and are sent to the jewellers of London and Quebec. It is supposed that they come from the conglomerate rocks on the Restigouche River.

Beyond Paspebiac are the shores of Hope, on which immense masses of caplin-fish are thrown up every spring. They are shovelled into wagons by the farmers and are used to fertilize the land. The next point of interest is the deep bay of *Port Daniel*, a safe and well-sheltered haven, on whose W. shore is a remarkable hill, 400 ft. high. Near the fishing-village up the harbor are deposits of oil-bearing shale. The steamer soon passes *Point Maquereau* (which some consider the N. portal of the Bay of Chaleur), with Point Miscou on the S. E.

At midnight on Oct. 15, 1838, the ship *Colborne* went ashore on Point Maquereau, and was soon broken to pieces. Her crew, consisting of 42 men, was lost. The cargo was composed of silks, wines, silver-plate, and specie, and was valued at over $ 400,000. The wreckers of Gaspé recovered rich treasures from the wreck.

Newport is 6 M. beyond Point Maquereau, and is inhabited by 200 Acadians, who are devoted to the fisheries and to the pursuit of the vast flocks of wild fowl which resort to these shores during the spring and autumn. Great and Little Pabos are seaside hamlets, 4 and 8 M. farther E. 4 M. beyond is *Grand River*, a large Acadian village clustered about the fishing-establishment of Robin & Co. It is 7 M. from this point to Cape Despair.

Cape Despair was named by the French *Cap d' Espoir*, or Cape Hope, and the present name is either an Anglicized pronunciation of this French word, or else was given in memory of the terrible disaster of 1711. During that year Queen Anne sent a great fleet, with 7,000 soldiers, with orders to capture Quebec and occupy Canada. The fleet was under Admiral Sir Hovenden Walker, and the army was commanded by Gen. Hill. During a black fog, on the 22d of August, a violent storm arose and scattered the fleet in all directions, hurling 8 large ships on the terrible ledges of Egg Island (see page 233) and Cape Despair, where they were lost with all on board. Fragments of the wrecks, called *Le Naufrage Anglais*, were to be seen along the shores until a recent date; and there was a wild superstition among the fishermen to the effect that sometimes, when the sea was quiet and calm, vast white waves would roll inward from the Gulf, bearing a phantom ship crowded with men in ancient military costumes. An officer stands on the bow, with a white-clad woman on his left arm, and as the maddened surge sweeps the doomed ship on with lightning speed, a tremendous crash ensues, the clear, agonized cry of a woman swells over the great voice of despair, — and naught is seen but the black cliffs and the level sea.

Just beyond Cape Despair is the prosperous fishing-station of *Cape Cove*, 9 M. from Percé. The traveller should now be on the lookout for the Percé Rock and Bonaventure Island. The steamer runs in between the Rock and the Island, affording fine views of both.

The * **Percé Rock** is 288 ft. high, rising with precipitous walls directly from the waves; and is about 500 ft. long. This citadel-like cliff is pierced by a lofty arch, through which the long levels of the sea are visible. Small boats sometimes traverse this weird passage, under the immense Gothic arch of rock. There was formerly another tunnel, near the outer point of the Rock, but its roof fell in with a tremendous crash, and left a great obelisk rising from the sea beyond.

The summit of the Percé Rock covers about two acres, and is divided into two great districts, one of which is inhabited by the gulls, and the cormorants dwell on the other. If either of these trespasses on the other's territory (which occurs every fifteen minutes, at least), a battle ensues, the shrill cries of hundreds or thousands of birds rend the air, great clouds of combatants hover over the plateau, and peace is only restored by the retreat of the invader. When the conflict is between large flocks, it is a scene worthy of close notice, and sometimes becomes highly exciting. The Rock is at right angles with Mt. Joli, and is of new red sandstone. The top is covered with fine grass.

Many years ago the Rock was ascended by two fishermen, and the way once being found, scores of men clambered up by ropes and carried away the eggs and young birds, finding the older ones so tame that they had to be lifted off the nests. This vast aviary would have been depopulated long ere this, but that the Percé magistrates passed a law forbidding the ascent of the Rock. There are numerous quaint and weird legends attached to this place, the strangest of which is that of *Le Génie de l' Ile Percée*, a phantom often seen over the plateau. "It is likely that the foundation for this legend can be traced to the vapory or cloud-like appearance the vast flocks of water-fowl assume when seen at a distance, wheeling in every fantastic shape through the air, previous to alighting on the summit."

The harbor of Percé is very insecure, and is open to the N. E. winds. In earlier times this port was called *La Terre des Tempétes*, so frequent and disastrous were the storms. The village has about 400 inhabitants, most of whom follow the shore-fisheries in small boats. The town is visited every spring and summer by hundreds of stalwart Jersey lads, sent out by the Robins.

Percé consists of South Beach, where are the white-and-red buildings of the Robin establishment; and North Beach, where is the bulk of the population, with the court-house, jail, and Catholic church. The two sections are separated by Mount Joli, a lofty promontory which here approaches Percé Rock. The Episcopal church is a cosey little Gothic structure, accommodating 100 persons. Percé is "the Elysium of fishermen," and hence arises a circumstance which detracts from its value as a summer resort, — when the shore is covered with the refuse parts of codfish, producing a powerful and unpleasant odor. It is said that even the potatoes are found to contain fish-bones.

Back of Percé is the remarkable * **Mount St. Anne**, with its bold and massive square top rising 1,230 ft. above the sea, and visible for a distance of 70 M. over the water. This eminence may be ascended without great trouble, and from its summit is obtained one of the noblest views in the Maritime Provinces. It includes many leagues of the savage mountain-land of Gaspé, extending also along the coast from the Bay of Chaleur to Gaspé Bay and Ship Head. But the marine view is the most attractive, and embraces many leagues of the Gulf of St. Lawrence, with its great fishing-fleets and squadrons of small boats. It overlooks Bonaventure and Percé Rock. A fine view is also obtained from the highway near French Town, including a vast area of the Gulf, the bird-colonies on top of the Rock, Point St. Peter, and Barry Head, with its conspicuous Catholic church. The walk around the mountain to the corner of the beach is full of interest; and the road through the hills to Gaspé is picturesque, though rough, leading by Corny Beach and through a profound mountain-gorge. Mt. St. Anne is also known as Mt. Joli and the Table Roulante. Upon its red-sandstone slopes are found shell-fossils, jasper, agate, and fine quartz crystals.

* **Bonaventure Island** forms a great natural breakwater before the Percé shores, and is surrounded by deep channels. It is 2½ M. from the mainland, and the passage around the island in a small boat affords a pleasant excursion. Bonaventure is 2½ M. long and ¾ M. wide, and is a vast pile of red conglomerate rock, with a line of cliffs 3–500 ft. high, facing the Gulf over 50 fathoms of water. There are about 300 French Catholics on the shores, connected with the fishing-establishment of LeBoutillier Brothers. The island was formerly the property of Capt. Duval, a brave mariner of the Channel Isles, who, in the privateer *Vulture*, swept the coasts of France during the Napoleonic wars. He is buried on Mount Joli.

" Percé is one of the curiosities of the St. Lawrence. If one should believe all the fantastic stories, to which tradition adds its prestige, that rest about this formidable rock, thrown forward into a ceaselessly surging and often stormy sea, like a fearless defiance from the shoal to the abyss, it could only be approached with a mysterious dread mingled with anguish. Percé proper is a village of 200 firesides, established on a promontory that seems to guard the St. Lawrence: this promontory is not lofty, nor does it compare with our northern mountains; but it is wrinkled, menacing, full of a fierce grandeur ; it might be said that the long battle with the ocean has revealed to it its strength and the power which it holds from God to restrain the waves from passing their appointed bounds. It is an archer of the Middle Ages, covered with iron, immovable in his armor, and who receives, invulnerable, all the blows of the enemy. In face of the Atlantic, which has beaten it with tempests through thousands of centuries, trembling under the eternal shower of the waves, but immovable as a decree of heaven, gloomy, thoughtful, enduring without murmur the wrathful torrents that inundate it, bent downward like a fallen god who expiates in an eternity the arrogant pride of a single day, Percé fills us at once with a sorrowful admiration and a sublime pity." (ARTHUR BUIES.)

Percé was visited by Cartier in 1534, and thereafter became a celebrated fishing-station for the French fleets. The coast from Canso to Cape Rosier was granted soon after, and on its reversion to the Crown this site was bestowed on De Fronsac, who founded a permanent village here, while over 500 transient fishermen made it a summer rendezvous. Bishop Laval sent the Franciscans here in 1673 to look after the spiritual welfare of the people, and they erected a chapel at Percé and the Church of St. Claire on Bonaventure Island. In 1690 the place was taken, with all its vessels, by two British frigates, whose crews sacked and burnt all the houses at Percé and Bonaventure, destroyed the churches, and fired 150 gunshots through the picture of St. Peter. In 1711 another naval attack was made by the British, and the French ships *Héros* and *Vermandois* were captured in the harbor. In 1776 a desperate naval combat took place off Percé Rock, between the American privateers that had devastated the shores of the Bay of Chaleur and the British war-vessels *Wolf* and *Diligence*. Two of the American vessels were sunk within cannon-shot of the Rock.

After leaving her anchorage off Percé the steamship runs N. across the openings of Mal Bay, and at 9 M. out passes *Point St. Peter*, with its fishing-village. The course is next laid to the N. W. up Gaspé Bay, with the fatal strand of the Grand Grêve on the r. To the l. is *Douglastown*, on the broad lagoon at the mouth of the St. John River (famous for salmon). This town was laid out by Surveyor Douglas, and is inhabited by Irish and French people. The vessel now steams in through the narrow strait between the grand natural breakwater of Sandybeach and the N. shore, and enters the *Gaspé Basin. The bay is 20 M. long and 5 M. wide, and the basin is a secure and land-locked harbor at its head. As the steamer rounds the lighthouse on Sandybeach, beautiful views are presented of the broad haven, with the North River Mts. to the W.

"The mountains of Gaspé are fair to behold,
With their fleckings of shadow and gleamings of gold."

Gaspé (*Gulf House*) is a town of 800 inhabitants, beautifully situated between the mountains and the sea, and fronting on the S. W. arm of the basin. It is the capital of the county and a free port of entry, and is devoted to the fisheries, having several whaling-ships and a large fleet of schooners. The Gaspé codfish are preferred, in the Mediterranean ports, to the Newfoundland fish, because they are not so salty. The chief establishment here is that of the LeBoutilliers, who have also a fine mansion near the village. Petroleum has been found here, and wells 7 – 800 ft. deep have been sunk by two companies. Gaspé is visited by 2 – 300 city people every summer, for the sake of its picturesque scenery, cool and sparkling air, and the conveniences for yachting and for fishing. The York and Dartmouth Rivers empty into the basin, and are famous for their game-fish. The adjacent shores are fertile and are thickly settled, and the town itself is rapidly advancing in importance. On a hill to the S. is *Fort Ramsay*, a line of guns among the trees. This is the first point N. of Newcastle where the steamer is moored to a wharf. Monthly mail-packets run from Gaspé to Esquimaux Bay, on the Labrador coast (see page 230).

"What a glorious sight! Imagine a bay 20 M. long ending in a basin where a fleet of a thousand vessels could be sheltered. On right and left, two rivers, which are parted by the port, sweep around the amphitheatrical shores; hills here and there of savage outline or covered with rounded lawns; below, a little line of piers, fishing-vessels, schooners and some brigs swinging their slackened sails in the light breeze which blows from the shores ; something wild, fresh, and vigorous, like the first spring of a great creation. The Gaspé Basin has traits of the giant and of the infant; it astonishes and charms; it has a harmony at once delicate and striking." (ARTHUR BUIES.)

The Indians of Gaspé were distinguished, in a remote age, for unusual advances in civilization. They knew the points of the compass, traced maps of their country, observed the positions of the stars, and worshipped the symbol of the cross. They informed the early Jesuit missionaries that in far distant ages they were scourged by a fatal pestilence, until a venerable man landed on their shore, and arrested the progress of the disease by erecting the cross (see PÈRE LECLERC'S *Nouvelle Relation de la Gaspésie*, 1676). It is supposed that this mysterious visitor was a Norseman. The name *Gaspé* means "land's end," one of its component parts being found also in the aboriginal words Mala-gash, Tracadie-gash, etc. The warlike tribes on this shore were formerly distinguished for their fierce and victorious forays into the remote lands of the Montaignais and Esquimaux.

Prof. Rafn, the great Danish archæologist, has advanced a theory to the effect that Gaspé was a fishing-station of the Norse vikings in the 11th, 12th, and 13th centuries. It is supposed that it was visited in 1506 by the Spanish mariner Velasco, who ascended the St. Lawrence for 200 leagues, or else by Stefano Gomez, who was sailing from Spain to Cuba in 1525, but was blown far from his course, and entered the Gulf of St. Lawrence. There is an old Castilian tradition that the gold-seeking Spaniards, finding no precious metals here, said, "*Aca náda*" ("There is nothing here"). This oft-repeated phrase became fixed in the memory of the Indians, though it was not comprehended; and when Cartier came, they supposed him to be of the same people as the previous European visitors, and endeavored to excite his interest by repeating the words, "Aca náda, Aca náda." He thought that they were giving him the name of their nation or country, and so, according to this puerile tradition, arose the name of CANADA. Another theory of the derivation of the name was given by the early New-Englanders: "New England is by some affirmed to be an island, bounded on the north with the River Canada (so called from Monsieur Cane)." (JOSSELYN'S *New England's Rarities Discovered*, 1672.) "From this lake northwards is derived the famous River of Canada, so named of Monsieur de Cane, a French Lord, who first planted a colony of French in America." (MORTON'S *New English Canaan*, 1632.)

The generally received account of the origin of the name Canada is that it is an Indian compound word. *Caugh-na-waugh-a* means "the village of the rapid," its first syllable being similar to that of the Indian word *Caugh-na-daugh*, "village of huts" (also of *Caugh-yu-ga*, or Cayuga, and *Caugh-na-daugh-ga*, now Canandaigua), which has been euphonized into "Canada." When Brant, the Mohawk chieftain, translated the Gospel of St. Matthew into his own language, he always put *Canada* for "a village."

In April, 1534 (being in his fortieth year), the bold and sagacious Jaques Cartier set sail from ancient St. Malo ("thrust out like a buttress into the sea, strange and grim of aspect, breathing war from its walls and battlements of ragged stone, — a stronghold of privateers, the home of a race whose intractable and defiant independence neither time nor change has subdued"). He was under the patronage of Philippe de Brion-Chabot, Admiral of France, and was sent forth to reconnoitre a new route to Cathay, for the great advantage of European commerce. It was also thought that in the new realms beyond the sea the Catholic Church might make such conquests as would requite her for the great schisms of Luther and Calvin and the Anglican Church. The result has nearly justified the hope.

The intrepid voyager traversed the Strait of Belle Isle, and stretched across to the *Baie des Chaleurs*, which was entered on the 9th of July, and received its name from the intense heats which the mariners encountered there. He then landed at Gaspé, and took possession of the country in the name of his Church and King by erecting a cross, 30 ft. high, adorned with the fleur-de-lis. Here he met a company of warriors from Quebec, campaigning against the natives of this region, and carried two of them to France. They were introduced to all the splendors of Paris and the court of Francis I., and in the following year returned with Cartier and piloted his fleet up the St. Lawrence to their home at Stadacona (Quebec).

"Twenty vessels were laden with stores, food, building implements, guns, and ammunition; nearly 150 pieces of ordnance were stowed away in the different holds, to be mounted upon the walls of Quebec and other forts; the decks were crowded with emigrants, male and female; priests were there, burning with religious zeal; and everything looked hopeful for their success. The whole fleet was put under the command of M. de Roquemont, a French Admiral; and full of hope and expectation they set sail from France in the month of April, 1627." This stately fleet was overtaken by a storm in the Gulf, and took refuge in Gaspé Bay, where they were boldly attacked by Captain Kirke's English squadron of 3 vessels. Kirke summoned the immensely superior French fleet to surrender, but De Roquemont, though unprepared for battle, and hampered with freight and non-combatants, sent back a spirited refusal. The Kirkes then sailed boldly into the hostile fleet, and after raking the Admiral's ship, carried it by boarding. The French resisted but feebly, and the whole squadron fell into the bold Briton's hands. He burnt 10 vessels, and freighted the others with the grand train of artillery and the other stores, with which he returned to England. Champlain was left in despair, at Quebec; and the Kirkes were burnt in effigy in the Place de Grève, at Paris.

Gaspé was honored, in 1663, by the sojourn of the brave old Baron Dubois d'Avaugour, some time Governor of New France. From this point he sent his celebrated memorial to Colbert, the French Prime Minister, after he had been deposed from office through the influence of Bishop Laval and the Jesuits. Hence he sailed to France, and soon met a soldier's death in the Croatian fortress of Zrin, which he was defending against the Turks.

In the year 1760 Commodore Byron's powerful fleet entered Gaspé Basin and captured the village. The French frigate *La Catharina* was in the harbor, but was soon taken and destroyed by fire. Many years ago the Gaspésian peninsula was erected into a province, and the seat of government was located at this town. But the number of inhabitants was not enough to warrant the expense of a vice-regal court, and the peninsula was reannexed to Quebec.

In leaving Gaspé Basin the steamship passes the beaches of the N. shore, lined with whale-huts and fish-stages, and then runs to the S. E. down Gaspé Bay. *Cape Gaspé is 7½ M. N. of Point St. Peter, and fronts the Gulf with a line of sandstone cliffs 692 ft. high. Off the S. E. point there was formerly a statue-like rock 100 ft. high, called *La Vieille* (the Old Woman), but it has been thrown down by the sea. The Indians named this rock *Gasepion*, whence the name *Gaspé*, which is now applied to the great peninsula between the Bay of Chaleur and the St. Lawrence River. Two leagues beyond Cape Gaspé the steamship passes **Cape Rosier**, and enters the St. Lawrence River.

67. The Lower St. Lawrence.

"The most interesting object in Canada to me was the River St. Lawrence, known far and wide, and for centuries, as the Great River. Cartier, its discoverer, sailed up it as far as Montreal in 1535, nearly a century before the coming of the Pilgrims; and I have seen a pretty accurate map of it so far, containing the city of 'Hochelaga' and the river 'Saguenay,' in Ortelius's *Theatrum Orbis Terrarum*, printed at Antwerp in 1575, in which the famous cities of 'Norumbega' and 'Orsinora' stand on the rough-blocked continent where New England is to-day, and the fabulous but unfortunate Isle of Demons, and Frisland, and others, lie off and on in the unfrequented sea, some of them prowling near what is now the course of the Cunard steamers. It was famous in Europe before the other rivers of North America were heard of, notwithstanding that the mouth of the Mississippi is said to have been discovered first, and its stream was reached by De Soto not long after; but the St. Lawrence had attracted settlers to its cold shores long before the Mississippi, or even the Hudson, was known to the world. The first explorers declared that the summer in that country was as warm as France, and they named one of the bays in the Gulf of St. Lawrence the Bay of Chaleur, or warmth: but they said nothing about the winter being as cold as Greenland. In the MS. account of Cartier's second voyage it is called 'the greatest river, without comparison, that is known to have ever been seen.' The savages told him that it was the '*Chemin du Canada*' (the highway to Canada), 'which goes so far that no man hath ever been to the end, that they had heard.' The Saguenay, one of its tributaries, is described by Cartier in 1535, and still more particularly by Jean Alphonse in 1542, who adds: 'I think that this river comes from the sea of Cathay, for in this place there issues a strong current, and there runs here a terrible tide.' The early explorers saw many whales and other sea-monsters far up the St. Lawrence. Champlain, in his map, represents a whale spouting in the harbor of Quebec, 360 M. from what may be called the mouth of the river; and Charlevoix took his reader to the summit of Cape Diamond to see the 'porpoises, white as snow,' sporting on the surface of the harbor of Quebec. In Champlain's day it was commonly called 'the Great River of Canada.' More than one nation has claimed it. In Ogilby's 'America of 1670,' in the map *Novi Belgi*, it is called 'De Groote Rivier van Niew Nederlandt' It rises near another father of waters, the Mississippi, issuing from a remarkable spring far up in the woods, called Lake Superior, 1,500 M. in circumference; and several other springs there are thereabouts which feed it. It makes such a noise in its tumbling down at one place as is heard all round the world. Bouchette, the Surveyor-General of the Canadas, calls it 'the most splendid river on the globe'; says that it is 2,000 M. long (more recent geographers make it 4-500 M. longer); that at the Rivière du Sud it is 11 M. wide; at the Paps of Matane, 25; at the Seven Islands, 73; and at its mouth, from Cape Rosier to the Mingan Settlements in Labrador, 96 M. wide. It has much the largest estuary, regarding both length and breadth, of any river on the globe. Perhaps Charlevoix describes the St. Lawrence truly as the most *navigable* river in the world. Between Montreal and Quebec it averages 2 M. wide. The tide is felt as far up as Three Rivers, 432 M., which is as far as from Boston to Washington. The geographer Guyot observes that the Maranon is 3,000 M. long, and gathers its waters from a surface of 1,500,000 square M.; that the Mississippi is also 3,000 M. long, but its basin covers only 8-900,000 square M.; that the St. Lawrence is 1,800 M. long, and its basin covers 1,000,000 square M.; and speaking of the lakes, he adds: 'These vast freshwater seas, together with the St. Lawrence, cover a surface of nearly 100,000 square M., and it has been calculated that they contain about one half of all the fresh water on the surface of our planet.' Pilots say there are no soundings till 150 M. up the St. Lawrence. McTaggart, an engineer, observes that 'the Ottawa is larger than all the rivers in Great Britain, were they running in one.' The traveller Grey writes: 'There is not perhaps in the whole extent of this immense continent so fine an approach to it as by the river St. Lawrence. In the Southern States you have, in general, a level country for many miles inland; here you are introduced at once into a majestic scenery, where everything is on a grand scale, — mountains, woods, lakes, rivers, precipices, waterfalls.' We have not yet the data for a minute comparison of the St. Lawrence with the South American rivers; but it is obvious that, taking it in connection with its lakes, its estuary, and its falls, it easily bears off the palm from all the rivers on the globe." (Freely condensed from THOREAU'S *A Yankee in Canada*.)

" Bien loin de ses gourbis, sous l'ombre des platanes,
L'Arabe au blanche burnous qui suit les caravanes
Sur les sables errant
Découvre moins joyeux son oasis humide,
Que les Canadiens sous la saison torride
Leur fleuve Saint-Laurent.

" A nous ses champs d'azur et ses fraîches retraites,
Les îlots couronnés de mourantes aigrettes,
Les monts audacieux.
Les arômes piquants que la mer y dépose
Et son grand horizon où votre œil se repose
Comme l étoile aux cieux."
L. J. C. FISET.

" Sur ces bords enchantés, notre mère, la France,
A laissé de sa gloire un immortel sillon,
Précipitant ses flots vers l'océan immense,
Le noble Saint-Laurent redit encor son nom.

" Salut, ô ma belle patrie !
Salut, ô bords du Saint-Laurent
Terre que l'étranger envie,
Et qu'il regrette en la quittant.
Heureux qui peut passer sa vie,

Toujours fidèle à te servir;
Et dans tes bras, mère chérie,
Peut rendre son dernier soupir.

" Salut, ô ciel de ma patrie !
Salut, ô noble Saint-Laurent !
Ton nom dans mon âme attendrie
Répand un parfum enivrant.
O Canada, fils de la France,
Qui te couvrit de ses bienfaits,
Toi, notre amour, notre esperance,
Qui pourra t'oublier jamais ?"
O. CRÉMAZIE.

Cape Rosier, "the Scylla of the St. Lawrence," is 6 M. beyond Cape Gaspé, and is the S. portal of the St. Lawrence River, whose mouth at this point is 96 M. wide. At the end of the cape is a stone lighthouse tower, 112 ft. high, with a fixed light (visible 16 M.) and a fog-horn and cannon. The hamlets of Grand Grêve, Griffin's Cove, and Cape Rosier are in this vicinity, and are inhabited by French people, who are dependent on the fishing-establishment of Wm. Fruing & Co., of the Isle of Jersey.

"The coast between Cape Rosier and Cape Chatte is high and bold, free from dangers, and destitute of harbors," and is lined with a majestic wall of mountains composed of slate and graywacke. They are covered with forests, and afford successions of noble views, sometimes of amphitheatrical coves, sometimes of distant vistas of blue peaks up the long gorges of the rivers.

"How can it be that men inhabit this harsh, arid, rough, almost hateful country, which extends from Cape Chatte to the Gaspé Basin? One can scarcely imagine. Yet, as you see, here and there appear parcels of tilled land, houses scattered along the banks, and little churches at various points."

"The peninsula of Gaspé, the land's end of Canada towards the E., from its geological formation of shale and limestone, presenting their upturned edges toward the sea and dipping inland, forms long ranges of beetling cliffs running down to a narrow strip of beach, and affording no resting-place even to the fishermen, except where they have been cut down by streams, and present little coves and bays opening back into deep glens, affording a view of great rolling wooded ridges that stand rank after rank behind the great sea-cliff, though with many fine valleys between."

7 M. N. W. of Cape Rosier the settlement at *Griffin's Cove* is passed; and 5 M. farther on is *Fox River* (Cloridorme), a settlement of 500 persons, with one of the Isle-of-Jersey fishing-establishments, a large Catholic church, and a court-house. The cod and mackerel fisheries are followed in the adjacent waters, and large American fleets are often seen off the port. The grand highway from Quebec ends here, but a rugged road runs down to Gaspé in 17 M. The inhabitants are nearly all French. 16 M. farther W. is the haven called Great Pond, 24 M. beyond which is **Cape Magdelaine** (red-and-white revolving light, visible 15 - 20 M.) at the mouth of the River Magdelaine, the home of some of the wildest legends of this region.

"Where is the Canadian sailor, familiar with this coast, who has not heard of the plaintive sounds and doleful cries uttered by the *Braillard de la Magdelaine*? Where would you find a native seaman who would consent to spend a few days by himself in this locality, wherein a troubled spirit seeks to make known the torments it endures? Is it the soul of a shipwrecked mariner asking for Christian burial for its bones, or imploring the prayers of the church for its repose? Is it the voice of the murderer condemned to expiate his crimes on the very spot which witnessed its commission? . . . . For it is well known that Gaspé wreckers have not always contented themselves with robbery and pillage, but have sometimes sought concealment and impunity by making away with victims, — convinced that the tomb is silent and reveals not its secrets." The Abbé Casgrain attributes these weird sounds to the fate of a priest who refused to christen a child who afterwards was lost by dying unbaptized. The conscience-stricken priest faded away to a skeleton, and the sound of his moaning has ever since been heard off these dark shores. Another legend tells that a terrible shipwreck occurred at this point, and that the only soul that reached the shore was a baby boy, who lay wailing on the beach throughout the stormy night. "Where La Magdelaine runs into the Gulf, horizontal layers of limestone, fretted away all around their base by the action of the tides and waves, assume the most fantastic shapes, — here representing ruins of Gothic architecture, there forming hollow caverns into which the surf rolling produces a moaning sound, like an unquiet spirit seeking repose." The strange wailing which is heard at certain seasons along this shore is otherwise referred to the rush of the wind through the pine-trees on the cape, whose trunks grate together with a harsh creaking.

Pleurese Point is 12 M. from Cape Magdelaine, and is near the remote hamlet of Mont Louis. Lines of wild cliffs front the shore for the next 28 M., to Cape St. Anne, near which is the French Catholic village of *St. Anne des Monts*, which has 250 inhabitants and a consulate of Italy. The adjacent waters abound in mackerel, cod, halibut, and herring, and great quantities of salmon and trout are caught in the River St. Anne. The stately peaks of the * **St. Anne Mountains** are seen on the S., commencing 12 M. S. W. of Cape St. Anne and running in a S. W. course for 40 M., nearly parallel with the river and 20-25 M. inland. These mountains are the most lofty in Canada, and are visible for 80-90 M. at sea, in clear weather. The chief peak is 14 M. from Cape Chatte, and is 3,973 ft. high.

"All those who come to New France know well enough the mountains of Notre Dame, because the pilots and sailors being arrived at that part of the great river which is opposite to those high mountains, baptize ordinarily for sport the new passengers, if they do not turn aside by some present the inundation of this baptism which is made to flow plentifully on their heads." (LALEMANT, 1648.)

Cape Chatte is 15 M. N. W. of Cape St. Anne, and sustains a white flashing light which is visible for 18 M.

Cape Chatte was named in honor of the officer who sent out the expedition of 1603, under Pontgravé and Lescarbot. His style was Eymard de Chaste, Knight of Malta, Commander of Lormetan, Grand Master of the Order of St. Lazarus, and Governor of Dieppe.
Somewhere in this broad reach of the river occurred the chivalrous naval battle between the English war-vessel *Abigail* and the French ship of Emery de Caen (son of Lord de la Motte). The *Abigail* was commanded by Capt. Kirke, and was sailing against Tadousac, when she was attacked (June, 1629) by De Caen. A running fight of several hours ensued, until a fortunate cannon-shot from the *Abigail* cut away a mast on the French vessel and compelled her to surrender. The loss on each ship was considerable.

The reach of the St. Lawrence next entered is about 35 M. wide, and on the N. shore is *Point de Monts* (see page 233). It is 33 M. from Cape Chatte to Matane, in which the steamer passes the hamlets of Dalibaire and St. Felicité. In 1688 the Sieur Riverin established a sedentary fishery at Matane, devoted to the pursuit of codfish and whales. Sometimes as many as 50 whales were seen at one time from the shore. This branch of the fisheries has now greatly declined. **Matane** is a village of 300 inhabitants, devoted to fishing and lumbering, and is visited by Canadian citizens on account of the facilities for sea-bathing on the fine sandy beach. There is also good fishing for trout and salmon on the Matane River. The remarkable peaks called the *Paps of Matane* are to the S.W., in the great Gaspésian wilderness. In clear weather, when a few miles E. of Matane, and well out in the river, *Mt. Camille* may be seen, 40 M. distant, S. W. by W. ½ W., like an island on the remote horizon.

The shore is now low, rocky, and wooded, and runs S. W. 22 M. to *Petit Métis*, which was populated with Scottish families by its seigneur. 4 M. from this point is the station of *St. Octave*, on the Intercolonial Railway. **Métis** is a little way W., and is occupied by 250 French Catholics

11*

and Scotch Presbyterians. It has a long government wharf; and the people are engaged also in the pursuit of black whales, which are sought by schooners equipped with harpoons, lances, etc. N. of Métis, across the river, is the great peninsula of *Manicouagan*, at the mouth of the rivers Manicouagan and Outarde, abounding in cascades.

The steamship comes to off **Father Point**, where there is a lighthouse and telegraph-station (for news of the shipping), and a hamlet of 100 inhabitants. Here the outward-bound vessels discharge their pilots. Near this place are the hamlets of St. Luce and St. Donat, and at **St. Flavie**, 15 M. N. E., the Intercolonial Railway reaches the St. Lawrence (see page 70). A few miles S. E. is *Mt. Camille*, which is 2,036 ft. high. Father Point (*Pointe au Père*) was so named because the priest Henri Nouvel wintered there in 1663. Canada geese, ducks, and brant are killed here in great numbers during the long easterly storms.

St. Germain de Rimouski (*Hotel St. Laurent; Rimouski Hotel*) is 6 M. from Father Point, and is an incorporated city, an important station on the Intercolonial Railway, and the capital of Rimouski County and of a Roman-Catholic diocese. It has 1,200-1,500 inhabitants, with a handsome cathedral, a Catholic college, convent, episcopal palace, court-house, and other public buildings. The Canadian government has built a large and substantial wharf out to the deep channel, and a prosperous future is expected for the young city. Many summer visitors come to this place, attracted by its cool air and fine scenery.

Rimouski was founded in 1688, and in 1701 a missionary was sent here, who founded a parish which has now grown into a strong bishopric. "Rimouski, the future metropolis of the Lower St. Lawrence, a little city full of promise and furrowed already by the rails of the Intercolonial, will have its harbor of refuge where the great ocean-steamers will stop in passing, and will attract all the commerce of the immense region of the Metapedia, the future granary of our country." The Rimouski River is famous for its abundance of trout.

Barnaby Island is low and wooded, and 3 M. long, sheltering the harbor of Rimouski. It was known by its present name in 1629, when the fleet of the Kirkes assembled here. From 1723 to 1767 it was the home of a pious French hermit, who avoided women and passed most of his time in his oratory. Some say that he was wrecked off these shores, and vowed to Heaven to abide here if he was saved; others, that he had been disappointed in love. In his last hours he was visited by people from Rimouski, who found him dying, with his faithful dog licking his chilling face.

Bic Island was formerly called *Le Pic*, but was named *St. Jean* by Cartier, who entered its harbor in 1535, on the anniversary of the decapitation of St. John. It was included in the scheme of D'Avaugour and Vauban (in the 17th century) for the defence of Canada, and was intended to have been made an impregnable maritime fortress, sheltering a harbor of refuge for the French navy. But this Mont St. Michel of the New World never received its ramparts and artillery. The place was taken by Wolfe's British fleet of 200 ships, June 18, 1759; and when the *Trent* affair threatened to involve the United States and Great Britain in war, in 1861, British troops were landed at Bic from the ocean-steamship *Persia*, and were carried hence in sleighs to Rivière du Loup. Near this point is *L'Islet au Massacre*, where, according to tradition, 200 Micmac Indians were once surprised at night by the Iroquois, while slumbering in a cavern. The vengeful enemy silently filled the cave's mouth with dry wood and then set it on fire, shooting the unfortunate Micmacs as they leaped through the flames. 195 of the latter were slain, and it is claimed that their bones strewed the islet until within a few years.

TROIS PISTOLES. *Route 67.* 251

Ste.-Cécile du Bic (two boarding-houses) is a prosperous French village of 600 inhabitants, with a good harbor and a large and ugly church. It is 9 M. from Rimouski, and is surrounded by fine scenery. The Bay of Bic is "large enough to be majestic, small enough to be overlooked in one glance; a shore cut into deep notches, broken with flats, capes, and beaches; a background of mountains hewn prodigally from the world's material, like all the landscapes of our Canada." The Intercolonial Railway was carried through this region at a vast expense, and sweeps around the flank of the mountain, 200 ft. above the village, affording beautiful views. Wonderful mirages are seen off this port, and out towards Point de Monts. The highlands immediately over Bic are nearly 1,300 ft. high; and the bay receives two rivers, which descend in cascades and rapids from the neighboring gorges. As the steamship passes the lighthouse on *Biquette Island*, the remarkable and varied peaks of the mountains to the S. will attract the attention by their fantastic irregularity. Between Bic and Trois Pistoles, but not visible from the river, are the new French villages of St. Fabien, among the mountains; St. Matthieu, with its great quarries of red stone for the Intercolonial Railway; and St. Simon, near a pretty highland lake.

The rocky islets of Rosade are 2 M. off the shore of Notre Dame des Anges, and are decorated with a large cross, in memory of a marvellous escape. Some 30 years ago the St. Lawrence froze for 6 M. out from the parish, and many hundreds of seals were discovered on the ice. The people gathered and went out to slay these strange visitors, but the ice suddenly broke adrift and was whirled away down the stream. There appeared no hope of escape for the 40 men on the outer floes, which were now ½ M from the shore. Their families and friends bade them an eternal farewell, and the village priests, standing at the water's edge, gave them final absolution in preparation for the approaching catastrophe. But even while they were kneeling on the ice, a bold mariner launched a tiny skiff from the shore and crossed the widening belt of tumultuous waters, touched the crumbling edges of the floes, and, after many trips back and forth, succeeded in landing every one of the men upon the isle of Rosade. Thence they passed easily to the mainland, and afterwards erected a cross on Rosade, as a token of their gratitude.

Trois Pistoles (two good hotels) is a thriving village of 650 inhabitants, situated inside of Basque Island (5 M. from the Rosades), and near valuable deposits of limestone. There are two Catholic churches here, whose construction involved a litigious contest which is still remembered in Lower Canada. The beauty of the marine scenery in this vicinity has induced several Quebec gentlemen to build summer cottages here.

There is a well-founded tradition that in the year 1700 a traveller rode up to the bank of the then unsettled and unnamed river and asked the Norman fisherman, who was tending his nets near his rude hut, what he would charge to ferry him across. "Trois pistoles" (three ten-franc pieces), said the fisher. "What is the name of this river?" asked the traveller. "It has no name; it will be baptized at a later day." "Well, then," said the traveller, "name it *Trois Pistoles.*" The river is now famous for its fine trout-fishing.

"That portion of the St. Lawrence extending between the Saguenay River and Goose Island is about 20 M. wide. The spring tides rise and fall a distance of 18 ft. The water is salt, but clear and cold, and the channel very deep. Here may be seen abundantly the black seal, the white porpoise, and the black whale." The white porpoise yields an oil of the best quality, and its skin makes good leather.

The Gulf-Ports steamship does not stop between Father Point and Quebec, but the villages described in this itinerary may be visited from Quebec; those on the S. shore by railway, and St. Paul's Bay, Murray Bay, Rivière du Loup, and Rimouski by river-steamers. The N. shore from Cape Tourmente to the Saguenay is described in Route 72.

The vessel steams up by *Green Island*, which is 6 – 7 M. long, and shelters the large manufacturing village of *Isle Verte*, whence fine butter is sent to Quebec. On the r. is *Red Island*, with its tall stone lighthouse, off which is a lightship. **Cacouna** and Rivière du Loup (see Route 72) are next passed, on the l., and the vessel runs W. with the three steep islets called the **Brandy Pots** (*Pots-à-l'eau-de-vie*) on the r. The S. islet bears a fixed light; the N. islet is 150 ft. high, of vesiculated conglomerate in which almond-shaped bits of quartz are imbedded. In war-time merchant-ships wait off the Brandy Pots for their convoying frigates. N. of these islets is *Hare Island*, which is about 10 M. long, and has extensive salt marshes, on which herds of cattle are kept. On the l. are now seen the five remarkable islets called **The Pilgrims**, about 1½ M. from the S. shore and 4½ M. in aggregate length. The *Long Pilgrim* is 300 ft. high and partially wooded, and is marked by a lighthouse, 180 ft. above the river. The *Kamouraska Islands* are 6 M. farther W., and over them is seen the pretty village of **Kamouraska** (*Albion Hotel*), with its great Church of St. Louis and Congregational Convent. The river-water at this point is as salt as the sea, and the village was the chief summer resort on the St. Lawrence before Cacouna arose.

"Who does not know Kamouraska? Who does not know that it is a charming village, bright and picturesque, bathing its feet in the crystal of the waters of the river like a naiad, and coquettishly viewing the reflections of its two long ranges of white houses, ... so near the river that from all the windows the great waves may be contemplated and their grand voices heard? On all sides, except towards the S., the horizon extends as far as the eye can reach, and is only bounded by the vast blue curtain of the Laurentides. At the N. E. the eye rests on a group of verdant isles, like a handful of emeralds dropped by the angel of the sea. . . . . These isles are the favorite resort of the strangers who visit Kamouraska. There they fish, or bathe, or seek other amusements. *Le pique-nique* is much in vogue there, and the truest joys are felt."

St. Paschal (700 inhabitants) is 5 M. from Kamouraska, on the Grand Trunk Railway.

"Bel endroit, Saint-Paschal, par sa croupe onduleuse,
Ses coteaux, ses vallons, sa route sinueuse!
C'est la Suisse ou l'Auvergne avec leurs gais chalets,
Leurs monts, leurs près en pente et leurs jardins coquets."

Beyond Kamouraska the steamer passes Cape Diable, and on the N. shore, 22 M. distant, are the bold mountains about Murray Bay (see Route 72). On the level plains to the S. is seen the tall *Church of St. Denis*, with its attendant village; and beyond Point Orignaux is the village of *Rivière Ouelle*, famous for its porpoise-fisheries. Near this point is the quaint Casgrain manor-house, now over a century old.

This parish is named for Madam Houel, wife of Comptroller-General Houel, who was captured here by Indians in the 17th century. Near the beach is a rock which bears the plain impress of three snow-shoes, and formerly had the marks of human feet and hands. In 1690 the priest of Rivière Ouelle led his parishioners, and drove back the New-Englanders of Sir William Phipps's fleet. Back among the hills are the hamlets of *St. Onésime* and *St. Pacome.*

St. Anne de la Pocatière (two hotels) is a large and prosperous town, 72 M. below Quebec, with 3,000 inhabitants, a weekly paper (*La Gazette des Campagnes*), and a convent. "Nature has given to St. Anne charming shores, laden with foliage and with melody, ravishing points of view, and verdant thickets, fitted for places of meditation." *St. Anne's College* is a stately pile of buildings with pleasant surroundings and a sumptuous chapel. It has 30 professors (ecclesiastics) and 230 students, and is maintained in a high state of efficiency. The parks cover several acres, and the museum is well supplied. St. Anne's Agricultural School and Model-Farm is connected with the college, and has 5 professors (zoötechny, rural law, etc.). The view from the dome of the college is of great extent and beauty.

As the steamer passes St. Anne the frowning mass of Mt. Éboulements is seen on the N. shore. A few miles beyond St. Anne the hamlet of *St. Roch-des-Aulnaies* is passed, on the l., and still farther to the W. is *St. Jean-Port-Joli*, a pretty little village about which is laid the scene of De Gaspé's popular romance, "Les Anciens Canadiens." The Isle aux Coudres is far away towards the N. shore. The course is laid in by the islet called the *Stone Pillar*, on which there is a lighthouse, and 1½ M. farther W. is the insulated rock of the Wood Pillar. The large and prosperous village of **L'Islet** (1,000 inhabitants) is seen on the l. *Goose Island* is passed on the r., and is connected with Crane Island (*L'Isle aux Grues*) by a long alluvial meadow, which produces rich hay, the total length being 11 M. Fine sporting is enjoyed here in the spring and autumn, when great flocks of snipe, plover, and wild geese visit these shores for a breeding-place. There is a settlement of about 150 persons on Crane Island, whence are obtained noble views of Cape Tourmente.

During the French régime these islands (*Les Isles de Ste.-Marguerite*) were erected into a seigniory and granted to an officer of France. He built a massive stone house on Crane Island, and was afterwards kept there, in rigorous captivity, by Madame de Granville. She claimed that she was his sister, and that he was insane; but this report was doubted by the people of the S. shore, and the island was regarded with dread. She kept him in close durance for many years, until at last he died.

Beyond the S. shore village of *Cap St. Ignace* (400 inhabitants) the steamer passes *St. Thomas*, the capital of Montmagny County. This town has 1,650 inhabitants, and carries on a large local trade. The College Montmagny is located here, and there is also a convent and a large and conspicuous church. The broad white band of a cascade is seen at the foot of the cove, where the Rivière du Sud falls 30 ft. On the r., beyond St. Thomas, is seen a cluster of picturesque islets, over which the massive Cape Tourmente frowns.

> "At length they spy huge Tourmente, sullen-browed,
> Bathe his bald forehead in a passing cloud;
> The Titan of the lofty capes that gleam
> In long succession down the mighty stream;
> When, lo! Orleans emerges to the sight,
> And woods and meadows float in liquid light;
> Rude Nature doffs her savage mountain dress,
> And all her sternness melts to loveliness.
> On either hand stretch fields of richest green,
> With glittering village spires and groves between,
> And snow-white cots adorn the fertile plain."

Grosse Isle formerly appertained to the Ursulines, and is 2½ M. long. On its graywacke ledges is the great Quarantine of Canada, where emigrant-ships are detained until thoroughly inspected and purified. The island is a vast tomb, so many have been the emigrants who have reached these shores only to die, poisoned in the filthy and crowded ships, poorly fed and rarely ministered unto. The Quarantine-station is occupied by medical and police forces, and is under a rigid code of rules.

The next town is *Berthier*, an ancient French parish of 400 inhabitants, W. of which is Bellechasse Island, composed of high, steep, and bare graywacke rocks. On the N. are Reaux Island (150 ft. high) and Madame Island, both of which are covered with trees. *St. Valier* is beyond Bellechasse, and is a place of 200 inhabitants, near which large deposits of bog iron-ore have been found. The **Isle of Orleans** (see Route 71) is now approached, on the r., and over it is seen the peak of Mt. St. Anne. Nearly opposite St. John (on the Orleans shore) is *St. Michel*, a lumber-working town of 700 inhabitants, in whose spacious church are some paintings for which a high value is claimed: St. Clara, by *Murillo* (?); St. Jerome, *Boucher*; the Crucifixion, *Romanelli*; the Death of the Virgin, *Gouly*; St. Bruno, *Philippe de Champagne*; the Flagellation, *Chally*. 6 M. beyond St. Michel is **Beaumont**, a village of 600 inhabitants, opposite Patrick's Hole, on the Orleans shore. The settlements now grow thicker on either shore, and in about 6 M. the steamship passes the W. end of the island of Orleans, and opens the grandest ** view on the route. On the r. is the majestic Montmorenci Fall, on the l. the rugged heights of Point Levi and St. Joseph, and in front the stately cliffs of **Quebec**, crowned with batteries, and flowering into spires.

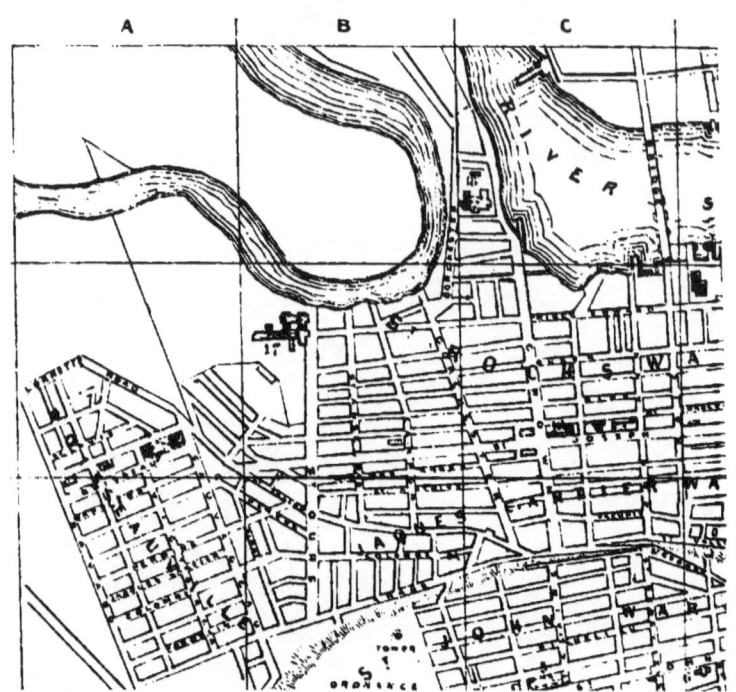

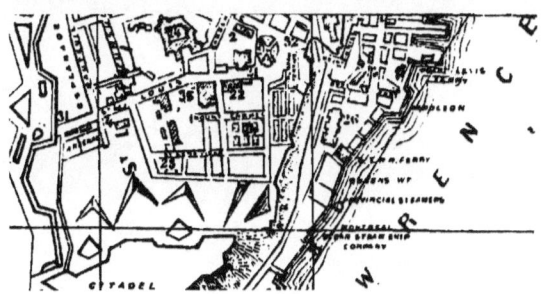

1 Catholic Cathedral   E.3.
2 Anglican   E.4.
3 Wesleyan Church... E.3.
4 Presbyterian "   E.4.
5 St. John (Cath.)   C.3.
6 St. Matthew   D.3.
7 St. Saveur   A.2.
8 St. Roch   C.2.
9 Notre Dame des
     Victoires.   E.4.
10 Archbishops Palace E.3.
11 Seminary   E.3.
12 Laval University.   E.3.
13 Hotel Dieu Convent E.3.
14 Ursuline   E.4.
15 Gray Sisters   D.3.
16 Congregational   C.2.
17 General Hospital   B.2.
18 Marine   C.1.
19 Morrin College   E.3.
20 Parliament House   E.3.
21 Court House   E.4.
22 Crown Lands Dep.   E.4.
23 High School   E.4.
24 Governor's Garden   E.4.
25 Custom House   E.3.
26 Champlain Market   E.4.
27 Jail   B.5.
28 Wolfe's Monument   B.5.
29 American Consulate E.3.
30 St. John's Gate   B.3.
31 St. Louis   D.4.
32 Prescott   E.E.4.
33 Hope   E.3.
34 Palace   E.3.

35 St. Louis Hotel   E.4.
36 Stadacona   E.3.
37 Jesuit Buildings   E.3.

68. Quebec.

Arrival. — If the traveller has much baggage, it is best to take a carriage or the hotel omnibus to the Upper Town. The *calèche* is not adapted for carrying luggage.

Hotels. — The *St. Louis Hotel is a large house near the Durham Terrace, kept by Willis Russell, an American gentleman. It accommodates 500 guests, and charges $ 2.50 – 3.50 a day. The Russell House is a large modern hotel, near the St. Louis, and under the same management. Its terms are lower than those of the St. Louis. The Albion Hotel is on Palace St., and charges $ 2.50 a day. Henchey's Hotel (on St. Anne St., opposite the Anglican Cathedral) is quiet and moderate, for gentlemen travelling *en garçon*. The Mountain-Hill House, on Mountain-Hill St., and Blanchard's Hotel, in the Lower Town, opposite Notre Dame des Victoires, are second-class houses, charging about $ 1.50 a day.

There are several good boarding-houses in the Upper Town, among which are those of the Misses Leonard, 3 St. Louis St.; Mrs. McDonell, 12 St. Louis St; Miss Lane, 44 St. Anne St.; Mrs. Boyce, 1 Garden St. Comfortable quarters may be obtained at these houses for about $ 10 a week.

Carriages in every variety may be procured at the livery-stables, and large numbers of them are kept at the stands near the St. Louis Hotel, in front of the Cathedral, and beyond St. John's Gate. The carriages in the Lower Town are less elegant and much less expensive than those within the walls. The rates for excursions in the suburbs in summer are from $ 3 to $ 4 for 1 – 3 persons (to Montmorenci Falls, Lorette, Cap Rouge, etc.). During the autumn the rates are reduced. The *calèche*-drivers of the Lower Town usually demand $ 2 for carrying 1 – 2 persons to the outer suburban resorts. The *calèche* is a singular and usually very shabby-looking vehicle, perched on two high wheels, with the driver sitting on a narrow ledge in front. It is drawn by a homely but hardy little horse, and is usually driven by a French Canadian, who urges the horse forward by the sharp dissyllabic cry, "*Marche-donc!*"

Horse-Cars run between St. Ours, St. Sauveur, and the Champlain Market, every 15 minutes, traversing St. Joseph, St. Paul, and St. Peter Sts. The fare is 5c.

Reading-Rooms. — The elegant library of the Quebec Literary and Historical Society (in Morrin College) is courteously opened to the visits of strangers. The Library of Parliament is also accessible, and is finely arranged. The *Institut Canadien* is at 11 St. John St., and the Y. M. C. Association has rooms at 24 Fabrique St. (near the Cathedral).

Post-Office at the corner of Buade and Du Fort Sts. According to the new rules of the Canadian postal service, stamps are not sold at the post-offices, but are kept on sale by the booksellers.

The most attractive shops are on Fabrique and St. John Sts., and in the vicinity of the French Cathedral.

Railways. — The Grand Trunk Railway has its terminal station at Point Levi, 317 M. from Portland, 425 M. from Boston, and 586 M. from New York. Passengers take the Grand Trunk ferry-steamer near the Champlain Market. The North Shore Railway is now being built from Quebec to Montreal along the N. shore of the St. Lawrence. The Quebec & Gosford Railway is of most primitive construction, and runs occasional trains from its terminal station in the Banlieue for 20 – 25 M. up the valley of the St. Charles.

Steamships. — The steamships of the Allan Line leave every Saturday for Liverpool (fares, $ 80, $ 70, and $ 25); also, once weekly in summer, to Glasgow (fares, $ 60 and $ 24). The Dominion Line sends a weekly steamship to Liverpool (fare, $ 60 and $ 24), and the Temperley Line despatches a fortnightly steamship to London (fares, $ 60 and $ 24). The vessels of the Quebec & Gulf Ports S. S. Co. leave every week for Father Point, Gaspé, Percé, the Bay of Chaleur, the Miramichi ports, Shediac, and Pictou (see page 239). Steamers for the Lower St. Lawrence (see Route 72) and for the Saguenay River (see Route 73) leave several times a week. The *Portneuf* leaves on Tuesday and Saturday for Cap Santé, Platon, Portneuf, St. Emelie, and St. Jean Deschaillons. The *Montmorenci* leaves semi-weekly for Chateau Richer, St. Famille, St. Anne, and Grande Rivière. Steamers run to the Isle of Orleans three times daily, and the Point-Levi ferry-boats cross the river every 15 minutes.

QUEBEC, "the Gibraltar of America," and the second city in the Dominion of Canada, is situated on a rocky promontory at the confluence of the St. Lawrence and St. Charles Rivers, 180 M. from Montreal, and over 400 M. from the Gulf of St. Lawrence. It has about 75,000 inhabitants, with 6 banks, 6 Masonic lodges, and numerous newspapers in the French and the English languages. The chief business of the city is in the handling and exportation of lumber, of which $5-7,000,000 worth is sent away annually. There are long lines of coves along the St. Lawrence shore, above the city, arranged for the reception and protection of the vast rafts which come down from the northern forests. A very considerable export trade in grain is also done here, and the various supplies of the populous counties to the N. and E. are drawn from this point. Shipbuilding is a leading industry, and many vessels of the largest size are launched every year from the shipyards on the St. Charles. Of late years several important manufactories have been established in the Lower Town, and the city is expected to derive great benefit from the convergence here of several lines of railway, connecting with the transatlantic steamships, and making it a depot of immigration and of freighting. The introduction of an abundant and powerful water supply from Lake St. Charles and the establishment of a fire-brigade and alarm-telegraph have preserved the city, during late years, from a recurrence of the terrible fires with which it was formerly scourged.

Quebec is built nearly in the form of a triangle, bounded by the two rivers and the Plains of Abraham, and is divided into the Upper Town and Lower Town, the former standing on an enwalled and strongly fortified bluff 350 ft. high, while the latter is built on the contracted strands between the cliffs and the rivers. The streets are narrow, crooked, and often very steep, and the houses are generally built of cut stone, in a style of severe simplicity. It is the most quaint, picturesque, and mediæval-looking city in America, and is surrounded by beautiful suburbs.

"Take mountain and plain, sinuous river, and broad, tranquil waters, stately ship and tiny boat, gentle hill and shady valley, bold headland and rich, fruitful fields, frowning battlement and cheerful villa, glittering dome and rural spire, flowery garden and sombre forest, — group them all into the choicest picture of ideal beauty your fancy can create, arch it over with a cloudless sky, light it up with a radiant sun, and lest the sheen should be too dazzling, hang a veil of lighted haze over all, to soften the lines and perfect the repose, — you will then have seen Quebec on this September morning." (ELIOT WARBURTON.)

"Quebec recalls Angoulême to my mind: in the upper city, stairways, narrow streets, ancient houses on the verge of the cliff; in the lower city, the new fortunes, commerce, workmen; — in both, many shops and much activity." (M. SAND.)

"The scenic beauty of Quebec has been the theme of general eulogy. The majestic appearance of Cape Diamond and the fortifications, — the cupolas and minarets, like those of an Eastern city, blazing and sparkling in the sun, — the loveliness of the panorama, — the noble basin, like a sheet of purest silver, in which might ride with safety a hundred sail of the line, — the graceful meandering of the river St. Charles, — the numerous village spires on either side of the St. Lawrence, — the fertile fields dotted with innumerable cottages, the abodes of a rich and moral peasantry, — the distant Falls of Montmorenci, — the park-like scenery of Point Levi, — the beauteous Isle of Orleans, — and more distant still, the frowning Cape Tourmente, and the lofty range of purple mountains of the most picturesque forms which bound the prospect, unite to form a *coup d'œil*, which, without exaggeration, is scarcely to be surpassed in any part of the world." (HAWKINS.)

"I rubbed my eyes to be sure that I was in the nineteenth century, and was not entering one of those portals which sometimes adorn the frontispiece of old black-letter volumes. I thought it would be a good place to read Froissart's Chronicles. It was such a reminiscence of the Middle Ages as Scott's novels.

"Too much has not been said about the scenery of Quebec. The fortifications of Cape Diamond are omnipresent. You travel 10, 20, 30 M. up or down the river's banks, you ramble 15 M. among the hills on either side, and then, when you have long since forgotten them, perchance slept on them by the way, at a turn of the road or of your body, there they are still, with their geometry against the sky..... No wonder if Jaques Cartier's pilot exclaimed in Norman-French, *Que bec!* ('What a peak!') when he saw this cape, as some suppose. Every modern traveller involuntarily uses a similar expression..... The view from Cape Diamond has been compared by European travellers with the most remarkable views of a similar kind in Europe, such as from Edinburgh Castle, Gibraltar, Cintra, and others, and preferred by many. A main peculiarity in this, compared with other views which I have beheld, is that it is from the ramparts of a fortified city, and not from a solitary and majestic river cape alone that this view is obtained..... I still remember the harbor far beneath me, sparkling like silver in the sun, — the answering headlands of Point Levi on the S. E., — the frowning Cape Tourmente abruptly bounding the seaward view far in the N. E., — the villages of Lorette and Charlesbourg on the N., — and farther W. the distant Val Cartier, sparkling with white cottages, hardly removed by distance through the clear air, — not to mention a few blue mountains along the horizon in that direction. You look out from the ramparts of the citadel beyond the frontiers of civilization. Yonder small group of hills, according to the guide-book, forms 'the portal of the wilds which are trodden only by the feet of the Indian hunters as far as Hudson's Bay.'" (THOREAU.)

"There is no city in America more famous in the annals of history than Quebec, and few on the continent of Europe more picturesquely situated. Whilst the surrounding scenery reminds one of the unrivalled views of the Bosphorus, the airy site of the citadel and town calls to mind Innspruck and Edinburgh. Quebec may be best described by supposing that an ancient Norman fortress of two centuries ago had been encased in amber, transported by magic to Canada, and placed on the summit of Cape Diamond."

"Quebec, at least for an American city, is certainly a very peculiar place. A military town, containing about 20,000 inhabitants; most compactly and permanently built, — stone its sole material; environed, as to its most important parts, by walls and gates, and defended by numerous heavy cannon; .... founded upon a rock, and in its highest parts overlooking a great extent of country; 3–400 miles from the ocean, in the midst of a great continent, and yet displaying fleets of foreign merchantmen in its fine, capacious bay, and showing all the bustle of a crowded seaport; its streets narrow, populous, and winding up and down almost mountainous declivities; situated in the latitude of the finest parts of Europe, exhibiting in its environs the beauty of a European capital, and yet in winter smarting with the cold of Siberia; governed by a people of different language and habits from the mass of the population, opposed in religion, and yet leaving that population without taxes, and in the enjoyment of every privilege, civil and religious: such are the prominent features which strike a stranger in the city of Quebec. A seat of ancient Dominion, — now hoary with the lapse of more than two centuries, — formerly the seat of a French empire in the west, — lost and won by the blood of gallant armies, and of illustrious commanders, — throned on a rock, and defended by all the proud defiance of war! Who could approach such a city without emotion! Who in Canada has not longed to cast his eyes on the water-girt rocks and towers of Quebec." (PROF. SILLIMAN; in 1820.)

"Few cities offer so many striking contrasts as Quebec. A fortress and a commercial city together, built upon the summit of a rock like the nest of an eagle, while her vessels are everywhere wrinkling the face of the ocean; an American city inhabited by French colonists, governed by England, and garrisoned by Scotch regiments; a city of the Middle Ages by most of its ancient institutions, while it is subject to all the combinations of modern constitutional government; a European city by its civilization and its habits of refinement, and still close by the remnants of the Indian tribes and the barren mts. of the North; a city with about the same latitude as Paris, while successively combining the torrid climate of southern regions with the severities of an hyperborean winter; a city at the same time Catholic and Protestant, where the labors of our (French) missions are still uninterrupted alongside of the undertakings of the Bible Society, and where the Jesuits, driven out of our own country, find refuge under the ægis of British Puritanism." (X. MARMIER'S *Lettres sur l'Amérique*, 1860.)

" Leaving the citadel, we are once more in the European Middle Ages. Gates and posterns, cranky steps that lead up to lofty, gabled houses, with sharp French roofs of burnished tin, like those of Liége; processions of the Host; altars decked with flowers; statues of the Virgin; sabots; blouses; and the scarlet of the British linesman,—all these are seen in narrow streets and markets that are graced with many a Cotentin lace cap, and all within 40 miles of the down-east, Yankee State of Maine. It is not far from New England to Old France..... There has been no dying out of the race among the French Canadians. They number twenty times the thousands that they did 100 years ago. The American soil has left their physical type, religion, language, and laws absolutely untouched. They herd together in their rambling villages, dance to the fiddle after mass on Sundays,—as gayly as once did their Norman sires,—and keep up the *fleur-de-lys* and the memory of Montcalm. More French than the French are the Lower Canada *habitans*. The pulse-beat of the continent finds no echo here." (SIR CHARLES DILKE.)

"Curious old Quebec! of all the cities of the continent of America the most quaint! It is a peak thickly populated! a gigantic rock, escarped, echeloned, and at the same time smoothed off to hold firmly on its summit the houses and castles, although according to the ordinary laws of matter they ought to fall off like a burden placed on a camel's back without a fastening. Yet the houses and castles hold there as if they were nailed down. At the foot of the rock some feet of land have been reclaimed from the river, and that is for the streets of the Lower Town. Quebec is a dried shred of the Middle Ages, hung high up near the North Pole, far from the beaten paths of the European tourists, ... a curiosity without parallel on this side of the ocean. We traversed each street as we would have turned the leaves of a book of engravings, containing a new painting on each page. ..... The locality ought to be scrupulously preserved antique. Let modern progress be carried elsewhere! When Quebec has taken the pains to go and perch herself away up near Hudson's Bay, it would be cruel and unfitting to dare to harass her with new ideas, and to speak of doing away with the narrow and tortuous streets that charm all travellers, in order to seek conformity with the fantastic ideas of comfort in vogue in the 19th century." (HENRY WARD BEECHER.)

" On l'a dit, Québec est un promontoire, c'est avant tout une forteresse remarquable. La citadelle s'élève au-dessus de la ville et mire dans les eaux du fleuve ses créneaux béants. Le voyageur s'étonne, après avoir admiré les bords verdoyants et fleuris du Saint-Laurent, les forêts aux puissantes ramures pleines de mystères et d'ombre, les riantes vallées pleines de bruits et de rayons, de rencontrer tout à coup cette ville qui semble venir d'Europe et qui serait moins étrange sur les bords du Rhin aux dramatiques légendes. Mais Québec n'est pas une ville où l'étranger vienne se distraire et chercher d'oubli un théâtre à grands luxes, à grands spectacles. .... C'est peut-être la seule ville du monde où les gens aient droit de se plaindre et où ils ne se plaignent pas. J'ai écrit que Québec est une forteresse remarquable; elle élève son front superbe et se cambre avec fierté dans sa robe de pierre. Elle a conservé un air des temps chevaleresques, elle a soutenu des sièges, elle a reçu son baptême du feu. En longeant ces vieux murs, en admirant cette forteresse élevée comme un nid d'aigle sur un roc sourcilleux, on se croirait dans une ville du moyen âge, au temps des factions et des guerres civiles, une de ces villes accoutumées aux bruits des armes, aux fanfares et aux hymnes guerriers, mais tout est silencieux dans la nuit sereine, et vous n'entendez même pas le pace cadencé d'une sentinelle. Dans cette ville et aux alentours, que d'événements ont été accompli! Quelle lutte pleine de poësie héroïque! Que de vicissitudes! et quel courage! En quelque lieu que vous alliez, à la basse-ville, sur le chemin Saint-Louis ou Sainte-Foye, sur les rives de la rivière Saint-Charles, tout respire un parfum historique, tout parle à vos yeux, tout a une voix qui exprime quelque chose de grand et de triste, et les pierres mêmes sont autour de vous comme les fantômes qui refléchissent le passé."

QUEBEC. *Route 68.* 259

The **Durham Terrace** is on the riverward edge of the Upper Town, and stands on the buttresses and platform formerly occupied by the Chateau of St. Louis, which was built by Champlain in 1620. The old Chateau was a massive stone structure, 200 ft. long, used for a fortress, prison, and governor's palace, and it stood until 1834, when it was ruined by fire. The terrace is 200 ft. above the river, and commands a * view of surpassing beauty. Immediately below are the sinuous streets of the Lower Town, with its wharves projecting into the stream. On one side are the lofty fortified bluffs of Point Levi, and on the other the St. Charles River winds away down its peaceful valley. The white houses of Beauport stretch off to the vicinity of the Montmorenci Falls, while beyond are seen the farms of L'Ange Gardien, extending towards the heights of St. Fereol. Vessels of all classes and sizes are anchored in the broad basin and the river, and the rich and verdant Isle of Orleans is in mid-stream below. Beyond and over all are the bold peaks of the Laurentian range, with Cape Tourmente towering over the river far in the distance. The Terrace is the favorite promenade of the citizens, and presents a pleasant scene in the late afternoon or on pleasant Sundays. At the upper end of the Terrace is a plain stone structure called the Old Chateau (built in 1779, for the British governors), which is now occupied by the Laval Normal and Model School (5 professors). In its gateway is a large stone bearing carvings of a Maltese cross and the date 1647, done by order of Gov. Montmagny, a Knight of Malta, in 1647.

" There is not in the world a nobler outlook than that from the Terrace at Quebec. You stand upon a rock overhanging city and river, and look down upon the guard-ships' masts. Acre upon acre of timber comes floating down the stream above the city, the Canadian boat-songs just reaching you upon the heights; and beneath you are fleets of great ships, English, German, French, and Dutch, embarking the timber from the floating docks. The Stars and Stripes are nowhere to be seen." (SIR CHARLES DILKE.)

" On a summer evening, when the Terrace is covered with loungers, and when Point Levi is sprinkled with lights and the Lower Town has illuminated its narrow streets and its long dormer-windows, while the lively murmur of business is ascending and the eye can discern the great shadows of the ships beating into port, the scene is one of marvellous animation. It is then, above all, that one is struck with the resemblance between Quebec and the European cities ; it might be called a city of France or Italy transplanted ; the physiognomy is the same, and daylight is needed to mark the alteration of features produced by the passage to America."

" At a later era, when, under the protection of the French kings, the Provinces had acquired the rudiments of military strength and power, the Castle of St. Louis was remarkable as having been the site whence the French governors exercised an immense sovereignty, extending from the Gulf of St. Lawrence, along the shores of that noble river, its magnificent lakes, and down the course of the Mississippi to its outlet below New Orleans. The banner which first streamed from the battlements of Quebec was displayed from a chain of forts which protected the settlements throughout this vast extent of country, keeping the English Colonies in constant alarm, and securing the fidelity of the Indian nations. During this period the council chamber of the castle was the scene of many a midnight vigil, many a long deliberation and deep-laid project, to free the continent from the intrusion of the ancient rival of France, and assert throughout the supremacy of the Gallic lily. At another period, subsequent to the surrender of Quebec to the British arms, and until the recognition of the independence of the United States, the extent of empire of which the Castle of Quebec was the principal seat comprehended the whole American continent north of Mexico." (HAWKINS.)

The **Anglican Cathedral** occupies the site of the ancient Recollet Convent and gardens, and is a plain and massive building, 135 ft. long, with a spire 152 ft. high. It was built by the British government in 1803–4, and received its superb communion-service, altar-cloths, and books as a present from King George III. There is a chime of 8 bells in the tower, which makes pleasant music on Sundays; and the windows are of rich stained glass. The interior is plain and the roof is supported on Corinthian pillars and pilasters, while over the chancel hang the old Crimean colors of the 69th Regiment of the British army. Under the altar lie the remains of Charles Lennox, Duke of Richmond, Lennox, and Aubigny, and Governor-General of Canada, who died of hydrophobia in 1819. There are numerous mural monuments in the cathedral, and in the chancel are the memorials to the early Anglican Bishops of Quebec, Jacob Mountain and Charles James Stewart. The former consists of a bust of the Bishop, alongside of which is a statue of Religion, both in relief, in white marble, on a background of black marble.

Dr. Mountain was in the presence of King George, when he expressed a doubt as to whom he should appoint as bishop of the new See of Quebec. Said the doctor, "If your Majesty had faith, there would be no difficulty." "How so?" said the king. Mountain answered, "If you had faith, you would say to this Mountain, Be thou removed into that See, and it would be done." It was.

Between the cathedral and the Durham Terrace is a pretty little park called the *Place d'Armes*, beyond which are the ruins of the court-house, which was recently destroyed by fire. Beyond the court-house (on St. Louis St.) is the *Masonic Hall*, opposite which are the old-time buildings of the St. Louis Hotel and the Commissariat and Crown Lands Departments. The latter is known as the *Kent House*, from the fact that Prince Edward, the Duke of Kent (father of Queen Victoria), dwelt here during his long sojourn at Quebec. Opposite the St. Louis Hotel is a quaint little building (now used as a barber-shop), in which Montcalm held his last council of war. St. Louis St. runs out through the ramparts, traversing a quiet and solidly built quarter, and is prolonged beyond the walls as the Grand Allée.

The * **Market Square** is near the centre of the Upper Town, and presents a curious and interesting appearance on market mornings, when the French peasantry bring in their farm-produce for sale.

"A few steps had brought them to the market-square in front of the cathedral, where a little belated traffic still lingered in the few old peasant-women hovering over baskets of such fruits and vegetables as had long been out of season in the States, and the housekeepers and servants cheapening these wares. A sentry moved mechanically up and down before the high portal of the Jesuit Barracks, over the arch of which were still the letters I. H. S. carved long ago on the keystone; and the ancient edifice itself, with its yellow stucco front and its grated windows, had every right to be a monastery turned barracks in France or Italy. A row of quaint stone houses — inns and shops — formed the upper side of the square, while the modern buildings of the Rue Fabrique on the lower side might serve very well for

QUEBEC. *Route 68.* 261

that show of improvement which deepens the sentiment of the neighboring antiquity and decay in Latin towns. As for the cathedral, which faced the convent from across the square, it was as cold and torpid a bit of Renaissance as could be found in Rome itself. A red-coated soldier or two passed through the square; three or four neat little French policemen lounged about in blue uniforms and flaring havelocks; some walnut-faced, blue-eyed old citizens and peasants sat upon the thresholds of the row of old houses and gazed dreamily through the smoke of their pipes at the slight stir and glitter of shopping about the fine stores of the Rue Fabrique. An air of serene disoccupation pervaded the place, with which the drivers of the long rows of calashes and carriages in front of the cathedral did not discord. Whenever a stray American wandered into the square, there was a wild flight of these drivers towards him, and his person was lost to sight amidst their pantomime. They did not try to underbid each other, and they were perfectly good-humored. As soon as he had made his choice, the rejected multitude returned to their places on the curbstone, pursuing the successful aspirant with inscrutable jokes as he drove off, while the horses went on munching the contents of their leathern head-bags, and tossing them into the air to shake down the lurking grains of corn." (HOWELLS's *A Chance Acquaintance.*)

On the W. side of this Square is the great pile of buildings which were begun in 1646 for the **Jesuits' College.** For some years this structure has been deserted, and in a state of dilapidation; and it is thought that it will be levelled and that on its site and in the spacious grounds adjacent will be founded a new market-house, although a movement has been made to erect here a superb Parliament Building for the Province of Quebec. The present structure is a parallelogram 224 ft. long by 200 ft. wide, and 3 stories high, whose quadrangle is entered by the lofty archway on the Square.

The Jesuits' College was founded in 1637, one year before Harvard College, and performed a noble work in its day. It was suspended in 1759 by Gen. Murray, who quartered his troops here, and in 1809 the property reverted to the crown, on the death of the last of the Jesuit Fathers. The buildings were used as barracks until the British armies evacuated Canada. " From this seat of piety and learning issued those dauntless missionaries, who made the Gospel known over a space of 600 leagues, and preached the Christian faith from the St. Lawrence to the Mississippi. In this pious work many suffered death in the most cruel form; all underwent danger and privation for a series of years, with a constancy and patience that must always command the wonder of the historian and the admiration of posterity."

The * **Basilica of Quebec** is on the E. side of the Market Square, and was known as the Cathedral of Notre Dame until 1874, when it was elevated by Pope Pius IX. to the rank of a basilica. It was founded in 1666 by Bishop Laval, and was destroyed by the bombardment from Wolfe's batteries in 1759. The present building dates from the era of the Conquest, and its exterior is quaint, irregular, and homely. From its towers the Angelus bells sound at 6 o'clock in the morning and 6 in the evening. The interior is heavy, but not unpleasing, and accommodates 4,000 persons. The High Altar is well adorned, and there are several chapels in the aisles. The most notable pictures in the Basilica are, ** the Crucifixion, by *Van Dyck* (" the Christ of the Cathedral"); the finest painting in Canada), on the first pillar l. of the altar; the Ecstasy of St. Paul, *Carlo Maratti;* the Annunciation, *Restout;* the Baptism of Christ, *Hallé;* the Pentecost, *Vignon;* Miracles of St. Anne, *Plamondon;* Angels waiting on Christ, *Restout* (in the choir); the Nativity, copy from *Annibale Caracci*; Holy Family, *Blanchard*.

The Basilica occupies the site of the ancient church of Notre Dame de la Recouvrance, built in 1633 by Champlain, in memory of the recovery of Canada by France. Within its walls are buried Bishops Laval and Plessis; Champlain, the heroic explorer, founder and first Governor of Quebec; and the Count de Frontenac, the fiery and chivalric Governor of Canada from 1688 to 1698. After his death his heart was enclosed in a leaden casket and sent to his widow, in France, but the proud countess refused to receive it, saying that she would not have a dead heart, which, while living, had not been hers. The noble lady (" the marvellously beautiful Anne de la Grand-Trianon, surnamed The Divine") was the friend of Madame de Sévigné, and was alienated from Frontenac on account of his love-affair with the brilliant Versaillaise, Madame de Montespan.

Most of the valuable paintings in the Basilica, and elsewhere in Canada, were bought in France at the epoch of the Revolution of 1793, when the churches and convents had been pillaged of their treasures of art. Many of them were purchased from their captors, and sent to the secure shores of New France.

Back of the Basilica, on Port Dauphin St., is the extensive palace of the Archbishop, surrounded by quiet gardens. To the E. are the Parliament Building and the Grand Battery.

The *Seminary of Quebec adjoins the Cathedral on the N., and covers several acres with its piles of quaint and rambling buildings and quiet and sequestered gardens. It is divided into *Le Grand Seminaire* and *Le Petit Seminaire*, the former being devoted to Roman-Catholic theology and the education of priests. The Minor Seminary is for the study of literature and science (for boys), and the course extends over nine years. Boarders pay $150 a year, exclusive of washing, music, and drawing. The students may be recognized in the streets by their peculiar uniform. The quadrangle, with its old and irregular buildings; the spotless neatness of the grounds; the massive walls and picturesquely outlined groupings, will claim the interest of the visitor.

"No such building could be seen anywhere save in Quebec, or in some ancient provincial town in Normandy. You ask for one of the gentlemen (priests), and you are introduced to his modest apartment, where you find him in his *soutane*, with all the polish, learning, and *bonhommie* of the nineteenth century." Visitors are conducted over the building in a courteous manner.

The **Seminary Chapel** has some fine paintings (beginning at the r. of the entrance): the Saviour and the Samaritan Woman, *La Gren.e;* the Virgin attended by Angels, *Dieu;* the Crucifixion, *Monet;* the Hermits of the Thebaid, *Guillot;* the Vision of St. Jerome, *D'Hullin;* the Ascension, *Philippe de Champagne;* the Burial of Christ, *Hulin;* (over the altar) the Flight into Egypt, *Vanloo;* above which is a picture of Angels, *Lebrun;* the Trance of St. Anthony, *Parrocel d'Avignon;* the Day of Pentecost, *P. de Champagne;* St. Peter freed from Prison, *De la Fosse;* The Baptism of Christ, *Hallé;* St. Jerome Writing, *J. B. Champagne;* Adoration of the Magi, *Bonnieu.* "The Chapel on the r. of the chief altar contains the relics of St. Clement; that on the l. the relics of St. Modestus."

The Seminary of Quebec was founded in 1663 by M de Laval, who endowed it with all his great wealth. The first buildings were erected in 1666, and the present Seminary is composed of edifices constructed at different dates since that time. In 1865 a large part of the quadrangle was burnt, but it has since been restored. In 1704 there were 54 teachers and students; in 1810 there were 110; and there are now over 400 (exclusive of the University students). "When we awake its departed shades, they rise upon us from their graves, in strange romantic guise. Men steeped in antique learning, pale with the close breath of the cloister, here spent the noon and evening of their lives, ruled savage hordes with a mild paternal sway, and stood

QUEBEC.   *Route 68.*   263

serene before the direst shapes of death. Men of courtly natures, heirs to the polish of a far-reaching ancestry, here with their dauntless hardihood put to shame the boldest sons of toil."

The * **Laval University** is between the Seminary gardens and the ramparts, and may be reached from St. Famille St. The main building is 280 ft. long and 5 stories high, is built of cut stone, and cost $ 225,000. The roof is a flat sanded platform, securely enrailed, where the students promenade and enjoy the grand * view of the city, the river, and the Laurentian Mts. Visitors are admitted to the collections of the University on application to the janitor. The reception-rooms contain the great picture of the Madonna of Quebec, a portrait of Pius IX., by *Pasqualoni*, and other paintings. The large hall of convocation has seats for 2,000, with galleries for ladies. The chemical laboratory is a fire-proof chamber, modelled after that of King's College, London; and the dissecting-room is spacious and well arranged. The * mineral museum was prepared by the late Abbé Haüy, an eminent scientist, and contains specimens of the stones, ores, and minerals of Canada, with a rare and valuable collection of crystals. It fills a long series of apartments, from which the visitor is ushered into the ethnological and zoölogical cabinets. Here are a great number of Indian remains, implements, and weapons, and other Huron antiquities; with prepared specimens of Canadian animals and fish. The Library contains 70,000 volumes (about half of which are French), arranged in two spacious halls, from whose windows delightful views are obtained. The * *Picture-Gallery* has lately been opened to the public, and is the richest in Canada. The works are mostly copies from the old masters, though there are several undoubted originals. It is by far the finest gallery N. of New York, and should be carefully studied. The visitor should also see the brilliant collection of Canadian birds; and the costly philosophical and medical apparatus, imported from Paris. The extensive dormitories occupy substantial stone buildings near the University, over the gardens.

The Seminary was founded in 1663 by François de Montmorenci Laval, first Bishop of Quebec, and has been the central power of the Catholic Church in this Province for over two centuries. The Laval University was founded in 1852, and has had the privileges of a Catholic University accorded to it by Pope Pius IX. The processes of study are modelled on those of the University of Louvain. The department of arts has 14 professors, the law has 6, divinity has 5, and medicine has 8. There are also 24 professors in the Minor Seminary.

The **Parliament Building** is on the site of Champlain's fort and the old Episcopal Palace, and is an extensive but plain building, whose glory has departed since the decapitalization of Quebec. The Legislative Council of the Province meets in a pleasant hall, upholstered and carpeted in crimson, with a very large throne, over which is a canopy surmounted by the arms of the United Kingdom. There are spacious galleries for visitors. The hall of the House of Assembly is on the front of the building, and is upholstered in green. Back of the speaker's chair is a line of Corinthian pilasters upholding a pediment on which are the Royal Arms. The * *Li-*

brary occupies a large and quiet apartment on the first floor, and is rich in French-Canadian literature. It also has copies of the costly volumes of Audubon's "Birds of America," Dugdale's "*Monasticon Anglicanum*," "The Antiquities of Italy," and the "*Acta Sanctorum*" (54 volumes, in vellum).

Mountain-Hill St. descends by the place of the Prescott Gate, to the Lower Town, winding down the slope of the cliff. On the r., about ⅓ of the way down, are the * **Champlain Steps**, or Côte la Montagne, a steep, crowded, and picturesque stairway leading down to Notre Dame des Victoires (see page 271). Near the foot of the steps is a grating, over the place where the remains of Champlain were recently found, in the vault of an ancient chapel. The Côte la Montagne has reminded one author of Naples and Trieste, another of Venice and Trieste, and another of Malta.

The new **Post-Office** is a handsome stone building at the corner of Buade and Du Fort Sts. In its front wall is a figure of a dog, carved in the stone and gilded, under which is the inscription:—

"Je suis un chien qui ronge l'os ;   ("I am a dog gnawing a bone.
En le rongeant je prend mon repos.   While I gnaw I take my repose.
Un temps viendra qui n'est pas venu   The time will come, though not yet,
Que je mordrais qui m'aura mordu."   When I will bite him who now bites me.")

This lampoon was aimed at the Intendant Bigot by M. Philibert, who had suffered wrong from him, but soon after the carved stone had been put into the front of Philibert's house, that gentleman was assassinated by an officer of the garrison. The murderer exchanged into the East Indian army, but was pursued by Philibert's brother, and was killed, at Pondicherry, after a severe conflict.

The Post-Office occupies the site of the Grand Place of the early French town, on which encamped the Huron tribe, sheltered by the fort from the attacks of the pitiless Iroquois. Here afterwards lived the beautiful Miss Prentice, with whom Nelson fell in love, so that he had to be forced on board of his ship to get him away. "How many changes would have ensued on the map of Europe ! how many new horizons in history, if Nelson had deserted the naval service of his country in 1782 ! Without doubt, Napoleon would have given law to the entire world. His supremacy on the sea would have consolidated his rule over the European continent ; and that because an amorous young naval officer was seized by a passion for a bewitching Canadian girl !" Near this place the Duke of Clarence, then a subaltern of the fleet, but afterwards King William IV. of England, followed a young lady home in an unseemly manner, and was caught by her father and very soundly horsewhipped.

The * **Ursuline Convent** is entered from Garden St., and is a spacious pile of buildings, commenced in 1686, and covering 7 acres with its gardens and offices. There are 40 nuns, who are devoted to teaching girls, and also to working in embroidery, painting, and fancy articles. The parlors and chapel may be visited by permission of the chaplain (whose office is adjacent); and in the latter are some valuable paintings : * Christ at the Pharisee's House, by *Philippe de Champagne;* Saints Nonus and Pelagius, *Prudhomme;* the Saviour Preaching, *P. de Champagne;* the Miraculous Draught of Fish, *Le Dieu de Jouvenet;* Captives at Algiers, *Restout;* St.

Peter, *Spanish School;* and several others. In the shrines are relics of St. Clement Martyr, and other saints from the Roman catacombs. Within a grave made by a shell which burst in this chapel during the bombardment of 1759 is buried "the High and Mighty Lord, Louis Joseph, Marquis of Montcalm," and over his remains is the inscription, "Honneur à Montcalm! Le destin en lui dérobant de la victoire l'a récompensé par une mort glorieuse." Montcalm's skull is carefully preserved under glass, and is shown as an object worthy of great veneration.

The first Superior of the Ursuline Convent was Mother Marie de l'Incarnation, who was "revered as the St. Teresa of her time." She mastered the Huron and Algonquin languages, and her letters to France form one of the most valuable records of the early days of Canada. The convent was founded in 1639, when the first abbess landed in Quebec amid the salutes of the castle-batteries ; and the special work of the nuns was that of educating the Indian girls. The convent was burnt down in 1650, and again in 1686, when the Ursulines were sheltered by the Hôpitaliéres. The Archbishop has recently ordered that the term of profession shall be for seven years, instead of for life.

Morrin College occupies a massive stone building at the corner of St. Anne and Stanislas Sts., and is the only non-Episcopal Protestant college in the Province. It was founded by Dr. Morrin, and has 5 professors, but has had but little success as an educational institution. The building was erected by the Government in 1810, for a prison; and occupied the site of an ancient fort of Champlain's era. It was used as a prison until the new Penitentiary was built, on the Plains of Abraham, and in the N. wing are the "sombre corridors that not long ago resounded with the steps of the jailers, and the narrow cells that are never enlivened by a ray of light."

The * Library of the Quebec Literary and Historical Society is in the N. wing of Morrin College, and contains a rare collection of books relating to Canadian history and science, in the French and English languages. This society is renowned for its valuable researches in the annals of the old St. Lawrence Provinces, and has published numerous volumes of records. It includes in its membership the leading literati of Eastern Canada. There is a small but interesting museum connected with the library-hall.

St. Andrew's Church, with its school and manse, occupy the triangle at the intersection of St. Anne and Stanislas Sts. It is a low, quaint building, erected in 1809 on ground granted by Sir James Craig. Previously, from the time of the Conquest of Canada, the Scottish Presbyterians had worshipped in the Jesuits' College. The *Wesleyan Church* is a comfortable modern building, just below Morrin College; beyond which, on Dauphin St., is the chapel of the Congregationalists (Roman Catholic). At the corner of St. John and Palace Sts. (second story) is a statue of Wolfe, which is nearly a century old, and bears such a relation to Quebec as does the Mannikin to Brussels. It was once stolen at night by some roystering naval officers, and carried off to Barbadoes, whence it was returned many months after, enclosed in a coffin.

The *Hôtel-Dieu Convent and Hospital is the most extensive pile of buildings in Quebec, and is situated on Palace St. (r. side) and the Rampart. E. of the long ranges of buildings (in which 650 sick persons can be accommodated) are pleasant and retired gardens. The convent-church is entered from Charlevoix St., and contains valuable pictures: the Nativity, by *Stella*; the Virgin and Child, *Coypel*; the Vision of St. Teresa, *Menageot*; St. Bruno in Meditation, *Le Sueur* (called "the Raphael of France"); the *Praying Monk, by *Zurbaran* (undoubted); and fine copies of the Twelve Apostles, by *Raphael*, and the Descent from the Cross, by *Rubens* (over the high altar).

The Hôtel Dieu was founded by the Duchesse d'Aguillon (niece of Cardinal Richelieu) in 1639. In 1654 one of the present buildings was erected, and most of it was built during the 17th century, while Talon, Baron des Islets, completed it in 1762. There are 30-40 cloistered nuns of the order of the Hôpitalières, and the hospital is open freely to the sick and infirm poor of whatever sect, with attendance by the best doctors of the city. The singing of the nuns during the Sunday services will interest the visitor.

The most precious relic in the Hôtel-Dieu is a silver bust (in life size) of Brébeuf, in whose base is preserved the skull of that heroic martyr. Jean de Brébeuf, a Norman Jesuit of noble blood, arrived at Quebec with Champlain in 1633, and went to the Huron country the next year. Here he had frequent celestial visions, and labored successfully in the work of converting the nation. He often said: " *Sentio me vehementer impelli ad moriendum pro Christo* "; and his wish was gratified when his mission-town of St. Ignace was stormed by the Iroquois (in 1649). He was bound to a stake and scorched from head to foot; the savages cut away his lower lip, and thrust a red-hot iron down his throat; hung around his neck a necklace of red-hot collars ("but the indomitable priest stood like a rock"); poured boiling water over his head and face, in demoniac mockery of baptism; cut strips of flesh from his limbs, and ate them before his eyes; scalped him; cut open his breast, and drank his living blood; filled his eyes with live coals; and after four hours of torture, a chief tore out his heart and devoured it. "Thus died Jean de Brébeuf, the founder of the Huron mission, its truest hero, and its greatest martyr. He came of a noble race, — the same, it is said, from which sprang the English Earls of Arundel; but never had the mailed barons of his line confronted a fate so appalling with more prodigious a constancy. To the last he refused to flinch, and ' his death was the astonishment of his murderers.'" The delicate and slender Lalemant, Brébeuf's colleague on the mission, was tortured for seventeen hours, with the most refined and exquisite varieties of torment. "It was said that, at times, he seemed beside himself; then, rallying, with hands uplifted, he offered his sufferings to Heaven as a sacrifice." The bones of Lalemant are preserved at the Hôtel Dieu.

Around the Ramparts.

*The Citadel is an immense and powerful fortification, covering 40 acres of ground, and is situated on the summit of Cape Diamond (so called from the glittering crystals found in the vicinity), which is said to be "the coldest place in the British Empire." Since the evacuation of Canada by the Imperial troops, the Citadel has been garrisoned by Provincial volunteers, and visitors are usually permitted to pass around the walls under the escort of a soldier. The **view from the most northerly bastion (which contains an immense Armstrong gun) surpasses that from the Durham Terrace, and is one of the most magnificent in the world. The

St. Charles is seen winding through a beautiful undulating plain, and the spires of Beauport, Charlesbourg, and Lorette, with the white cottages around them, form pleasing features in the landscape. On the S. of the parade are the officers' quarters and the bomb-proof hospital, while barracks and magazines are seen in advance. The armory contains a great number of military curiosities, but is not always accessible to visitors. The Citadel is separated from the town by a broad glacis, which is broken by three ravelins; and the wall on that side contains a line of casemated barracks. The entrance to the Citadel is by way of a winding road which leads in from St. Louis St. through the slope of the glacis, and enters first the outer ditch of the ravelin, beyond the strong Chain Gate. Thence it passes, always under the mouths of cannon, into the main ditch, which is faced with masonry, and at this point opens into a narrow parade, overlooked by the retiring angles of the bastion. The curious iron-work of the Chain Gate being passed, the visitor finds himself in an open triangular parade, under the loopholes of the Dalhousie Bastion.

" Such structures carry us back to the Middle Ages, the siege of Jerusalem, and St. Jean d'Acre, and the days of the Buccaniers. In the armory of the Citadel they showed me a clumsy implement, long since useless, which they called a Lombard gun. I thought that their whole Citadel was such a Lombard gun, fit object for the museums of the curious..... Silliman states that 'the cold is so intense in the winter nights, particularly on Cape Diamond, that the sentinels cannot stand it more than one hour, and are relieved at the expiration of that time; and even, as it is said, at much shorter intervals, in case of the most extreme cold.' I shall never again wake up in a colder night than usual, but I shall think how rapidly the sentinels are relieving one another on the walls of Quebec, their quicksilver being all frozen, as if apprehensive that some hostile Wolfe may even then be scaling the Heights of Abraham, or some persevering Arnold about to issue from the wilderness; some Malay or Japanese, perchance, coming round by the N. W. coast, have chosen that moment to assault the Citadel. Why I should as soon expect to see the sentinels still relieving one another on the walls of Nineveh, which have so long been buried to the world. What a troublesome thing a wall is! I thought it was to defend me, and not I it. Of course, if they had no walls they would not need to have any sentinels." (THOREAU.)

The Citadel was formerly connected with the Artillery Barracks, at the farther end of the city, by a bomb-proof covered way 1,837 yards long. These fortifications are 345 feet above the river, and considerably higher than the Upper Town. The rock on which they are founded is of dark slate, in which are limpid quartz-crystals.

The picturesque walls of Quebec are of no defensive value since the modern improvements in gunnery; and even the Citadel could not prevent dangerous approaches or a bombardment of the city. Skilful military engineers have therefore laid out a more extensive system of modern fortifications, including lines of powerful detached forts on the heights of Point Levi, and at Sillery. The former were begun in 1867, and are nearly completed; but the Sillery forts are not yet commenced.

The spirit of utilitarianism, which has levelled the walls of Frankfort and Vienna and is menacing Boston Common, has been attacking the ramparts of Quebec for many years. The people of the Upper Town and the extra-mural wards are doubtless much incommoded by this broad wall of separation, which has also become useless in a military point of view. However much it may be deplored by antiquarians and men of culture, the day is at hand when the mediæval fortifications of Quebec will be sacrificed to the spirit of the times. There are not wanting reverent American Ruskins to cry out against such demolition, but the wishes of the indigenous population will probably prevail against these ideas. Already the picturesque old gates are gone. The St. Louis and Prescott Gates were taken down in 1871, and the Palace and Hope Gates were removed in 1873.

The **Esplanade** extends to the r. from the St. Louis Gate (within), and the tourist is recommended to walk along the ramparts to St. John's Gate, viewing the deep fosse, the massive outworks, and the antiquated ordnance at the embrasures. On the r. are the Stadacona Club, the Congregational (Catholic) Church, and the National School; and Montcalm's Ward is on the l. * **St. John's Gate** is the only remaining gate of the city, and is a strong and graceful structure which was erected in 1869. While rallying his soldiers outside of this gate, the Marquis de Montcalm was mortally wounded; and Col. Brown (of Massachusetts) attacked this point while Arnold and Montgomery were fighting in the Lower Town. To the l. is St. John's Ward (see page 269); and the road to St. Foy passes below. The ramparts must be left at this point, and D'Auteuil and St. Helene Sts. follow their course by the *Artillery Barracks*, amid fine grounds at the S. W. angle of the fortifications. The French garrison erected the most important of these buildings (600 ft. long) in 1750, and the British Government has since made large additions; but the barracks are now unoccupied and are closed up. On and near St. Helene St. are several churches, — St. Patrick's (Irish Catholic), Trinity (Anglican), the Baptist, and the Congregational.

After crossing the wide and unsightly gap made by the removal of the Palace Gate, the rambler may follow the course of the walls from the Hôtel Dieu (see page 266) to the Parliament Building. They occupy the crest of the cliff, and command fine views over the two rivers and the Isle of Orleans and Laurentian Mts. The walls are thin and low, but are furnished with lines of loopholes and with bastions for artillery. The walk takes an easterly course beyond the angle of the convent-buildings, and passes between the battlements and the high walls of the Hôtel-Dieu gardens for nearly 500 ft.

The streets which intersect the Rampart beyond this point are of a quaint and pleasing character. One of them is thus described by Howells: "The thresholds and doorsteps were covered with the neatest and brightest oilcloth; the wooden sidewalk was very clean, like the steep, roughly paved street itself; and at the foot of the hill down which it sloped was a breadth of the city wall, pierced for musketry, and, past the corner of one of the houses, the half-length of cannon showing. It had all the charm of those ancient streets, dear to Old-World travel, in which the past and present, decay and repair, peace and war, have made friends in an effect that not only wins the eye, but, however illogically, touches the heart; and over the top of the wall it had a stretch of landscape as I know not what European street can command: the St. Lawrence, blue and wide; a bit of the white village of Beauport on its bank; then a vast breadth of pale green, upward-sloping meadows; then the purple heights; and the hazy heaven above them."

Since Prescott Gate fell, there was "nothing left so picturesque and characteristic as Hope Gate, and I doubt if anywhere in Europe there is a more mediæval-looking bit of military architecture. The heavy stone gateway is black with age, and the gate, which has probably never been closed in our century, is of massive frame, set thick with mighty bolts and spikes. The wall here sweeps along the brow of the crag on which the city is built, and a steep street drops down, by stone-parapeted curves and angles from the Upper to the Lower Town, where, in 1775, nothing but a narrow lane bordered the St. Lawrence. A considerable breadth of land has since been won from the river, and several streets and many piers now stretch between this alley and the water; but the old Sault au Matelot still crouches and creeps along under the shelter of the city wall and the overhanging rock, which is thickly bearded with weeds and grass, and trickles with abundant moisture. It must be an ice-pit in winter, and I should think it the last spot on the continent for the summer to find; but when the summer has at last found it, the old Sault au Matelot puts on a vagabond air of Southern leisure and abandon, not to be matched anywhere out of Italy. Looking from that jutting rock near Hope Gate, behind which the defeated Americans took refuge from the fire of their enemies, the vista is almost unique for a certain scenic squalor and gypsy luxury of color: sag-roofed barns and stables, weak-backed and sunken-chested workshops of every sort lounge along in tumble-down succession, and lean up against the cliff in every imaginable posture of worthlessness and decrepitude; light wooden galleries cross to them from the second stories of the houses which look back on the alley; and over these galleries flutters, from a labyrinth of clothes-lines, a variety of bright-colored garments of all ages, sexes, and conditions; while the footway underneath abounds in gossiping women, smoking men, idle poultry, cats, children, and large indolent Newfoundland dogs." (HOWELLS'S *A Chance Acquaintance*.)

Passing the ends of these quiet streets, and crossing the gap caused by the removal of Hope Gate, the Rampart promenade turns to the S., by the immense block of the Laval University (see page 263) and its concealed gardens. The course is now to the S., and soon reaches the * **Grand Battery**, where 22 32-pounders command the river, and from whose terrace a pleasing view may be obtained. The visitor is then obliged to leave the walls near the Parliament Building (see page 263) and the site of the Prescott Gate. A short détour leads out again to the Durham Terrace (see page 259). Des Carrières St. runs S. from the Place d'Armes to the *Governor's Garden*, a pleasant summer-evening resort, with a monument 65 ft. high, erected in 1827 to the memory of Wolfe and Montcalm, and bearing the elegant and classic inscription:

> MORTEM. VIRTUS. COMMUNEM.
> FAMAM. HISTORIA.
> MONUMENTUM. POSTERITAS.
> DEDIT.

In the lower garden is a battery which commands the harbor. Des Carrières St. leads to the inner glacis of the Citadel, and by turning to the r. on St. Denis St., its northern outworks and approaches may be seen. Passing a cluster of barracks on the r., the *Chalmers Church* is reached. This is a symmetrical Gothic building occupied by the Presbyterians, and its services have all the peculiarities of the old Scottish church. Beyond this point is St. Louis St., whence the circuit of the walls was begun.

The Montcalm and St. John Wards extend W. on the plateau, from the city-walls to the line of the Martello Towers. The population is mostly French, and the quarter is entered by passing down St. John St. and through St. John's Gate. Glacis St. leads to the r., just beyond the walls, to the *Convent of the Gray Sisters*, which has a lofty and elegant chapel. There are about 70 nuns, whose lives are devoted to teaching and to visiting the sick. This building shelters 136 orphans and infirm persons, and the sisters teach 700 female children. It overlooks the St. Charles valley, commanding fine views. Just above the nunnery is the Convent of the Christian Brothers, facing on the glacis of the rampart. A short distance out St. John St. is St. Matthew's Church (Episcopal); beyond which is the stately *Church of St. John* (Catholic), whose twin spires are seen for many leagues to the N. and W. The interior is lofty and light, and contains 12 copies from famous European paintings, executed by *Plamondon*, a meritorious Canadian artist. Claire-Fontaine St. leads S. from this church to the Grande Allée, passing just inside the line of the Martello Towers; and Sutherland St., leading into the Lower Town, is a little way beyond. The St. Foy toll-gate is about ½ M. from St. John's Church.

" Above St. John's Gate, at the end of the street of that name, devoted entirely to business, there is at sunset one of the most beautiful views imaginable. The river St. Charles, gambolling, as it were, in the rays of the departing luminary, the light still lingering on the spires of Lorette and Charlesbourg, until it fades away beyond the lofty mountains of Bonhomme and Tonnonthuan, presents an evening scene of gorgeous and surpassing splendor." (HAWKINS.)

" A sunset seen from the heights above the wide valley of the St. Charles, bathing in tender light the long undulating lines of remote hills, and transfiguring with glory the great chain of the Laurentides, is a sight of beauty to remain in the mind forever." (MARSHALL.)

The **Montcalm Ward** may also be reached by passing out St. Louis St., through the intricate and formidable lines of ravelins and redoubts near the site of the St. Louis Gate. On the r. is the skating-rink, beyond which are the pleasant borders of the Grand Allée. The Convent of the Good Shepherd is in this ward, and has, in its church, a fine copy of Murillo's "Conception," by Plamondon. There are 74 nuns here, 90 penitents, and 500 girl-students. The dark and heavy mediæval structure on the Grand Allée was built for the *Canada Military Asylum*, to take care of the widows and orphans of British soldiers who died on the Canadian stations. Near the corner of De Salaberry St. is *St. Bridget's Asylum*, connected with St. Patrick's Church. The *Ladies' Protestant Home* is nearly opposite, and is a handsome building of white brick, where 70 old men and young girls are kept from want by the bounty of the ladies of Quebec.

The **Martello Towers** are four in number, and were built outside the extra-mural wards in order to protect them and to occupy the line of heights. They were erected in 1807-12, at an expense of $60,000, and are arranged for the reception of 7 guns each. They are circular in form, and have walls 13 ft. thick toward the country, while on the other side they are 7 ft. thick. The new Jail is about ½ M. in advance of the towers, and is a massive stone building, with walls pierced for musketry. Near this point (turning to the l. from the Grand Allée beyond the toll-gate), and on the edge of the Plains of Abraham (extending to the S.), is a monument consisting of a tall column, decked with trophies, and rising from a square base, on which is the inscription:

HERE DIED
WOLFE
VICTORIOUS.
SEPT. 13.
1759.

"The horror of the night, the precipice scaled by Wolfe, the empire he with a handful of men added to England, and the glorious catastrophe of contentedly terminating life where his fame began. . . . . Ancient story may be ransacked, and ostentatious philosophy thrown into the account, before an episode can be found to rank with Wolfe's." (WILLIAM PITT.)

The Lower Town.

The most picturesque approach from the Upper to the Lower Town is by the *Champlain Steps* (see page 264). This route leads to the busiest and most crowded part of the old river wards, and to the long lines of steamboat wharves. **Notre Dame des Victoires** is in the market square in the Lower Town, and is a plain old structure of stone, built on the site of Champlain's residence. It was erected in 1690, and was called Notre Dame *des Victoires* to commemorate the deliverance of the city from the English attacks of 1690 and 1711, in honor of which an annual religious feast was instituted. A prophecy was made by a nun that the church would be destroyed by the conquering British; and in 1759 it was burned during the bombardment from Wolfe's batteries. S. of Notre Dame is the spacious *Champlain Market*, near an open square on whose water-front the river-steamers land. The narrow Champlain St. may be followed to the S., under Cape Diamond and by the point where Montgomery fell, to the great timber-coves above.

St. Peter St. runs N. between the cliffs and the river, and is the seat of the chief trade of the city, containing numerous banks, public offices, and wholesale houses. The buildings are of the prevalent gray stone, and are massive and generally plain. The parallel lane at the foot of the cliff is the scene of the final discomfiture of the American assault in 1775. It is named *Sault au Matelot*, to commemorate the leap of a dog from the cliff above, near the Grand Battery. Leadenhall St. leads off on the r. to the great piers of Pointe à Garcy and to the imposing classic building of the ***Custom-House**, which is at the confluence of the St. Lawrence and St. Charles Rivers. *St. Paul St.* runs W. from near the end of St. Peter St., along the narrow strip between the St. Charles and the northern cliffs, and passes the roads ascending to the Hope and Palace Gates.

The *Queen's Fuel-Yard* (l. side) is beyond the Palace Market, and occupies the site of an immense range of buildings erected by M. Begon, one of the later Royal Intendants of New France. Here also lived Bigot in all the feudal splendor of the old French *noblesse*, on the revenues which he extorted from the oppressed Province. In 1775 the palace was captured by Arnold's Virginia riflemen, who so greatly annoyed the garrison that the buildings were set on fire and consumed by shells from the batteries of the Upper Town.

St. Paul St. is prolonged by *St. Joseph St.*, the main thoroughfare of this quarter, and the boundary between the Jaques Cartier and St. Roch Wards. The latter is occupied chiefly by manufactories and shipyards (on the shores of the St. Charles); and the narrow and plank-paved streets of Jaques Cartier, toward the northern walls, are filled with quaint little houses and interesting *genre* views about the homes of the French-Canadian artisans. **St. Roch's Church** is a very spacious building, with broad interior galleries, and contains several religious paintings. The *Convent of Notre Dame* is opposite St. Roch's, and has 70 nuns (black costume), who teach 725 children.

The * **Marine Hospital** is a large and imposing modern building, in Ionic architecture, situated in a park of six acres on the banks of the St. Charles River. The **General Hospital** and the monastery of Notre Dame des Anges form an extensive pile of buildings, on St. Ours St., near the St. Charles. They were founded by De Vallier, second bishop of Quebec (in 1693), for invalids and incurables. He spent 100,000 crowns in this work, erecting the finest building in Canada (at that time). It is now conducted by a superior and 45 nuns of St. Augustine. The conventchurch of Notre Dame des Anges has 14 paintings by *Légaré*, with an Assumption (over the high altar) dating from 1671.

Pointe aux Liévres, or Hare Point, is beyond the General Hospital, on the meadows of the St. Charles. It is supposed to be the place where the pious Franciscan monks founded the first mission in Canada. Jaques Cartier's winter-quarters in 1536 were here, and on leaving this point he carried off the Indian king, Donnacona, who was afterwards baptized with great pomp in the magnificent cathedral of Rouen. On this ground, also, the army of Montcalm tried to rally after the disastrous battle on the Plains of Abraham.

The suburb of the *Banlieue* lies beyond St. Ours St., and is occupied by the homes of the lower classes, with the heights toward St. Foy rising on the S. *St. Sauveur's Church* is the only fine building in this quarter.

In May, 1535, Jaques Cartier with his patrician officers and hardy sailors attended high mass and received the bishop's blessing in the Cathedral of St. Malo, and then departed across the unknown western seas. The largest of his vessels was of only 120 tons' burden, yet the fleet crossed the ocean safely, and ascended the broad St. Lawrence. Having passed the dark Saguenay cliffs and the vine-laden shores of the Isle of Orleans, he entered a broad basin where "a mighty promontory, rugged and bare, thrust its scarped front into the raging current. Here, clothed in the majesty of solitude, breathing the stern poetry of the wilderness, rose the cliffs now rich with heroic memories, where the fiery Count Frontenac cast defiance at his foes, where Wolfe, Montcalm, and Montgomery fell. As yet all was a nameless barbarism, and a cluster of wigwams held the site of the rock-built city of Quebec. Its name was Stadaconé, and it owned the sway of the royal Donnacona."

It is held as an old tradition that when Cartier's Norman sailors first saw the promontory of Cape Diamond, they shouted "*Quel bec!*" ("What a beak!") which by a natural elision has been changed to Quebec. Others claim that they named the place in loving memory of Caudebec, on the Seine, to which its natural features bear a magnified resemblance. But the more likely origin of the name is from the Indian word *kebec*, signifying a strait, and applied to the comparative narrowing of the river above the Basin. It is, however, held in support of the Norman origin of the name that the seal of William de la Pole, Earl of Suffolk in the 15th century, bears the title of Lord of Quebec. This noble had large domains in France, and was the victor at Crevant and Compeigne, and the conqueror of Joan of Arc, but was impeached and put to death (as narrated by Shakespeare, King Henry VI, Part II., Act IV., Scene 1) for losing the English provinces in France after 34 arduous campaigns.

When Cartier went to Montreal his men built a fort and prepared winter-quarters near the St. Charles River. Soon after his return an intense cold set in, and nearly every man in the fleet was stricken down with the scurvy, of which many died in great suffering. In the springtime, Cartier planted the cross and fleur-de-lis on the site of Quebec, and returned to France, carrying King Donnacona and several of his chiefs as prisoners. These Indians were soon afterwards received into the Catholic Church, with much pomp and ceremony, and died within a year, in France. In 1541 Cartier returned with 5 vessels and erected forts at Cap Rouge, but the Indians were suspicious, and the colony was soon abandoned. Soon afterwards Roberval, the Viceroy of New France, founded another colony on the same site, but after a long and miserable winter it also was broken up.

In the year 1608 the city of Quebec was founded by the noble Champlain,[1] who erected a fort here, and laid the foundations of Canada. A party of Franciscan monks arrived in 1615, and the Jesuits came in 1644. In 1628 Sir David Kirke vainly attacked the place with a small English fleet, but in 1629 he was more successful, and, after a long blockade, made himself master of Quebec. It was restored to France in 1632; and in 1635 Governor Champlain died, and was buried in the Lower Town. Champlain's successor was Charles de Montmagny, a brave and devout Knight of Malta, on whom the Iroquois bestowed the name of *Onontio* ("Great Mountain"). The work of founding new settlements and of proselyting the Hurons and combating the Iroquois was continued for the next century from the rock of Quebec.

After the king had erected his military colonies along the St. Lawrence, he found that another element was necessary in order to make them permanent and progressive. Therefore, between 1665 and 1673 he sent to Quebec 1,000 girls, most of whom were of the French peasantry; though the Intendant, mindful of the tastes of his officers, demanded and received a consignment of young ladies ("*demoiselles bien choisies*"). These cargoes included a wide variety, from Parisian vagrants to Norman ladies, and were maliciously styled by one of the chief nuns, "mixed goods" (*une marchandise mêlée*). The government provided them with dowries; bachelors were excluded by law from trading, fishing, and hunting, and were distinguished by "marks of infamy"; and the French Crown gave bounties for children (each inhabitant who had 10 children being entitled to a pension of from 400 to 800 livres). About the year 1664 the city indulged in extraordinary festivities on the occasion of the arrival of the bones of St. Flavien and St. Felicité, which the Pope had presented to the cathedral of Quebec. These honored relics were borne in solemn procession through the streets, amid the sounds of martial music and the roaring of saluting batteries, and were escorted by the Marquis de Tracy, the Intendant Talon, and the valiant Courcelles, behind whom marched the royal guards and the famous Savoyard regiment of Carignan-Salières, veterans of the Turkish campaigns. The diocese of Quebec was founded in 1674, and endowed with the revenues of the ancient abbeys of Maubec and Benevent. In the same ship with Bishop Laval came Father Hennepin, who explored the Mississippi from the Falls of St. Anthony to the Gulf of Mexico, and the fearless explorer La Salle.

In 1672 the Count de Frontenac was sent here as Governor, and in 1690 he bravely repulsed an attack by Sir Wm. Phipps's fleet (from Boston), inflicting severe damage by a cannonade from the fort. Besides many men, the assailants lost their admiral's standard and several ships. In 1711 Sir Hovenden Walker sailed from Boston against Quebec, but he lost in one day eight vessels and 884 men by shipwreck on the terrible reefs of the Egg Islands. Strong fortifications were built soon after; and in 1759 Gen. Wolfe came up the river with 8,000 British soldiers. The Marquis de Montcalm was then Governor, and he moved the French army into fortified lines on Beauport Plains, where he defeated the British in a sanguinary action. On the night of Sept. 12, Wolfe's army drifted up stream on the rising tide, and succeeded in scaling the steep cliffs beyond the city. They were fired upon by the French outposts; but before Montcalm could bring his forces across the St. Charles the Brit-

---

[1] Champlain was born of a good family in the province of Saintonge, in 1570. He became a naval officer, and was afterward attached to the person of King Henri IV. In 1603 he explored the St. Lawrence River up to the St. Louis Rapids, and afterward (until his death in 1635) he explored the country from Nantucket to the head-waters of the Ottawa. He was a brave, merciful, and zealous chief, and held that "the salvation of one soul is of more importance than the founding of a new empire." He established strong missions among the Hurons, fought the Iroquois, and founded Quebec.

ish lines were formed upon the Plains of Abraham; and in the short but desperate battle which ensued both the generals were mortally wounded. The English lost 664 men, and the French lost 1,500. The French army, which was largely composed of provincial levies (with the regiments of La Guienne, Royal Roussilon, Bearn, La Sarre, and Languedoc), gave way, and retreated across the St. Charles, and a few days later the city surrendered.

In April, 1760, the Chevalier de Levis (of that Levis family — Dukes of Ventadour — which claimed to possess records of their lineal descent from the patriarch Levi) led the reorganized French army to St. Foy, near Quebec. Gen. Murray, hoping to surprise Levis, advanced (with 3,000 men) from his fine position on the Plains of Abraham; but the French were vigilant, and Murray was defeated and hurled back within the city gates, having lost 1,000 men and 20 cannon. Levis now laid close siege to the city, and battered the walls (and especially St. John's Gate) from three heavy field-works. Quebec answered with an almost incessant cannonade from 132 guns, until Commodore Swanton came up the river with a fleet from England. The British supremacy in Canada was soon afterwards assured by the Treaty of Paris, and Voltaire congratulated Louis XV. on being rid of "1,500 leagues of frozen country." The memorable words of Gov. Shirley before the Massachusetts Legislature (June 28, 1746), "*Canada est delenda,*" were at last verified, but the campaigns had cost the British Government $400,000,000, and resulted in the loss of the richest of England's colonies. For the attempted taxation of the Americans, which resulted in the War of Independence, was planned in order to cover the deficit caused in the British Treasury by the Canadian campaigns.

In the winter of 1775-6 the Americans besieged the city, then commanded by Gen. Guy Carleton (afterwards made Lord Dorchester). The provisions of the besiegers began to fail, their regiments were being depleted by sickness, and their light guns made but little impression on the massive city walls; so an assault was ordered and conducted before dawn on Dec. 31, 1775. In the midst of a heavy snow-storm Arnold advanced through the Lower Town from his quarters near the St. Charles River, and led his 800 New-Englanders and Virginians over two or three barricades. The Montreal Bank and several other massive stone houses were filled with British regulars, who guarded the approaches with such a deadly fire that Arnold's men were forced to take refuge in the adjoining houses, while Arnold himself was badly wounded and carried to the rear. Meanwhile Montgomery was leading his New-Yorkers and Continentals N. along Champlain St. by the river-side. The intention was for the two attacking columns, after driving the enemy from the Lower Town, to unite before the Prescott Gate and carry it by storm. A strong barricade was stretched across Champlain St. from the cliff to the river; but when its guards saw the great masses of the attacking column advancing through the twilight, they fled. In all probability Montgomery would have crossed the barricade, delivered Arnold's men by attacking the enemy in the rear, and then, with 1,500 men flushed with victory, would have escaladed the Prescott Gate and won Quebec and Canada, — but that one of the fleeing Canadians, impelled by a strange caprice, turned quickly back, and fired the cannon which stood loaded on the barricade. Montgomery and many of his officers and men were stricken down by the shot, and the column broke up in panic, and fled. The British forces were now concentrated on Arnold's men, who were hemmed in by a sortie from the Palace Gate, and 426 officers and men were made prisoners. A painted board has been hung high up on the cliff over the place in Champlain St. where Montgomery fell. Montgomery was an officer in Wolfe's army when Quebec was taken from the French 15 years before, and knew the ground. His mistake was in heading the forlorn hope. Quebec was the capital of Canada from 1760 to 1791, and after that it served as a semi-capital, until the founding of Ottawa City. In 1845, 2,900 houses were burnt, and the place was nearly destroyed, but soon revived with the aid of the great lumber-trade, which is still its specialty.

In September, 1874, Quebec was filled with prelates, priests, and enthusiastic people, and the second centennial of the foundation of the diocese was celebrated with great pomp. Nine triumphal arches, in Latin, Byzantine, Romanesque, Classic, and Gothic architecture, were erected over the streets of the Upper Town, and dedicated to the metropolitan dioceses of North America; an imposing procession passed under them and into the Cathedral, which was endowed on that day with the name and privileges of a basilica; and at evening the city was illuminated, at a cost of $30,000. In the pageant was borne the ancient flag of Ticonderoga (*Le Drapeau de Carillon*), which floated over Montcalm's victorious army when he defeated Abercrombie on Lake Champlain (July 8, 1758), and is now one of the most esteemed trophies of Quebec.

The annals of the Church contain no grander chapter than that which records the career of the Canadian Jesuits. Unarmed and alone, they passed forth from Quebec and Montreal, and traversed all the wide region between Labrador and the remote West, bravely meeting death in its most lingering and horrible forms at the hands of the vindictive savages whom they came to bless. Their achievements and their fate filled the world with amazement. Even Puritan New England, proudly and sternly jealous of her religious liberty, received their envoy with honors; Boston, Plymouth, and Salem alike became his gracious hosts; and the Apostle Eliot entertained him at his Roxbury parsonage, and urged him to remain. "'To the Jesuits the atmosphere of Quebec was wellnigh celestial. 'In the climate of New France,' they write, 'one learns perfectly to seek only one God, to have no desire but God, no purpose but for God.' And again: 'To live in New France is in truth to live in the bosom of God.' 'If,' adds Le Jeune, 'any one of those who die in this country goes to perdition, I think he will be doubly guilty.'"
"Meanwhile from Old France to New came succors and reinforcements to the missions of the forest. More Jesuits crossed the sea to urge on the work of conversion. These were no stern exiles, seeking on barbarous shores an asylum for a persecuted faith. Rank, wealth, power, and royalty itself smiled on their enterprise, and bade them God-speed. Yet, withal, a fervor more intense, a self-abnegation more complete, a self-devotion more constant and enduring, will scarcely find its record on the pages of human history. . . . . It was her nobler and purer part that gave life to the early missions of New France. That gloomy wilderness, those hordes of savages, had nothing to tempt the ambitious, the proud, the grasping, or the indolent. Obscure toil, solitude, privation, hardship, and death were to be the missionary's portion. . . . .
"The Jesuits had borne all that the human frame seems capable of bearing. They had escaped as by miracle from torture and death. Did their zeal flag or their courage fail? A fervor intense and unquenchable urged them on to more distant and more deadly ventures. The beings, so near to mortal sympathies, so human, yet so divine, in whom their faith impersonated and dramatized the great principles of Christian faith, — virgins, saints, and angels, — hovered over them, and held before their raptured sight crowns of glory and garlands of immortal bliss. They burned to do, to suffer, and to die; and now, from out a living martyrdom, they turned their heroic gaze towards an horizon dark with perils yet more appalling, and saw in hope the day when they should bear the cross into the blood-stained dens of the Iroquois.
In 1647, when the powerful and bloodthirsty Iroquois were sweeping over Canada in all directions, the Superior of the Jesuits wrote: "Do not imagine that the rage of the Iroquois, and the loss of many Christians and many catechumens, can bring to naught the mystery of the cross of Jesus Christ and the efficacy of his blood. We shall die; we shall be captured, burned, butchered: be it so. Those who die in their beds do not always die the best death. I see none of our company cast down. On the contrary, they ask leave to go up to the Hurons, and some of them protest that the fires of the Iroquois are one of their motives for the journey."
"The iron Brébeuf, the gentle Garnier, the all-enduring Jogues, the enthusiastic Chaumonot, Lalemant, Le Mercier, Chatelain, Daniel, Pijart, Rogueneau, Du Peron, Poncet, Le Moyne, — one and all bore themselves with a tranquil boldness, which amazed the Indians and enforced their respect. . . . . When we look for the result of these missions, we soon become aware that the influence of the French and the Jesuits extended far beyond the circle of converts. It eventually modified and softened the manners of many unconverted tribes. In the wars of the next century we do not often find those examples of diabolic atrocity with which the earlier annals are crowded. The savage burned his enemies alive, it is true, but he seldom ate them; neither did he torment them with the same deliberation and persistency. He was a savage still, but not so often a devil." (PARKMAN.)
The traveller who wishes to study more closely this sublime episode in the New-World history may consult the brilliant and picturesque historical narratives of Mr. Francis Parkman: "The Jesuits of North America," "The Pioneers of France in the New World," and "The Discovery of the Great West."

69. The Environs of Quebec.

This district is famed for its beauty, and is filled with objects of interest to the tourist. The suburban villages can be visited by pedestrian tours; but in that case it is best to cut off communication with the city, and to sweep around on the great curve which includes the chief points of attraction. The village inns furnish poor accommodations. Such a walking tour should be taken only after a season of dry weather, else the roads will be found very muddy. But all the world goes about in carriages here, and a *calèche* and driver can be hired at very low rates (see page 255). The drivers' statements of distances can seldom be relied on, for they generally err on the side of expansion.

"I don't know whether I cared more for Quebec or the beautiful little villages in the country all about it. The whole landscape looks just like a dream of 'Evangeline.' . . . . But if we are coming to the grand and beautiful, why, there is no direction in which you can look about Quebec without seeing it; and it is always mixed up with something so familiar and homelike that my heart warms to it." (HOWELLS's *A Chance Acquaintance*.)

** The Falls of Montmorenci are 7 M. from the Dorchester Bridge, which is about 1 M. from the Upper-Town Market Square. The route usually taken leads down Palace St. and by the Queen's Fuel-Yard (see page 271) and St. Roch's Church. As the bridge is being crossed, the Marine Hospital is seen on the l., and on the r. are the shipyards of St. Roch's Ward and the suburb of St. Charles. The road is broad and firm, and leads across a fertile plain, with fine retrospective views. The *Beauport Lunatic Asylum* is soon reached, near which is the villa of Glenalla. The asylum formerly consisted of two large buildings, one for each sex; but the female department was destroyed by fire in January, 1875, and several of its inmates were burnt with it. **Beauport** is $3\frac{1}{2}$ - 5 M. from Quebec, and is a long-drawn-out village of 1,300 inhabitants, with a tall and stately church whose twin spires are seen from a great distance. There are several flour and barley mills in the parish, and a considerable lumber business is done. The seigniory was founded in 1634 by the Sieur Giffard, and along its plains was some of the heaviest fighting of the war of the Conquest of Canada.

It is "in that part of Canada which was the first to be settled, and where the face of the country and the people have undergone the least change from the beginning, where the influence of the States and of Europe is least felt, and the inhabitants see little or nothing of the world over the walls of Quebec." The road from Quebec to St. Joachim is lined by a continuous succession of the quaint and solid little Canadian houses of whitewashed stone, placed at an angle with the street in order to face the south. The farms are consequently remarkably narrow (sometimes but a few yards wide and $\frac{1}{4}$ M. long), and the country is bristling with fences. In 1664 the French king forbade that the colonists should make any more clearings, "except one next to another"; but in 1745 he was obliged to order that their farms should be not less than $1\frac{1}{2}$ arpents wide. These narrow domains arose from the social character of the people, who were thus brought close together; from their need of concentration as a defence against the Indians; and from the subdivision of estates by inheritance. The Latin Catholicism of the villagers is shown by roadside crosses rising here and there along the way.

So late as 1827 Montmorenci County (which is nearly as large as Massachusetts) had but 5 shops, 30 artisans, 2 schools, 5 churches (all Catholic), and 5 vessels (with an aggregate of 59 tons). There has been but little change since. In 1861, out of 11,136 inhabitants in the county, 10,708 were of French origin, of whom but a few score understand the English language.

M. Rameau ("*La France aux Colonies*") has proved, after much labor and research, that the colonists who settled the Cote de Beaupré and Beauport were from the ancient French province of La Perche; adding that Montreal was colonized from the province of Anjou, the Isle of Orleans from Poitou, and Quebec, Trois Rivières, and the Richelieu valley from Normandy.

Beyond the church of Beauport the road continues past the narrow domains on either hand, and runs along the side of the Haldimand estate. The Montmorenci River is crossed, and the traveller stops at the *Montmorenci Restaurant*, where lunch may be obtained. At this point admission is given to the grounds about the Falls (fee, 25c.); and the tourist should visit not only the pavilion near the brink (which commands a charming view of Quebec), but also the small platform lower down (and reached by a long stairway), whence the best front-view is obtained. The descent to the basin below is difficult, and will hardly repay the labor of the return. A short distance below the Falls is the confluence of the Montmorenci with the St. Lawrence, and immense saw-mills are located there, employing 7 - 800 men and cutting up 2,500 logs a day. Near the Falls is *Haldimand House*, formerly occupied by the Duke of Kent, Queen Victoria's father; and on the cliffs by the river are seen the towers of a suspension-bridge which fell soon after its erection, hurling three persons into the fatal abyss below. At the foot of these Falls an immense ice-cone (sometimes 200 ft. high) is formed every winter, and here the favorite sport of tobogginning is carried on. The * **Natural Steps** are $1\frac{1}{2}$ M. above the Falls, where the Montmorenci is contracted into a narrow limit and rushes down with great velocity, having cut its bed down through successive strata and leaving step-like terraces on either side. Fine specimens of trilobites have been found in this vicinity.

The road running on beyond the Montmorenci Restaurant leads to Ange Gardien and St. Anne (see Route 70). The views on the way back to Quebec are very beautiful.

The old French *habitans* call the Montmorenci Fall, *La Vache* (" The Cow"), on account of the resemblance of its foaming waters to milk. Others attribute this name to the noise like the lowing of a cow which is made by the Fall during the prevalence of certain winds. Immediately about the basin and along the Montmorenci River, many severe actions took place during Wolfe's siege of Quebec. This river was for a time the location of the picket-lines of the British and French armies.

"It is a very simple and noble fall, and leaves nothing to be desired. . . . . It is a splendid introduction to the scenery of Quebec. Instead of an artificial fountain in its square, Quebec has this magnificent natural waterfall to adorn one side of its harbor." (THOREAU.)

"The effect on the beholder is most delightful. The river, at some distance, seems suspended in the air, in a sheet of billowy foam, and, contrasted, as it is, with the black frowning abyss into which it falls, it is an object of the highest interest. It has been compared to a white ribbon, suspended in the air; this comparison does justice to the delicacy, but not to the grandeur of the cataract." (SILLIMAN.)

"A safe platform leads along the rocks to a pavilion on a point at the side of the fall, and on a level with it. Here the gulf, nearly 300 ft. deep, with its walls of chocolate-covered earth, and its patches of emerald herbage, wet with eternal spray, opens to the St. Lawrence. Montmorenci is one of the loveliest waterfalls. In its general character it bears some resemblance to the Pisse-Vache, in Switzerland, which, however, is much smaller. The water is snow-white, tinted, in the heaviest portions of the fall, with a soft yellow, like that of raw silk. In fact, broken as it is by the irregular edge of the rock, it reminds one of masses of silken, flossy skeins, continually overlapping one another as they fall. At the bottom, dashed upon a pile of rocks, it shoots far out in star-like radii of spray, which share the regular throb or pulsation of the falling masses. The edges of the fall flutter out into lace-like points and fringes, which dissolve into gauze as they descend." (BAYARD TAYLOR.)

"The Falls of Montmorenci present the most majestic spectacle in all this vicinity, and even in the Province. The river, in its course through a country which is covered with an almost unbroken forest, has an inconsiderable flow of water except when swelled by the melting of the snow or the autumnal rains, until it reaches the precipice, where it is 8-10 fathoms wide. Its bed, being inclined before arriving at this point, gives a great velocity to the current, which, pushed on to the verge of a perpendicular rock, forms a large sheet of water of a whiteness and a fleecy appearance which resembles snow, in falling in a chasm among the rocks [251] ft. below. At the bottom there rises an immense foam in undulating masses, which, when the sun lights up their brilliant prismatic colors, produces an inconceivably beautiful effect." (BOUCHETTE.)

"For those who go from Montmorenci to Quebec, the time to be on the road is about sunset. The city, climbing up from the great river to the heights, on which stands the castle, looks especially beautiful in the warm light that then falls full upon it, and the level rays, striking on the quaint old metal-sheathed roofs and on all the westward-facing windows, light up the town with a diamond-like sparkling of wonderful brilliancy." (WHITE'S *Sketches from America*.)

* **Indian Lorette** (small inn) is 9 M. from Quebec, by the Little River Road. It is an ancient village of the Hurons ("Catholics and allies of France"), and the present inhabitants are a quiet and religious people in whom the Indian blood predominates, though it is never unmixed. The men hunt and fish, the women make bead-work and moccasons, and the boys earn pennies by dexterous archery. There are 60 Huron families here, and their quaint little church is worthy of notice. The population of the parish is 3,500, and the district is devoted to farming. The * **Lorette Falls** are near the mill, and are very pretty.

The best description of Lorette is given in Howells's *A Chance Acquaintance* (Chap. XIII.), from which the following note is extracted: "The road to Lorette is through St. John's Gate, down into the outlying meadows and rye-fields, where, crossing and recrossing the swift St. Charles, it finally rises at Lorette above the level of the citadel. It is a lonelier road than that to Montmorenci, and the scattering cottages upon it have not the well-to-do prettiness, the operatic repair, of stone-built Beauport. But they are charming, nevertheless, and the people seem to be remoter from modern influences..... By and by they came to Jeune-Lorette, an almost ideally pretty hamlet, bordering the road on either hand with galleried and balconied little houses, from which the people bowed to them as they passed, and piously enclosing in its midst the village church and churchyard. They soon after reached Lorette itself, which they might easily have known for an Indian town by its unkempt air, and the irregular attitudes in which the shabby cabins lounged along the lanes that wandered through it..... The cascade, with two or three successive leaps above the road, plunges headlong down a steep, crescent-shaped slope, and hides its foamy whiteness in the dark-foliaged ravine below. It is a wonder of graceful motion, of iridescent lights and delicious shadows; a shape of loveliness that seems instinct with a conscious life."

Charles Marshall says, in his "Canadian Dominion" (London, 1871): "For picturesque beauty the environs of Quebec vie with those of any city in the world. .... It is not too much to say that the Lorette cascades would give fame and fortune to any spot in England or France; yet here, dwarfed by grander waters, they remain comparatively unknown."

When the French came to Canada the Hurons were a powerful nation on the shores of Lakes Huron and Simcoe, with 32 villages and 20-30,000 inhabitants. They received the Jesuit missionaries gladly, and were speedily converted to Christianity. Many of them wore their hair in bristling ridges, whence certain astonished Frenchmen, on first seeing them, exclaimed " *Quelles hures!* " (" What boars' heads ! ") and the name of *Huron* supplanted their proper title of *Ouendat* or *Wyandot*. The Iroquois, or Five Nations (of New York), were their mortal foes, and after many years of most barbarous warfare, succeeded in storming the Christian Huron towns of St. Joseph, St. Ignace, and St. Louis. The nation was annihilated: a few of its people fled to the far West, and are now known as the Wyandots; multitudes were made slaves among the Iroquois villages; 10,000 were killed in battle or in the subjugated towns; and the mournful remnant fled to Quebec. Hundreds of them were swept away from the Isle of Orleans by a daring Iroquois raid; the survivors encamped under the guns of the fort for 10 years, then moved to St. Foy; and, about the year 1673, this feeble fragment of the great Huron nation settled at Ancienne Lorette. It was under the care of the Jesuit Chaumonot, who, while a mere boy, had stolen a small sum of money and fled from France into Lombardy. In filth and poverty he begged his way to Ancona, and thence to Loretto, where, at the Holy House, he had an angelic vision. He went to Rome, became a Jesuit, and experienced another miracle from Loretto; after which he passed to the Huron mission in Canada, where he was delivered from martyrdom by the aid of St. Michael. He erected at Ancienne Lorette a chapel in exact fac-simile of the Holy House at Loretto; and here he claimed that many miracles were performed. In 1697 the Hurons moved to New Lorette, "a wild spot, covered with the primitive forest, and seamed by a deep and tortuous ravine, where the St. Charles foams, white as a snow-drift, over the black ledges, and where the sunshine struggles through matted boughs of the pine and the fir, to bask for brief moments on the mossy rocks or flash on the hurrying waters. On a plateau beside the torrent, another chapel was built to Our Lady, and another Huron town sprang up; and here to this day, the tourist finds the remnant of a lost people, harmless weavers of baskets and sewers of moccasons, the Huron blood fast bleaching out of them, as, with every generation, they mingle and fade away in the French population around." (PARKMAN.)

Visitors to Lorette are recommended to return to Quebec by another road from that on which they went out. Ancienne Lorette may be reached from this point, and so may the lakes of Beauport and St. Charles. 1½ days' journey to the N. is Lac Rond, famous for its fine hunting and fishing.

Charlesbourg (Huot's boarding-house) is 4 M. from Quebec, on a far-viewing ridge, and is clustered about a venerable convent and old church (with copies of the Last Communion of St. Jerome and the Sistine Madonna over its altars). It is the *chef-lieu* of the seigniory of Notre Dame des Anges, and its products are lumber and oats. To this point (then known as *Bourg Royal*) retired the inhabitants of the Isle of Orleans, in 1759, when ordered by Montcalm to fall back before the British. They were 2,500 in number, and were led by their curés. Pleasant roads lead from Charlesbourg to Lorette, Lake St. Charles, Lake Beauport, and Château Bigot.

Lake St. Charles is 11 M. from Quebec, and 6 M. from Lorette. It is 4 M. long, and its waters are very clear and deep. The red trout of this lake are of delicate flavor. There is a remarkable echo from the shores.

" On arriving at the vicinity of the lake, the spectator is delighted by the beauty and picturesque wildness of its banks. . . . . Trees grow immediately on the borders of the water, which is indented by several points advancing into it, and forming little bays. The lofty hills which suddenly rise towards the N., in shapes singular and diversified, are overlooked by mountains which exalt, beyond them, their more distant summits." (HERIOT.)

Château Bigot is about 7 M. from Quebec, by way of Charlesbourg, where the traveller turns to the r. around the church, and rides for 2 M. along a ridge which affords charming views of the city on the r. "It is a lovely road out to Château Bigot. First you drive through the ancient suburbs of the Lower Town, and then you mount the smooth, hard highway, between pretty country-houses, towards the village of Charlesbourg, while Quebec shows, to your casual backward glance, like a wondrous painted scene, with the spires and lofty roofs of the Upper Town, and the long, irregular wall wandering on the verge of the cliff; then the thronging gables and chimneys of St. Roch, and again many spires and convent walls." The ruins of the Château are only reached after driving for some distance through a narrow wheel-track, half overgrown with foliage. There remain the gables and division-wall, in thick masonry, with a deep cellar, outside of which are heaps of débris, over which grow alders and lilacs. The ruins are in a cleared space over a little brook where trout are found; and over it is the low and forest-covered ridge of *La Montagne des Ormes.*

This land was in the *Fief de la Trinité*, which was granted about the year 1640 to M. Denis, of La Rochelle. The château was built for his feudal mansion by the Royal Intendant Talon, Baron des Islets, and was afterwards occupied by the last Royal Intendant, M. Bigot, a dissolute and licentious French satrap, who stole $2,000,000 from the treasury. The legend tells that Bigot used this building for a hunting-lodge and place of revels, and that once, while pursuing a bear among the hills, he got lost, and was guided back to the château by a lovely Algonquin maiden whom he had met in the forest. She remained in this building for a long time, in a luxurious boudoir, and was visited frequently by the Intendant; but one night she was assassinated by some unknown person, — either M. Bigot's wife, or her own mother, avenging the dishonor to her tribe (see "Château Bigot," by J. M. LeMoine, sold at the Quebec bookstores for 10c.; also Howells's *A Chance Acquaintance*, Chap. XII.).

Sillery (or *St. Colomb*) is 3 M. from Quebec, by the Grand Allée and the Cap-Rouge Road (see page 270). After passing Wolfe's Monument, the road leads across the *Plains of Abraham*, on which were fought the sanguinary battles of 1759 and 1760. Sillery is a parish of 3,000 inhabitants, on whose river front are 17 coves, where most of the lumber of Quebec is guarded. The *Convent of Jesus-Maria* is a new building of great size and imposing architecture; opposite which is the handsome Gothic school-house which was given to this parish by Bishop Mountain. In the vicinity of Sillery are several fine villas, amid ornamental grounds: *Marchmont*, once the home of Sir John Harvey and Bishop Stewart; *Spencer Wood*, "the most beautiful domain of Sillery, or, it can be said, of Canada," with a park of 80 acres, formerly the home of the Earl of Elgin and other Canadian governors; *Woodfield*, founded by the Bishop of Samos *in partibus infidelium*; *Spencer Grange*, where lives J. M. LeMoine, the author and antiquarian; *Bardfield*, Bishop Mountain's former home; *Cataracouy*, where the British princes, Albert Edward and Alfred, sojourned; *Benmore*, Col. Rhodes's estate; and several others. The beautiful cemetery of Mount

Hermon, which was laid out by Major Douglas, the planner of Greenwood Cemetery, is in this vicinity, and is adorned by the graceful chapel of St. Michael. The people of Sillery have recently (1870) erected a monument, sustaining a marble cross, near the place where Father Massé was buried, in 1646, in the ancient Church of St. Michael (which has long since disappeared). The old Jesuit Residence still remains, and is a massive building of stone.

The Chevalier Noel Brulart de Sillery, Knight of Malta, and formerly a high officer at the court of Queen Marie de Médicis, having renounced the world, devoted his vast revenues to religious purposes. Among his endowments was the foundation of a Christian Algonquin village just above Quebec, which the Jesuits named *Sillery*, in his honor. Here the Abenaquis of Maine learned the elements of Catholicism, which was afterwards unfolded to them in their villages on the Kennebec, by Father Druilletes. This worthy old clergyman followed them in their grand hunts about Moosehead Lake and the northern forests, " with toil too great to buy the kingdoms of this world, but very small as a price for the Kingdom of Heaven." From the mission-house at Sillery departed Jogues, Brébeuf, Lalemant, and many other heroic missionaries and martyrs of the primitive Canadian Church. "It was the scene of miracles and martyrdoms, and marvels of many kinds, and the centre of the missionary efforts among the Indians. Indeed, few events of the picturesque early history of Quebec left it untouched; and it is worthy to be seen, no less for the wild beauty of the spot than for its heroical memories. About a league from the city, where the irregular wall of rock on which Quebec is built recedes from the river, and a grassy space stretches between the tide and the foot of the woody steep, the old mission and the Indian village once stood; and to this day there yet stands the stalwart frame of the first Jesuit Residence, modernized, of course, and turned to secular uses, but firm as of old, and good for a century to come. All around is a world of lumber, and rafts of vast extent cover the face of the waters in the ample cove, — one of many that indent the shore of the St. Lawrence. A careless village straggles along the roadside and the river's margin; huge lumber-ships are loading for Europe in the stream; a town shines out of the woods on the opposite shore; nothing but a friendly climate is needed to make this one of the most charming scenes the heart could imagine."

Cap Rouge is 9 M. from Quebec, and may be reached by the road which passes through Sillery. It is a village of 800 inhabitants, with a timber-trade and a large pottery; and is connected with Quebec by semi-daily stages. The cape forms the W. end of the great plateau of Quebec, which, according to the geologists, was formerly an island, around which the St. Lawrence flowed down the St. Charles valley. It is thought that the Grand Trunk Railway will throw a suspension-bridge over the St. Lawrence at this point, being forced to bring its trains into Quebec by the competition of the North-Shore Railway. The mansion of *Redclyffe* is on the cape, and is near the site where Jaques Cartier and Roberval passed the winters of 1541 and 1542. On the same point batteries were erected by Montcalm and Murray.

In returning from Cap Rouge to the city, it may be well to turn to the l. at St. Albans and gain the St. Foy road. The village of **St. Foy** is 5 M. from Quebec, and contains many pleasant villas and mansions. To the N. is the broad and smiling valley of the St. Charles, in which may be seen *Ancienne Lorette* (two inns), a lumbering village of 3,000 inhabitants, on the Gosford Railway, 4½ M. from St. Foy. Beyond the Church of St.

Foy is the * monumental column, surmounted by a statue of Bellona (presented by Prince Napoleon), which marks the site of the fiercest part of the Second Battle of the Plains, in which De Levis defeated Murray (1760). The monument was dedicated with great pomp in 1854, and stands over the grave of many hundreds who fell in the fight. Passing now the handsome Finlay Asylum and several villas, the suburb of St. John is entered.

Point Levi (or *Levis*) is on the S. shore of the St. Lawrence, opposite Quebec, with which it is connected by ferry-boats running every 15 minutes. It has about 10,000 inhabitants, with a large and increasing trade, being the terminus of the Quebec branch of the Grand Trunk Railway and of the (as yet uncompleted) Levis & Kennebec Railway. On the lofty plateau beyond the town are the great forts which have been erected to defend Quebec from a second bombardment from this shore. They are three in number, 1 M. apart, solidly built of masonry and earth, with large casemates and covered ways; and are to be armed with Moncrieff guns of the heaviest calibre. It is said that these forts cost $15,000,000, — a palpable exaggeration, — but they have been a very expensive piece of work, and are said to be more nearly like Cherbourg, the best of modern European fortifications, than any others in America. The batteries with which Gen. Wolfe destroyed Quebec, in 1759, were located on this line of heights.

St. Joseph is 2½ M. from Point Levi, and transacts a large business in wood and timber. *South Quebec* is above Point Levi, and is closely connected with it. The Liverpool steamers stop here, and there are great shipments of lumber from the harbor. The town has 3,000 inhabitants, and is growing rapidly.

St. Romuald (or *New Liverpool*) is 5 M. from Quebec, and adjoins S. Quebec. It has several factories and mills and a large lumber-trade, and is connected with Quebec by semi-daily steamers. The * **Church of St. Romuald** is "the finest on the Lower St. Lawrence," and is celebrated for its paintings (executed in 1868 – 9 by *Lamprech* of Munich).

In the choir are the Nativity, Crucifixion, and Resurrection of Christ; in the Chapel of St. Joseph, the Marriage of St. Joseph, the Flight into Egypt, Nazareth, Jesus and the Doctors, the Death of St. Joseph; in the Chapel of the Virgin, the Annunciation, the Visitation, the Adoration of the Magi, and the Presentation in the Temple. Above are eight scenes from the life of St. Romuald, from his Conversion to his Apotheosis. There are 16 medallions on a gold ground, representing Sts. Peter and Paul, the Four Evangelists, and five doctors of the Greek Church and five of the Latin Church. The altars were designed by *Schneider* of Munich, and the statues were carved in wood by *Rudmiller* of Munich.

The * **Chaudière Falls** are 4½ M. beyond St. Romuald, and over 9 M. from Quebec. They can only be reached by walking a considerable distance through the bordering fields. "The deep green foliage of the woods overhanging, the roar of the cataract, and the solitude of the place, especially as you emerge suddenly from the forest fastnesses on the scene, produce a strong and vivid impression, not soon to be forgotten." Some visitors even prefer this fall to that of Montmorenci. The Chaudière descends from Lake Megantic, near the frontier of Maine, traversing the Canadian gold-fields. Arnold's hungry and heroic army followed the course of this river from its source to its mouth in their arduous winter-march, in 1775. The Chaudière Falls are 3 M. from its confluence with the St. Lawrence, and at a point where the stream is compressed into a breadth of 400 ft. The depth of the plunge is about 135 ft., and the waters below are continually in a state of turbulent tossing. At the verge of the fall the stream is divided by large rocks, forming three channels, of which that on the W. is the largest. The view from the E. shore is the best. "The wild diversity of rocks, the foliage of the overhanging woods, the rapid motion, the effulgent brightness and deeply solemn sound of the cataracts, all combine to present a rich assemblage of objects highly attractive, especially when the visitor, emerging from the wood, is instantaneously surprised by the delightful scene."

70. Quebec to La Bonne Ste. Anne.—The Côte de Beaupré.

The steamer *Montmorenci* runs from Quebec to St. Anne twice a week. A better route is that by land, through the mediæval hamlets of the Côte de Beaupré. Three days should be devoted to the trip,—one to go and one to return, and the other to the Falls of St. Anne and St. Feréol. Gentlemen who understand French will find this district very interesting for the scene of a pedestrian tour. The inns at St. Anne and along the road are of a very humble character, resembling the wayside *auberges* of Brittany or Normandy; but the people are courteous and well-disposed.

Distances. — Quebec to the Montmorenci Falls, 7 M.; Ange Gardien, 10; Château Richer, 15; St. Anne, 22 (St. Joachim, 27; St. Feréol, 30).

The Seigniory of the Côte de Beaupré contains several parishes of the N. shore, and is the most mountainous part of the Province. It was granted in 1636, and is at present an appanage of the Seminary of Quebec. No rural district N. of Mexico is more quaint and mediæval than the Beaupré Road, with its narrow and ancient farms, its low and massive stone houses, roadside crosses and chapels, and unprogressive French population. But few districts are more beautiful than this, with the broad St. Lawrence on the S., and the garden-like Isle of Orleans; the towers of Quebec on the W., and the sombre ridges of Cape Tourmente and the mountains of St. Anne and St. Feréol in advance. "In the inhabitant of the Côte de Beaupré you find the Norman peasant of the reign of Louis XIV., with his annals, his songs, and his superstitions." (ABBÉ FERLAND.)

"Though all the while we had grand views of the adjacent country far up and down the river, and, for the most part, when we turned about, of Quebec, in the horizon behind us,—and we never beheld it without new surprise and admiration, —yet, throughout our walk, the Great River of Canada on our right hand was the main feature in the landscape, and this expands so rapidly below the Isle of Orleans, and creates such a breadth of level surface above its waters in that direction, that, looking down the river as we approached the extremity of that island, the St. Lawrence seemed to be opening into the ocean, though we were still about 325 M. from what can be called its mouth." (THOREAU.)

Quebec to the Montmorenci Falls, see page 276.

Beyond the Falls the road passes on over far-viewing and breezy hills, and between the snug estates of the rural farmers with their great barns and exposed cellars (*caves*). The village of **Ange Gardien** is guarded at each end by roadside oratories, and lies in a sheltered glen near the river. It is clustered about a venerable old church, in which are paintings of the Annunciation and the Adoration of the Magi. On its front is a large sundial. This dreamy old parish has 1,500 inhabitants, and dates from 1678, when it was founded by Bishop Laval. In 1759 it was overrun and occupied by the famous British corps of the Louisbourg Grenadiers.

After ascending out of the glen of Ange Gardien, the road crosses elevated bluffs, and on the r. are rich and extensive intervales, cut into narrow strips by walls. They extend to the margin of the river, beyond which are the white villages and tin-clad spires of the Isle of Orleans.

Château Richer is a compact and busy village of 2,000 inhabitants, over which, on a bold knoll, is the spacious parish-church. The views from the platform of this edifice are very pretty, including a large area of the parish, the village of St. Pierre on the Isle of Orleans, and the distant promontory of Cape Diamond. During the hunting season the Château-Richer marshes are much frequented by Quebec sportsmen, who shoot great numbers of snipe, ducks, and partridges. The upland streams afford good trout-fishing.

On a rocky promontory near Château Richer was the site of the ancient Franciscan monastery. This massive stone building was erected about the year 1695, and was occupied by a community of peaceful monks. When the British army was fighting the French near the Falls of Montmorenci, a detachment was sent here to get provisions; but the French villagers, under the influence of their spiritual guides, refused to give aid, and fortified themselves in the monastery. The reduction of this impromptu fortress gave Gen. Wolfe considerable trouble, and it was only accomplished by sending against it the valiant Louisbourg Grenadiers and a section of artillery. The monks surrendered after their walls were well battered by cannon-shot, and were dispossessed by the troops. Before the bombardment the parish priest met the English officers, and told them that they fought for their king, and he should be as fearless in defending his people. The villagers made a fierce sortie from the convent during the siege, but were repulsed with the loss of 30 killed. The site of the monastery is now occupied by the school of the Sisters of Le Bon Pasteur, and part of its walls still remain.

The little roadside *auberge* called the Hôtel Campagne is about 1 M. beyond Château Richer. The *Sault à la Puce* is about 2 M. beyond the village, and is visited by leaving the road where it crosses the Rivière à la Puce, and ascending to the l. by the path. The stream leaps over a long cliff, falling into the shadows of a bowery glen, and has been likened to the Cauterskill Falls.

"This fall of La Puce, the least remarkable of the four which we visited in this vicinity, we had never heard of until we came to Canada, and yet, so far as I know, there is nothing of the kind in New England to be compared with it. Most travellers in Canada would not hear of it, though they might go so near as to hear it." (THOREAU.) There are other pretty cascades farther up the stream, but they are difficult of access.

"The lower fall is 112 ft. in height, and its banks, formed by elevated acclivities, wooded to their summits, spread around a solemn gloom, which the whiteness, the movements, and the noise of the descending waters combine to make interesting and attractive..... The environs of this river display, in miniature, a succession of romantic views. The river, from about one fourth of the height of the mountain, discloses itself to the contemplation of the spectator, and delights his eye with varied masses of shining foam, which, suddenly issuing from a deep ravine hollowed out by the waters, glide down the almost perpendicular rock, and form a splendid curtain, which loses itself amid the foliage of surrounding woods. Such is the scene which the fall of La Puce exhibits." (HERIOT.)

La Bonne St. Anne (otherwise known as St. Anne du Nord and St. Anne de Beaupré) is 7 M. beyond Château Richer, and is built on a level site just above the intervales. It has about 1,200 inhabitants, and is supported by the thousands of pilgrims who frequent its shrine, and by supplying brick to the Quebec market. Immense numbers of wild fowl (especially pigeons) are killed here every year. There are numerous small inns in the narrow street, all of which are crowded during the season of pilgrimage. On the E. of the village is the new Church of St. Anne, a massive and beautiful structure of gray stone, in classic architecture, which will probably be completed in 1876. The old building of the * *Church of St. Anne* is on the bank just above, and is probably the most highly venerated shrine in Anglo-Saxon America. The relics of St. Anne are guarded in a crystal globe, and are exhibited at morning mass, when their contemplation is said to have effected many miraculous cures. Over the richly adorned high altar is a *picture of St Anne, by the famous French artist, *Le Brun* (presented by Viceroy Tracy); and the side altars have paintings (given by Bishop Laval) by the Franciscan monk *Lefrançois* (who died in 1685). There are numerous rude *ex-voto* paintings, representing marvellous deliverances of ships in peril, through the aid of St. Anne; and along the cornices and in the sacristy are great sheaves of crutches, left here by cripples and invalids who claimed to have been healed by the intercession of the saint. Within the church is the tomb of Philippe Réné de Portneuf, priest of St. Joachim, who was slain, with several of his people, while defending his parish against the British troops (1759).

"Above all, do not fail to make your pilgrimage to the shrine of St. Anne. . . . . Here, when Aillebout was governor, he began with his own hands the pious work, and a *habitant* of Beaupré, Louis Guimont, sorely afflicted with rheumatism, came grinning with pain to lay three stones in the foundation, in honor probably of St. Anne, St. Joachim, and their daughter, the Virgin. Instantly he was cured. It was but the beginning of a long course of miracles continued more than two centuries, and continuing still. Their fame spread far and wide. The devotion to St. Anne became a distinguishing feature of Canadian Catholicity, till at the present day at least thirteen parishes bear her name. . . . . Sometimes the whole shore was covered with the wigwams of Indian converts who had paddled their birch canoes from the farthest wilds of Canada. The more fervent among them would crawl on their knees from the shore to the altar. And, in our own day, every summer a far greater concourse of pilgrims, not in paint and feathers, but in cloth and millinery, and not in canoes, but in steamboats, bring their offerings and their vows to the 'Bonne St. Anne.'" (PARKMAN.)

According to the traditions of the Roman Church, St. Anne was the mother of the Blessed Virgin, and after her body had reposed for some years in the cathedral at Jerusalem, it was sent by St. James to St. Lazare, first bishop of Marseilles. He, in turn, sent it to St. Auspice, bishop of Apt, who placed it in a subterranean chapel to guard it from profanation in the approaching heathen inroads. Barbarian hordes afterwards swept over Apt and obliterated the church. 700 years later,

Route 70. THE FALLS OF ST. ANNE.

Charlemagne visited the town, and while attending service in the cathedral, several marvellous incidents took place, and the forgotten remains of St. Anne were recovered from the grotto, whence a perpetual light was seen and a delicious fragrance emanated. Ever since that day the relics of the saint have been highly venerated in France. The colonists who founded Canada brought with them this special devotion, and erected numerous churches in her honor, the chief of which was St. Anne de Beaupré, which was founded in 1658 by Gov. d'Aillebout on the estate presented by Etienne Lessart. In 1668 the cathedral-chapter of Carcasson sent to this new shrine a relic of St. Anne (a bone of the hand), together with a lamp and a reliquary of silver, and some fine paintings. The legend holds that a little child was thrice favored with heavenly visions, on the site of the church; and that, on her third appearance, the Virgin commanded the little one to tell the people that they should build a church on that spot. The completion of the building was signalized by a remarkable miracle. The vessels ascending the St. Lawrence during the French domination, always fired off a saluting broadside when passing this point, in recognition of their delivery from the perils of the sea. Bishop Laval made St. Anne's Day a feast of obligation; and rich *ex-voto* gifts were placed in the church by the Intendant Talon, the Marquis de Tracy, and M. d'Iberville, "the Cid of New France." For over two centuries the pilgrimages have been almost incessant, and hundreds of miraculous cures have been attributed to *La Bonne St. Anne*. Between June and October, 1874, over 20,000 pilgrims visited the church, some of whom came from France and some from the United States. An extract from a Lower-Canada newspaper of October, 1874, describes one of the latest of these curious phenomena, the curing of a woman who had been bedridden for 4 years: "She was placed in the Church of St. Anne, on a portable bed, at 6 o'clock on Wednesday morning. After low mass she was made to venerate the relics of St. Anne. A grand mass was chanted a few minutes afterwards. Toward the middle of the divine office the patient moved a little. After the elevation she sat up. At the termination of the mass she got up and walked and made the circuit of the church."

The Côte de Beaupré and the site of St. Anne were granted by the Compagnie des Cents Associés, in 1636, to the Sieur Cheffault de la Regnardière, who, however, made but little progress in settling this broad domain, and finally sold it to Bishop Laval. In 1661, after the fall of Montreal, this district was ravaged by the merciless Iroquois, and in 1682 St. Anne was garrisoned by three companies of French regulars. On the 23d of August, 1759, St. Anne was attacked by 300 Highlanders and Light Infantry and a company of Rangers, under command of Capt. Montgomery. The place was defended by 200 villagers and Indians, who kept up so hot a fire from the shelter of the houses that the assailants were forced to halt and wait until a flanking movement had been made by the Rangers. Many of the Canadians were slain during their retreat, and all who fell into the hands of the British were put to death. The victors then burnt the village, saving only the ancient church, in which they made their quarters. A tradition of the country says that they set fire to the church three times, but it was delivered by St. Anne. The following day they advanced on Château Richer and Ange Gardien, burning every house and barn, and cutting down the fruit trees and young grain. They were incessantly annoyed by the rifles of the countrymen, and gave no quarter to their prisoners.

The * **Falls of St. Anne** are visited by passing out from St. Anne on the road to St. Joachim, as far as the inn, "like an *auberge* of Brittany," at the crossing of the St. Anne River. Thence the way leads up the river-bank through dark glens for 3–4 M., and the visitor is conducted by a guide. In descending from the plateau to the plain below, the river forms seven cascades in a distance of about a league, some of which are of rare beauty, and have been preferred even to the Trenton Falls, in New York. The lower fall is 130 ft. high.

"A magnificent spectacle burst upon our sight. A rapid stream, breaking its way through the dark woods, and from pool to pool among masses of jagged rock, suddenly cleaves for itself a narrow chasm, over which you may spring if you have an iron nerve, and then falls, broken into a thousand fantastic forms of spray along the steep face of the rock, into a deep gorge of horrid darkness. I do not know the volume of water; I forgot to guess the height,—it may be two hundred feet. Figures are absurd in the estimate of the beauty and grandeur of a scene like this. I only know that the whole impression of the scene was one of the most intense I have ever experienced. The disposition of the mass of broken waters is the most graceful conceivable. The irresistible might of the rush of the fall, the stupendous upright masses of black rock that form the chasm; the heavy fringe of dark woods all around; the utter solitariness and gloom of the scene,—all aid to impress the imagination. An artist might prefer this spot to Niagara." (MARSHALL.)

"Here the river, 1-200 ft. wide, comes flowing rapidly over a rocky bed out of that interesting wilderness which stretches toward Hudson's Bay and Davis's Straits. Ha Ha Bay, on the Saguenay, was about 100 M. N. of where we stood. Looking on the map, I find that the first country on the N. which bears a name is that part of Rupert's Land called East Main. This river, called after the Holy Anne, flowing from such a direction, here tumbles over a precipice, at present by three channels, how far down I do not know, but far enough for all our purposes, and to as good a distance as if twice as far. . . . . The falling water seemed to jar the very rocks, and the noise to be ever increasing. The vista was through a narrow and deep cleft in the mountain, all white suds at the bottom." From the bed of the stream below "rose a perpendicular wall, I will not venture to say how far, but only that it was the highest perpendicular wall of bare rock that I ever saw. . . . . This precipice is not sloped, nor is the material soft and crumbling slate as at Montmorenci, but it rises perfectly perpendicular, like the side of a mountain fortress, and is cracked into vast cubical masses of gray and black rock shining with moisture, as if it were the ruin of an ancient wall built by Titans. . . . . Take it altogether, it was a most wild and rugged and stupendous chasm, so deep and narrow where a river had worn itself a passage through a mountain of rock, and all around was the comparatively untrodden wilderness." (THOREAU.)

The base of the **St. Anne Mts.** is reached by a road running up the valley for 3-5 M. The chief peak is 2,687 ft. high, but the view thence is intercepted by trees. The *Valley of St. Feréol* is 8 M. from St. Anne, and is surrounded by beautiful scenery. It contains 1,100 inhabitants, and in the vicinity are several lofty and picturesque cascades. *St. Tite des Caps* is a village of 800 inhabitants, 5 M. from the river, between Cape Tourmente and the St. Feréol Mts. The trouting in these glens is very good, and rare sport is found at *Lake St. Joachim*, several miles beyond.

St. Joachim is 5 M. beyond St. Anne, and is a village of 1,000 inhabitants, situated near the river, and opposite St. François d'Orleans. 2 M. beyond this point is the Château Bellevue and the farm of the Quebec Seminary. The summit of **Cape Tourmente** is about 3 M. from the château, and is sometimes ascended for the sake of its superb * view. The Seminarians have kept a cross upon this peak for the last half-century; and in 1869, 44 Catholic gentlemen, led by the Archbishop of Quebec, erected a new one, 25 ft. high, and covered with tin.

The *Château Bellevue* is a long and massive building of limestone, situated near the foot of Cape Tourmente, and surrounded by noble old forests, in which are shrines of St. Joseph and the Virgin. The château is furnished with reading and billiard rooms, etc., and is occupied every summer by about 40 priests and students from the Seminary of Quebec. The neat Chapel of St. Louis de Gonzaga (the protector of youth) is S. of the château.

Near this point Jaques Cartier anchored in 1535, and was visited by the Indians, who brought him presents of melons and maize. In 1623 Champlain came hither from Quebec and founded a settlement, whose traces are still seen. This post was destroyed by Sir David Kirke's men in 1628, and the settlers were driven away.

St. Joachim was occupied in August, 1759, by 150 of the 78th Highlanders, who had just marched down the Isle of Orleans, through St. Pierre and St. Famille. They were engaged in the streets by armed villagers, and had a sharp skirmish before the Canadians were driven into the forest, after which the Scottish soldiers fortified themselves in the priest's house, near the church.

The site of the seminary was occupied before 1670 by Bishop Laval, who founded here a rural seminary in which the youth of the peasantry were instructed. They were well grounded in the doctrine and discipline of the Church, and were instructed in the mechanic arts and in various branches of farming. This was the first "agricultural college" in America. The broad seigniory of the Côte de Beaupré, which lies between St. Joachim and Beauport, was then an appanage of Bishop Laval, and was more populous than Quebec itself. "Above the vast meadows of the parish of St. Joachim, that here border the St. Lawrence, there rises like an island a low flat hill, hedged round with forests, like the tonsured head of a monk. It was here that Laval planted his school. Across the meadows, a mile or more distant, towers the mountain promontory of Cape Tourmente. You may climb its woody steeps, and from the top, waist-deep in blueberry-bushes, survey, from Kamouraska to Quebec, the grand Canadian world outstretched below; or mount the neighboring heights of St. Anne, where, athwart the gaunt arms of ancient pines, the river lies shimmering in summer haze, the cottages of the *habitants* are strung like beads of a rosary along the meadows of Beaupré, the shores of Orleans bask in warm light, and far on the horizon the rock of Quebec rests like a faint gray cloud; or traverse the forest till the roar of the torrent guides you to the rocky solitude where it holds its savage revels..... Game on the river; trout in lakes, brooks, and pools; wild fruits and flowers on the meadows and mountains; a thousand resources of honest and healthful recreation here wait the student emancipated from books, but not parted for a moment from the pious influence that hangs about the old walls embosomed in the woods of St. Joachim. Around on plains and hills stand the dwellings of a peaceful peasantry, as different from the restless population of the neighboring States as the denizens of some Norman or Breton village." (PARKMAN.)

71. The Isle of Orleans.

Steam ferry-boats leave Quebec three times daily for the Isle of Orleans. The trip gives beautiful views of the city and its marine environs, and of the Montmorenci Falls and the St. Anne Mts.

The island is traversed by two roads. The N. shore road passes from West Point to St. Pierre, in 5 M.; St. Famille, 14 M.; and St. François, 20 M. The S. shore road runs from West Point to Patrick's Hole, in 6 M.; St. Laurent, 7½; St. John, 13½; St. François, 21. A transverse road crosses the island from St. Laurent to St. Pierre.

The Isle of Orleans is about 3½ M. from Quebec, and contains 70 square miles (47,923 acres) of land, being 20 M. long and 5½ M. wide. The beautiful situation of the island, in the broad St. Lawrence, its picturesque heights and umbrageous groves, its quaint little hamlets and peaceful and primitive people, render Orleans one of the most interesting districts of the Lower Province, and justify its title of "the Garden of Canada."

The island was called *Minigo* by the Indians, a large tribe of whom lived here and carried on the fisheries, providing also a place of retreat for the mainland tribes in case of invasion. In 1535 Cartier explored these shores and the hills and forests beyond, being warmly welcomed by the resident Indians and feasted with fish, honey, and melons. He speaks of the noble forests, and adds: "We found there great grape-vines, such as we had not seen before in all the world; and for that we named it the Isle of Bacchus." A year later it received the name of the Isle of Orleans, in honor of De Valois, Duke of Orleans, the son of Francis I. of France. The popular name was *L'Isle des Sorciers* (Wizards' Island), either on account of the marvellous skill of the natives in foretelling future storms and nautical events, or else because the superstitious colonists on the mainland were alarmed at the nightly movements of lights along the insular shores, and attributed to demons and wizards the dancing fires which were carried by the Indians in visiting their fish-nets during the night-tides.

The island was granted in 1620 to the Sieur de Caen by the Duke de Montmorenci, Viceroy of New France. In 1675 this district was formed into the Earldom of St. Laurent, and was conferred on M. Berthelot, who assumed the title of the Count of St. Lawrence. In 1651 the N. part was occupied by 600 Christian Hurons, who had taken refuge under the walls of Quebec from the exterminating Iroquois. In 1656 the Iroquois demanded that they should come and dwell in their country, and upon their refusal fell upon the Hurons with a force of 300 warriors, devastated the island, and killed 72 of the unfortunate Christians. Two tribes were compelled soon after to surrender and be led as captives into the Iroquois country, while the Tribe of the Cord left the island and settled at Lorette. The Isle was overrun by Iroquois in 1661, and in an action with them at Rivière Maheu, De Lauzon, Seneschal of New France, and all his guards were killed, preferring to die fighting than to surrender and be tortured. The great cross of Argentenay was carried away and raised in triumph at the Iroquois village on Lake Onondaga (New York).

For nearly a century the Isle enjoyed peace and prosperity, until it had 2,000 inhabitants with 5,000 cattle and rich and productive farms. Then came the advance of Wolfe's fleet; the inhabitants all fled to Charlesbourg; the unavailing French troops and artillery left these shores; Wolfe's troops landed at St. Laurent, and erected camps, forts, and hospitals on the S. E. point; and soon afterward the British forces systematically ravaged the deserted country, burning nearly every house on the Isle, and destroying the orchards.

The Isle is now divided into two seigniories, or lordships, whose revenues and titles are vested in ancient French families of Quebec. The soil is rich and diversified, and its pretty vistas justify Charlevoix's sketch (of 1720): "We took a stroll on the Isle of Orleans, whose cultivated fields extend around like a broad amphitheatre, and gracefully end the view on every side. I have found this country beautiful, the soil good, and the inhabitants very much at their ease." The agricultural interest is now declining, owing to the antique and unprogressive ideas of the farmers, who confine themselves to small areas and neglect alternation of crops. The farms are celebrated for their excellent potatoes, plums, apples, and for a rare and delicious variety of small cheeses. The people are temperate, generous, and hospitable, and, by reason of their insular position, still preserve the primitive Norman customs of the early settlers under Champlain and Frontenac. The Isle and the adjacent shore of Beaupré have been called the nursery of Canada, so many have been the emigrants from these swarming hives who have settled in other parts of the Provinces.

St. Pierre is the village nearest to Quebec (9 M.), and is reached by ferry-steamers, which also run to *Beaulieu*. It has about 700 inhabitants, and is beautifully situated nearly opposite the Montmorenci Falls and Ange Gardien. The first chapel was erected here in 1651 by Père Lalemant, and was used by the Hurons and French in common. In 1769 the present church of St. Pierre was erected. On this shore, in 1825, were built the colossal timber-ships, the *Columbus*, 3,700 tons, and the *Baron Renfrew*, 3,000 tons, the largest vessels that the world had seen up to that time.

The convent of *St. Famille* was founded in 1685, by the Sisters of the Congregation, and since that time the good nuns have educated the girls of the village, having generally about 70 in the institution. The nunnery is seen near the church, and was built in 1699, having received additions from time to time as the village increased. Its cellar is divided into narrow and contracted cells, whose design has been long forgotten. The woodwork of the convent was burned by Wolfe's foragers in 1759, but was restored in 1761, after the Conquest of Canada. The first church of St. Famille was built in 1671, and the present church dates from 1745. The village is nearly opposite Château Richer, and commands fine views of the Laurentian Mts.

The Parish of *St. François* includes the domain of the ancient fief of Argentenay, and was formed in 1678. In 1683 the first church was built, and the present church dates from 1736, and was plundered by Wolfe's troops in 1759. The view from the church is very beautiful, and includes the St. Lawrence to the horizon, the white villages of the S. coast, and the isles of Madame, Grosse, and Réaux. On the N. shore, at the end of the island, are the broad meadows of Argentenay, where wild-fowl and other game are sought by the sportsmen of Quebec. This district looks across the N. Channel upon the dark and imposing ridges of the St. Anne Mts. and the peaks of St. Feréol; and the view from the church is yet more extensive and beautiful.

The church of *St. John* was built in 1735, near the site of a chapel dating from 1675, and contemporary with the hamlet. This parish is famous for the number of skilful river-pilots which it has furnished. It has about 1,300 inhabitants, and is the most important parish on the island. It is nearly opposite the S. shore village of St. Michel (see page 254).

St. Laurent is 7 M. from St. Jean, upon the well-settled royal road. The parish is entered after crossing the Rivière Maheu, where the Seneschal of New France fell in battle. The Church of St. Laurent is a stately edifice of cut stone with a shining tin roof, and is 113 ft. in length. It replaced churches of 1675 and 1697, and was consecrated in 1861. The *Route des Pretres* runs N. from St. Laurent to St. Pierre, and was so named 50 years ago, when this church had a piece of St. Paul's arm-bone, which was taken away to St. Pierre, and thence was stolen at night by the St. Laurent people. After long controversy, the Bishop of Quebec ordered that each church should restore to the other its own relics, which was done along this road by large processions, the relics being exchanged at the great black cross midway on the road. 1½ M. W. of St. Laurent is the celebrated haven called *Trou St. Patrice* (since 1689), or *Patrick's Hole*, where vessels seek shelter in a storm, or outward-bound ships await orders to sail. The river is 1¼ M. wide here, and there are 10 – 12 fathoms of water in the cove. 2 M. W. of this point is the *Caverne de Bontemps*, a grotto about 20 ft. deep cut in the solid rock near the level of the river.

72. Quebec to Cacouna and the Saguenay River.—The North Shore of the St. Lawrence.

The St. Lawrence Steam Navigation Company has several first-class steamers plying on the lower reaches of the river. The time-table below is that of 1874; but if any changes have been made, they may be seen in the Quebec newspapers, or at the ticket-office, opposite the St. Louis Hotel.

At 7 A. M., on Tuesday and Friday, the *Saguenay* leaves Quebec for St. Paul's Bay, Les Éboulements, Murray Bay, Rivière du Loup (Cacouna), Tadousac, Ha Ha Bay, and Chicoutimi; reaching Quebec again on Thursday and Monday mornings.

On Wednesday, Thursday, and Saturday the *Union* or the *St. Lawrence* leaves Quebec at 7 A. M., for Murray Bay, Rivière du Loup, Tadousac, and Ha Ha Bay; reaching Quebec the second morning after.

On Saturday the *St. Lawrence* leaves Quebec, at noon, for Murray Bay, Rivière du Loup, and Rimouski; reaching Quebec again on Tuesday morning.

Distances.—Quebec to St. Laurent, 12 M.; St. John (Orleans), 17; Isle Madame, 23; Cape Tourmente, 28; St. François Xavier, 45; St. Paul's Bay, 55; Les Éboulements, 66; Murray Bay, 82; Rivière Du Loup, 112 (Cacouna, 118); Tadousac, 134 (Chicoutimi, 235).

The S. shore is described in Route 67 (pages 246–255), and the Isle of Orleans in Route 71. As the steamer moves down across the Basin of Quebec, beautiful * views are afforded on all sides, including a fascinating retrospect of the lofty fortress.

" Behind us lay the city, with its tinned roofs glittering in the morning sunshine, and its citadel-rock towering over the river; on the southern shore, Point Levi, picturesquely climbing the steep bank, embowered in dark trees; then the wooded bluffs with their long levels of farm-land behind them, and the scattered cottages of the *habitants*, while northward the shore rose with a gradual, undulating sweep, glittering, far inland, with houses, and gardens, and crowding villages, until it reached the dark stormy line of the Laurentian Mts. in the N. E. . . . . The sky, the air, the colors of the landscape, were from Norway; Quebec and the surrounding villages suggested Normandy,—except the tin roofs and spires, which were Russian, rather; while here and there, though rarely, were the marks of English occupancy. The age, the order, the apparent stability and immobility of society, as illustrated by external things, belonged decidedly to Europe. This part of America is but 70 or 80 years older than New England, yet there seems to be a difference of 500 years." (BAYARD TAYLOR.)

After running for 17 M. between the populous shores and bright villages of Orleans and Bellechasse (see page 254), the steamer turns to the N. E., when off St. John, and goes toward Cape Tourmente, passing between Isle Madame and the Isle of Orleans. Then St. François is passed, on the l., and the meadows of Argentenay are seen, over which is St. Joachim. As the N. Channel is opened, a distant view of St. Anne de Beaupré may be obtained, under the frowning St. Anne Mts. Cape Tourmente (see page 287) is now passed, beyond which are the great Laurentian peaks of *Cape Rouge* and *Cape Gribaune*, over 2,000 ft. high, and impinging so closely on the river that neither road nor houses can be built. These mountains are of granite, and are partially wooded. 3 M. N. E. of Cape Tourmente is a lighthouse, 175 ft. above the water, on the rugged slope of Cape Rouge. A few miles to the E. is the *Sault au Cochon*, under the crest of a mountain 2,370 ft. high.

ST. PAUL'S BAY.

Boucher asserted, in 1663, that the shore between Cape Tourmente and Tadousac was uninhabitable, " being too lofty, and all rocky and escarped." But the French Canadians, hardy and tireless, and loving the St. Lawrence more than the Normans love the Seine, have founded numerous hamlets on the rocks of this iron shore. The coast between St. Joachim and St. François Xavier is as yet unoccupied.

" We ran along the bases of headlands, 1,000 to 1,500 ft. in height, wild and dark with lowering clouds, gray with rain, or touched with a golden transparency by the sunshine, — alternating belts of atmospheric effect, which greatly increased their beauty. Indeed, all of us who saw the Lower St. Lawrence for the first time were surprised by the imposing character of its scenery." (BAYARD TAYLOR.)

Beyond Abattis and the high cliffs of Cape Maillard the steamer passes the populous village of *St. François Xavier*, extending up the valley of the Bouchard River. On the S. a long line of picturesque islets is passed (see page 254). Beyond Cape Labaie the steamer lies to off St. Paul's Bay, whose unique and beautiful scenery is seen from the deck.

St. Paul's Bay (two small inns) is a parish of 4,000 inhabitants, situated amid the grandest scenery of the N. shore. The people are all French, and the village is clustered about the church and convent near the Gouffre River. In the vicinity are found iron, plumbago, limestone, garnet-rock, and curious saline and sulphurous springs. It is claimed that "no parish offers so much of interest to the tourist, the poet, or the naturalist." The wild and turbulent streams that sweep down the valley have carried away all the bridges which have been erected by the people. Passengers who wish to land at this point are transferred from the steamer to a large sail-boat.

The vistas up the valleys of the Gouffre and the Moulin Rivers show distant ranges of picturesque blue mountains, with groups of conical Alpine peaks. In 1791 it is claimed that the shores of the bay were shaken by earthquakes for many days, after which one of the peaks to the N. belched forth great volumes of smoke and passed into the volcanic state, emitting columns of flame through several days. The peaks are bare and white, with sharp precipices near the summit. The valley of the Gouffre has been likened to the Vale of Clwyd, in Wales, and is traversed by a fair road along the r. bank of the rapid river. 10 - 12 M. from the bay are the extensive deposits of magnetic iron-ore which were explored by order of Intendant Talon, a century and a half ago. In the upper part of the valley, 9 M. from St. Paul's Bay, is *St. Urbain*, a French Catholic village of about 1,000 inhabitants. By this route the tri-weekly Royal mail-stages cross to Chicoutimi, on the upper Saguenay (see page 300). *St. Placide* (Clairvaux) is also back of St. Paul's Bay, and has 400 inhabitants.

" In all the miles of country I had passed over, I had seen nothing to equal the exquisite beauty of the Vale of Baie St. Paul. From the hill on which we stood, the whole valley, of many miles in extent, was visible. It was perfectly level, and covered from end to end with little hamlets, and several churches, with here and there a few small patches of forest..... Like the Happy Valley of Rasselas, it was surrounded by the most wild and rugged mountains, which rose in endless succession one behind the other, stretching away in the distance, till they resembled a faint blue wave in the horizon." (BALLANTYNE.)

" Nothing can be more picturesque than the landscape which may be viewed from the crest of Cap au Corbeau. Have you courage to clamber up the long slopes of Cap au Corbeau; to see the white-sailed schooners at the entrance of the bay; to comprehend the thousand divers objects at your feet; the sinuous course of the Marée and of the serpentine Gouffre; on the S. the old mansions and rich pastures; to see the church and convent and the village, the Cap à la Rey, the bottom of the bay; and, farther away, the shores of St. Antoine Perou, St. Jerome, St. John, St. Joseph, and St. Flavien?" (TRUDELLE.)

The Bay was settled early in the 17th century, and has always been noted for its earthquakes and volcanic disturbances. In October, 1870, it felt such a severe shock that nearly every house in the valley was damaged. In 1759 the village was destroyed by Gorham's New-England Rangers, after the inhabitants had defended it for two hours.

"Above the Gulph I have just mentioned is the Bay of *St. Paul*, where the Habitations begin on the North Side; and there are some Woods of Pine-Trees, which are much valued; Here are also some red Pines of great Beauty. Messrs. of the Seminary of *Quebec* are Lords of this Bay. Six Leagues higher, there is a very high Promontory, which terminates a Chain of Mountains, which extend above 400 Leagues to the West; It is called Cape *Tourmente*, probably because he that gave it this Name, suffered here by a Gust of Wind." (CHARLEVOIX.)

The W. promontory of St. Paul's Bay is Cape Labaie; that on the E., opposite the Isle aux Coudres, is *Cape Corbeau*. "This cape has something of the majestic and of the mournful. At a little distance it might be taken for one of the immense tombs erected in the middle of the Egyptian deserts by the vanity of some puny mortal. A cloud of birds, children of storm, wheel continually about its fir-crowned brow, and seem, by their sinister croaking, to intone the funeral of some dying man."

Between St. Paul's Bay and the Isle aux Coudres is the whirlpool called *Le Gouffre*, where the water suddenly attains a depth of 30 fathoms, and at ebb-tide the outer currents are repulsed from Coudres to Corbeau in wide swirling eddies. It is said that before the Gouffre began to fill with sand schooners which were caught in these eddies described a series of spiral curves, the last of which landed them on the rocks. It was the most dreaded point on this shore, and many lives were lost here; but its navigation is now safe and easy.

The **Isle aux Coudres** is 5½ M. long and 2½ M. wide, and is a charming remnant of primitive Norman life. It has about 500 inhabitants, engaged in farming, and more purely mediæval French than any other people in Canada. The houses are mostly along the lines of the N. W. and S. E. shores; and the Church of St. Louis is on the S. W. point. The island is still owned by the Seminary of Quebec, to which it was granted in 1687. Large numbers of porpoises are caught between this point and the Rivière Ouelle, on the S. shore. Bayard Taylor says: "The Isle aux Coudres is a beautiful pastoral mosaic in the pale emerald setting of the river."

Off the Isle aux Coudres, and between that point and Rivière Ouelle, great numbers of white whales are caught, in fish-pounds made for the purpose. These fish (often taken for porpoises) live in the Lower St. Lawrence from April to October, when they migrate to the Gulf and the Arctic Ocean. They are from 14 to 22 ft. in length, and yield 100-120 gallons of fine oil, which is much used for lighthouse purposes, because it does not freeze in winter. A valuable leather is made from their skins.

When Cartier was advancing up the St. Lawrence in 1535, under the direction of the Quebec Indians whom he had abducted from Gaspé, he landed on this island, and, marvelling at the numerous hazel-trees upon the hills, named it *L'Isle aux Coudres* (Hazel-tree Island). This point he made the division between the country of Saguenay and that of Canada. "In 1663 an Earthquake rooted up a Mountain, and threw it upon the Isle of *Coudres*, which was made one half larger than before, and in the Place of the Mountain there appeared a Gulf, which it is not safe to approach."

The island was deserted by its inhabitants in the summer of 1759, when great British fleets were anchored off the shores, but several boats' crews were driven from the strand by rangers. Three British officers landed on the isle, carrying a flag which they were about to raise on the chief eminence before the fleet; but they were cut off by a small party of Canadians, and were led prisoners to Quebec. Admiral Durell first reached the island, with 10 frigates, and captured 3 French vessels bearing 1,800 barrels of powder.

The steamer runs S. E. for several miles in the narrow channel between the Isle aux Coudres and the mountains of the N. Shore. At 11 M. from St. Paul's Bay it rounds in at the pier (920 ft. long) of the parish of Les Éboulements, a farming district of 2,400 inhabitants. "High on the crest of the Laurentides, old as the world, the tourist sees on the N., on landing at the Éboulements pier, the handsome parish-church." The situation of this village is one of the most quaint and charming on the river, and overlooks the St. Lawrence for many leagues. The white houses are grouped snugly about the tall Notre Dame Church, above which the dark peak of Mt. Éboulements rises to the height of 2,547 ft.

In the vicinity of Les Éboulements are visible the tracks of the great land-slides of 1663, in that season when so many marvellous phenomena were seen in Canada. The St. Lawrence ran "white as milk," as far down as Tadousac; ranges of hills were thrown down into the river, or were swallowed up in the plains; earthquakes shattered the houses and shook the trees until the Indians said that the forests were drunk; vast fissures opened in the ground; and the courses of streams were changed. Meteors, fiery-winged serpents, and ghastly spectres were seen in the air; roarings and mysterious voices sounded on every side; and the confessionals of all the churches were crowded with penitents, awaiting the end of the world.

The steamer now rounds the huge mass of Mt. Éboulements, passing the rugged spurs called Goose Cape and Cape Corneille. On the E. slope is seen the large village of *St. Irénée*, where 900 French people preserve their ancient customs and language. A few miles farther E. the steamer rounds in at Murray Bay.

Murray Bay is the favorite summer resort of the N. Shore, and has fine facilities for boating and bathing, with a long firm beach. It is also one of the best fishing-centres in the Province, and sportsmen meet with success in the waters of the beautiful Murray River, or the Gravel and Petit Lakes. The steamer stops at the long wharf at *Point à Pique*, near which are the hotels, — Duberger's, the Lorne, and Warren's. A new hotel, of 300 ft. front, is being built for the summer of 1875. There are also summer cottages about the base of *Cap à l'Aigle*. The tourists occupy Point à Pique with their hotels, and make excursions to the lakes and the falls. The French town is at the bridge over the Murray River, and is clustered about the great church and the court-house of Charlevoix County. It has 3,000 inhabitants.

"Of all the picturesque parishes on the shore of our grand river, to which innumerable swarms of tourists go every summer to take the waters, none will interest the lover of sublime landscapes more than Malbaie. One must go there to enjoy the rugged, the grandeur of nature, the broad horizons. He will not find here the beautiful wheat-fields of Kamouraska, the pretty and verdurous shores of Cacouna or Rimouski, where the languorous citizen goes to strengthen his energies during the dog-days; here is savage and unconquered nature, and view-points yet more majestic than those of the coasts and walls of Bic. Precipice on precipice; impenetrable gorges in the projections of the rocks; peaks which lose themselves in the clouds, and among which the bears wander through July, in search of berries; where the caribou browses in September; where the solitary crow and the royal eagle make their nests in May; in short, alpine landscapes, the pathless highlands of Scotland, a Byronic nature, tossed about, heaped up in the North, far from the ways of civilized men, near a volcano that from time to time awakens and shakes the country in a manner to frighten, but not to endanger, the romantic inhabitants. According to some, in order to enjoy all the fulness of these austere beauties, one must be at the privileged epoch of life. If then you wish to taste, in their full features, the dreamy solitudes of the shores, the grottos, the great forests of Point à Pique or Cap à l'Aigle, or to capture by hundreds the frisking trout of the remote Gravel Lake, you must have a good eye, a well-nerved arm, and a supple leg." (LEMOINE.)

This district was formerly known as the King's Farm, and had 30 houses at the conquest of Canada. It was then granted to the Scottish officers, Major Nairn and Malcom Fraser, who soon promoted its settlement. It was explored in June, 1608, by Champlain, who named it Malle Baie, on account of "the tide which runs there marvellously, and, even though the weather is calm, the bay is greatly moved." It is still generally known as Malbaie, though the English use the name Murray Bay, given in honor of the general who granted it to the Scots. The Scotch families brought out by Fraser and Nairn are now French in language and customs. A depot for American prisoners-of-war was established here in 1776, near the Nairn manor-house, and the barracks were built by the captives themselves.

The great French settlement of *St. Agnes*, with 1,600 inhabitants, is 9 M. W. of Murray Bay, up the valley, and on the verge of the wide wilderness of the Crown Lands. A rugged road follows the N. shore from Murray Bay to the Saguenay River, a distance of about 40 M., passing the romantic St. Fidèle (9 M. out ; 1,000 inhabitants), the lumbering village of Port au Persil, the hamlets of Black River, Port aux Quilles, St. Simeon, and Callière, back of which are mountains where many moose and caribou are found. Still farther E. is Baie des Rochers, on an island-studded bay.

The steamer now stretches out across the river in a diagonal course of 30 M., the direction being about N. E. The river is about 20 M. wide, and the steamer soon comes in sight of the Kamouraska Islands (see page 252), on the l., and then passes between Hare Island (l.) and the Pilgrims. The vessel soon reaches the long pier at *Point à Beaulieu*, 3 M. from Rivière du Loup.

Rivière du Loup (*La Rochelle House;* and several large summer boarding-houses) is a prosperous village of 1,200 inhabitants, occupying a fine position on a hillside near the mouth of the river. There are some pretty villas in the vicinity, and the great church in the centre of the town is a prominent landmark for miles. About 3 M. up the river are the famous **Rivière-du-Loup Falls*, near the new and massive bridge of the Intercolonial Railway. The stream here plunges over a cliff about 80 ft. high, and then rests quietly in a broad pool below. The views of the river and its islands and shipping, from the streets of the village, are broad and beautiful; and many summer visitors pass their vacations here, finding comfortable accommodations in the boarding-houses. The Grand-Portage road runs S. E. from this point into New Brunswick, crossing numerous trout-streams and leading through a desolate region of hills. Its first point of interest is the long *Temiscouata Lake* (see page 58).

Rivière du Loup will soon be one of the chief railway-centres of Canada. It has been the E. terminus of the Grand Trunk line for years. The Intercolonial is now nearly (or quite) completed from this point to St. John and Halifax, and the New-Brunswick Railway is being pushed hitherward up the St. John Valley (see page 49).

This domain was granted by the Compagnie des Indes Occidentales to the Sieur de la Chesnoye in 1673. It is said that its name is derived from the fact that in former years great droves of seals (*loups-marins*) frequented the shoals at the mouth of the river, making a remarkable uproar at night.

A persistent attempt has been made to call this town *Fraserville*, in honor of the Frasers, who are its seigniors. The numerous Frasers of this Province met at Quebec in 1868 to re-form their ancient Scottish clan organization, and to name Provincial, county, and parish chieftains. The head-chief is entitled *The* Fraser, and is the Hon. John Fraser de Berry, "58th descendant of Jules de Berry, a rich and powerful lord, who gave a sumptuous feast to the Emperor Charlemagne and his numerous suite, at his castle in Normandy, in the 8th century." The solemn Scots maintain that De Berry then regaled Charlemagne with strawberries (*fraises*, in the French language), and that the Emperor was so greatly pleased that he ordered that he should thenceforth be known as *Fraiser de Berry*, and from him the Clan Fraser traces its name and descent.

Cacouna is 6 M. from Rivière du Loup, and is the chief summer resort of Canada. The * *St. Lawrence Hall* is the most fashionable hotel, and accommodates 600 guests, at $2.50-3 a day. The *Mansion House* charges $2 a day. There are several summer boarding-houses whose rates are still lower. The traveller who visits Cacouna from Rivière du Loup must be on his guard against the extortions of the carriage-drivers, who frequently demand exorbitant fares.

Twenty years ago Cacouna was nothing; it is now filled with great hotels and boarding-houses, and adorned with many summer cottages. It is visited by thousands of Canadians, and also by many Americans "fuyant le ciel corrosif de New-York." Here may be seen the Anglo-Canadian girls, who are said to combine the physical beauty and strength of the English ladies with the vivacity and brilliancy of the Americans. The amusements of the village are like those of similar places farther S.,—sea-bathing and fishing, driving, and balls which extend into the small hours. The beach is good, and the river-views from the heights are of famed beauty. There is a pretty lake back among the hills, where many trout are found.

The great specialties of Cacouna are its pure cool air and brilliant northern scenery. It is sometimes found too cold, even in August, during rainy weather, for the American visitors, who then hurry away in crowds. The peninsula of Cacouna is a remarkable mass of rock, nearly 400 ft. high, which is connected with the mainland by a low isthmus. Its name was given by the Indians, in allusion to its form, and signifies "the turtle." The village is French, and has 400 inhabitants and 3 churches. 4½ M. distant is the populous parish of *St. Arsène*, and 8 M. S. is *St. Modeste*.

From Rivière du Loup the steamer runs across to the Saguenay River, passing within 3-4 M. of Cacouna, and running between the Brandy Pots (l.) and Red Island (see page 252).

The **Saguenay River**, see Route 73.

73. The Saguenay River.

Steamers leave Quebec for Chicoutimi, the farthest port on the Saguenay, on Tuesday and Friday, at 7 A. M. (see page 291); and for Ha Ha Bay on Wednesday, Thursday, and Saturday. They reach Tadousac by nightfall, and start on the return from Chicoutimi the next morning.

Distances. — Quebec to Tadousac, 134 M.; Tadousac to Rivière St. Marguerite, 15; St. Louis Islets, 19; Rivière aux Canards, 23; Little Saguenay River, 27; St. John's Bay, 32; Eternity Bay, 41; Trinity Bay, 48; Cape Rouge, 56; Cape East, 63; Cape West, 65; St. Alphonse, 72; St. Fulgence, 95; Chicoutimi, 100. This itinerary is based on that of the steamship company and is not correct, but will be useful in marking approximations to the relative distances between the points on the river. There is no other table of distances accessible. Imray's *Sailing Directions* (precise authority) says that it is 65 M. from the St. Lawrence to Chicoutimi.

The ** **Saguenay River** is the chief tributary of the Lower St. Lawrence, and is the outlet of the great Lake St. John, into which 11 rivers fall. For the last 50 M. of its course the stream is from 1 to 2½ M. wide, and is bordered on both sides by lofty precipices of syenite and gneiss, which impinge directly on the shores, and are dotted with stunted trees. Along their slopes are the deep lines of glacial striations, telling of the passage of formidable icebergs down this chasm. The bed of the river is 100 fathoms lower than that of the St. Lawrence, a difference which is sharply marked at the point of confluence. The shores were stripped of their forests by a great fire, in 1810, but there are large numbers of hemlock and birch trees in the neighboring glens. The river is frozen from the St. Louis Isles to Chicoutimi during half the year, and snow remains on the hills until June. The awful majesty of its unbroken mountain-shores, the profound depth of its waters, the absence of life through many leagues of distance, have made the Saguenay unique among rivers, and it is yearly visited by thousands of tourists as one of the chief curiosities of the Western World.

"The Saguenay is not, properly, a river. It is a tremendous chasm, like that of the Jordan Valley and the Dead Sea, cleft for 60 M. through the heart of a mountain wilderness. . . . . No magical illusions of atmosphere enwrap the scenery of this northern river. Everything is hard, naked, stern, silent. Dark-gray cliffs of granitic gneiss rise from the pitch-black water; firs of gloomy green are rooted in their crevices and fringe their summits; loftier ranges of a dull indigo hue show themselves in the background, and over all bends a pale, cold, northern sky. The keen air, which brings out every object with a crystalline distinctness, even contracts the dimensions of the scenery, diminishes the height of the cliffs, and apparently belittles the majesty of the river, so that the first feeling is one of disappointment. Still, it exercises a fascination which you cannot resist. You look, and look, fettered by the fresh, novel, savage stamp which nature exhibits, and at last, as in St. Peter's or at Niagara, learn from the character of the separate features to appreciate the grandeur of the whole. . . . . Steadily upwards we went, the windings of the river and its varying breadth — from ½ M. to nearly 2 M. — giving us a shifting succession of the grandest pictures. Shores that seemed roughly piled together out of the fragments of chaos overhung us, — great masses of rock, gleaming duskily through their scanty drapery of evergreens, here lifting long irregular walls against the sky, there split into huge, fantastic forms by deep lateral gorges, up which we saw the dark-blue crests of loftier mountains in the rear. The water beneath us was black as night, with a pitchy glaze on its surface; and the only life in all the savage solitude was, now and then, the back of a white porpoise, in some of the deeper coves. . . . . The river is a reproduction — truly on a contracted scale — of the fiords of the Norwegian

13*

coast..... The dark mountains, the tremendous precipices, the fir forests, even the settlements at Ha Ha Bay and L'Anse à l'Eau (except that the houses are white instead of red) are as completely Norwegian as they can be. The Scandinavian skippers who come to Canada all notice this resemblance, and many of them, I learn, settle here." (BAYARD TAYLOR.)

"From Ha Ha right down to the St. Lawrence, you see nothing but the cold, black, gloomy Saguenay, rolling between two straight lines of rocky hills that rise steeply from the water's edge. These hills, though steep, are generally roughly rounded in shape, and not abrupt or faced with precipices. This makes the scenery differ from that with which it has been often compared, the boldest of the fiords of Norway. Over the rugged hills of the Saguenay there is generally enough of earth here and there lodged to let the gray rock be dotted over with a dark-green sprinkling of pine-trees. Perhaps there is hardly a spot on the Saguenay, which, taken by itself, would not impress any lover of wild nature by its grandeur, and even sublimity; but after sailing for 70 miles downwards, passing rocky hill after rocky hill, rising one beyond the other in monotonously straight lines alongside of you; .... after vainly longing for some break in these twin imprisoning walls, which might allow the eye the relief of wandering over an expanse of country,—you will begin to compare the Saguenay in no kindly spirit to the Rhine..... It is a cold, savage, inhuman river, fit to take rank with Styx and Acheron; and, into the bargain, it is dull. For the whole 70 miles, you will not be likely to see any living thing on it or near it, outside of your own steamer, not a house, nor a field, nor a sign of any sort that living things have ever been there." (WHITE.)

"Sunlight and clear sky are out of place over its black waters. Anything which recalls the life and smile of nature is not in unison with the huge naked cliffs, raw, cold, and silent as the tombs. An Italian spring could effect no change in the deadly, rugged aspect; nor does winter add one iota to its mournful desolation. It is with a sense of relief that the tourist emerges from its sullen gloom, and looks back upon it as a kind of vault,—Nature's sarcophagus, where life or sound seems never to have entered. Compared to it the Dead Sea is blooming, and the wildest ravines look cosey and smiling. It is wild without the least variety, and grand apparently in spite of itself; while so utter is the solitude, so dreary and monotonous the frown of its great black walls of rock, that the tourist is sure to get impatient with its sullen dead reverse, till he feels almost an antipathy to its very name. The Saguenay seems to want painting, blowing up, or draining,—anything, in short, to alter its morose, quiet, eternal awe. Talk of Lethe or the Styx,—they must have been purling brooks compared with this savage river; and a picnic on the banks of either would be preferable to one on the banks of the Saguenay." (*London Times.*)

On Sept. 1, 1535, Tadousac was visited by the wonder-loving Cartier, with three vessels. He saw the Indians fishing off shore, and reported that, "in ascending the Saguenay, you reach a country where there are men dressed like us, who live in cities, and have much gold, rubies, and copper." The river was visited by Roberval in 1543, and part of the expedition was lost. Thenceforward the country of the Saguenay was explored by the fur-traders and the fearless Jesuits. In 1603 Tadousac was visited by Champlain, around whose vessel the natives crowded in their canoes in order to sell or barter away their peltries. Seven years later a solemn and beautiful scene occurred at Point la Boule (the immense promontory which is seen 5 M. up-stream), when Champlain and Lescarbot attended the great council of the Montaignais. They were received with dignified courtesy by the Sagamore Anadabijou, and conducted to the meeting of the warriors, where several grave and eloquent speeches were made while the pipe of peace was passed around. The Montaignais at that time numbered 9 tribes, 2 of which dwelt along the river, and the other 7 occupied the vast area towards Hudson's Bay and the land of the Esquimaux. Their last Sagamore, Simeon, died in 1849, and had no successor, and the poor remnant of the nation now obtains a precarious living by beggary, or has withdrawn into the fastnesses of the North. The present name of the river is a modification of the original Indian word *Saggishsékuss*, which means "a river whose banks are precipitous."

In 1671 the heroic and self-abnegating Jesuit, Père de Crepieul, founded the mission at Tadousac, where he remained for 26 years, passing the winters in the wretched huts of the savages. Before this time (in 1661) the Fathers Druillettes and

TADOUSAC.   *Route 73.*   299

Dablon had ascended the river to Lake St. John and there had baptized many Indians, and founded the mission of St. François Xavier. The Montaiguais are still in the Catholic faith, and each family has its prayer-book and breviary, in which they are able to read. In 1671 Father Albanel ascended the Saguenay from Tadousac, by order of Intendant Bigot, and passed N. to Hudson's Bay by way of the great lakes of St. John and Mistassini. The country about the Upper Saguenay was then well known to the zealous churchmen, but after the decline of the missions it was forgotten. About 50 years ago the Canadian government had it re-explored by efficient officers, and this remote region is now being occupied by French-Canadian hamlets. The chief business on the river is the exportation of lumber, which is shipped from Chicoutimi in immense quantities.

Tadousac is a small village, prettily situated on a semicircular terrace surrounded with mountains and fronting on a small harbor, deep and secure. The St. Lawrence is here about 24 M. wide, and the mountains of the S. shore are visible, while on clear days the view includes the white villages of Cacouna and Rivière du Loup. The * *Tadousac Hotel* ($ 2.50 a day) is a spacious establishment on the bluff over the beach. It was founded in 1865 by a joint-stock company, and has been successful. The sea-bathing is very good, although the water is cold, and sea-trout are caught off the shore. The old buildings of the Hudson's Bay Company are near the hotel, and on the lawn before them is a battery of antiquated 4-pounders. E. of the hotel is the old * chapel of the Jesuit mission, which was erected in 1746 on the site of a still more ancient church. The summer cottages are near the shore, and are cheerful little buildings. The Earl of Dufferin, Governor-General of Canada, has erected a handsome house here. The scenery of the landward environs is described in the Indian word *Tadousac*, which means knobs or mamelons.

"Tadousac is placed, like a nest, in the midst of the granite rocks that surround the mouth of the Saguenay. The chapels and the buildings of the post occupy the edge of a pretty plateau, on the summit of an escarped height. So perched, these edifices dominate the narrow strip of fine sand which sweeps around at their feet. On the r. the view plunges into the profound waters of the sombre Saguenay ; in front, it is lost in the immense St. Lawrence. All around are mountains covered with fir-trees and birches. Through the opening which the mighty river has cut through the rock, the reefs, the islands, and south shores are seen. It is a delicious place" (TACHÉ.)

4 M. E. of Tadousac is the harbor of *Moulin à Baude*, where are large beds of white marble. Charlevoix anchored here in the *Chameau* (in 1700), and was so enthusiastic over the discovery that he reported that "all this country is full of marble." *Pointe Rouge*, the S. E. promontory before Tadousac, is composed of an intensely hard red granite. The shore extends to the N. E. to the famous shooting-grounds of Mille Vaches, the trout-stream of the Laval River, and the Hudson's Bay post of Betsiamitis (see page 233).

In the year 1599 a trading-post was established at Tadousac by Pontgravé and Chauvin, to whom this country had been granted. They built storehouses and huts, and left 16 men to gather in the furs from the Indians, but several of these died and the rest fled into the forest. Two subsequent attempts within a few years ended as disastrously. In 1628 the place was captured by Admiral Kirke, and in 1632 his brother died here. In 1658 the lordship of this district was given to the Sieur Demaux, with the dominion over the country between Eboulements and Cape Cormorant. Three years later the place was captured by the Iroquois, and the garrison was massacred. In 1690 three French frigates, bearing the royal treasure to Quebec, were chased in here by Sir William Phipps's New-England fleet. They formed batteries on the Tadousac shores, but the Americans were unable to get their vessels up through the swift currents, and the French fleet was saved. The trading-post and mission were kept up with advantage. Charlevoix visited the place in 1720, and says: " The greatest Part of our Geographers have here placed a Town, but where there never was but one *French* house, and some huts of Savages who came there in the Time of the Trade and who carried away their Huts or Booths, when they went away; and this was the whole matter. It is true that this Port has been a long Time the Resort of all the Savage Nations of the North and East, and that the *French* resorted thither as soon as the Navigation was free both from *France* and *Canada;* the Missionaries also made Use of the Opportunity, and came to trade here for Heaven. And when the Trade was over, the Merchants returned to their Homes, the Savages took the Way to their Villages or Forests, and the Gospel Labourers followed the last, to compleat their Instructions."

The steamer leaves Tadousac during the evening, and ascends the river by night, when, if the sky is unclouded, there are beautiful effects of starlight or moonlight on the frowning shores. The return trip down the river is made the next day, and the full power of the scenery is then felt. This description of the river begins, therefore, at the head of navigation, and follows the river downward, detaching the détour into Ha Ha Bay, for the sake of continuity.

Chicoutimi (good hotel) is the capital of Chicoutimi County, and has 700 inhabitants. It is situated at the head of navigation on the Saguenay, and is the great shipping-point of the lumber districts. Over 40 ships load here every year, most of them being squarely built Scandinavian vessels. The trade amounts to $500,000 a year, and is under the control of Senator Price of Quebec, who has fine villas at Chicoutimi and Tadousac, and is known as " The King of the Saguenay." The powerful house of Price Brothers & Co. owns most of the Saguenay country, and has establishments on the Lower St. Lawrence and in England. Their property in mills, buildings, and vessels is of immense value. Over the steamboat-pier is the new college, which is being built of stone, about an open quadrangle. Near by are the church and the convent of the Good Shepherd. Beyond the village the court-house is seen, on the dark slope of a high hill; and the white ribbon of the * *Chicoutimi Falls* is visible to the l. The Chicoutimi River here falls 40–50 ft., just before entering the Saguenay. This stream affords fine sport for the fisherman, and contains great numbers of fish resembling the land-locked salmon, or grilse.

Chicoutimi signifies "deep water," and was so called by the Northern Indians who here first encountered the profound depths of the Saguenay. There is fine fishing about the falls and the adjacent rapids (permission must be obtained, and is often granted in courtesy to strangers). The ancient Jesuit chapel and the Hudson's Bay Company's post were situated near the confluence of the two rivers, and within the chapel (which remained until recently) was the tomb of Father Cocquart, the last of the Jesuit missionaries. A strong mission was founded here in 1727, by Father Labrosse, and many Indians were converted.

St. Anne du Saguenay is a village of 200 inhabitants, on the high bank of the river opposite Chicoutimi. Lake St. John is about 60 M. W. of Chicoutimi, and is reached by a good road, which passes through Jonquière, Kenogami, and Hebertville (1,200 inhabitants). The *Rapids of Terres Rompues*, on the Saguenay River, are 9 M. above Chicoutimi.

"These rapids extend 3 M.; then there are 3 M. of smooth water; then a second rapid of terrific strength; then 10 M. of still water; then 2 M. of rapids; then ¾ M. of still water. Finally, there succeed the mighty rush and uproar of the Grand Décharge, mingling with the foam and tumult of the Petit Décharge. These empty the waters of the Grand St. John Lake, and sweeping around a rugged island with terrific and unnatural force, unite, and rage, contend, and finally melt and settle down into the quiet mood of the still water below." In this part of the river is found the winninish, or Northern charr, a game-fish whose pink meat is considered a greater delicacy than brook-trout or salmon.

Lake St. John was discovered in 1647 by Father Duquen, the missionary at Tadousac, who was the first European to ascend the Saguenay to its source. It was then called by the Indians *Picouagami*, or Flat Lake. Several Jesuit missionaries soon passed by this route to the great Nekouba, where all the northern tribes were wont to meet in annual fairs; and in 1672 Father Albanel advanced from Tadousac, by Lake St. John and Lake Mistassini, to the *Mer du Nord*, or Hudson's Bay. A Catholic mission was founded on the lake, at Metabetchuan, and posts of the Hudson's Bay Company were also established here. The lake is of great area, and receives the waters of 8 large rivers, the chief of which is the Mistassini, flowing down 250 M. from *Lake Mistassini*, which is 75 × 30 M. in area. The water is shallow, and is agitated into furious white waves by the N. W. winds. To the N. and W. is a vast region of low volcanic mountains and dreary lands of low spruce forests. The soil along the lake-shores is said to be a fertile alluvium, capable of nourishing a dense population; but the winters are long and terrible. 20 years ago there were no settlements here except the Hudson's Bay posts; now there are numerous villages, the chief of which are Roberval, Rivière à l'Ours, and St. Jerome.

Mr. Price, M. P., states that a missionary has recently discovered, high upon the Saguenay (or on the Mistassini), an ancient French fort, with intrenchments and stockades. On the inside were two cannon, and several broken tombstones dating from the early part of the 16th century. It is surmised that these remote memorials mark the last resting-place of the Sieur Roberval, Governor-General of Canada, who (it is supposed) sailed up the Saguenay in 1543, and was never heard from afterwards. The Robervals were favorites of King Francis I., who called one of them "the Petty King of Viemen," and the other, "the Gendarme of Hannibal." They were both lost on their last expedition to America.

In descending the Saguenay from Chicoutimi to Ha Ha Bay, the scenery is of remarkable boldness, but is less startling than the lower reaches of the river. Soon after leaving the village the steamer passes the pretty villa and the Anglican church pertaining to Senator Price. Below this point is a line of hills of marly clay; and Cape St. François soon rears its dark crest on the l. bank. The river widens rapidly, and the hamlet of *St. Fulgence* is seen on the l., near Pointe Roches. Beyond the ponderous walls of *High Point* is another broad reach, with small islets under the l. bank. The steamer now runs between the frowning promontories of Cape East and Cape West, and passes the entrance to Ha Ha Bay.

* **Ha Ha Bay** runs 7 M. S. W. from the Saguenay, and is ascended between lofty and serrated ridges, bristling with sturdy and stunted trees. So broad and stately is this inlet that it is said that the early French explorers ascended it in the belief that it was the main river, and the name originated from their exclamations on reach'ng the end, either of amusement at their mistake or of pleasure at the beautiful appearance of the meadows. After running for several miles between the terraced cliffs of Cape West (on the r.) and the opposite ridges, the steamer enters a wide haven whose shores consist of open intervale-land, backed by tall blue heights. The entrance is 4 M. long, 1 M. wide, and 100 fathoms deep, and the haven can be reached by ships of the line without difficulty. It is expected that this bay will be the great port of "the hyperborean Latin nation" which is fast settling the Upper Saguenay and Lake St. John country. Large quantities of lumber are loaded here upon British and Scandinavian ships, and a flourishing trade is carried on in the autumn by sending farm-produce and blueberries to Quebec, — the latter being packed in coffin-shaped boxes and sold for 10 – 20 cents a bushel.

The steamer is moored to the wharf at St. *Alphonse* (Bagotville), near which is the church and a village of 250 inhabitants. Calashes are found at the pier, on which the passengers can ride up over the hills or to *St. Alexis* (Grande Baie), a village of 300 inhabitants, 3 M. distant on the S. shore of the bay. The mail-road is prolonged from this point, through the uninhabited wilderness of the Crown Lands, to St. Urbain and St. Paul's Bay (see page 292). The Rivière à Mars, emptying into the bay between St. Alphonse and St. Alexis, is famous for its salmon-fisheries.

"The long line of sullen hills had fallen away, and the morning sun shone warm on what in a friendlier climate would have been a very lovely landscape. The bay was an irregular oval, with shores that rose in bold but not lofty heights on one side, while on the other lay a narrow plain with two villages clinging about the road that followed the crescent beach, and lifting each the slender tin-clad spire of its church to sparkle in the sun. At the head of the bay was a mountainous top, and along its waters were masses of rocks, gayly painted with lichens and stained with metallic tints of orange and scarlet." (HOWELLS.)

21 M. from Ha Ha Bay is *Lac d la Belle Truite*, famous for its immense red trout, and beyond is the Great Ha Ha Lake, among the mountains, with bold capes encircling forests, and a pretty island. 6 M. from Belle Truite is the Little Ha Ha Lake, on whose shore is a stupendous cliff nearly 2,000 ft. high. The blue peaks of the *St. Margaret Mts.* are about 30 M. from Ha Ha Bay, and sweep from Lake St. John to Hudson's Bay. Carriages may be taken from St. Alphonse to Chicoutimi (12 M.), and for longer excursions toward Lake St. John.

After passing the dark chasm of Ha Ha Bay, Cape East is seen on the l., throwing its serrated ledges far out into the stream, and cutting off the retrospective view. Rugged palisades of syenite line the shores on both sides. "The procession of the pine-clad, rounded heights on either shore began shortly after Ha Ha Bay had disappeared behind a curve, and it hardly ceased, save at one point, before the boat re-entered the St. Lawrence. The shores of the river are almost uninhabited. The hills rise from the water's edge; and if ever a narrow vale divides them, it is but to open drearier solitudes to the eye." Just before reaching Cape Rouge (l. bank) the ravine of *Descente des Femmes* opens to the N., deriving its singular name from a tradition that a party of Indians were starving, in the back-country, and sent their squaws for help, who descended to the river through this wild gorge and secured assistance.

ETERNITY BAY.   *Route 73.*   303

On the r. bank is *** Le Tableau**, a cliff 900 ft. high, whose riverward face contains a broad sheet of dark limestone, 600 × 300 ft. in area, so smooth and straight as to suggest a vast canvas prepared for a picture. Still farther down (r. bank) is

"***Statue Point**, where, at about 1,000 feet above the water, a huge, rough Gothic arch gives entrance to a cave, in which, as yet, the foot of man has never trodden. Before the entrance to this black aperture, a gigantic rock, like the statue of some dead Titan, once stood. A few years ago, during the winter, it gave way, and the monstrous statue came crashing down through the ice of the Saguenay, and left bare to view the entrance to the cavern it had guarded perhaps for ages."

The steamer soon passes Cape Trinity on the r. bank, and runs in close to ****Eternity Bay**, which is a narrow cove between the majestic cliffs of Cape Trinity and Cape Eternity. The water is 150 fathoms deep, and the cliffs descend abruptly into its profoundest parts. ***Cape Trinity** consists of three vast superimposed precipices, each of which is 5 – 600 ft. high, on whose faces are seen two remarkable profiles. The echo in the bay is wonderful, and is usually tested by discharging a gun or blowing a whistle. (In recent maps and descriptions the name of Eternity has been given to the N. cape, and Trinity to the other. This is not correct, for the N. cape was named *La Trinité* by the Jesuits on account of its union of three vast sections into one mountain. It is known by that name among the old pilots and river-people. The Editor has substituted the correct names in the ensuing quotations.*)*

" The masterpiece of the Saguenay is the majesty of its two grandest bulwarks, — Cape Trinity and Cape Eternity, — enormous masses of rock, 1,500 feet high, rising sheer out of the black water, and jutting forward into it so as to shelter a little bay of the river between their gloomy portals. In the sublimity of their height and steepness, and in the beautiful effect against the rock of the pine-trees which here and there gain a dizzy foothold, nestling trustfully into every hollow on the face of the tremendous precipice, these capes can hardly be surpassed by any river-scene in the world." (WHITE.)

"Suddenly the boat rounded the corner of the three steps, each 500 ft. high, in which Cape Trinity climbs from the river, and crept in under the naked side of the awful cliff. It is sheer rock, springing from the black water, and stretching upward with a weary, effort-like impulse, in long impulses of stone marked by deep seams from space to space, till, 1,500 ft. in air, its vast brow beetles forward, and frowns with a scattering fringe of pines. . . . . The rock fully justifies its attributive height to the eye, which follows the upward rush of the mighty acclivity, steep after steep, till it wins the cloud-capt summit, when the measureless mass seems to swing and sway overhead, and the nerves tremble with the same terror that besets him who looks downward from the verge of a lofty precipice. It is wholly grim and stern ; no touch of beauty relieves the austere majesty of that presence. At the foot of Cape Trinity the water is of unknown depth, and it spreads, a black expanse, in the rounding hollow of shores of unimaginable wildness and desolation, and issues again in its river's course around the base of Cape Eternity. This is yet loftier than the sister cliff, but it slopes gently backward from the stream, and from foot to crest it is heavily clothed with a forest of pines. The woods that hitherto have shagged the hills with a stunted and meagre growth, showing long stretches scarred by fire, now assume a stately size, and assemble themselves compactly upon the side of the mountain, setting their serried stems one rank above another, till the summit is crowned with the mass of their dark green plumes, dense and soft and beautiful ;

so that the spirit, perturbed by the spectacle of the other cliff, is calmed and assuaged by the serene grandeur of this." (HOWELLS'S *A Chance Acquaintance.*)

"These awful cliffs, planted in water nearly a thousand feet deep, and soaring into the very sky, form the gateway to a rugged valley, stretching inland, and covered with the dark primeval forest of the North. I doubt whether a sublimer picture of the wilderness is to be found on this continent. . . . The wall of dun-colored syenitic granite, ribbed with vertical streaks of black, hung for a moment directly over our heads, as high as three Trinity spires atop of one another. Westward, the wall ran inland, projecting bastion after bastion of inaccessible rock, over the dark forests in the bed of the valley." (BAYARD TAYLOR.)

"The wild scenery of the river culminates at a little inlet on the right bank between Capes Trinity and Eternity. Than these two dreadful headlands nothing can be imagined more grand and impressive. For one brief moment the rugged character of the river is partly softened, and, looking back into the deep valley between the capes, the land has an aspect of life and mild luxuriance which, though not rich, at least seems so in comparison with the grievous awful barrenness. Cape Eternity on this side towards the landward opening is pretty thickly clothed with fir and birch mingled together in a color contrast which is beautiful enough, especially where the rocks show out among them, with their little cascades and waterfalls like strips of silver shining in the sun. But Cape Trinity well becomes its name, and is the reverse of all this. It seems to frown in gloomy indignation on its brother for the weakness it betrays in allowing anything like life or verdure to shield its wild, uncouth deformity of strength. Cape Trinity certainly shows no sign of relaxing in this respect from its deep savage grandeur. It is one tremendous cliff of limestone, more than 1,500 feet high, and inclining forward more than 200 feet, brow-beating all beneath it, and seeming as if at any moment it would fall and overwhelm the deep black stream which flows so cold and motionless down below. High up, on its rough gray brows, a few stunted pines show like bristles their scathed white arms, giving an awful weird aspect to the mass, blanched here and there by the tempests of ages, stained and discolored by little waterfalls in blotchy and decaying spots. Unlike Niagara, and all other of God's great works in nature, one does not wish for silence or solitude here. Companionship becomes doubly necessary in an awful solitude like this." (*London Times.*)

When the *Flying Fish* ascended the river with the Prince of Wales and his suite, one of her heavy 68-pounders was fired off near Cape Trinity. " For the space of half a minute or so after the discharge there was a dead silence, and then, as if the report and concussion were hurled back upon the decks, the echoes came down crash upon crash. It seemed as if the rocks and crags had all sprung into life under the tremendous din, and as if each was firing 68-pounders full upon us, in sharp, crushing volleys, till at last they grew hoarser and hoarser in their anger, and retreated, bellowing slowly, carrying the tale of invaded solitude from hill to hill, till all the distant mountains seemed to roar and groan at the intrusion."

St. John's Bay (r. bank) is 6 M. below Eternity Bay, and is shallow enough to afford an anchorage for shipping. It is 2 M. wide and 3 M. long, and receives the St. John River. At its end is a small hamlet, situated in a narrow valley which appears beautiful in contrast with the surrounding cliffs. Far inland are seen the blue peaks of distant mountains. In the little cove opposite is the white thread of a lofty cascade.

The *Little Saguenay River* (r. bank) is 4 M. below, and flows down out of a bristling wilderness where are famous Indian hunting-grounds and pools filled with trout. A short distance below are the islets at the mouth of the Rivière aux Canards. The steamer then sweeps by the **St. Louis Isle**, a granite rock, ½ M. long, covered with firs, spruces, and birch-trees. There is 1,200 ft. depth of water around this islet, in which are multitudes of salmon-trout. On the r. bank are the massive promontories of Cape Victoria and Cape George. The *retrospect from this point affords one of the grandest views on the river. 2 M. below (l. bank) is seen the intervales of the *St. Marguerite River*, the chief tributary of the Saguenay, descending from a lake far N. of Chicoutimi, and famous for its salmon-fisheries (leased). It is a swift stream, flecked with rapids, but is navigable for 20 M. by canoes; and flows from a valuable region of hard-wood trees. There are huts along the strand at its mouth, and vessels are usually seen at anchor here; while far inland are bare and rugged ridges. The tall promontory beyond this river is seamed with remarkable trap-dikes, of a color approaching black; opposite which is the mouth of the St. Athanase.

Beyond Point Crepe (r. bank) is the deep cove of *St. Etienne Bay*, affording an anchorage, and bordered with narrow strips of alluvial land. The steamer now sweeps rapidly down, between immense cliffs, and with but narrow reaches of the river visible ahead and astern. Beyond the *Passe Pierre Isles* (r. bank) it approaches a castellated crag on the r., opposite which is the frowning promontory called * **Pointe la Boule**, a vast granite mountain which narrows the channel to very close confines. From Pointe la Boule to Tadousac, the river flows between escarped cliffs of feldspathic granite, with an appearance resembling stratification dipping to the S. E. Their lofty rounded summits are nearly barren, or at most support a thin fringe of low trees; and the sheer descent of the sides is prolonged to a great depth beneath the water.

The vessel calls at *L'Anse à L'Eau*, the little cove near Tadousac (see page 299); and soon afterwards steams out into the broad St. Lawrence, in the darkness of evening. The next morning, the traveller awakes at or near Quebec.

74. Quebec to Montreal.—The St. Lawrence River.

The steamboats of the Richelieu Company, the *Quebec* and the *Montreal*, are among the largest and most elegant river-boats in America. They leave Quebec every evening, arriving at Montreal early the next morning. Day-boats are sometimes put on during the summer season, and should be preferred by the tourist, as enabling him to see the river and its villages. The Union Line entered into competition with the Richelieu boats, in 1874, with the fine steamers *Athenian* and *Corinthian*. The prices on both lines were reduced to the following figures, but will probably be raised again in 1875:—

Fares.—Quebec to Batiscan, 75c.; to Three Rivers, $1; to Sorel, $1.25; to Montreal, $2 (supper and berth included).

Distances.— Quebec to Batiscan, 69 M.; Three Rivers, 90; Sorel, 135; Montreal, 180.

The Grand Trunk Railway runs two trains daily between Quebec and Montreal. *Stations.* — Quebec (Point Levi); Hadlow, 2 M.; Chaudière Curve, 8; Craig's Road, 15; Black River, 20; Methot's Mills, 28; Lyster, 37; Becancour, 41; Somerset, 49; Stanfold, 55; Arthabaska, 64; Warwick, 71; Danville, 84; Richmond, 96; New Durham, 106; Acton, 118; Upton, 124; Britannia Mills, 130; St. Hyacinthe, 137; Soixante, 144; St. Hilaire, 150; St. Bruno, 157; St. Hubert, 162; St. Lambert, 167; Montreal, 172.

" It could really be called a village, beginning at Montreal and ending at Quebec, which is a distance of more than 180 M.; for the farm-houses are never more than five arpents apart, and sometimes but three asunder, a few places excepted." (KALM, the Swedish traveller, in 1749.) In 1684 La Hontau said that the houses along these shores were never more than a gunshot apart. The inhabitants are simple-minded and

T

primitive in their ways, tenaciously retaining the Catholic faith and the French language and customs. Emery de Caen, Champlain's contemporary, told the Huguenot sailors that "Mouseigneur, the Duke de Ventadour (Viceroy), did not wish that they should sing psalms in the Great River." When the first steamboat ascended this river, an old Canadian *voyageur* exclaimed, in astonishment and doubt, "Mais croyez-vous que le bon Dieu permettra tout cela!"

As the steamboat swings out into the stream a fine series of views are afforded, including Quebec and the Basin, the bold bluffs of Point Levi, and the dark walls of the Citadel, almost overhead. As the river is ascended, the villas of Sillery and Cap Rouge are seen on the r., and on the l. are the wharves and villages of South Quebec and New Liverpool, beyond which are the mouths of the Etchemin and Chaudière Rivers. *St. Augustin* is on the N. shore, 15 M. above Quebec, and has a *Calvaire*, to which many pilgrimages are made, and a statue of the Guardian Angel, erected on a base of cut stone in front of the church, and commemorating the Vatican Council of 1870.

Near the village is a ruined church dating from 1720, at whose construction the Devil is said to have assisted, in the form of a powerful black stallion who hauled in the blocks of stone, until his driver unbridled him at a watering-place, when he vanished in a cloud of sulphur-smoke. In front of St. Augustine the French frigate *Atalante* surrendered to the British fleet in 1760, after a heroic but hopeless battle; and in the same waters the steamer *Montreal* was burned in 1857, and 200 passengers lost their lives.

Pointe aux Trembles is 3 M. above St. Augustin (N. shore, and is a shipbuilding village of 700 inhabitants. Here many of the ladies of Quebec took refuge during Wolfe's siege (1759), and were captured by his Grenadiers. Here also the American armies of Arnold and Montgomery united their forces (Dec. 1, 1775) before the disastrous assault on Quebec. Passing the hamlet of St. Antoine de Tilly, on the S. shore, the village of *Les Ecureuils* is seen on the N., 7 M. above Pointe aux Trembles. This is near the mouth of the **Jacques Cartier River**, famous for its remarkable scenery and for its fine trout-fishing (on the upper waters). On the heights near the mouth of the river was *Fort Jacques Cartier*, to which 10,000 French troops retreated after the defeat of Montcalm. Nearly a year later (June, 1760) the fort was held by the Marquis d'Albergotti, and was bombarded and taken by Fraser's Highlanders.

6 M. above Les Ecureuils is *St. Croix* (S. shore), a village of 750 inhabitants, with a black nunnery and the public buildings of Lotbinière County. 8 M. beyond (N. shore) is *Portneuf*, a prosperous little town with paper-mills and a large country trade. This seigniory was granted to M. Le Neuf by the Cent Associés in 1647, and was completely desolated by the famishing French cavalry in 1759. Beyond this point the scenery becomes less picturesque, and the bold ridges of the Laurentian Mts. sink down into level lowlands. *Deschambault* (N. shore) has 500 inhabitants, with a trade in lumber and flour. *Lotbinière* (S. shore) is a town of 2,500 inhabitants, with a Convent of the Bon Pasteur and two stove-foundries. *Grondines* (N. shore) is 3 M. beyond Deschambault, and

has 400 inhabitants; and *St. Jean Deschaillons* (S. shore) is noted for its brickyards. *St. Anne de la Perade* (N. shore) has a great church, and is situated at the mouth of the St. Anne River, which is here crossed by a bridge 1,500 ft. long. Beyond *St. Pierre les Becquets* (S. shore) is the busy little port of *Batiscan* (N. shore), with its two lighthouses; *Gentilly* (S. shore) has 600 inhabitants and the Convent of the Assumption; and *Champlain* (N. shore) has 400 inhabitants.

Three Rivers (*British American Hotel*) is a city of 9,000 inhabitants, midway between Quebec and Montreal, and at the head of tide-water on the St. Lawrence River. It was founded in 1618, under the name of Trois Rivières, and played an important part in the early history of Canada. The chief buildings are the stately Catholic Cathedral, the Court-House, the Ursuline Convent, St. Joseph's College, and the Episcopal and Wesleyan churches. The city has a bank, 2 Masonic lodges, and 4 semi-weekly and weekly newspapers (2 of which are French). Besides the daily boats of the Richelieu Line, there are 5 steamboats plying from this port to the adjacent river-villages. It is connected with Quebec and Montreal by the Three-Rivers Branch of the Grand Trunk Railway and by the North-Shore Railway, and is building a new line up the St. Maurice Valley to Grand Piles. There are large iron-works and machine-shops here, and stoves and car-wheels are made in great numbers from bog-iron ore. The chief industry is the shipment of lumber, which comes down the St. Maurice River. The Canadian government has expended $200,000 in improving the navigation on the St. Maurice, and over $1,000,000 has been invested in mills and booms above.

The **St. Maurice River** waters a district of immense (and unknown) extent, abounding in lakes and forests. Portions of this great northern wilderness have been visited by the lumbermen, who conduct rafts to Three Rivers, where the lumber is sawed. About 22 M. above the city are the noble *Falls of the Shawanegan*, where the great river plunges over a perpendicular descent of 150 ft. between the lofty rocks called *La Grand' Mère* and *Le Bonhomme*. A few miles above are the Falls of the *Grand' Mère*. These falls are visited by engaging canoes and guides at Three Rivers, while hunting-parties conducted by Canadian *voyageurs* or Algonquin Indians sometimes pass thence into the remote northern forests in pursuit of the larger varieties of game. The head-waters of the St. Maurice are interlocked with those of the Saguenay.

Across the St. Maurice is the thriving village of *Cap de la Magdelaine;* and on the S. shore are *Becancour*, the capital of Nicolet County, and *St. Angel de Laval* (Doucett's Landing), the terminus of a branch of the Grand Trunk Railway.

The steamer soon enters **Lake St. Peter,** a shallow widening of the river 22 M. long and 8 M. broad. It has a deep and narrow channel (partly artificial), which is marked out by buoys and poles, and is used by large vessels. Immense lumber-rafts are often seen here, drifting downward like floating islands, and bearing streamers, sails, and the rude huts of the lumbermen. In stormy weather on the lake these rafts sometimes come to pieces. The inlets along the low shores afford good duck-shooting; and enormous quantities of eels and pike are taken from the waters. Near the

E. end of the lake, at the mouth of the Nicolet River, is the populous town of **Nicolet,** famous for its flour and lumber trade and for its noble college, with its 250 students and a library of 10,000 volumes. The buildings are surrounded by attractive parks and gardens. On the N. shore is Rivière du Loup *en haut,* near which are the celebrated **St. Leon Springs** (reached by daily stage from Three Rivers, in 24 M.; fare, $1.50; Gilman's Hotel, and others). *St. François du Lac* is a pretty village on the S. W. shore, at the mouth of the great St. Francis River.

On leaving Lake St. Peter, the steamer threads her way through an archipelago of low islands, and soon reaches **Sorel** (four hotels), a city of 7,500 inhabitants, with 3 weekly papers (2 French), a Catholic college, several shipyards and foundries, and a large country trade. It is at the mouth of the great River Richelieu, the outlet of Lake George and Lake Champlain, whose head-waters are interlocked with those of the Hudson. Navigation is kept up between this point and the Lake-Champlain ports by the Chambly Canal, and a railway is being built to meet the Grand Trunk line at Upton. The town is regularly laid out, and its broad streets are adorned with trees. In the centre is the Royal Square, whose fine old elms are much admired.

Fort Richelieu was built on this site in 1641, and was re-constructed and enlarged by Capt. Sorel, of the Carignan Regiment, under orders from Gov. de Tracy (1665). In November, 1775, it was occupied by Col. Easton, with a strong force of Continental troops and a flotilla, and this detachment captured 11 sail of vessels, containing Gen. Prescott and the British garrison of Montreal. Sorel was for many years the summer residence of the Canadian governors, and on being visited by Prince William Henry of England (afterward King William IV.) an abortive attempt was made to change its name to *William Henry.*

Berthier en haut is 6 M. above Sorel, on the N. shore (semi-daily steamers), and is an important manufacturing town of 1,700 inhabitants, situated amid rich farming lands. It was the birthplace of M. Faribault, long time a N. W. Commissioner, and founder of Faribault, Minnesota. Back of Berthier are the populous towns of St. Cuthbert, St. Norbet, St. Felix de Valois, and St. Elizabeth. *Lanoraie* is 9 M. above Berthier (N. Shore), and is the terminus of the St. Lawrence & Industry Railway, which runs N. W. 12 M. to St. Thomas and Joliette, and thence into Montcalm County. 15 M. above Sorel (S. shore) is Contrecœur, noted for its maple-sugar; and Lavaltrie is 15 M. above Berthier (N. shore), and has 2 lighthouses. 6 M. above is St. Sulpice (N. shore), beyond which is L'Assomption (Hotel Richard), a prosperous village of 2,600 inhabitants. Above the N. shore village of Repentigny the N. branch of the Ottawa River (Rivière des Prairies) flows into the St. Lawrence, having diverged from the Ottawa at the Lake of the Two Mountains.

Varennes is a pretty village on the S. shore, opposite Isle St. Therese, and connected by a ferry with Bout de l'Isle, and with Montreal (15 M. distant) by a daily steamer. It has 825 inhabitants, and manufactures many

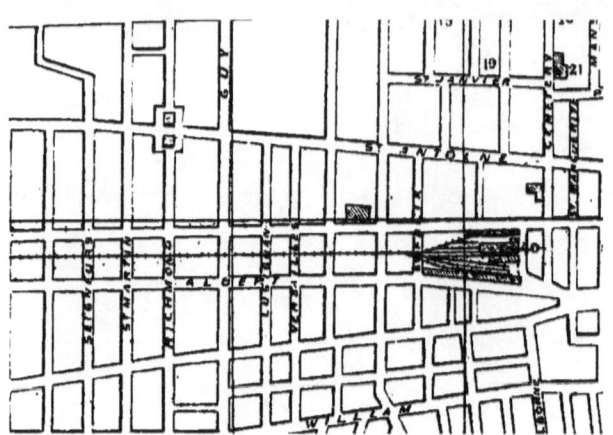

1. City Hall and
    cours
2. Post Office.
3. Court House.
5. Papineau Mar
6. St. Ann's
7. St. Patrick's Ha
8. Military School
9. Crystal Palace

Churche
10. Catholic Cathe
11. Notre Dame.
12. Christ Church C
13. Gesu (Jesuits).
14. St. Patrick's (C
15. American.
16. Trinity.
17. Notre Dame de Bon
18. St. James (Cath)
19. St. George (Epis.
20. St. Paul (Pres.)

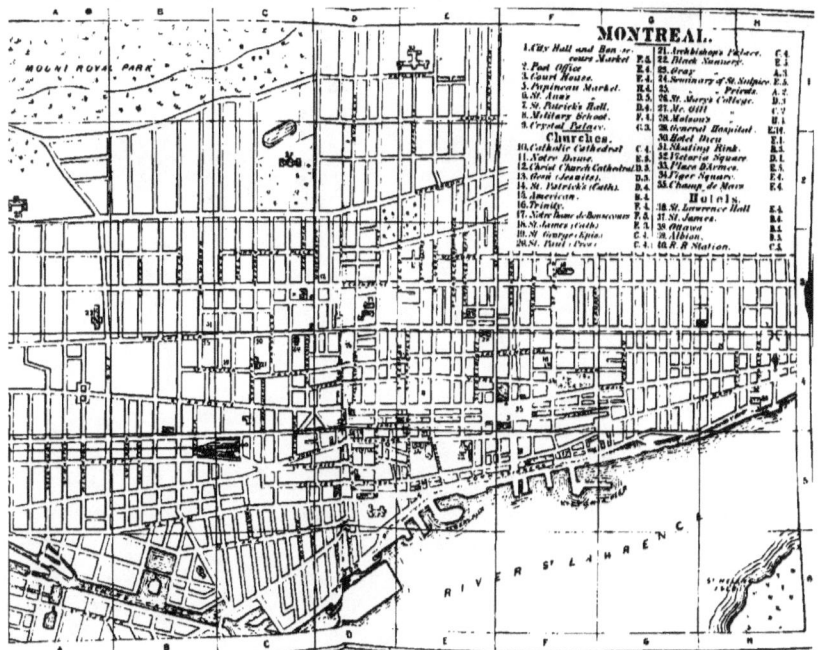

carriages. The church is a large and stately building, with two conspicuous towers. 1 M. from the village are the celebrated *Varennes Springs*, which are saline in character and possessed of valuable medicinal properties. One of them emits great quantities of carbonated hydrogen gas, and the other yields 2 – 3 gallons a minute, and is much visited by invalids. Arrangements are being made to establish a first-class summer resort at this point. Above Varennes is *Boucherville*, the birthplace of Chief Justice Sir Louis Hippolyte Lafontaine. The low and marshy islands off this shore are famous for duck-shooting, and for the ice-dams which form here at the close of the winter. *Pointe aux Trembles* is to the N., on the Island of Montreal, and is an ancient village dating from 1674.

"We were gliding past Longueuil and Boucherville on the (left), and *Pointe aux Trembles*, ' so called from having been originally covered with aspens,' on the (right). I repeat these names not merely for want of more substantial facts to record, but because they sounded singularly poetic in my ears. There certainly was no lie in them. They suggested that some simple and perchance heroic human life might have transpired there." (THOREAU.)

Clustering villages are now seen on either shore, and the river is strewn with low islands. At 9 M. above Pointe aux Trembles the steamer reaches her pier at **Montreal**, with the magnificent Victoria Bridge spanning the river in front.

75. Montreal.

Hotels. — * St. Lawrence Hall, 139 Great St. James St., accommodating 500 guests, $ 3.50 a day ; * Ottawa Hotel, 246 Great St. James St., $3.50 a day ; Montreal Hou·e, Custom-House Square, $ 2.50 a day ; Donnegana Hotel, Notre Dame St., $ 2.50 a day ; Albion Hotel, 141 McGill St., $ 2 a day ; Canada Hotel, St. Gabriel St., $ 2 a day (frequented by French Canadians) ; the American, 22 and 26 St. Joseph St., $ 2 a day. There are also numerous French hotels of the second class, among which are the Hôtel Richelieu, 45 St. Vincent St. ; the Morrisseau, St. Paul St ; Hôtel du Nord, 131 St. Paul St. ; and the Hôtel Lepine, 151 St. Paul St. Montreal needs a modern first-class hotel.

Shops. — The most attractive are on Great St. James and Notre Dame Sts. Among the chief houses for *clothing* are Henry & Wilson, 236 St. James St. ; W. Walsh & Co., 463 Notre Dame St. *Dry goods and gloves*, Brown & Clagget, corner Notre Dame and St. Helen Sts. ; Wm. McDunnough, 280 Notre Dame St. (*laces* a specialty) ; Thomas Mussen, 257 and 259 Notre Dame St. ; Ste. Marie Brothers, 454 Notre Dame St. *Furs and hats*, A. Brahadi, corner St. Lambert and Notre Dame Sts. *Jewelry*, Savage, Lyman, & Co., 228 St. James St. ; E. G. Mellor, 285 Notre Dame St.; Alex. D. Daly, 426 and 428 Notre Dame St. The parlors of W. Notman, the celebrated photographer (very high prices), are at 17 Bleury St. *Turkish baths*, Swedish movement and health-lift, at 140 St. Monique St., near the Crystal Palace.

Amusements. — Theatre Royal, 19 Cotté St., open usually during the summer. *Lectures* are given at the Association Hall, corner of Craig St. and Victoria Square. Lectures and other entertainments are also given at the hall of the Mechanics' Institute, 204 Great St. James St. *Billiards*, at Nordheimer's Hall, Great St. James St. The Victoria Skating Rink, Drummond and Dorchester Sts., is famous for its winter carnivals. The Thistle Rink is near the Crystal Palace.

Reading-Rooms. — Young Men's Christian Association, Victoria Square ; Merchants' Exchange, 11 St. Sacrament St. ; Mechanics' Institute, 204 Great St. James St. ; Institut Canadien, 111 Notre Dame St. ; Œuvre des Bons Livres, 327 Notre Dame St. There is a circulating library at 666 Dorchester St. (Mrs. Hill's).

Consuls. — United States, 145 Great St. James St. ; Germany, 61 St. Sulpice St.; France, 75 Notre Dame St. ; Austro-Hungary, 61 St. Sulpice St. ; Italy, 158 Fortification Lane ; Belgium, 873 Sherbrooke St.

310   *Route 75.*   MONTREAL.

Post-Office, on Great St. James St., near St. François Xavier St.   *Telegraph*, central office of the Montreal Telegraph Company, corner of St. Sacrament and St. François Xavier Sts.

Carriages.—(One-horse) For 1-2 persons, within a city division, 15 cents ; between two points in the city, 25 c. ; by the hour, 50 c., and 20 c. for each additional ½ hour. For 3-4 persons, within a city division, 25 c. ; between two points in the city, 40 c. ; by the hour, 70 c., and 30 c. for each additional ½ hour. (Two-horse carriages.) For 1-2 persons, within a city division, 30 c. ; between two points in the city, 40 c. ; by the hour, 75 c., and 30 c. for each additional ½ hour. For 3-4 persons, within a city division, 40 c. ; between two points in different divisions, 50 c. ; by the hour, $ 1 ; and 40 c. for every additional ½ hour. The first division includes the district between McGill St., Craig St., the Quebec-Gate Barracks, and the river ; the second, the wards S. and W. of the first and of St. Lawrence Main St. ; the third the wards N. and E. of the first and of St. Lawrence Main St. (approximately).

Horse-cars run across the city on Craig, Bleury, and St. Catherine Sts. ; also on St. Mary, Notre Dame, and St. Joseph Sts. ; and out St. Lawrence Main St. to St. Jean Baptiste.

Railways. — To Boston by way of St. Albans, Concord, and Lowell, in 384 M. ; or by way of Fitchburg, in 344 M. ; or by the new route, the Southeastern Railway. To New York, by Rutland and Albany, 365 M. (by Lake Champlain, 405 M.); to Quebec, 172 M. (in 7 hrs.); to Plattsburg, 63 M. ; to Rouse's Point, 50 M. ; to Toronto, 333 M. (14-15 hrs.); to Detroit (861 M.) and Chicago (1,145 M.) ; to Ottawa, 164 M.

Stages run out from Montreal in all directions, daily. To St. Césaire, Marieville, and Chambly ; St. Eustache, St. Augustin, St. Scholastique, St. Columban, and St. Canut ; New Glasgow, Kilkenny, St. Jérome. Stanbridge, St. Lin, St. Hippolyte, St. Agathe des Monts, St. Adèle, St. Janvier, St. Thérèse de Blainville, St. Sophie ; St. Vincent de Paul, Mascouche, Terrebonne, and St. Sauveur ; Pointe aux Trembles, Sault au Recollet, and St. Martin.

Steamships. — The first-class ocean steamships of the Allan Line and the Dominion Line leave Montreal 2-3 times weekly during the season of navigation, for Liverpool and Glasgow. The Richelieu Line and the Union Line each run daily steamers to the lower river-ports and Quebec. The morning and evening trains to Lachine connect with the steamboats for Ottawa, by way of the Ottawa River. The vessels of the Canadian Navigation Company ascend the St. Lawrence and Lake Ontario, from Montreal to the upper river-ports, Toronto and Hamilton. The *St. Hélène* and *Ottawa* make semi-weekly trips to the Bay of Quinté. The Quebec & Gulf Ports S. S. Co. despatch a weekly steamer from Montreal to Percé, Charlottetown, and Pictou. The *Chambly* runs semi-weekly from Montreal to Verchères, Contrecœur, Sorel, St. Ours, St. Denis, St. Antoine, St. Charles, St. Marc, St. Hilaire, Beloeil, St. Matthias, and Chambly (90 M.). The *Three Rivers* runs semi-weekly to Verchères, Sorel, Maskinongé, Rivière du Loup *en haut*, Yamachiche, Port St. Francis, Champlain, and Three Rivers. The *Berthier* runs semi-weekly to Repentigny, St. Sulpice, Lavaltrie, Lanoraie, and Berthier. The *Terrebonne* runs daily to Boucherville, Varennes, Bout de l'Isle, Lachenaie, L'Assomption, and Terrebonne (24 M.). Ferry steamers cross the river at frequent intervals to La Prairie, St. Lambert, and Longueuil.

MONTREAL, the metropolis of the Dominion of Canada, and "the Queen of the St. Lawrence," is one of the most beautiful cities on the continent. It is situated on an island (at the confluence of the Ottawa and St. Lawrence Rivers) containing 197 square miles, and which, from its fertility, has been called the Garden of Canada. The St. Lawrence is 1½ M. wide opposite the city, and the river-front is lined for over 1 M. with lofty and massive walls, quays, and terraces of gray limestone, unequalled elsewhere in the world, except at Liverpool, Paris, and St. Petersburg. The commercial buildings of the city are generally of stone, in plain and substantial architecture, and the number of fine public buildings is very large. Three fourths of the population are Catholics, most of whom are French, and the bright suburban villages are almost entirely inhabited by Frenchmen.

Although Montreal is 800 M. from the sea, it is the port which receives the greater part of the importations to Canada; and its manufacturing interests are extensive and important. The admirable systems of railway and steamboat communication of which Montreal is the centre have made it the commercial emporium of the North; and new lines of traffic and internal railways are being built from year to year, binding all the St. Lawrence counties to this city. Montreal forms the Metropolitical See of the Anglican Church in Canada, and is the capital of a Roman Catholic diocese. The water-supply, street-lamps, paving, and fire department are similar to those of American cities of the first rank.

The population of Montreal was 107,225, at the census of 1871, and it now probably contains 160,000 inhabitants (including the populations of the closely connected suburbs). In 1870 its assessed valuation was $ 47,679,000; its imports, $ 25,680,814; and its exports, $ 19,100,413. In the same year 602 vessels arrived here from the sea, and the customs revenue was $ 4,128,052. The city has 19 banks, 62 churches, more than 30 newspapers and magazines (in English and French), and scores of societies of freemasons, antiquarians, sportsmen, and cricket-players. There are numerous charitable and benevolent organizations, 6 building associations, 3 musical clubs, and societies for the English, Scotch, Irish, French, German, and New-England residents.

The **Victoria Square** is a public ground at the intersection of McGill and Great St. James Sts., ornamented with a fountain and a bronze statue of Queen Victoria. On its S. side is the elegant Gothic building which pertains to the Young Men's Christian Association, the oldest society of that name in America. On the lower side of the Square are the stately *Albert Buildings*, devoted to commerce; and on the N. are the ruins of the great Irish headquarters, *St. Patrick's Hall*.

Passing to the N. E. along **Great St. James St.**, the visitor sees many fine stores, and the attractive buildings of * Molson's Bank (of Ohio stone and Scotch granite), the Merchants' Bank, the stately new * Post-Office, and other symmetrical and solidly constructed edifices. This street is the Broadway of Montreal. *St. Peter St.* runs to the S. E. by the stately Caverhill Buildings (of cut limestone in Italian Palazzo architecture) to *St. Paul St.*, the seat of an extensive wholesale trade. The Central Wesleyan Church, on Great St. James St., has a fine organ. In this vicinity are the chief hotels of the city, the St. Lawrence and the Ottawa.

Opposite the beautiful Corinthian colonnade of the Bank of Montreal (beyond St. François Xavier St., the Wall St. of Montreal) the *Place d'Armes* is seen. This square was so named because it was the parade-ground of Montgomery's American army in 1775. Here is the lofty front of the * **Church of Notre Dame**, the largest church on the continent, with seats for 8,000 persons on the floor and 2,000 in the galleries. It is 255½ ft. long and 144½ ft. wide, and has a chancel window of stained glass

64 × 32 ft. in size. The interior is not striking, and the pictures are poor. There are two towers on the front, each 220 ft. high, and, like the church, in the simplest form of mediæval Gothic architecture. One tower has a chime of bells, and in the other hangs "Gros Bourdon," the largest bell in America, weighing nearly 15 tons. The tower is generally open (fee of 25 c. to the door-keeper), and affords from its summit a noble *view of the city and its environs (especially of the city and river, the Victoria Bridge, and the islands). The suburbs of Laprairie, Longueuil, and St. Lambert, the Lachine Rapids, and the blue mountains of Vermont, are seen from this point. Alongside the church is the ancient *Seminary of St. Sulpice*, on the site of the Seminary of 1657, as the church is near the site of the Notre Dame of 1671. The present church was built in 1824–9, and was consecrated by the Bishop of Telmesse *in partibus*. The seminary consists of low and massive buildings, surrounded with gardens and court-yards of spotless neatness. It has 24 priests connected with its various works.

"I soon found my way to the Church of Notre Dame. I saw that it was of great size and signified something..... Coming from the hurrahing mob and the rattling carriages, we pushed back the listed door of this church, and found ourselves instantly in an atmosphere which might be sacred to thought and religion, if one had any..... It was a great cave in the midst of a city; and what were the altars and the tinsel but the sparkling stalactics, into which you entered in a moment, and where the still atmosphere and the sombre light disposed to serious and profitable thought? Such a cave at hand, which you can enter any day, is worth a thousand of our churches which are open only Sundays." (THOREAU.)

Fronting on the Place d'Armes are the elegant Ontario Bank and the hall of the Grand Lodge of the Masons of Canada. A short distance to the E., on Notre Dame St., an archway on the r. admits one to the extensive and secluded Convent of the Black Nuns (founded in 1657). Farther on, the *Court House is seen on the l.,—a stately stone building in Ionic architecture (300 × 125 ft.), back of which is the *Champ de Mars*, or Parade Ground, an open space covering 50,000 square yards, and ample enough for the display of 3,000 troops. The great structure fronting across Craig St. was built for the Dominion Military School, which is now established at Kingston. The Museum of the Canadian Geological Survey is on St. Gabriel St., opposite the Champ de Mars, and was founded by Sir William Logan. It contains a large collection of ores, building-stones, rare minerals, and one of the best palæontological museums in America. The new City Hall is to be erected on the E. side of the Champ de Mars. Just beyond the Court House the Jacques Cartier Square opens off Notre Dame St., and is encumbered with a dilapidated monument to Nelson. The Jacques Cartier Normal School (in the ancient French government building) and the Institut Canadien (with a fine library) are near the head of this square.

By the next side-street (St. Claude) to the r., the *Bonsecours Market may be visited. This market is unrivalled in America, and is built of stone, in quasi-Doric architecture, at a cost of $300,000. It is three stories high, has a lofty dome, and presents an imposing front to the river. The curious French costumes and language of the country people who congregate here on market-days, as well as some peculiarities of the wares offered for sale, render a visit very interesting. Alongside of the market is the *Bonsecours Church* (accommodating 2,000 persons), which was built in 1658. A short distance beyond are the extensive Quebec-Gate Barracks, on Dalhousie Square; and the Victoria Pier makes out into the stream towards *St. Helen's Isle* (a fortified depot of ammunition and war *matériel*), which was named by Champlain in honor of his wife. To the N., on Craig St., is the attractive Viger Garden, with a small conservatory and several fountains, fronting on which is *Trinity Church* (Episcopal), built of Montreal stone, in early English Gothic architecture, and accommodating 4,000 persons. N. of Trinity, and also on St. Denis St., is St. James Church (Catholic), in the pointed Gothic style, with rich stained glass. Some distance E. of Dalhousie Square, on St. Mary St., are Molson's College (abandoned) and St. Thomas Church (Episcopal), with the great buildings of Molson's brewery and the Papineau Market and Square (on which are the works of the Canadian Rubber Co.). The suburb of Hochelaga (see page 318) is about 1 M. beyond the Papineau Square.

McGill St. is an important thoroughfare leading S. from Victoria Square to the river. Considerable wholesale trade is done here and in the intersecting St. Paul St. The Dominion and Albert Buildings are rich and massive, and just beyond is St. Ann's Market, on the site of the old Parliament House. In 1849 the Earl of Elgin signed the obnoxious Rebellion Bill, upon which he was attacked by a mob, who also drove the Assembly from the Parliament House, and burnt the building. On account of these riots, Montreal was decapitalized the same year. Commissioners' St. leads E. by St. Ann's Market and the elegant *Custom-House* to the broad promenades on the river-walls. Ottawa St. leads W. to the heavy masonry of the Lachine-Canal Basins and the vicinity of the Victoria Bridge.

Radegonde St. and Beaver-Hall Hill run N. from Victoria Square, passing *Zion Church*, where the Gavazzi riots took place in 1853. The armed congregation repulsed the Catholic assailants twice, and then the troops restored order, 40 men having been killed or badly wounded. Just above is the Baptist Church, overlooked by the tall Church of the Messiah (Unitarian), with St. Andrew's Presbyterian Church on the r. A few steps to the r., Lagauchetière St. leads to *St. Patrick's Church*, a stately Gothic building 240 × 90 ft., accommodating 5,000 persons, and adorned with a spire 225 ft. high. The nave is very lofty, and the narrow lancet-windows are filled with stained glass. Near by, on Bleury St., are the massive stone buildings of St. Mary's College (Jesuit; 9 professors) and the *Church of the Gesù. The nave of the church (75 ft. high) is bounded by rich composite columns; and the transepts are 144 ft. long, and are adorned with fine frescos in chiaroscuro.

Over the High Altar is the Crucifixion, and the Adoration of the Spotless Lamb, above which is the Nativity. Against the columns at the crossing of the nave and transepts are statues of St. Mark with a lion, St. Matthew with an ox, St. Luke with a child, and St. John with an eagle. On the ceiling of the nave are frescos of St. Thomas Repentant, the Bleeding Lamb, and the Virgin and Child amid Angelic Choirs. Medallions along the nave contain portraits of eight saints of the Order of Jesus. Over the Altar of the Virgin, in the l. transept, is a fresco of the Trinity, near which is a painting of St. Aloysius Gonzaga receiving his first communion from St. Charles Borromeo, Cardinal Archbishop of Milan. To the r. is a fresco of St. Ignatius Loyola in the Grotto of Manresa, and on the l. is Christ's Appearance to him near Rome, while above is Christ blessing Little Children. Over St. Joseph's Altar, in the r. transept, is a painting of the Eternal Father; on the r. of which is another picture, St. Stanislaus Kostka receiving Communion from Angels. On the l. is a fresco of the Martyrdom of the Jesuits at Nagasaki (Japan); on the r. is the Martyrdom of St. Andrew Bobola, in Poland; and above is the Raising of Lazarus. On the ceiling is the Holy Family at Work.

Turning now to the W. on St. Catherine St., one soon reaches * **Christ Church Cathedral,** the best representative of English Gothic architecture in America. It is built of Montreal and Caen stone, and is 112 ft. long, and 100 ft. wide at the transepts. A stately stone spire springs from the intersection of the nave and transepts, and attains a height of 224 ft. The choir is 46 ft. long, is paved with encaustic tiles, and contains a fine stained-glass window. On either side are elaborately carved stalls for the clergy; and the pointed roof of the nave (67 ft. high) is sustained by columns of Caen stone whose capitals are carved to represent Canadian plants. In front of the cathedral is a monument to Bishop Fulford, and on the N. is a quaint octagonal chapter-house, where the diocesan library is kept. The residence of the Lord Bishop (and Metropolitan of Canada) is near this building. One square E. of the cathedral (corner of Cathcart and University Sts.) is the large and interesting *Natural-History Museum*, which is open to the public (fee, 25 c.). The Ferrier Collection of Egyptian Antiquities and the cases of Canadian birds are of much interest. Farther out on St. Catherine St. is the *Crystal Palace.*

McGill University is about $\frac{1}{2}$ M. from the cathedral, at the foot of Mount Royal. It was endowed in 1814 and opened in 1828, and has faculties of Arts (9 professors), Medicine (10 professors), and Law (8 professors). The Medical School is N. of the main building, and the museum is worthy of a visit. The University is under the charge of Dr. J. W. Dawson (see page 138), and is the most flourishing institution of the kind in Lower Canada. The reservoir for the water-supply of Montreal is back of the University, 200 ft. above the river, and has a capacity of 15,000,000 gallons. The water is taken from the St. Lawrence, 1½ M. above the Lachine Rapids, flows for 5 M. in an open canal, and is then forced up to the reservoir by powerful machinery. A pleasant view of the city may be obtained from this terrace, and on the W. is *Ravenscray*, the mansion of Sir Hugh Allan.

The * **Great Seminary** of St. Sulpice and the *Montreal College* are ¾ M. S. W. of the University, and front on the same street (Sherbrooke). They occupy a portion of the broad ecclesiastical domain which is known as the Priests' Farm. The incongruous towers in front of the main building pertained to the ancient college of the 17th century, and were at that time loopholed and held as a part of the defences of the town against the Iroquois Indians. The Seminary is for the education of Roman Catholic priests, and has 4 professors and 112 students. The Montreal College is for the education of Canadian youth, and has 10 ecclesiastics for professors and 260 students. It was founded in 1773 by the Sulpicians, who still remain in charge. The Seminary chapel is worthy of a visit, and the gardens about the buildings are said to be the finest in Canada. Sherbrooke St. and the environs of Mount Royal contain many elegant residences.

Dorchester St. runs S. W. from Beaver-Hall Square, soon crossing University St., on whose r. corners are the High School and the St. James Club. This street leads, on the l., to the Normal and Model Schools; and on the r. to the Natural-History Museum and the Cathedral. Dorchester St. passes on by St. Paul's Church (l. side) and the Knox Church (r. side) to **Dominion Square**, which occupies the site of a cemetery. In this vicinity are several fine churches, — the Wesleyan Methodist, a graceful building in the English Gothic style; the American Presbyterian, an exact copy of the Park Church in Brooklyn, N. Y.; and St. George's Church (Episcopal), an elegant edifice in decorated Gothic architecture, with deep transepts, costly stained windows, a timber roof, and fine school-buildings attached.

The new Roman Catholic **Cathedral of St. Peter** is being erected at the corner of Dorchester and Cemetery Sts. It is 300 ft. long and 225 ft. wide at the transepts; and is to be surmounted by a stone dome 250 ft. high, supported on 4 piers (each of which are 36 ft. thick) and 32 Corinthian columns. 4 minor domes are to surround this noble piece of architecture. The portico is to resemble that of the Roman St. Peter's, surmounted also by colossal statues of the Apostles; and gives entrance to the vestibule, which is 200 ft long and 30 ft. wide. The interior colonnades support lines of round arches; and there are 20 minor chapels. The exterior walls are very massive, but extremely plain and rough. This building is to supply the place of the Cathedral on St. Denis St., which was burned in 1852. The design was conceived by Bishop Bourget, who secured the land, and after inspecting numerous plans in different styles, determined to erect a cathedral like St. Peter's (though smaller). The architects went to Rome and studied the Vatican Basilica carefully, and the work was soon begun. At present strenuous exertions are being made by the clergy, monks, and nuns to procure the needful funds to finish the building, and it is expected that it will be completed by the year 1884.

The *Bishop's Palace* is on the E. of Dominion Square; and Cemetery St. runs thence to St. Joseph's Church and the Bonaventure station of the Grand Trunk Railway. Beyond this point is the populous *St. Ann's Ward*, toward the great basins of the Lachine Canal.

The * **Gray Nunnery** is nearly ½ M. S. W. of Dominion Square, near Dorchester St., and occupies an immense pile of stone buildings. This convent (*L'Hôpital Général des Sœurs Grises*) was founded in 1747, and contains 202 nuns, 116 on mission, 42 novices and postulants, and over 600 patients. It takes care of aged and infirm men and women, orphans and foundlings, and has large revenues from landed estates. Over 600 foundlings are received every year, of whom more than seven eighths die, and the remainder are kept in the convent until they reach the age of 12 years. Opposite the nunnery is *Mont Ste. Marie*, a large building which was erected for a Baptist college, but has become a ladies' boarding-school (169 students) under the Congregational Nuns of the Black Nunnery, who have, in the city, 57 schools and 12,000 pupils. This order was founded by Marguerite Bourgoys in 1659.

The *Nazareth Asylum for the Blind* is N. of the Gesù, on St. Catherine St., and has also an infant school with over 400 pupils. The chapel is built in a light and delicate form of Romanesque architecture, and is richly decorated and frescoed. On the same square are the handsome stone buildings of the Catholic Commercial Academy. To the E. (on Dorchester St.) is the *General Hospital*, with 150 beds; the Hospice of St. Vincent de Paul (30 brethren) and the *Asile de la Providence* (122 nuns) are near Labelle St.; and numerous other convents and asylums are found throughout this singular city, which is both British and French, commercial and monastic, progressive and mediæval, — combining American enterprise with English solidity and French ecclesiasticism.

The * *Hôtel Dieu de Ville Marie* is about 1 M. N. W. of Great St. James St., and is one of the largest buildings in Canada. The chapel is a spacious hall over which is a dome 150 ft. high, frescoed with scenes from the life of the Holy Family. This institution was founded in 1859, and is conducted by about 80 cloistered nuns of the Order of St. Joseph. There are generally about 500 persons in the building, consisting of the nuns and their charges, old and infirm men and women, orphans, and about 200 sick people. To the N. is the populous French suburb of *St. Jean Baptiste* (5,000 inhabitants), which is connected with the city by horse-cars on St. Lawrence Main St.

* **Mount Royal** is S. W. of the Hôtel Dieu and W. of the city, and is a long wooded ridge 750 ft. high. The larger part of this picturesque eminence has been purchased by the municipal government, and is being formed into a public park, whose natural attractions are certainly great. The *Mount Royal Cemetery* is on the W. slope, and is locally famous for its pleasant drives and artistic monuments (among which are those of the Molson family).

Point St. Charles is beyond the Lachine-Canal Basins, and is traversed by the tracks of the Grand Trunk Railway. Near the Victoria Bridge is a great bowlder, surrounded by a railing, commemorating the place where were buried 6,500 Irish immigrants, who died here of ship-fever in the summer of 1847. The * **Victoria Bridge** is the longest and most costly bridge in Canada. It consists of 23 spans of 242 ft. each (the central one

330 ft.), resting on 24 piers of blue limestone masonry, cemented and iron-riveted, with sharp wedge-faces to the down-current. The tubes containing the track are 19 × 16 ft. and the bridge is approached by abutments 2,600 ft. long and 90 ft. wide, which, with the 6,594 ft. of iron tubing, makes a total length of 9,194 ft. from grade to grade and over 1¼ M. from shore to shore. The bridge was commenced in 1854, and finished in 1859; it contains 250,000 tons of stone and 8,000 tons of iron, and cost $ 6,300,000. There is a beautiful view of the city from the central tube.

In the early autumn of 1535 Jaques Cartier heard, from the Indians of Quebec, of a greater town far up the river. The fearless Breton chief took 2 boats and 50 men, and ascended the St. Lawrence to the Iroquois town of Hochelaga, occupying the present site of the metropolis of Canada. " Before them, wrapped in forests painted by the early frosts, rose the ridgy back of the Mountain of Montreal, and below, encompassed with its cornfields, lay the Indian town," surrounded with triple palisades arranged for defence. The French were admitted within the walls and rested on the great public square, where the women surrounded them in curiosity, and the sick and maimed were brought to them to be healed, " as if a god had come down among them." The warriors sat in grave silence while he read aloud the Passion of our Saviour (though they understood not a word); then presents were given to all the people, and the French trumpeters sounded a warlike melody. The Indians then guided their guests to the summit of the adjacent mountain, whence scores of leagues of unbroken forest were overlooked. Cartier gave to this fair eminence the name of *Mont Royal*, whence is derived the present name of the city.

In 1603 this point was visited by the noble Champlain, but Hochelaga had disappeared, and only a few wandering Algonquins could be seen in the country. The Iroquois of the great town had been driven to the S. by the powerful Algonquins (such is the Mohawk tradition).

At a later day a tax-gatherer of Anjou and a priest of Paris heard celestial voices, bidding them to found a hospital (Hôtel Dieu) and a college of priests at Mont Royal, and the voices were followed by apparitions of the Virgin and the Saviour. Filled with sacred zeal, and brought together by a singular accident, these men won several nobles of France to aid their cause, then bought the Isle of Mont Royal, and formed the Society of Notre Dame de Montreal. With the Lord of Maisonneuve and 45 associates, in a solemn service held in the Cathedral of Notre Dame de Paris, they consecrated the island to the Holy Family under the name of " Ville Marie de Montreal " (Feb., 1641). May 18, 1642, Maisonneuve and his people landed at Montreal and raised an altar, before which, when high mass was concluded, the priest said, " You are a grain of mustard-seed that shall arise and grow until its branches overshadow the land. You are few, but your work is the work of God. His smile is on you, and your children shall fill the land." The Hôtel Dieu was founded in 1647, and in 1657 the Sulpicians of Paris established a seminary here. In 1689, 1,400 Iroquois Indians stormed the western suburbs, and killed 200 of the inhabitants, and a short time afterwards Col. Schuyler destroyed Montreal with troops from New York, leaving only the citadel, which his utmost efforts could not reduce. In 1760 Lord Amherst and 17,000 men captured the city, which then had 4,000 inhabitants, and was surrounded by a wall with 11 redoubts and a citadel. In 1775 Ethan Allen attacked Montreal with a handful of Vermonters, and was defeated and captured, with 100 of his men. Gen. Prescott sent them to England as " banditti," and Allen was imprisoned in Pendennis Castle. In the fall of 1775 the city was taken by the American army under Gen. Montgomery. With the close of the War of 1812, a brisk commerce set in, and the city grew rapidly, having, in 1821, 18,767 inhabitants. The completion of the Grand Trunk Railway greatly benefited this place, and its increase has for many years been steady, substantial, and rapid. In 1832 the cholera destroyed 1,843 persons, out of a population of 30,000; and in 1852 a large part of the city was burned. 80 years ago vessels of over 300 tons could not reach Montreal, but a ship-channel has been cleared by the exertions of the merchants (headed by Sir Hugh Allan), and now the city is visited regularly by ocean steamships of 4,000 tons, and by the largest vessels of the merchant-marine.

76. The Environs of Montreal.

Montreal is situated on the S. E. side of the island of Montreal, which is 28 M. long, 10 M. wide, and 70 M. around. It is divided into 10 parishes, and is composed of fertile and arable soil, supporting a dense population. The favorite drive is that * "**Around the Mountain,**" a distance of 9 M. The road passes out by the Hôtel Dieu and through the suburb of St. Jean Baptiste (whence a road runs E. to the limestone-quarries at *Côté St. Michel*). At *Mile-End* the carriage turns to the l. and soon passes the avenue which leads (to the l.) to the Mount Royal Cemetery. The road ascends to higher grades, and beautiful views open on the N. and W., including 13 villages, the distant shores of the Isle of Jesus, and the bright waters of Lake St. Louis and the Lake of the Two Mountains. On a clear day the spires of the Catholic College of *St. Thérèse* are seen, several leagues to the N., beyond the Rivière aux Chiens. The village of *Côté des Neiges* (three inns) has an antique church, and is occupied by 1,200 inhabitants. It was first settled by families from Côté des Neiges in France, which derived its name from a legend that a miraculous cruciform fall of snow took place there in August, marking the place on which a pious citizen afterwards built the Church of Notre Dame des Neiges. From this village the inter-mountain road leads E. to Montreal. On the lower slope of Mount Royal a platform has been built on the wall of the Seminary grounds, from which a beautiful * view is obtained. (The usual charges for the ride around the mountain are $1.50 for 2 - 3 persons, in a cab, or $2 for 4 persons; for a two-horse carriage, $4, for 1 - 4 persons.)

A road turns to the r. from Côté des Neiges and passes around the bold highlands S. of Mount Royal, through fair rural scenery. Beyond the hamlet of *Côté St. Luc* it reaches *Côté St. Antoine*, the seat of the fine building and grounds formerly known as Monklands, when the home of Governor-General Lord Elgin. It is now called **Villa Maria**, and is occupied by the black nuns as a boarding-school. There are 25 sisters and 172 pupils, most of whom are from the United States. Opposite Villa Maria is the Church of St. Luc. The short road from this point to the city is made interesting by beautiful views and fair villas, and for ¼ M. after passing the toll-gate it skirts the Seminary grounds.

The **Sault au Recollet** is 7 M. W. of Montreal, on the Rivière des Prairies, and is frequently visited for the sake of its picturesque rapids. Picnic parties occupy the forest-covered *Priests' Island*, whence the descent of rafts may be observed. The Convent of the Sacred Heart is beautifully situated amid pleasant grounds near the river. Opposite Sault au Recollet is the **Isle Jesus**, which is nearly 25 M. long, and contains the villages of St. Martin, St. Rose de Lima, and St. Vincent de Paul (near which is the Provincial Reformatory Prison).

Hochelaga is at the N. E. end of the Montreal horse-car line, and is

LACHINE RAPIDS. *Route 76.* 319

the point where the Northern-Colonization and North-Shore Railways are to terminate. It has a good harbor on the St. Lawrence, below the Rapid of St. Mary. There are several fine villas here, and the * *Convent of the Sacred Name of Jesus and Mary* is the most extensive monastic institution in Canada. Hochelaga is 3 M. from the Victoria Bridge; and 3-4 M. farther E. is Longue Point, near which the late Sir George E. Cartier resided. The river-road gives views of Longueuil, Boucherville, and Varennes, on the S. shore.

Lachine (three hotels) is 9 M. S. W. of Montreal, and is a favorite summer-resort of the citizens. The river-road is very picturesque; and the upper road runs through the manufacturing town called *Tannery West*, which has over 4,000 inhabitants. Visitors usually go out on one road and return by the other. Lachine is at the foot of Lake St. Louis, and is noted for its annual regattas. It was so named by Champlain in 1613, because he believed that beyond the rapids the river led to China (*La Chine*). In 1689 the Iroquois Indians destroyed the French town here, with all its inhabitants, 200 of whom were burnt at the stake. Opposite Lachine is the populous village of *Caughnawaga*, inhabited by about 500 of the orderly and indolent descendants of the Iroquois Indians, who are governed by a council of seven chiefs.

The * **Lachine Rapids** may be visited by taking the 7 A. M. train (at the Bonaventure station) to Lachine, where a steamer is in waiting, by which the tourist returns through the rapids to Montreal. After taking a pilot from Caughnawaga, the steamer passes out.

" Suddenly a scene of wild grandeur bursts upon the eye. Waves are lashed into spray and into breakers of a thousand forms by the submerged rocks which they are dashed against in the headlong impetuosity of the river. Whirlpools, a storm-lashed sea, the chasm below Niagara, all mingle their sublimity in a single rapid. Now passing with lightning speed within a few yards of rocks, which, did your vessel but touch them, would reduce her to an utter wreck before the crash could sound upon the ear; did she even diverge in the least from her course, — if her head were not kept straight with the course of the rapid, — she would be instantly submerged and rolled over and over. Before us is an absolute precipice of waters; on every side of it breakers, like dense avalanches, are thrown high into the air. Ere we can take a glance at the scene, the boat descends the wall of waves and foam like a bird, and in a second afterwards you are floating on the calm, unruffled bosom of ' below the rapids.' "

The steamer then passes under the central arch of the Victoria Bridge (see page 316), and opens an imposing panoramic * view of the city. (Tickets for the round-trip cost 50 c.; and the tourist gets back to Montreal about 9.30 A. M.)

The **Belœil Mountain** may be visited in a day by taking the Grand Trunk Railway to Belœil (22 M.), whence the mountain is easily ascended, passing a pretty little lake. On this peak (1,400 ft. above the St. Lawrence) the Bishop of Nancy erected an oratory surmounted by a huge tin-covered cross which was visible for over 30 M. The cross was blown down, several years ago. The * view from Belœil includes a radius of 60 M. over the fertile and thickly settled plains of the St. Lawrence Valley, with the blue mountains of Vermont far away in the S. E. The *Boucherville Mountain* is reached from St. Bruno, a station on the Grand Trunk Railway, and commands fine views. There are 10 lakes on this ridge, one of which, the *Manor Lake*, is on a level with the top of the towers of Notre Dame, in Montreal.

St. Anne (*du Bout de l'Isle*) is 21 M. S. W. of Montreal, and may be reached in an hour by the Grand Trunk Railway. It is a village of 1,000 inhabitants, with two inns, and has an ancient church which is much revered by the Canadian boatmen and *voyageurs*. Many of the people of Montreal visit this place during the summer. The village is at some distance from the railway, between Lake St. Louis (of the St. Lawrence) and the Lake of the Two Mountains (of the Ottawa River). The Ottawa is here crossed by a fine railway-bridge, resting on 16 stone piers; and the famous Rapids of St. Anne are flanked by a canal. Here Tom Moore wrote his Canadian Boat-Song, beginning: —

> "Faintly as tolls the evening chime,
> Our voices keep tune, and our oars keep time.
> Soon as the woods on shore look dim
> We'll sing at St. Anne's our parting hymn.
> Row, brothers, row; the stream runs fast,
> The Rapids are near, and the daylight's past.
>
> "Uttawa's tide! this trembling moon
> Shall see us float o'er thy surges soon.
> Saint of this green isle! hear our prayers;
> O, grant us cool heavens and favoring airs!
> Blow, breezes, blow; the stream runs fast,
> The Rapids are near, and the daylight's past."

Steamers run daily up the Ottawa River to **Ottawa** (*Russell Hotel*), the capital of Canada. The Canadian **Parliament House** is situated on a lofty bluff over the Ottawa River, and is the finest specimen of Italian Gothic architecture in America or the world. The great *Victoria Tower in the centre of the façade is imposing in its proportions; and the polygonal structure of the *Dominion Library* is in the rear of the buildings. The halls of the Senate and Chamber of Commons are worthy of a visit, and are adorned with stained-glass windows and marble columns. In the Senate is a statue of Queen Victoria, and near the vice-regal throne are busts of the Prince and Princess of Wales. The departmental buildings which flank the Parliament House are stately structures, in harmonious architecture, and of the same kinds of stone. The *Cathedral of Notre Dame* and the nunneries of the lower town are interesting; also the new churches of the middle town (which, like the rest of the city, is still undergoing a formative process). The **Chaudière Falls** are just above the city, where the broad Ottawa River plunges down over long and ragged ledges. In this vicinity are immense lumber-yards, with the connected industries which support the French Canadians, who form the majority of the citizens here. S. of the city are the pretty *Rideau Falls*. Steamers depart frequently for Montreal, and for the remote forests of the N.

The river and city of Ottawa are fully described in the companion to this hand-book, Osgood's *Middle States* ("with the Northern Frontier from Niagara Falls to Montreal; also, Baltimore, Washington, and Northern Virginia"), which was published in 1874 and revised in 1875. It also includes descriptions of the Upper St. Lawrence and Lake Ontario, Lake Champlain and Lake George, and the routes from New York to Montreal.

Osgood's *New England* contains also descriptions of Northern Vermont and New Hampshire, and the routes between Boston and Montreal or Quebec (published in 1873, and revised in 1874 and 1875).

INDEX.

Abattis, P. Q. 292.
Advocate Harbor 103, 80.
Agulquac River 54.
Ainslie Glen 167, 169.
Albert Bridge, C. B. 154.
Albert Mines, N. B. 72.
Alberton, P. E. I. 179.
Albion Mines, N. S. 136.
Aldouin River, 60.
Alemek Bay, N. B. 63.
Alexander Point 63.
Alexis River 225.
Allagash River, Me. 58.
Allandale, N. B. 52.
Alright Id. 184.
Alston Point, N. B. 65.
Amherst, N. S. 78, 74.
Amherst Id. 183.
Ancienne Lorette 281, 279.
Andover, N. B. 54.
Ange Gardien, P. Q. 283.
Annandale, P. E. I. 182.
Annapolis Basin, N. S. 84.
Annapolis Royal 85.
Annapolis Valley 88.
Anticosti 234.
Antigonish, N. S. 138.
Apohaqui, N. B. 71, 48.
Apple River, N. S. 80.
Apsey Cove, N. F. 210.
Aquafort, N. F. 198.
Ardoise Mt., N. S. 93.
Argentenay, P. Q. 290.
Argyle, N. S. 116, 125.
Arichat, C. B. 145.
Arisaig, N. S. 139.
Aroostook Valley, Me. 55.
Arthurette, N. B. 54.
Aspotogon Mt., N. S. 127.
Aspy Bay, C. B. 160.
Athol, N. B. 80.
Atlantic Cove, C. B. 160.
Aulac, N. B. 74.
Avalon, N. F. 198, 209.
Avonport, N. S. 91.
Aylesford, N. S. 89.
Aylesford Lakes 90.

Baccalieu Id., N.F. 201, 205.

Baccaro Point, N. S. 123.
Baddeck, C. B. 162.
Baddeck River, 167.
Bagotville, P. Q. 302.
Baie des Rochers, P. Q. 295.
Baie St. Paul, P. Q. 292.
Baie Verte, N. S. 74.
Ballard Bank, The 199.
Ballyhaly Bog, N. F. 195.
Bangor, Me. 39.
Barachois, N. B. 59.
Bareneed, N. F. 207.
Barnaby Id., P. Q. 250.
Barra, Strait of 164.
Barr'd Ids., N. F. 210.
Barrow, N. F. 214.
Barrow Harbor 203.
Barton, N. S. 112.
Basin of Minas 101, 108.
Basque Harbor 183.
Basque Island 251.
Bass River 81.
Bathurst, N. B. 65, 61.
Batiscan, P. Q. 307.
Batteau Harbor 225.
Battery Point, N. B. 68.
Battle Id., Lab. 224, 200, 206.
Bay, Argyle, N. S. 116.
Belleisle, N. B. 42.
Bonavista, N. F. 203.
Bonne, Lab. 219.
Bradore, Lab. 230.
Bulls, N. F. 194, 197.
Canada, N. F. 221.
Cardigan, P. E. I. 175.
Conception, N.F. 195, 206.
De Grave, N. F. 207.
Du Vin, N. B. 61.
East, C. B. 147.
Esquimaux, Lab 230.
Eternity, P. Q. 303.
Fortune, N. F. 214.
Garia, N. F. 215.
Ha Ha, P. Q. 301.
Hall's, N. F. 211.
Hermitage, N. F. 215.
Hillsborough 174, 175.
Ingornachoix 219.
Kennebecasis 40.

Bay, Little, N. F. 215.
Mahone, N. S. 118, 127.
Miramichi, N. B. 61.
Oak, N. B. 34.
of Chaleur 64, 240.
of Despair 215.
of Fair and False 203.
of Fundy 31, 83.
of Islands 218.
of Notre Dame, N. F. 210.
of St. John 219.
Placentia, N. F. 212.
Richmond, P. E. I. 178.
Roberts, N. F. 207.
St. Anne's, C. B. 158
St. George's, N. F. 217.
St. John's, P. Q. 304.
St. Margaret's 126, 113.
St. Mary's 112, 213.
Sandwich, Lab. 225.
Trinity, N. F. 208, 201.
Verd, N. F. 201, 208.
White, N. F. 221.
Beach, The 206.
Bear Cove 93.
Bear Point 143.
Bear River 85.
Beaubair's Id., N. B. 63.
Beaulieu, P. Q. 289.
Beaumont, P. Q. 254.
Beauport, P. Q. 276.
Beaver Bank, N. S. 93
Beaver Harbor, C. B. 162.
Beaver Harbor, N. B. 31.
Beaver Harbor, N. S. 132.
Beaver River 114.
Becancour, P. Q. 307.
Bedeque Bay, P. E. I. 174.
Bedford Basin, N. S 100.
Bellechasse Id. 254.
Belledune, N. B. 66.
Belle Isle 220, 206.
Belleisle Bay, N. B. 42.
Bell Isle, N. F. 221.
Belleorem, N. F. 214.
Belliveau Cove, N. S. 112.
Belliveau Village 73.
Beloeil Mt., P. Q. 319.
Benacadie Point 165.

INDEX.

Benmore 280.
Bersimis River 233.
Berthier *en bas* 254.
Berthier *en haut* 308.
Berwick, N. S. 90.
Bic Id., P. Q. 251.
Big Loran, C. B. 154.
Big Tancook Id. 128.
Biquette, P. Q. 251.
Birch Point 64.
Birchtown, N. S. 121.
Bird Island Cove 202.
Bird Isles 184.
Bird Rock 161.
Black Bay 228.
Black Brook 61.
Blackhead 196.
Blackhead Cove 210.
Black Point, N. S. 122.
Black River, N. F. 212.
Black River, P. Q. 295.
Blancherotte, C. B. 147.
Blanc Sablon, Lab 229.
Blandford, N. B. 27.
Blind Lake, N. S. 126.
Bliss Id., N. B. 31.
Blissville, N. B. 49.
Blockhouse Mines 153.
Blomidon, Cape 102, 103.
Bloody Bay, N. F. 203.
Bloody Bridge 79
Bloody Brook, N. S. 89.
Blow-me-Down Head 207.
Blue Mts., N. S. 90, 115.
Blue Pinion, N. F. 214.
Blue Rocks, N. S. 118.
Boar's Back, N. S. 82.
Boar's Head, N. B. 40.
Boiestown, N. B. 47, 62.
Boisdale 162.
Bonami Point 67.
Bonaparte Lake 36.
Bonaventure Id. 243.
Bonavista Bay, N. F. 203.
Bonhomme, Le 307.
Bonne Bay 219.
Bonne Esperance Bay 230.
Bonny, Lab. 230.
Bon Portage Id. 124.
Bonshaw, P. E. I. 174.
Bothwell, P. E I. 182.
Boucherville, P. Q. 309.
Boularderie, C. B. 161.
Bout de l'Isle 308.
Bradford's Cove 29.
Bradore Bay, Lab. 230.
Brae, P. E. I. 179.
Braha, N. F. 221.
Branch, N. F. 212.
Brandies, The 201.
Brandy Pots 252, 296.
Bras d'Or, The 161.
Breton, Cape 149, 154.
Bridgeport, C. B. 152.

Bridgetown, N. S. 88.
Bridgeton, P. E. I. 182.
Bridgewater 128, 119.
Brigg's Corner 49.
Brighton, N. S. 112.
Brigus, N. F. 207.
Bristol, N. B. 51.
Broad Cove, N. B. 29.
Broad Cove, N. F. 203.
Broad Cove, N. S. 120.
Broad Cove Intervale 169.
Brookfield, N. S. 82, 130.
Brooklyn, N. S. 93.
Brookvale, N. B. 48.
Broyle Harbor 197.
Brucker's Hill 26.
Brule Harbor 81.
Brunet Id. 214.
Bryant's Cove 207.
Buctouche, N. B. 59.
Bull Arm, N. F. 209.
Bull Moose Hill 41.
Burgeo, N. F. 215.
Burgoyne's Ferry 51.
Burin, N. F. 214, 212.
Burlington, N. S. 93.
Burnt Church 62, 63.
Burnt Head 207.
Burnt Ridge 202.
Burton, N. B. 43.
Burying Place 211.
Butter Pots, The 199.

Cacouna, P. Q. 296, 252.
Calais, Me. 35.
Caledonia Corner 130.
Callière, P. Q. 295.
Calvaire, Miq. 185.
Calvaire, P. Q. 306.
Cambridge, N. B. 42.
Cambriol, N. F. 214.
Campbell River 55.
Campbellton, N. B. 68.
Camille, Mt. 249.
Campobello Id. 25.
Canaan River 72.
Canada Bay 221.
Canada Creek 90.
Canning, N. B. 43.
Canning, N. S. 91.
Canso 142.
Canterbury 37, 52.
Cap à l'Aigle 294.
   au Corbeau 292.
   de la Magdelaine 307.
   de Meule 184.
   Rouge 281.
   St. Ignace 253.
Cape Alright 184.
   Anguille, N. F. 217.
   Ballard, N. F. 213.
   Bauld, N. F. 220.
   Bear 175, 181.
   Blomidon, 91, 102, 103.

Cape Bluff, Lab. 225.
   Breton, 149, 154.
   Broyle, N. F. 197.
   Canso, N. S. 134, 142.
   Chapeau Rouge 214, 189.
   Chatte, P. Q. 249.
   Chignecto, N. S. 104.
   Cove, N. S. 114.
   Cove, P. Q. 241.
   Colombier, P. Q. 233.
   Corneille, 294.
   Dauphin 158, 161.
   Desolation 226.
   Despair, P. Q. 241.
   Diable, P. Q. 252.
   d'Or, N. S. 103.
   East, P. Q. 301.
   Egmont, P. E. I. 174, 179.
   English, N. F. 213.
   Enragé, N. B. 72.
   Eternity, P. Q. 303.
   Fogo, N. F. 204, 210.
   Fourchu, N. S. 125.
   Freels, N. F. 203, 213.
   Gaspé, P. Q. 246.
   George, P. Q. 304.
   Goose 294.
   Grand Bank 214.
   Gribaune 291.
   Jourimain 59, 73.
   Kildare 180.
   Labaie 292.
   Lahave, N. S. 120.
   La Hune 215.
   Largent 202.
   Mabou, C. B. 168.
   Magdelaine 248.
   Maillard 292.
   Marangouin 73.
   Morien, C. B. 153.
   Negro, N. S. 122.
   Norman, N. F. 220.
   North, C. B. 160.
   Perry, C. B. 153.
   Pine, N. F. 213.
   Porcupine, N. S. 144.
   Race, N. F 199, 189.
   Ray, N. F. 217, 216.
   Rhumore, C. B. 147.
   Ridge, N. F. 203.
   Roseway, N. S. 121.
   Rosier 247, 246.
   Rouge 291.
   Sable, N. S. 123.
   St. Anne 249.
   St. Francis 201, 225, 301.
   St. George 218
   St. Lawrence 160, 170.
   St. Michael 225.
   St. Nicholas 233.
   Sambro 118, 93.
   Smoky, C. B. 159.
   Spear, N. F. 189, 196.
   Spencer 104, 83.

INDEX.   323

Cape Split, N. S. 104.
  Tourmente 287, 253.
  Tourmentine 59, 73, 174.
  Traverse 174.
  Trinity, P. Q. 303.
  Tryon, P. E. I. 178.
  Victoria, P. Q. 304.
  West 302.
  Whittle, Lab. 230.
  Wolfe 179.
Caplin Cove 198.
Caraquette 66, 62.
Carbonear, N. F. 208.
Cardigan, N. B. 50.
Cardigan, P. E. I. 181.
Caribacon 145.
Caribou Id. 175, 224.
Caribou Plains 80.
Caribou Point 233.
Carleton, N. B. 24.
Carleton, P. Q. 239.
Carrousel Id. 233.
Cascapediac Bay 240.
Cascumpec 180.
Castle Id., Lab. 227.
Catalina, N. F. 201.
Catalogue, C. B. 154.
Cataracouy 280.
Cat Cove 221.
Caughnawaga 319.
Cavendish, P. E. I. 178.
Caverne de Bontemps 290.
Cawee Ids. 233.
Central Falmouth 91.
Centre Hill 209.
Chaleur, Bay of, 64, 240.
Chamcook Mt. 33.
Champlain, P. Q. 307.
Chance Harbor 31.
Change Ids. 205, 210.
Channel, N. F. 216.
Chapel Id., C. B. 147.
Charlesbourg, P. Q. 279.
Charlottetown, P. E. I. 175.
Château Bay, Lab. 227.
Château Bellevue 287.
Château Bigot 280.
Château Richer 284.
Chatham, N. B. 61, 66.
Chaudière Falls 232.
Chebucto Head 93.
Chedabucto Bay 143.
Chester, N. S. 127, 90.
Cheticamp, C. B. 170.
Cheticamp, N. S. 114.
Chezzetcook, N. S. 131.
Chicoutimi, P. Q. 300.
Chignecto, Cape, 104.
Chignecto Peninsula 79.
Chimney Tickle 227.
Chiputneticook Lakes, N. B. 38, 46.
Chivirie 93, 102, 106.
Chouse Brook 221.

Ciboux Ids. 161.
Clairvaux, P. Q. 292.
Clare, N. S. 113.
Clarendon, N. B. 33.
Clementsport, N. S. 85.
Clementsvale 85.
Clifton, N. B. 66, 71.
Clode Sound 203.
Cloridorme 248.
Clouds, The, 221.
Clyde River, N. S. 124.
Coacocho River 231.
Cobequid Mts., N. S. 80.
Cocagne, N. B. 59.
Colebrooke, N. B. 55.
Cole's Id. N. B. 47.
Colinet, N. F. 213.
Columbe 215
Conception Bay 195, 206.
Conche, N. F. 221.
Contrecœur, P. Q. 308.
Corbin, N. F. 214.
Cornwallis Valley, N. S. 90, 103, 107.
Corny Beach 243.
Côté de Beaupré, 283.
  des Neiges 318.
  St. Antoine 318.
  St. Luc 318.
  St. Michel 318.
Cottel's Id. 203.
Coudres, Isle aux 293.
Country Harbor 133.
Covehead, P. E. I. 181.
Cow Bay 101, 150, 153.
Cox's Point 49.
Crabb's Brook 217.
Crane Id., P. Q. 253.
Crapaud, P. E. I. 174.
Creignish 168.
Croque, N. F. 221.
Cross Id., N. S. 118.
Cumberland Bay 49.
Cumberland Harbor 230.
Cupids, N. F. 207.

Dalhousie, N. B. 67.
Dalibaire, P. Q. 249.
Dark Cove, 30.
Dartmouth, N. S. 101.
Dauphiney's Cove 126.
Davis Strait 226.
Dead Ids. 216, 225.
Deadman's Isle 184.
Debec Junction 37.
Debert 80, 105.
Deep Cove 127.
Deerfield, N. S 115.
Deer Harbor 209.
Deer Isle, N. B. 25.
Deer Lake 37.
Deer Pond 219.
Demoiselle Hill 183.
Denys River, C. B. 165.

De Sable 174.
Descente des Femmes 302.
Deschambault 306.
D'Escousse, C. B. 145.
Despair, Bay of, 215.
Despair, Cape, 241.
Devil Id. 93.
Devil's Back, N. B. 41.
Devil's Goose-Pasture 90.
Devil's Head 34.
Diable Bay 228.
Digby, N. S. 84.
Digby Neck 116.
Dipper Harbor 31.
Distress Cove 212.
Dodding Head 214.
Dollannan Bank 202.
D'Or, Cape, N. S. 103.
Dorchester, N. B. 73.
Doucet's Id. N. B. 34.
Douglas Harbor 49.
Douglastown, N. B. 62.
Douglastown, P. Q. 244.
Douglas Valley 38.
Dumfries, N. B. 52.
Dundas, N. B. 59.
Dundas, P. E. I. 182.
Dunk River 174.

Earltown, N. S. 136.
East Bay 147, 165, 214.
Eastern Passage 93.
East Point 182.
Eastport, Me. 26.
East River 126, 225.
Éboulements, Les, 204.
Echo Lake 131.
Economy Point 105, 80.
Ecureuils, Les, 306.
Eddy Point 143.
Edmundston, N. B. 57.
Edoobekuk, C. B. 147.
Eel Brook 30.
Egg Ids., Lab. 233.
Ekum Sekum, N. S. 132.
Ellershouse, N. S. 93.
Elliot River 174.
Elmsdale, N. S. 82.
Elysian Fields, N. S. 79.
Enfield, N. S. 82.
English Harbor 201.
English Harbor West 214.
English Point 233.
Englishtown, C. B. 158.
Enniskillen, N. B. 38.
Entry Id. 184.
Escasoni, C. B. 148.
Escuminac Point 61.
Esquimaux Bay 230, 244.
Eternity Bay 303.
Exploits Id. 205, 210.
Exploits, River of 210.
Factory Dale, N S. 89.
Fairville, N. B. 37.

Fairy Lake, N. S. 130.
Falkland, N. S. 90, 93.
Falls, Chaudière 282, 320.
  Chicoutimi, P. Q. 300.
  Grand 55, 66.
  Grand, N. F. 210.
  Grande-Mère 307.
  Lorette, P. Q. 278.
  Magaguadavic 32.
  Manitousin 232.
  Montmorenci 277.
  Nictau, N. S. 89.
  North River 105.
  Pabineau, N. B. 66.
  Pokiok, N. B. 52.
  Pollett 72.
  Rideau, Ont. 320.
  Rivière du Loup 295.
  Rivière du Sud 253.
  St. Anne, P. Q. 286.
  Sault à la Puce 284.
  Shawanegan 307.
  Sissiboo, N. S. 112.
Falmouth, N. S. 91.
Farmington, N. S. 89.
Father Point, P. Q. 250.
Ferguson's Cove 101.
Fermeuse, N. F. 198.
Fern Ledges 24.
Ferryland, N. F. 198.
Fish Head 30.
Five Ids., N. S. 105, 80.
Flagg's Cove 29.
Fleurant Point 67.
Flint Id., C. B. 150, 153.
Florenceville, N. B. 53.
Flower Cove 219.
Fogo, N. F. 204.
Folly Pass, N. S. 80.
Forks, The 48, 54.
Fort Beaubassin 74, 78.
Fort Beausejour 74, 78.
Fort Cumberland 74, 78.
Forteau, Lab. 228.
Fort Fairfield, Me. 54.
Fort Ingalls, N. B. 58.
Fort Jaques Cartier 306.
Fort Kent, Me. 58.
Fort Lawrence 74, 78.
Fort Meductic, N. B. 52, 46.
Fort Nascopie, Lab. 226.
Fort Norwest, Lab. 226.
Fortune, N. F. 214.
Foster's Cove 54.
Fourchette, N. F. 221.
Fourchu, C. B. 147.
Fox Harbor, N. S. 103, 81.
Fox Harbor, Lab. 224.
Fox River 248.
Framboise, C. B. 147.
Frazer's Head 104.
Fredericton, N. B. 44.
Fredericton Junc. 33.
French Cross, N. S. 89.

French Fort Creek 180.
French Lake 48.
Frenchman's Cove 214.
French River 138.
French Shore, The 216.
French Village 151.
Frenchville, Me. 57.
Freshwater Bay 203.
Friar's Face 26.
Frozen Ocean 130.
Funk Id., N. F. 204.

Gabarus Bay 154, 149.
Gagetown, N. B. 42, 48.
Gairloch, N. S. 136.
Galantry Head 185.
Gambo Ponds 203.
Gander Bay 210.
Gannet Rock, N. B. 29.
Gannet Rock 184.
Garia Bay 215.
Garnish, N. F. 214.
Gaspé, P. Q. 244.
Gaspereaux Lake 90.
Gay's River, N. S. 82.
Gentilly, P. Q. 307.
George Id. 179.
George's Id., N. S. 98.
Georgetown, P. E. I. 181, 175.
Gibson, N. B. 49.
Gilbert's Cove 112.
Glace Bay 153, 150.
Glengarry, N. S. 136.
Goat Id., N. S. 85.
Godbout, Lab. 233.
Goldenville, N. S. 133.
Gold River 128.
Gondola Point 71.
Gooseberry Isles, 203.
Goose Id. 253.
Gouffre, Le 293.
Gowrie Mines 153.
Grand Anse, C. B. 145.
Grand Anse, N. B. 66.
Grand Banks, The 199.
Grand Bay 40.
Grand Digue 145.
Grande Baie 302.
Grande-Mère Falls 307.
Grand Falls, Lab. 226.
Grand Falls, N. B. 55.
Grand Grève, P. Q. 244.
Grand Harbor 29.
Grand Lake 36, 48.
Grand Lake Stream 35.
Grand Manan 28.
Grand Narrows 164.
Grand Pond 218, 211.
Grand Pré 107, 91, 101.
Grand River, C. B. 147.
Grand River, N. B. 56.
Grand River 241.
Grand-River Lake 147.
Grand Rustico 178.

Grandy's Brook, 215.
Grant Isle, Me. 57.
Granville, N. S. 86.
Great Bartibog 61.
Great Boule 233.
Great Bras d'Or 161, 164.
Great Codroy 217.
Great Ha Ha Lake 302.
Great Harbor Deep 221.
Great Meccatina 230.
Great Miquelon 186.
Great Pabos 241.
Great Pond 248.
Great Pubnico Lake 124.
Great St. Lawrence 214.
Great Shemogue 59.
Great Village 81.
Green Bay 211.
Greenfield 130.
Green Harbor 209.
Green Ids. 124, 214, 252.
Greenly Id. 229.
Green River 57.
Greenspond, N. F. 203.
Greenville 80.
Greenwich Hill 41.
Grenville Harbor 178.
Griffin's Cove 248.
Griguet, N. F. 221.
Grimross, N. B. 42.
Grindstone Id. 183.
Grondines, P. Q. 306.
Grosse Isle 254.
Grosses Coques 113.
Gull Rock 121.
Gut of Canso 142.
Guysborough 133.

Habitants Bay 143.
Ha Ha Bay, P. Q. 301.
Halifax, N. S. 93.
  Admiralty House 97.
  Cathedral 98.
  Citadel 96.
  Dalhousie Coll. 98.
  Gov't. House 98.
  Harbor 93.
  Hortic. Gardens 98.
  Museum 96.
  Parliament Building 95.
  Provincial Building 95.
  Queen's Dockyard 97.
  Y. M. C. A. 96.
Halifax, P. E. I. 179.
Hall's Bay 211, 218.
Hammond's Plains 100.
Hampton, N. B. 71.
Hampton, N. S. 89.
Hantsport, N. S. 91, 101.
Harbor Briton 214.
Harbor Buffet 212.
Harbor Grace, N. F. 207.
Harborville, N. S. 90.
Hare Bay, N. F. 221.

INDEX.

Hare Id., P. Q. 252.
Hare's Ears 198.
Hare's-Head Hills 218.
Harmony, P. E. I. 182.
Harvey, N. B. 38.
Harvey Corner 72.
Haulover Isthmus 146.
Havelock, N. S. 89.
Head of Amherst 78.
Heart Ridge, N. F. 210.
Heart's Content 208.
Heart's Delight 209.
Heart's Desire 209.
Heart's Ease, N. F. 209.
Hebertville, P. Q. 300.
Hebron, Lab. 226.
Heights of Land 226.
Hell Hill 197.
Hermitage Bay 215.
Herring Cove, N. S. 93.
High Beacon 227.
Highland Park 23.
Highland Village 81.
High Point 301.
Hillsborough, N. B. 72.
Hillsborough Bay 174.
Hillsborough River 180.
Hillsburn 86.
Hochelaga, P. Q. 318.
Hodge-Water River 213.
Holland Bay, 180.
Holyrood, N. F. 199.
Holyrood Pond 213.
Hooping Harbor 221.
Hope, P. Q. 241.
Hope All, N. F. 209.
Hopedale, Lab. 226.
Hopewell 136.
Hopewell Cape 72.
Horton Landing 91.
Houlton, Me. 37, 51.
Howe's Lake 23.
Hudson's Strait 226.
Humber River 219.
Hunter River 177, 178.

Indian Bay 167, 203.
Indian Beach 30.
Indian Gardens 130.
Indian Id., Lab. 225.
Indian Ids. 210.
Indian Lorette 278.
Indian Tickle 225.
Indiantown, N. B. 47.
Indian Village 51.
Ingonish, C. B. 159.
Intervale 133.
Ionclay Hill 197.
Irish Cove, C. B. 147.
Ironbound Cove, N. B 49.
Ironbound Id., N. S. 119.
Island, Alright 184.
　Amherst 183.
　Anticosti 234.

Island, Baccalieu, N. F. 201.
Barnaby, P. Q. 250.
Beaubair's 63.
Bellechasse 254.
Bic, P. Q. 250.
Blackbill 227.
Bonaventure 243.
Bon Portage 124.
Boughton 175.
Boularderie 161.
Brandy Pots 252.
Brier 117.
Brunet 214.
Bryon 184.
Campobello 25.
Cape Breton 141.
Cape Sable 123.
Caribou 175, 224.
Carrousel 233.
Castle, Lab. 227.
Caton's 41.
Cawee 233.
Chapel 147.
Cheticamp 170.
Cheyne 29.
Christmas 164.
Cobbler's 203.
Coffin 184.
Cole's 47.
Cottel's 203.
Crane, P. Q. 253.
Cross, N. S. 118.
Dead, N. F. 225.
Deer 203.
Devil, N. S. 93.
Egg, Lab. 233.
Entry 184.
Esquimaux, Lab. 231.
Exploits, N. F. 205, 210.
Fair, N. F. 203.
Fishflake 227.
Fly 225.
Fogo, N. F. 204, 210.
Foster's, N. B. 41.
Fox, N. B. 61.
Funk, N. F. 203.
George 179.
George's, N. S. 98.
Goat, N. S. 85.
Goose, P. Q. 253.
Governor's 175.
Grand Dune 61.
Grand Manan 28.
Grassy, N. B. 41.
Great Caribou 224.
Green 124, 201, 220, 252.
Grimross, N. B. 43.
Grindstone 72, 183.
Grosse 184.
Hare, P. Q. 252.
Henry 169.
Heron 67.
Horse 221.
Huntington 225.

Island, Indian 225.
Ireland, N. F. 215.
Ironbound 119.
Jaques Cartier 220.
Kamouraska 252.
Large 231.
Lennox, P. E. I. 179.
Little Miquelon 186.
Little Bay 211.
Locke's, N. S. 121.
Long 42, 101,107, 117, 212.
Lower Musquash 42.
McNab's, N. S. 101, 93.
Madame, P. Q. 254.
Mauger's 43.
Melville 101.
Merasheen 212.
Middle 48.
Miquelon 186.
Miscou 64.
Moose 26.
Nantucket 29.
Negro 122.
Newfoundland 187.
New World 205.
of Ponds 225.
Panmure, P. E. I. 175.
Park, P. E. I. 179.
Partridge, N. B. 15.
Partridge, N. S. 102, 103.
Penguin 203.
Pictou, N. S. 175.
Pilgrims 252.
Pincher's 203.
Pinnacle 105.
Pocksuedie 63.
Pool's 25.
Portage 61.
Priests' 318.
Prince Edward 172.
Quarry 231.
Quirpon 220.
Ram 121.
Random, N. F. 209.
Reaux, P. Q. 254.
Red 212, 218, 252.
Sable 134.
Saddle 217.
Sagona 214.
St. Barbe 221.
St. Paul's 160.
St. Pierre 185.
Sandous 46.
Seal, N. S. 124.
Sea-Wolf 169.
Sheldrake 61.
Shippigan 63.
Smith's 169.
Spencer's 103,104, 106.
Spotted, N. F. 225.
Square, Lab. 225.
Stone Pillar 253.
Sugar 60, 51.
Venison 225.

INDEX.

Island, Vin, N. B 61.
  White Head 29.
  White Horse 31.
  Wolf 184.
  Wood Pillar 253.
Islands, Battle 224.
  Burnt 215.
  Camp 227.
  Ciboux 161.
  Dead 215.
  Five 105.
  Little St. Modeste 228.
  Magdalen 183.
  Mingan 231.
  Mutton 124.
  Penguin 203.
  Ragged 212.
  Ramea 215.
  Ram's, N. F. 212.
  Red 147.
  Seal 225.
  Seven, Lab. 232.
  Tancook, N. S. 128.
  Tusket, N. S. 125.
Isle aux Chiens 185.
  aux Coudres 293.
  Bell, N. F. 221.
  Belle 206, 220.
  Deadman's 184.
  Deer, N. B. 25.
  Groais 221.
  Haute 104.
  Jesus 318.
  Madame 145.
  of Orleans 288.
  St. Louis 304.
  St. Therese 308.
  Verte, P. Q. 252.
Isles, Bird 184.
  Burgeo 215.
  de la Demoiselle 230.
  Gooseberry 203.
  Passe Pierre 305.
  Peterel 227.
  Twillingate 205.
  Wadham, N. F. 203.
  West, N. B. 25, 31.

Jackson's Arm 221.
Jacksonville, N. S. 90.
Jaques Cartier 306.
Jebogue Point 125.
Jeddore, N. S. 132.
Jemseg, N. B. 42, 48.
Jerseyman Id. 145.
Jesus, Isle 318.
Jeune-Lorette 278.
Joe Batt's Arm 210.
Joggins Shore 80.
Jolicœur, N. B. 73.
Joliette, P. Q.
Jonquière 300.
Judique, C. B. 168.
Julianshaab, Gr. 226.

Kamouraska, P. Q. 252.
Keels, N. F. 203.
Kegashka Bay 231.
Kempt Head 162.
Kempt, N. S. 115.
Kempt Lake, N. S. 90.
Kennebecasis Bay 40, 22.
Kenogami, P. Q. 300.
Kensington 178.
Kentville, N. S. 90.
Keswick Valley 50.
Keyhole, N. B. 49.
Kingsclear, N. B. 51.
King's Cove 203.
Kingston, N. B. 42.
Kingston, N. S. 89.
Kouchibouguac Bay 61.

La Bonne St. Anne 285.
Labrador 223.
Lac à la Belle Truite 302.
Lachine, P. Q. 320.
La Fleur de Lis 221.
Lahave River 128.
Lake Ainslie 167, 169.
  Bathurst 211.
  Bear 38.
  Beauport 279.
  Belfry 154.
  Ben Lomond 23.
  Blind 126.
  Catalogne, C. B. 154.
  Cedar, N. S. 115.
  Chamberlain, Me. 58.
  Chesuncook 58.
  Cleveland 57.
  Cranberry 38.
  Croaker's 211.
  Echo, N. S. 131.
  Fairy, N. S. 130.
  French, N. B. 48.
  Gabarus, C. B. 154.
  Gaspereaux 90.
  George 51, 90, 115.
  George IV. 211.
  Grand 48, 36, 82.
  Gravel 295.
  Great Ha Ha 302.
  Jones 23.
  Kempt 90.
  Lewey's, Me. 35.
  Lily, N. B. 22.
  Little Ha Ha 302.
  Long, P. Q. 58.
  Long, N. S. 82.
  Magaguadavic 38.
  Malaga, N. S. 129.
  Manor, P. Q. 319.
  Maquapit, N. B. 48.
  Metapedia 69.
  Mira, C. B. 154.
  Mistassini 301.
  Moosehead 58.
  Mount Theobald 71.

Lake Nepisiguit 55.
  Nictor, N. B. 55.
  Oromocto 38.
  Pechtaweekagomic 58.
  Pemgockwahen 58.
  Pockwock 100.
  Pohenagamook 58.
  Ponhook, N. S. 126.
  Porter's 131.
  Port Medway 130.
  Preble, Me. 57.
  Prince William 52.
  Queen's, N. B. 37.
  Quiddy Viddy 195.
  Robin Hood 37.
  Rocky, N. S. 82.
  Rossignol 130.
  St. Charles 279.
  St. Joachim 287.
  St. John, P. Q. 301.
  St. Peter, P. Q. 307.
  Sedgwick 57.
  Segum Sega 130.
  Sheogomoc 52.
  Shepody, N. B. 72.
  Sherbrooke 90.
  Sherwood, N. B. 37.
  Ship Harbor 132.
  S. Oromocto 38.
  Spruce, N. B. 24.
  Stream 49.
  Taylor's 23.
  Temiscouata 58, 295.
  Terra Nova 203.
  Tracy's, N. B. 71.
  Tusket, N. S. 115.
  Two-Mile 90.
  Utopia, N. B. 32.
  Vaughan, N. S. 115.
  Washademoak 47, 42.
  Welastookwangamis 58.
  Wentworth 113.
  Windsor, N. F. 195.
  Winthrop, Me. 58.
Lakes, Aylesford 90.
  Bras d'Or 161.
  Chiputneticook 38.
  Dartmouth 101.
  Eagle, Me. 58.
  Schoodic, Me. 35.
  Tusket, N. S. 115.
La Manche 197, 212.
Lance-au-Loup 228.
Lance Cove 206.
Land's End, 41.
Langley Id. 186.
Lanoraie, P. Q. 308.
L'Anse à l'Eau 305.
La Poîle, N. F. 215.
L'Archevêque 147.
L'Ardoise, C. B. 146.
Large Id. 231.
La Scie 221, 211.
L'Assomption, P. Q. 308.

INDEX.

Laval River 299.
Lavaltrie, P. Q. 308.
La Vieille 246.
Lawlor's Lake 70.
Lawrencetown 89, 131.
Lazaretto, Tracadie 62.
Ledge, The 35.
Leitchfield, N. S. 86.
Lennox Id. 179.
Lennox Passage 145.
Les Eboulements 294.
Les Ecureuils 306.
Les Escoumains 233.
L'Etang du Nord 184.
L'Etang du Savoyard 185.
L'Etang Harbor 31.
Letite Passage 32.
Levis, P. Q. 282.
Lewey's Id. 35.
Lewis Cove 47.
Lily Lake 22.
Lingan 152, 150.
Lion's Back 23.
Liscomb Harbor 132.
L'Islet, P. Q. 253.
L'Islet au Massacre 250.
Little Arichat 145.
Little Bay Id. 205, 211.
Little Bras d'Or 161.
Little Falls 57.
Little Glace Bay 153.
Little Ha Ha Lake 302.
Little Loran 154.
Little Miquelon 186.
Little Narrows 167.
Little Pabos 241.
Little Placentia 212.
Little River 22.
Little Rocher 72.
Little Saguenay 304.
Little St. Lawrence 214.
Little Seldom-come-by 210.
Little Shemogue 59.
Little Tancook 128.
Liverpool, N. S. 120, 130.
Lobster Harbor 221.
Loch Alva 37.
Loch an Fad 147.
Loch Lomond, C. B. 147.
Loch Lomond, N. B. 22.
Lochside, C. B. 147.
Loch Uist 147.
Locke's Id., N. S. 121.
Logie Bay 195, 200.
Londonderry 105.
Long Id. 40, 42, 101, 117.
Long Pilgrim 252.
Long Point 231.
Long Range 217.
Long Reach 41.
Long's Eddy 30.
Longue Point 319.
Lorette, Indian 278.
Lotbinière, P. Q 306.

Louisbourg, C. B. 154, 149.
Loup Bay 228.
Low Point 168.
Lower Canterbury 52.
Lower Caraquette 66.
Lower French Vill 51.
Lower Horton 107.
Lower Middleton 89.
Lower Prince William 51.
Lower Queensbury 51.
Lower Woodstock 52.
Lubec, Me. 26.
Ludlow, N. B. 47.
Lunenburg 118, 128.

Mabou, C. B. 169.
Mabou Valley 168.
McAdam Junc. 38.
Maccan, N. S. 80, 79.
Mace's Bay 31.
McNab's Id. 101, 93.
Madawaska 57.
Magaguadavic River 32.
Magdalen Ids. 183.
Magdelaine, Cape 248.
Maguacha Point 67, 239.
Magundy, N. B. 51.
Mahogany Road 24.
Mahone Bay 127, 118.
Main-à-Dieu 150.
Maitland 82, 105, 129.
Malaga Lake 130.
Malagawdatchkt 165.
Malbaie, P. Q. 294.
Mal Bay 244.
Malcolm Point 61.
Malignant Cove 139.
Malpeque Harbor 178.
Manchester, N. S. 133.
Manicouagan 233, 250.
Manitousin Falls 232.
Maquapit Lake 48.
Marchmont 280.
Margaree River 167.
Margaree Forks 170.
Margaretsville 89.
Maria, P. Q. 240.
Marie Joseph 132.
Marion Bridge 154.
Marshalltown 112.
Mars Head 117.
Mars Hill 54.
Marsh Road 22.
Marshy Hope 138.
Mascarene 32.
Masstown 81.
Matane, P. Q. 249.
Mattawamkeag 39, 58.
Maugerville, N. B. 43.
Mealy Mts. 225.
Meccatina, Lab. 230.
Medisco, N. S. 76.
Meductic Rapids 52.
Mejarmette Portage 40.

Melford Creek 143.
Melrose, N. S. 82.
Melvern Square 89.
Melville Id. 101.
Melville Lake 226.
Memramcook 73.
Merasheen Id. 212.
Merigomish 138.
Metapedia 69.
Meteghan, N. S. 113.
Métis, P. Q. 249.
Middle Musquodoboit 82.
Middle River 163, 167.
Middle Simonds, 53.
Middle Stewiacke 81.
Middleton, N. S. 89.
Milford, N. S. 129.
Milford Haven 133.
Milkish Channel 41.
Mill Cove, N. B. 49.
Mille Vaches 299.
Milltown, N. B. 35.
Mill Village 128.
Minas Basin 101, 108.
Mingan Ids., Lab. 231.
Ming's Bight 221.
Minister's Face 22.
Minudie, N. S. 79.
Miquelon 185, 214.
Mira Bay 150.
Mira Lake, C. B. 154.
Miramichi, N. B. 61.
Miscouche 179.
Miscou Id. 64.
Mispeck, N. B. 23.
Missiguash Marsh 79, 74.
Mission Point 68.
Mistanoque Id. 230.
Mistassini, Lake 301.
Moisic River 232.
Molasses Harbor 134.
Momozeket River 55.
Moncton, N. B. 72.
Money Cove 30.
Montague Bridge 181.
Montague Mines 101, 131.
Mont Joli 231.
Mont Louis 249.
Montmorenci Falls 277.
Montreal, P. Q. 309.
　Bonsecours Market 312.
　Champ de Mars, 312.
　Christ Ch. Cathed. 314.
　Court House 312.
　Dominion Sq. 315.
　Geolog. Museum 312.
　Gesù Church 313.
　Gray Nunnery 315.
　Great Seminary 315.
　Hôtel Dieu 316.
　Institut Canadien 312.
　McGill Univ. 314.
　Montreal Coll. 315.
　Mt. Royal 316.

INDEX.

Montreal, Nazareth Asyl. 316.
New Cathedral 315.
Notre Dame 311.
Place d'Armes 311.
Post-Office 311.
Seminary 312.
St. Helen's Isle 313.
Victoria Bridge 316.
Victoria Square 311.
Moose Harbor 120.
Moosepath Park 22.
Morden, N. S. 89.
Morrell, P. E. I. 182.
Morris Id. 116.
Morristown 90, 139.
Mosquito Cove 208.
Moss Glen 22.
Moulin à Baude 299.
Mount Aspotogon 127.
 Blair 32.
 Calvaire 186.
 Camille 250.
 Chapeau 186.
 Dalhousie 67.
 Denson 91.
 Éboulements 294, 253.
 Granville 146.
 Hawley 89.
 Hermon Cemet. 280.
 Joli 242.
 Nat 225.
 Pisgah 71.
 Royal 316, 318.
 St. Anne 242.
 Stewart, P. E. I. 181.
 Teneriffe, N. B. 55.
 Uniacke, N. S. 93.
Mountain, Ardoise, 93.
 Bald, 38, 55.
 Belœil 319.
 Boar's Back 132.
 Boucherville 319.
 Chamcook 33.
 North 84.
 Salt 168.
 South 84.
 Sugar-Loaf 159.
 Tracadiegash 67, 239.
Mountains, Antigonish 139.
 Baddeck 163.
 Blue 84, 90, 115, 130.
 Cobequid 80.
 Ingonish 161.
 Mealy 225.
 Notre Dame 249.
 St. Anne 287.
 St. Margaret 302.
 Scaumenac 68.
 Sporting 146.
Mull River 168.
Murray Bay 294.
Murray Harbor 181.
Mushaboon Harbor 132.

Musquash, N. B. 31.
Musquodoboit 131.
Mutton Ids. 124.

Nain, Lab. 226.
Napan Valley 61.
Narrows, The 47, 54.
Narrows, Grand 164.
Nashwaak 47.
Nashwaaksis 45.
Natashquan Point 231.
Natural Steps, The 277.
Necum Tench 132.
Negro Id., N. S. 122.
Negrotown Point 15.
Nelson, N. B. 63.
Nepisiguit Lake 55.
Nepisiguit River 65.
Nerepis Hills, N. B. 41.
Nerepis River 38.
Netsbuctoke 225.
Neutral Id., N. B. 34.
New Albany, N. S. 89.
New Bandon 66.
New Bay 211.
New Bonaventure 210.
New Brunswick 13.
Newburgh, N. B. 50.
New Canaan 48.
New Carlisle 240.
Newcastle 49, 62.
New Dublin 119.
New Edinburgh 112.
Newfoundland 187.
New Glasgow, N. S. 136.
New Glasgow, P. E. I. 178.
New Liverpool 282.
New London 178.
Newman Sound 203.
New Perlican 209.
Newport, N. S. 92, 101.
Newport, P. Q. 241.
New Richmond 240.
New Ross, N. S. 90.
New Tusket 113.
Niapisca Id. 231.
Nicolet, P. Q. 308.
Nictau Falls 89.
Nictor Lake 55.
Niger Sound 227.
Nimrod, N. F. 211.
Nipper's Harbor 205, 211.
Noel, N. S. 105.
North Bay 214.
Northern Head 30.
Northfield 129.
North Harbor 212.
North Joggins 73.
North Lake 182.
North Mt. 84.
North Point 180.
North River Falls 105.
North Rustico 178.
North Sydney 151.

Northumberland Strait 60, 174, 239.
Northwest Arm 100.
North Wiltshire 177.
Norton, N. B. 71, 42.
Norwest, Lab. 226.
Notre Dame Bay 210, 205.
Notre Dame du Lac 58.
Nova Scotia 75.
Nubble Id. 31.

Oak Bay, N. B. 34.
Oak Point 41, 61.
Ochre Pit Cove 208.
Offer Wadham 204.
Okkak, Lab. 226.
Old Barns 81.
Old Bonaventure 210.
Old Ferolle 219.
Old Fort Point 158.
Oldham Mines 82.
Old Maid 29.
Old Perlican 209, 201.
Oldtown, Me. 39.
Olomanosheebo 231.
Onslow 80.
Oromocto, N. B. 43.
Oromocto Lake 38.
Orono, Me. 39.
Otnabog, N. B. 42.
Ottawa, Ont. 320.
Outarde River 250.
Oxford, N. S. 80.
Ovens, the 119.

Pabineau Falls 66.
Pabos, P. Q. 241.
Painsec Junc. 72, 59.
Paps of Matane 249.
Paradise, N. F. 225.
Paradise, N. S. 89.
Parrsboro', N. S. 102.
Partridge Id., N. B. 15.
Partridge Id., N. S. 102.
Paspebiac, P. Q. 240.
Patrick's Hole 290.
Patten, Me. 58.
Penguin Ids. 203.
Penobscot River 39.
Penobsquis, N. B. 71.
Pentecost River 233.
Pepiswick Lake 131.
Percé, P. Q. 242.
Perroquets, The 232.
Perry, Me. 28.
Perth, N. B. 54.
Petitcodiac 72, 48.
Petit de Grat 145.
Petite Bergeronne 233.
Petite Passage 117.
Petit Métis 249.
Petty Harbor 197.
Piccadilly Mt. 71.
Pickwaakeet 42.

INDEX. 329

Pictou 137, 166.
Pictou Id. 175.
Pilgrims, The 252.
Pincher's Id. 203.
Pinnacle Id., N. S. 105.
Pirate's Cove 143.
Pisarinco Cove 31.
Placentia Bay 212.
Plains of Abraham 280.
Plaster Cove 143, 168.
Pleasant Bay 183.
Pleasant Point 27.
Pleureuse Point 249.
Plumweseep 71.
Pockmouche, N. B. 62.
Pockshaw, N. B. 66.
Point à Beaulieu 295
Point Aconi, C. B. 161.
  Amour, Lab. 228.
  à Pique 294.
  au Bourdo 69.
  de Monts 233, 249.
  du Chêne 59, 60.
  la Boule 305.
  Lepreau 31.
  Levi, P. Q. 282.
  Maquereau 241.
  Miscou, N. B. 64.
  Orignaux 252.
  Pleasant 40, 68, 100.
  Prim 175, 181.
  Rich, N. F. 219.
  St. Charles 316.
  St. Peter 244.
  Wolfe, N. B. 71.
Pointe à la Garde 68.
  à la Croix 68.
  aux Trembles 306, 309.
  Mille Vaches 233.
  Roches 301.
  Rouge 299.
Pokiok Falls 52.
Pollett River 72.
Pomquet Forks 139.
Pond, Deer, N. F. 219.
  Grand, N. F. 218.
  Red Indian 211.
  Quemo-Gospen 213.
Ponhook Lake 130, 126.
Port Acadie, N. S. 113.
Portage Road, N. B. 61.
Port au Basque, N. F. 216.
  au Choix 219.
  au Persil, P. Q. 295.
  au Pique 81.
  au Port, N. F. 218.
  aux Quilles, 295.
  Daniel 241.
  Elgin, N. B. 73.
  Porter's Lake, N. S. 131.
Port Greville, N. S. 103.
  Hastings, C. B. 143.
  Hawkesbury 143.
  Herbert, N. S. 121.

Port Hill, P. E. I. 179.
  Hood, C. B. 169.
  Joli, N. S. 121.
  Latour, N. S. 122.
  Medway, N. S. 120.
  Mouton 120.
  Mulgrave 143, 140.
Port Neuf, Lab. 233.
Portneuf, P. Q. 306.
Porto Nuevo Id. 149.
Portugal Cove 195, 206.
Port St. Augustine 230.
Port Williams 89, 91.
Powder-Horn Hills 212.
Pownal, P. E. I. 177.
Presque Isle, Me. 54.
Preston, N. S. 131.
Preston's Beach 61.
Prim Point 83.
Prince Edward Id. 172.
Princetown, P. E. I. 178.
Prince William 52.
Prince William St. 89.
Pubnico, N. S. 125.
Pugwash 81, 80.

Quaco, N. B. 71.
Quebec, P. Q. 255.
  Anglican Cathedral 260.
  Basilica 261.
  Cathedral 261.
  Citadel 266.
  Custom House 271.
  Durham Terrace 259.
  Esplanade 268.
  Gen. Hospital 272.
  Gov.'s Garden 269.
  Grand Battery 269.
  Hôtel Dieu 266.
  Jesuits' College 261.
  Laval University 263.
  Lower Town 271.
  Marine Hosp. 272.
  Market Sq. 260.
  Martello Towers 270.
  Montcalm Ward 270.
  Morrin College 265.
  N. D. des Victoires 271.
  Parliament Building 263.
  Post-Office 264.
  St. John Ward 269.
  St. Roch 272.
  Seminary 262.
  Ursuline Conv. 264.
Quemo Gospen 213.
Quiddy Viddy 195.
Quirpon, N. F. 220.
Quispamsis, N. B. 70.
Quoddy Head 26.
Ragged Harbor 201.
Ragged Ids. 212.
Ramea Ids. N. F. 215.
Ram Id. 121.
Ram's Ids. N. F. 212.

Random Sound 209.
Rankin's Mills, N. B. 37.
Rapide de Femme 56.
Rapids, Lachine 319.
  Meductic 52.
  St. Anne 320.
  St. Mary's 319.
  Terres Rompues 300.
Red Bay 228.
Red Cliffs, Lab. 220, 228.
Red Head, N. F. 200.
Red Hills, N. F. 199.
Red-Indian Pond 210, 211.
Red Ids. 147.
Red Point 182.
Red Rapids, N. B. 54.
Remsheg, N. S. 81.
Renewse, N. F. 198.
Renfrew, N. S. 82.
Repentigny, P. Q. 308.
Restigouche River 69, 56.
Richibucto, N. B. 60.
Richmond Bay 178.
Rigolette, Lab. 226.
Rimouski, P. Q. 250.
River, Avon, N. S. 91.
  Charlo, N. B. 66.
  Denys, C. B. 165.
  Gold, N. S. 128.
  Gouffre, P. Q. 292.
  Hillsborough 180.
  Humber, N. F. 219.
  John, N. B. 219.
  LaHave, N. S. 128.
  Louison, N. B. 66.
  Magaguadavic 32.
  Manitou, Lab. 232.
  Miramichi 61.
  Mistassini 301.
  Moisic, Lab. 232.
  Nepisiguit 65, 55.
  of Castors 219.
  Exploits 210.
  Ottawa 320.
  Petitcodiac 72.
  Philip, N. S. 80.
  Restigouche 69, 56.
  Saguenay 297, 233.
  St. Anne, P. Q. 286.
  St. Croix, N. B. 33.
  St. John, Lab 232.
  St. Lawrence 246, 305.
  St. Marguerite 305.
  St. Mary's, N. S. 133.
  St. Maurice 307.
Riversdale, N. S 136.
River, Tobique 54.
Rivière à l'Ours 301.
  à Mars 302.
  du Loup 295, 252.
  Maheu 290.
  Ouelle, P. Q. 252.
Robbinston, Me. 33.
Roberval, P. Q. 301.

Robinson's Point 48.
Rochette, N. B. 66.
Rock, Percé 242.
Rockland, N. B. 73.
Rockport 73.
Rocky Bay, N. F. 210.
Rocky Lake, N. S. 82.
Rollo Bay, P. E. I. 182.
Rosades, The 251.
Rose Bay 119.
Rose Blanche 215.
Rossignol Lake 130.
Rossway, N. S. 116.
Rothesay 22, 70.
Rough Waters 66.
Round Harbor 211.
Route des Pretres 290.
Royalty Junc. 177.
Rustico, P. E. I. 178.

Sabbattec Lake 127.
Sabimm Lake 124.
Sable Id. 134.
Sackville, N. B. 73.
Sacred Ids. 220.
Saddle Id. 227.
Sagona Id. 214.
Saguenay River 297.
St. Agnes, P. Q. 295.
St. Albans, P. Q. 281.
St. Alexis 69, 302.
St. Alphonse, P. Q. 302.
St. Andrews, N. B. 33, 28.
St. Andrews, P. E. I. 181.
St. Andrew's Channel 165.
St. Angel de Laval 307.
St. Anne (Bout de l'I.) 320.
St. Anne de Beaupré 285.
St. Anne de la Perade 307.
St. Anne de la Pocatière 253.
St. Anne des Monts 249.
St. Anne du Nord 285.
St. Anne du Saguenay 300.
St. Anne Mts. 287.
St. Aune's Bay 158.
St. Anthony 221.
St. Antoine de Tilly 306.
St. Antoine Perou 292.
St. Arsène 296.
St. Augustin 306.
St. Barbe 219.
St. Basil 57.
St. Bruno 319.
St. Cécile du Bic 251.
St. Charles Harbor 227.
St. Colomb 280.
St. Croix, P. Q. 306.
St. Croix Cove 89.
St. Croix River 33.
St. Cuthbert 308.
St. David's 178.
St. Denis, P. Q. 252.
St. Donat, P. Q. 250.
St. Eleanors, P. E. I. 179.

St. Elizabeth, P. Q. 308.
St. Esprit, C. B. 148.
St. Etienne Bay 305.
St. Fabien, P. Q. 251.
St. Famille, P. Q. 289.
St. Felicité, P. Q. 249.
St. Felix de Valois 308.
St. Feréol, P. Q. 287.
St. Fidèle, P. Q. 295.
St. Flavie 70, 250.
St. Foy, P. Q. 281.
St. Francis 58.
St. Francis Harbor 225.
St. François 290.
St. François du Lac 308.
St. François Xavier 292.
St. Fulgence 301.
St. Genevieve 219.
St. George, N. B. 32.
St. George's Bay 217.
St. George's Channel 165.
St. Germain de Rim. 250.
St. Irénée 294.
St. Ignace, Cap 253.
St. Jaques 214.
St. Jean Baptiste 318.
St. Jean Deschaillons 307.
St. Jean d'Orleans 290.
St. Jean-Port-Joli 253.
St. Jerome, P. Q. 301.
St. Joachim 287.
St. John, N. B. 15.
  Cathedral 18.
  Custom-House 17.
  Gen. Pub. Hosp. 18.
  Harbor 15.
  King Square 16.
  Post-Office 17.
  St. Paul's 19.
  Trinity 17.
  Valley, The 19.
  Wiggins Asyl. 17.
  Y. M. C. A. 16.
St. John, Lake 301.
St. John's, N. F. 189.
  Anglican Cathedral 191.
  Colonial Building 192.
  Gov't House 192.
  Harbor 189.
  Narrows 191.
  Roman-Catholic Cathedral 192.
  Signal Hill 193.
St. John's Bay 304.
St. Jones Harbor 209.
St. Joseph, N. B. 73.
St. Joseph P. Q. 282.
St. Laurent 290.
St. Lawrence Bay 160.
St. Lawrence River 246, 305.
St. Leonard, N. B. 56.
St. Leon Springs 308.
St. Lewis Sound 225.
St. Louis Isle 304.

St. Luce, P. Q. 250.
St. Lunaire 221.
St. Margaret River 233.
St. Margaret's Bay 219.
St. Margaret's Bay 126,118.
St. Marguerite River 305.
St. Martin, P. Q. 318.
St. Martin's, N. B. 71.
St. Mary's, N. B. 45.
St. Mary's, N. F. 213.
St. Mary's Bay, N. F. 213.
St. Mary's Bay, N. S. 112.
St. Mary's Bay, P. E. I. 181.
St. Maurice River 307.
St. Matthieu 251.
St. Michael's Bay 225.
St. Michel 254.
St. Modeste 296.
St Norbert 308.
St. Octave, P. Q. 249.
St. Onésime, P. Q. 253.
St. Pacome, P. Q. 253.
St. Paschal 252.
St. Patrick's Channel 167.
St. Paul's Bay 292.
St. Peter's, C. B. 146.
St. Peter's, N. B. 65.
St. Peter's, P. E. I. 182.
St. Peter's Bay 227.
St. Peter's Inlet 165.
St. Peter's Id. 174.
St. Peter, Lake 307.
St. Pierre 185, 214.
St. Pierre d'Orleans 289.
St. Pierre les Becquets 307.
St. Placide, P. Q. 292.
St. Roch-des-Aulnaies 253.
St. Romuald, P. Q. 282.
St. Rose de Lima 318.
St. Shot's, N. F. 213.
St. Simeon, 295.
St. Simon 251.
St. Stephen, N. B. 35.
St. Sulpice, P. Q. 308.
St. Thérèse 318.
St. Thomas, P. Q. 253.
St. Tite des Caps 287.
St. Urbain 292.
St. Valier, P. Q. 254.
St. Vincent de Paul, 318.
Salisbury, N. B. 72.
Salmon Cove 201.
Salmonier, N. F. 213.
Salmon River 49, 71, 114.
Salt Mt., C. B. 167.
Salutation Point 174.
Sambro Id. 117.
Sandwich Bay 225.
Sandwich Head 227.
Sandybeach 244.
Sandy Cove 116, 112.
Sandy Point 217.
Sault à la Puce 284.
Sault au Cochon 291.

INDEX. 331

Sault au Recollet 318.
Sault de Mouton 233.
Scatari, C. B. 150.
Schoodic Lakes 35.
Scotchtown, N. B. 48.
Scotch Village 93.
Sculpin Point 214.
Seal Cove, N. B. 29.
Seal Cove, N. F. 221.
Seal Id. N. S. 124.
Seal Ids. 225.
Sea-Trout Point 175.
Sea-Wolf Id. 169.
Seeley's Mills 71.
Segum-Sega Lakes 130.
Seldom-come-by 210.
Seven Ids., Lab. 232.
Shag Id. 230.
Shawanegan Falls 307.
Shecatica Bay 230.
Shediac 59, 60, 174.
Sheet Harbor 132.
Shelburne, N. S. 121.
Shepody Bay 73.
Shepody Mt. 72.
Sherbrooke 133, 132.
Sherbrooke Lake 90.
Shinimicas, N. S. 78.
Ship Harbor 132.
Shippigan Id. 63.
Shoe Cove 211, 221.
Shubenacadie 82.
Sillery, P. Q. 280.
Silver Falls, N. B. 22.
Sir Charles Hamilton's Sound, N. F. 203.
Sissiboo Falls 112.
Skye Glen 168.
Smith's Sound 209.
Smoky, Cape 159.
Sorel, P. Q. 308.
Souris, P. E. I. 182.
South Bay, N. B. 40.
South Mt. 84.
South Oromocto Lake 38.
Southport, P. E. I. 177.
South Quebec 282.
S. W. Head 29.
S. W. Miramichi 62.
Spaniard's Bay 207.
Spear Harbor 225.
Spectacle Id. 120.
Spencer's Id. 103, 104, 106.
Spencer Wood 280.
Spiller Rocks 202.
Split, Cape 104.
Split Rock, 31.
Spotted Id. 225.
Spout, The 197.
Spragg's Point 42.
Sprague's Cove 29.
Springfield, N. B. 42.
Springfield, N. S. 89.
Springhill, N. B. 51.

Spring Hill, N. S. 80.
Spruce Id. 31
Spruce Lake 24.
Spry Bay 132.
Stanley, N. B. 50.
Statue Point 303.
Steep Creek 143.
Stellarton, N. S. 136.
Stewiacke 82.
Stone Pillar 253.
Stormont, N. S. 133.
Strait of Barra 164.
Strait of Belle Isle 220, 227.
Strait of Canso 142.
Strait of Northumberland 60, 174, 239.
Strait Shore, N. F. 196.
Sugar Id 50, 51.
Sugar-Loaf, N. B. 68.
Sugar-Loaf, N. F. 200, 217.
Summerside, P E. I. 178.
Sunacadie, C. B. 164.
Sussex Vale, N. B. 71.
Swallow-Tail Head 29.
Sydney, C. B. 150.
Sydney Mines 152.

Tableau, Le 303.
Table Head 227.
Table Roulante 243.
Tabusintac 61, 62.
Tadousac, P. Q. 299.
Tangier, N. S. 132
Tannery West 319.
Tantramar Marsh 79, 74.
Tatamagouche, N. S. 81.
Tea Hill, P. E. I. 177.
Tedish, N. B. 59
Temiscouata Lake 58, 295.
Temple Bay, Lab. 227.
Tennant's Cove 42.
Thoroughfare, The 48.
Three Rivers 307.
Three Tides, P. E. I. 174.
Three Towers, N. F. 211.
Thrumcap Shoal 93.
Tickle Cove 203.
Tidnish, N. S. 78.
Tignish, P. E. I. 180.
Tilt Cove 205, 211.
Tilton Harbor 210.
Toad Cove 197.
Tobique, N. B. 54.
Tolt Peak 217.
Tomkedgwick River 69.
Topsail, N. F. 206.
Torbay, N. F. 195, 200.
Tor Bay, N. S. 134.
Tormentine, Cape 174.
Torrent Point 227.
Tracadie, N. B. 62.
Tracadie, N. S. 139.
Tracadie, P. E. I. 181.
Tracadiegash 67, 239.

Tracy's Lake 71.
Tracy's Mills, 38.
Traverse, Cape 174.
Tremont, N. S. 89.
Trepassey, N. F. 213.
Trinity, N. F. 201.
Trinity Bay 208, 201.
Trinity, Cape 303.
Trinity Cove 160.
Trois Pistoles 251.
Trois Rivières 307.
Trou St. Patrice 290.
Trouty, N. F. 210.
Truro, N. S. 81.
Tryon, P. E. I. 174.
Tusket Ids. 125, 115.
Tusket Lakes 115.
Tweednogie, C. B. 148.
Tweedside, N. B. 38.
Twillingate, N. F. 205.

Ungava Bay 226.
Upper Caraquette 66.
Upper Gagetown 43.
Upper Musquodoboit 82.
Upper Queensbury 52.
Upsalquitch River 69.
Utopia, Lake 32.

Van Buren, Me. 56.
Vanceboro, Me. 38.
Varennes, P. Q. 308.
Veazie, Me. 39.
Venison Id. 225.
Vernon River 181.
Victoria 53.
Victoria Line 168.
Victoria Mines 152.
Virginia Water 195.

Wallace Valley 80.
Walrus Id. 231.
Walton 106, 93.
Wapitagun Har. 230.
Wapskehegan River 54.
Ward's Harbor 211.
Washademoak Lake 47.
Wash-shecootai 231
Watagheistic Sound 230
Watchabaktchkt 164.
Watt Junc. 37.
Waverley Mines 82.
Waweig, N. B. 36.
Welchpool, N. B. 25.
Wellington 179.
Welsford, N. B. 38.
Wentworth, N. S. 80.
West Bay, C. B. 165.
Westchester, N. S. 80.
Westfield, N. B. 41.
West Isles 31.
West Point 179.
West Port, N. S. 117.
West River 225.

Weymouth, N. S. 112.
Whale Cove 29.
White Bay 221.
White Haven 134.
White Horse 31.
White's Cove 49.
Whycocomagh, C. B. 167.
Wickham, 42, 47.
Wicklow, N. B. 53.
Wiggins Cove 49.

William Henry 308.
Wilmot Springs 89.
Wilson's Beach 25.
Wilton Grove 210.
Windsor, N. S. 91, 101.
Windsor Junc. 82, 93.
Windsor Lake 195.
Wine Harbor 133.
Wiseman's Cove 221.

Witless Bay, N. F. 197.
Wolf River 231.
Wolfville 107, 91.
Wolves, The 25, 31.
Wood Pillar 253.
Woodstock 50, 37.

Yarmouth, N. S. 114, 125.
York River 174.

Index to Historical and Biographical Allusions.

Acadian Exiles 108, 113, 131
Annapolis Royal, N. S. 86.
Anticosti, P. Q. 234.
Aukpaque, N. B. 46.
Avalon, N. F. 198.
Bathurst, N. B. 65.
Bay Bulls, N. F. 197.
Bay of Chaleur 65.
Beaubassin and Beausejour 78.
Bic Island, P. Q. 250.
Bras d'Or, C. B. 165.
Br:beuf, Père 266.
Brest, Lab. 230.
Campobello Id., N. B. 26.
Canada, Lower 235.
Canada, the name of 245.
Canso, N. S. 144.
Cape Breton 149.
Cape Breton (old Province) 141.
Cape Broyle, N. F. 197.
Cape Chatte, P. Q. 249.
Cape Despair, P. Q. 241.
Cape d'Or, N. S. 104.
Cape Sable, N. S. 123.
Cape Sambro, N. S. 118.
Caraquette, N. B. 66.
Carbonear, N. F. 208.
Cartier's Voyages 193, 204, 245, 272, 293.
Caughnawaga, P. Q. 319.
Champlain, Samuel de 273.
Charlottetown, P. E. I. 176.
Château, Lab. 227.
Château Bigot, P. Q. 280.
Château Richer, P. Q. 284.
Chaumonot, Père 279.
Chezzetcook, N. S. 131.
Chicoutimi, P. Q. 300.
Clare Settlements, N. S. 113.
Conception Bay, N. F. 206.
Constitution and Guerrière 200.
Côté de Beaupré 276.
D'Aulnay and La Tour 19, 87, 122.

D'Avaugour, Baron 246.
Dawson, Dr. J. W. 138.
Dead Islands, N. F. 216.
Eastport, Me. 27.
Esquimaux, the 226.
Ferryland, N. F. 198.
Fort La Hève, N. S. 119.
Forts Lawrence and Cumberland 78.
Fort Meductic, N. B. 52.
Fredericton, N. B. 46.
Frontenac, Count de 262, 273.
Gaspé, P. Q. 244.
Gilbert, Sir Humphrey 135, 193.
Glooscap 19, 41, 102, 106, 120, 137, 144.
Goat Island, N. S. 85.
Grand Banks 199.
Grand Lake, N. B. 48.
Grand Manan 28.
Grand Pré, N. S. 108.
Guysborough, N. S. 134.
Haliburton, Judge 92.
Halifax, N. S. 99.
Huron Indians 279, 289.
Indian Lorette 279.
Ingonish, C. B. 159.
Isle aux Coudres 293.
Isle of Orleans 288.
Jemseg, N. B. 42.
Jesuits, the 261, 266, 275, 281.
King's College 92.
Labrador 222, 223.
Lachine, P. Q. 319.
Lake St. John, P. Q. 301.
Lake Utopia, N. B. 32.
Liverpool, N. S. 120.
Lord's-Day Gale 170, 153, 185.
Louisbourg, C. B. 154, 149.
Lunenburg, N. S. 118.
Madawaska, N. B. 57.
Magdalen Islands 184.
Mahone Bay, N. S. 128.

Maugerville, N. B. 43.
Micmac Indians 68, 147, 163, 244.
Mingan Ids., Lab. 231.
Miramichi District 63.
Miscou Id., N. B. 64.
Montreal, P. Q. 317.
Moravian Missions 226.
Murray Bay, P. Q. 295.
New Brunswick 14.
Newfoundland 187, 201, 202, 204, 222.
Norsemen, the 123, 204, 245.
Nova Scotia 76.
Oromocto, N B. 43.
Passamoquoddy Bay 27.
Penobscot Indians 39.
Percé, P. Q. 243.
Pictou, N. S. 137.
Placentia, N. F. 212.
Pleasant Point, Me. 27.
Port Latour, N. S. 122.
Port Mouton, N. S. 121.
Prince Edward Island 172.
Quebec 272.
Red Indians 210, 218.
Restigouche 69.
Richibucto Indians 60.
Rivière du Loup 296.
Rivière Ouelle 252.
Robervals, the 301.
Robin & Co. 240.
Sable Island 135.
Saguenay River 298.
St. Anne de Beaupré 285.
St. Anne's Bay, C. B 158.
St. Augustin, P. Q. 306.
St. Croix Island 34.
St. Joachim, P. Q. 287.
St. John, N. B. 19.
St. John River 40.
St. John's, N. F. 193.
St. Mary's Bay 112.
St. Paul's Bay 292.
St. Paul's Island 160.
St. Peter's, C. B. 146.
St. Pierre, Miq. 186

Scottish Migration 164.
Sillery, P. Q. 281.
Sorel, P. Q. 308.
Strait of Belle Isle 220.
Sydney, C. B. 151.
Sydney Coal-Mines 153.

Tadousac, P. Q. 298, 299.
Tilbury, Wreck of the 148.
Trepassey, N. F. 213.
Trois Pistoles, P. Q. 251.
Truro, N. S. 81.
Ursulines of Quebec 265.

Walker's Expedition 233, 241.
Wallis, Admiral 100.
Williams, Gen. 100.
Windsor, N S. 92.
Yarmouth, N. S. 114.

Index to Quotations.

Alexander, Sir J. E. 38, 58.
Baillie, T. 43.
Ballantyne, R. M. 292.
Beecher, Henry Ward 258.
Boucher 292.
Bouchette, R. 247, 278.
Bougainville 238.
Bonnycastle, Sir R. 67, 195, 218.
Brown, Richard 141, 154, 155, 157, 159, 166, 233.
Buies, Arthur 240, 243, 244, 248, 250.
Cartier, Jacques 204, 246, 288, 298.
Champlain 124, 273, 295.
Charlevoix 30, 77, 150, 158, 184, 204, 233, 238, 247, 289, 293, 299, 300.
Cozzens, F. S. 92, 96, 100, 111, 131, 140, 142, 147, 166.
Crémazie, O. 247.
Dawson, J. W. 102, 142.
De Costa, B. F. 28, 29, 30.
De Mille, Prof. 105.
Dilke, Sir Charles 258, 259.
Dufferin, Lord 237.
Ferland, Abbé 232, 248, 283.
Fiset, L. J. C. 247.
Gesner, Dr. A. B. 32, 36, 43, 56.
Gilpin, Dr. 134.
Gordon, Hon. Arthur 51, 52, 53, 55, 56, 62, 67.
Grey 247.
Haliburton, Judge 90, 91, 109, 111, 113.
Hallock, Charles 67, 78, 103, 126, 127, 128, 129, 130, 169, 170, 225, 227, 240, 301.
Hamilton, 88.
Hardy, Capt. 129, 130, 131.
Hawkins's *Quebec* 256, 259, 261, 272.
Heriot, George 279, 284.
Hind, Prof. 232, 233.
Howells, W. D. 260, 268, 276, 278, 280, 281, 302, 303.
Imray's *Sailing Directions* 73, 158, 169, 248.
Johnston, Prof. J. F. W. 28, 31, 45, 57, 71, 117.
Jukes, Prof. J. B. 189, 195, 196, 216, 218.
Kalm 305.
Kirke, Henry 245.
La Hontan, Baron 87, 212, 305.
Lalemant, Père 249.

Lanman, Charles 68.
Le Moine, J. M. 258, 264, 280, 294.
Lescarbot, M. 34, 85, 86, 201.
London Times 257, 298, 304.
Longfellow, H. W. 109, 110, 111, 113.
Lowell, R. T. S. 187.
McCrea, Lt.-Col. 193, 195, 197.
Marmier, X. 257.
Marshall, C. 278, 286.
Martin, M. 154.
M'Gregor, John 19, 42, 117, 128, 166.
Moore, Tom 184, 320.
Moorson, Capt. 116, 118, 122.
Murdoch, B. 75, 109, 122, 155, 156.
Noble, Rev. L. L. 30, 91, 103, 141, 160, 189, 193, 196, 204, 219, 221, 223, 224, 228.
Novus Orbis 125.
Parkman, Francis 237, 245, 262, 266, 275, 279, 285, 288.
Perley, M. H. 182.
Rameau, M. 238, 277.
Roosevelt, R. B. 66.
Routhier, A. B. 252.
Sagas of Iceland 123, 204.
Sand, Maurice 186, 256.
Scott, G. C. 8, 36, 200.
Shirley, Gov. 274.
Silliman, Prof 238, 257, 267, 277.
Stedman, R. H. 170.
Strauss, 231
Sutherland, Rev. George 178, 180.
Taché 251, 299.
Taylor's *Canadian Handbook* 242, 248, 251, 282, 319.
Taylor, Bayard 277, 291, 292, 293, 297, 304.
Thoreau, H. D. 237, 238, 246, 257, 267, 276, 277, 283, 284, 287, 309, 312.
Trudelle 292.
Voltaire 274.
Warburton, Eliot 190, 195, 234, 256.
Warner, Charles Dudley, 20, 25, 26, 84, 86, 91, 92, 95, 107, 138, 140, 158, 162, 165, 166, 167, 168, 175, 176, 179.
Whitburne, Capt. 187.
White, John, 278, 298, 303.
Whittier, John G. 21, 65, 209, 224, 230.

Index to Railways and Steamboat Lines.

European and North American 37.
Grand Trunk 305.
Intercolonial 70, 78.
New Brunswick 49.
New Brunswick and Canada 33.

Pictou Branch 136.
Prince Edward Island 177, 180, 182.
Quebec and Gosford 255.
Shediac Branch 59.
Windsor and Annapolis 83.

Basin of Minas 101.
Bras d'Or, 161.
Conception Bay (N. F.) 206.
Eastport 25.
Grand Lake 48.
Halifax to Sydney 148.
Labrador 224.
Magdalen Islands 183.
Moisic River (Labrador) 229.
Newfoundland 188, 148.
Northern Coastal (N. F.) 200.
North Shore (N. B.) 60.

Passamaquoddy Bay 25, 30.
Prince Edward Island 174, 175.
Quebec and Gulf Ports 238, 60.
Quebec to Cacouna 291.
Richelieu (St. Lawrence) 305.
Saguenay River 291, 297.
St. John River 39, 51, 53.
St. Pierre (Miq.) 185.
Union (St. Lawrence) 305.
Washademoak Lake 47.
Western Outports 213.
Yarmouth and Halifax 117.

Authorities Consulted in the Preparation of this Volume.

The Editor acknowledges his obligations to the officers of the Boston Athenæum, the Parliament Library at Halifax, the Colonial Library at Charlottetown, the Mechanics' Institute at St. John, and the libraries of Parliament, of the Laval University, of the *Institut Canadien*, and of the Literary and Historical Society, of Quebec.

New Brunswick, with Notes for Emigrants; by Abraham Gesner, M. D. (1847.)
Geology of New Brunswick, etc.; by Dr. Gesner.
New Brunswick and its Scenery; by Jno. R. Hamilton. (St. John, 1874.)
Account of New Brunswick; by Thomas Baillie. (London, 1832.)
Handbook for Emigrants to New Brunswick; by M. H. Perley. (St. John, 1854.)
Mount Desert; by B. F. De Costa. (New York.)
History of New Brunswick; by Cooney.
Nouveau Brunswick; by E. Regnault. (Paris.)

History of Maine; by James Sullivan, LL. D. (1795.)
History of Maine; by W. D. Williamson. (2 vols.; 1839.)
Transactions of the Maine Historical Society.

Letters from Nova Scotia; by Captain Moorson. (London, 1830.)
Travels in Nova Scotia and New Brunswick; by J. S. Buckingham, M. P.
Forest Life in Acadie; by Capt. Campbell Hardy. (London.)
The Fishing Tourist; by Charles Hallock. (New York, 1873.)
Acadia; or A Month among the Bluenoses; by Frederick S. Cozzens. (New York, 1859.)
The Neutral French; a Story of Nova Scotia.
The Lily and the Cross; by Prof. De Mille.
The Boys of Grand Pré School; by Prof. De Mille.
The Clock-Maker; by Judge T. C. Haliburton.
The Old Judge; by Judge T. C. Haliburton.
The Pre-Columbian Discovery of America; by B. F. De Costa. (New York.)
Acadian Geology; by J. W. Dawson, LL. D., F. R. S. (Halifax, 1855.)
On the Mineralogy and Geology of Nova Scotia; by Dr. A. Gesner.
An Historical and Statistical Account of Nova Scotia; by T. C. Haliburton, D. C. L., M. P. (2 vols.; Halifax, 1829.)
History of Nova Scotia, or Acadie; by Beamish Murdoch, Q. C. (3 vols.; Halifax, 1865.)
A General Description of Nova Scotia. (Halifax, 1823.)
Account of the Present State of Nova Scotia. (Edinburgh, 1786.)

INDEX.   335

A History of the Island of Cape Breton ; by Richard Brown, F. G. S., F. R. G. S. (London, 1869.)
Importance and Advantages of Cape Breton ; by Wm. Bollan. (London, 1746.)
Letters on Cape Breton ; by Thomas Pichon. (London, 1760.)
Baddeck, and that Sort of Thing ; by Charles Dudley Warner. (Boston, 1874.)

Prince Edward Island ; by Rev. George Sutherland. (Charlottetown, 1861.)
Progress and Prospects of Prince Edward Island ; by C. B. Bagster. (Charlottetown, 1861.)
Travels in Prince Edward Island ; by Walter Johnstone. (Edinburgh, 1824.)

A Concise History of Newfoundland ; by F. R. Page. (London, 1860.)
History of the Government of Newfoundland ; by Chief Justice John Reeve. (London, 1793.)
Catechism of the History of Newfoundland ; by W. C. St. John. (Boston, 1855.)
Pedley's History of Newfoundland.
Anspach's History of Newfoundland.
Newfoundland in 1842 ; by Sir R. H. Bonnycastle. (2 vols.; London, 1842.)
Voyage of H. M. S. *Rosamond;* by Lieut. Chappell, R. N. (London, 1818.)
Lost amid the Fogs ; by Lieut.-Col. McCrea, Royal Artillery. (London, 1869.)
The New Priest of Conception Bay ; by R. T. S. Lowell. (Boston, 1838.)
Excursions in and about Newfoundland by Prof. J. B. Jukes. (2 vols.; London, 1842.)
Geological Survey of Newfoundland for 1873 ; by Alex. Murray, F. G. S. (St. John's, 1874.)
After Icebergs with a Painter ; by Rev. L. L. Noble. (New York, 1860.)

A Voyage to Labrador ; by L'Abbé Ferland. (Quebec.)
Notes on the Coast of Labrador ; by Robertson. (Quebec.)
Explorations in the Interior of the Labrador Peninsula ; by Prof. H. Y. Hind, F. R. G. S. (2 vols.; London, 1863.)
Sixteen Years' Residence on the Coast of Labrador ; by George Cartwright. (3 vols.; Newark, 1792.)
A Summer Cruise to Labrador ; by Charles Hallock. In Harper's Magazine, Vol. XXII.

History and General Description of New France ; by Father P. F. X. Charlevoix. (6 vols.; in Shea's translation ; New York, 1872.)
Histoire de la Nouvelle France ; by Marc Lescarbot. (1609 ; Paris, 1866 ; 3 vols.)
Cours d'Histoire du Canada ; by L'Abbé Ferland.
Histoire de la Colonie Française en Canada ; by M. Faillon. (3 vols.; Ville-Marie [Montreal], 1865-6).
History of Canada ; by F. X. Garneau. (Bell's translation ; Montreal, 1866.)
History of Canada ; by John MacMullen. (Brockville, 1868.)
Novus Orbis ; by Johannes de Laet. (Leyden, 1633.)
Les Relations des Jesuits.
Lower Canada ; by Joseph Bouchette. (London, 1815.)
British Dominions in North America ; by Joseph Bouchette. (2 vols.; London, 1832.)
British America ; by John M'Gregor. (2 vols.; London, 1832.)
La France aux Colonies ; by M. Rameau. (Paris, 1859.)
Le Canada au Point de Vue Economique ; by Louis Strauss. (Paris, 1867.)
Hochelaga, or England in the New World ; by Eliot Warburton. (2 vols.; New York, 1846.)
The Conquest of Canada ; by Eliot Warburton. (2 vols.; London, 1849.)
The First English Conquest of Canada ; by Henry Kirke. (London, 1871.)
The Pioneers of France in the New World ; by Francis Parkman. (Boston, 1865.)
The Jesuits of North America ; by Francis Parkman.
The Old Régime in Canada ; by Francis Parkman. (Boston, 1874.)
Histoire du Canada ; by Gabriel Sagard. (4 vols. ; Paris, 1866.)
Sketches of Celebrated Canadians ; by Henry J. Morgan. (Montreal, 1865.)
Hawkins's New Picture of Quebec. (Quebec, 1834.)
Reminiscences of Quebec. (Quebec, 1858.)
Découverte du Tombeau de Champlain ; by Laverdière and Casgrain. (Quebec, 1866.)
Maple Leaves ; by J. M. Le Moine. (Quebec.)

INDEX.

Letters sur l'Amerique; by X. Marmier. (Paris.)
Account of a Journey between Hartford and Quebec; by Prof. B. Silliman. (1820.)
Taylor's Canadian Handbook. (Montreal.)
English America; by S. P. Day. (2 vols; London, 1864.)
Three Years in Canada; by John MacTaggart. (2 vols.; London, 1829.)
Western Wanderings; by W. H. G. Kingston. (2 vols.; London, 1856.)
Sketches of Lower Canada; by Joseph Sanson. (New York, 1817.)
The Canadian Dominion; by Charles Marshall. (London, 1871.)
Five Years' Residence in the Canadas; by E. A. Talbot. (2 vols.; London, 1824.)
Sketches from America; by John White. (London, 1870.)
Travels through the Canadas; by George Heriot. (London, 1807.)
British Possessions; by M. Smith. (Baltimore, 1814.)
Adventures in the Wilds of America; by Charles Lanman. (2 vols.; Philadelphia, 1856.)
Pine-Forests; by Lieut.-Col. Sleigh. (London, 1853.)
The travels of Hall, Lyell, Trollope, Dickens, Johnston, etc.
Bref Recit et Succincte Narration de la Navigation faite en MDXXXV. et MDXXXVI. par le Capitaine Jacques Cartier. (Paris, 1863.)
The Principal Navigations, Voyages, etc., of the English Nation; by Richard Hakluyt. (1589–1600.)
Les Voyages à la Nouvelle France, etc.; by Samuel de Champlain. (1632; Paris, 1830.)
Relation du Voyage au Port Royal; by M. Diéreville. (Amsterdam, 1710.)
Nouveaux Voyages, etc.; by the Baron La Hontan. (1703; London, 1785.)
Relation Originale du Voyage de Jacques Cartier. (Paris, 1867.)
Memoires, Relations, et Voyages de Découverte au Canada. (Quebec, 1838.)
Voyage to Canada; by Father Charlevoix. (London, 1763.)
Six Mille Lieues à Toute Vapeur; by Maurice Sand. (Paris.)
Greater Britain; by Sir Charles Dilke.
The Hudson's Bay Company; by R. M. Ballantyne.
Imray's *Sailing Directions*. (London)
Journal of a Voyage to the Coast of Gaspé; by L'Abbé Ferland. (Quebec.)
The Lower St. Lawrence; by Dr. W. J. Anderson. (Quebec, 1872.)
Le Chercheur de Trésors; by Ph. Aubert de Gaspé fils. (Quebec, 1863.)
Chroniques Humeurs et Caprices; by Arthur Buies. (Quebec, 1873.)
Les Anciens Canadiens; by Philippe Aubert de Gaspé. (Quebec, 1864.)
L'Album du Touriste; by J. M. Le Moine. (Quebec, 1872.)
The Blockade of Quebec; by Dr. W. J. Anderson. (Quebec, 1872.)
Journal of the Siege of Quebec; by Gen. James Murray. (Quebec, 1871)
The Expedition against Quebec; by "A Volunteer." (Quebec, 1872.)
Château Bigot; by J. M. Le Moine. (Quebec, 1874.)
A Chance Acquaintance; by W. D. Howells. (Boston, 1873.)
A Yankee in Canada; by Henry D. Thoreau. (Boston, 1862.)
La Littérature Canadienne. (2 vols.; Quebec, 1863–4.)
Soirées Canadiennes. (2 vols.; Quebec, 1861.)
Travels in New Brunswick; by Hon. Arthur Gordon. (In *Vacation Tourists* for 1862–3, London.)
Field and Forest Rambles; by A. Leith Adams. (London, 1873.)
L'Acadie, or Seven Years' Explorations in British North America; by Sir James E. Alexander. (2 vols.; London, 1849.)
Game-Fish of the North and the British Provinces; by R. B. Roosevelt. (New York, 1865.)
Fishing in American Waters; by Genio C. Scott. (New York.)
The American Angler's Guide; by Norris. (New York.)
Fish and Fishing; by H. W. Herbert ("Frank Forrester"). (New York, 1850.)
The Fishing Tourist; by Charles Hallock. (New York, 1873.)
Les Muses de la Nouvelle France; by Marc Lescarbot. (Paris, 1609.)
Evangeline, a Tale of Acadie; by Henry W. Longfellow. (Boston, 1847.)
The Poetical Works of John G. Whittier. (Boston.)
The St. Lawrence and the Saguenay; by Charles Sangster (Kingston.)
Essais Poétiques; by Léon Pamphile Le May. (Quebec, 1865.)
Mes Loisirs; by Louis Honoré Fréchette. (Quebec.)
The Poetical Works of O. Crémazie, J. Lenoir, and L. J. C. Fiset. (Quebec.)

www.ingramcontent.com/pod-product-compliance
Lightning Source LLC
Chambersburg PA
CBHW020300240426
43673CB00039B/652